Coastal Engineering

Historically, much harm has been done by well-meaning coastal engineering attempts, which seemed like good ideas on paper but which failed to allow for practical issues. For this reason, it is vital that theories and models are well grounded in practice.

This second edition brings the models and examples of practice up to date. It has expanded coverage of tsunamis and generating energy from waves to focus both on the great dangers and the great opportunities which the ocean presents to the coastal zone. With an emphasis on practice and detailed modelling, this is a thorough introduction to all aspects of coastal processes, morphology and design of coastal defences. It describes numerous case studies to illustrate the successful application of mathematical modelling to real world practice.

This is a must-have book for engineering students looking to specialise in coastal engineering and management.

Dominic Reeve is Professor of Coastal Engineering at Swansea University, UK.

Andrew Chadwick is Visiting Professor of Coastal Engineering at the University of Plymouth, UK.

Chris Fleming is a former Group Director of Halcrow.

Coastal Engineering

Processes, theory and design practice

Second edition

Dominic Reeve, Andrew Chadwick
and Chris Fleming

 Spon Press
an imprint of Taylor & Francis

LONDON AND NEW YORK

First published 2004
This edition published 2012
by Spon Press
2 Park Square, Milton Park, Abingdon, Oxon OX14 4RN

Simultaneously published in the USA and Canada
by Spon Press
711 Third Avenue, New York, NY 10017

Spon Press is an imprint of the Taylor & Francis Group

British Library Cataloguing in Publication Data
A catalogue record for this book is available
from the British Library

Library of Congress Cataloging in Publication Data
Reeve, Dominic.
    Coastal engineering : processes, theory and design practice / Dominic Reeve,
    Andrew Chadwick and Christopher Fleming.
        p. cm.
    Includes bibliographical references and index.
    ISBN 0–415–26840–0 (hb: alk. paper)—ISBN 0–415–26841–9 (pb: alk. paper)
    1. Coastal engineering. I. Chadwick, Andrew, 1960– II. Fleming,
    Christopher. III. Title.
    TC205.R44 2004
    6270.58—dc22

                                                                            2004001082

ISBN: 978–0–415–58352–7 (hbk)
ISBN: 978–0–415–58353–4 (pbk)
ISBN: 978–0–203–84907–1 (ebk)

Typeset in Sabon
by Swales & Willis Ltd, Exeter, Devon

100623315X

MIX
Paper from
responsible sources
FSC® C004839
www.fsc.org

Printed and bound in Great Britain by
CPI Antony Rowe, Chippenham, Wiltshire

Under heaven nothing is more soft
And yielding than water.
Yet for attacking the solid and the strong,
Nothing is better;
It has no equal.

Lao Tsu (6th Century BC)

Contents

Figures

Tables

Symbols

a_i	amplitude of the ith harmonic (obtained from astronomical theory)
a_b	wave amplitude at the bed
a_x, a_y, a_z	respective components of acceleration in the x, y and z directions
A	empirical coefficient related to sediment fall velocity
A, B	empirical coefficients (wave overtopping)
A_c	freeboard between crest of armour and top of crown wall
A_f	adjustment factor (wave overtopping)
b	exponent related to interaction between waves and revetment type
B_c	width of crown wall element
B_d	width of dry beach
c	wave celerity (phase speed)
c_0	offshore wave celerity
c_f	friction coefficient such that $\tau = \rho c_f |u|u$
c_g, c_g	group velocity
c_{go}	group velocity offshore
c_p	coefficients describing the temporal variation of the pth eigenfunction
C	sediment concentration
C_B	bottom friction coefficient
C_D	drag coefficient
C_I	inertia coefficient
$Cov(x,y)$	covariance of x and y
$C(x_1, x_2)$	autocovariance
C_r	crest berm reduction factor (wave overtopping)
C_w	width of permeable crest
d	depth below a fixed datum (e.g. still water level)
d_0	offshore depth
d_B	depth at wave breaking
d_c	depth of closure
d_s	depth of scour (general)
d_{sw}	depth of scour at a wall
D	pile or pipe diameter
D_{50}	median grain size
D_n	nominal rock diameter
D_{n50}	nominal rock diameter (50th percentile)
D_f	discharge factor

D_*	dimensionless particle size number
e	voids ratio
e_p	spatial eigenfunctions
E	total wave energy per unit area of ocean
E_K	kinetic energy of waves per wavelength per unit crest length
E_0	value of E offshore
Ep	potential energy of waves per wavelength per unit crest length
$E\{.\}$	the mean or expected value
$\mathrm{Erf}(x)$	Error function
$\mathrm{Erfc}(x)$	complementary error function
f	wave frequency
f_b	berm reduction factor (wave overtopping)
f_b	friction roughness reduction factor (wave overtopping)
f_o	angle of attack reduction factor (wave overtopping)
f_w	wall presence reduction factor (wave overtopping)
f_w	wave friction factor
$f_X(x)$	density function of X (= Pr (X = x))
F	fetch length
F	tidal ratio
F_D	drag force
F_{eff}	effective fetch
F_{gr}	particle mobility number
F_H	horizontal force on vertical structural element (e.g. crown wall)
F_I	inertia force
Fr	Froude number
$F_X(x)$	cumulative distribution function of X (= Pr(X ≤ x))
g	acceleration due to gravity
G	gravitational constant
$G(R,S)$	reliability function
$G(f,\theta)$	directional spreading function
G_{gr}	sediment transport parameter, which is based on the stream power concept
h	water depth
h_t	depth of water above toe bund
$h(x, y, t)$	seabed levels
H	wave height
H_0	wave height offshore
H_B	breaking wave height
H_c	mean height between wave crests
H_i	incident wave height
H_{max}	maximum difference between adjacent crest and trough
H_{rms}	root-mean-square wave height
H_s	significant wave height
H_{sb}	significant wave height (broken)
H_z	mean height between zero upward crossing
$H_{1/3}$	mean height of the highest one-third of the waves
$H_{1/10}$	mean height of the highest one-tenth of the waves

$H_{1/100}$	mean height of the highest one-hundredth of the waves
I_{ls}	longshore immersed weight sediment transport rate
k	wave number ($= 2\pi/L$)
k_s	seabed grain size
k_s	Nikuradse roughness
k_Δ	layer thickness coefficient
K	coastal constant
K	run-up constant for smooth plane surface
K_D	diffraction coefficient
K_D	Hudson's non-dimensional stability factor
$K_p(z)$	pressure attenuation factor
K_r	reflection coefficient
K_R	refraction coefficient
K_S	shoaling coefficient
L	wave length
$L(.,.,...,.)$	Likelihood function
L_{berm}	width of berm between +/- H_s
L_g	gap length (e.g. between offshore breakwaters)
L_m	model length scale
L_m	wave length of wave with period T_m
L_0	wave length offshore
L_p	wave length of wave with period T_p
L_p	prototype length scale
L_s	length of structure
L_{slope}	length of profile between +/- 1.5 H_s
m_n	nth spectral moment
M_n	nth sample moment
M_2, S_2, O_1, K_1	tidal constituents
n	Manning's n
n_w	volumetric porosity
N	number of waves during design storm
N_A	area scale ratio
N_L	length scale ratio
N_{od}	number of units displaced out of armour layer strip D_{n50} wide
N_V	volume scale ratio
O_r	wave incidence reduction factor (wave overtopping)
p	pressure
p_f	probability of failure
p_s	porosity
P	permeability coefficient
$P(x \geq X)$	probability that random variable x takes on a value greater than or equal to X
P	rate of transmission of wave energy (wave power)
P_0	value of P offshore
P_{ls}	longshore component of wave power per unit length of beach
$q_b,$	volumetric bedload transport rate per unit width
q_t	volumetric total load transport rate per unit width

Q_g	alongshore drift rate with groyne (m³/sec)
Q_{ls}	volumetric longshore transport rate
Q_m	mean wave overtopping rate (m³/m/sec)
Q_{max}	maximum permissible overtopping discharge rate (m³/m/sec)
Q_0	alongshore drift rate without groyne (m³/sec)
Q_p	spectral peakedness
Q_w	wave overtopping discharge over wave wall (m³/m/sec)
Q_*	dimensionless overtopping rate
r	correlation coefficient
r	roughness coefficient
R	strength function in reliability analysis
R	run-up level relative to still water level
R	overfill ratio (beach nourishment)
R_*	dimensionless run-up coefficient
R_c	crest level relative to still water level (freeboard)
$R(t)$	residual water level variation
$R(x_1, x_2)$	autocorrelation
Re	Reynolds' number
Re_*	Grain sized Reynolds' number
Re_w	local Reynolds' number
S	load function in reliability analysis
S_d	damage level parameter
$S(f)$	spectral energy density
$S(f,\theta)$	directional energy density
S_{om}	deep water wave steepness (mean wave period T_m)
S_{XX}, S_{XY}, S_{YY}	wave radiation stresses
t	time
t_a	thickness of armour layer
t_f	thickness of filter layer
t_u	thickness of underlayer
T	wave period
T_c	mean period between wave crests
T_m	mean wave period
T_p	peak period (=$1/f_p$ where f_p=frequency at the maximum value of the frequency spectrum)
T_s	significant wave period
T_z	mean period between zero upward (or downward) crossings
T_R	return period
u, v, w	respective components of velocity in the x, y and z directions
u_b	bottom orbital velocity
um	maximum near bed orbital velocity
u_*	friction velocity
U_r	Ursell number
w_S	particle fall velocity of a given grain size
W_{a50}	median armour layer weight
W_{cm}	arithmetic average weight of all blocks in a consignment
W_h	height of wave wall above crest level of armour

W_{u50}	median underlayer weight
W_{50}	median weight of armour rock
W'	submerged self weight
W_*	dimensionless wave wall height
x, y, z	ordinates in horizontal and vertical directions
X	offshore distance from original shoreline of breakwater or contour line
X_g	maximum distance from salient shoreline to original shoreline
X_{gi}	maximum distance from salient bay to initial beach fill shoreline
X_i	offshore distance from initial beach fill shoreline
z_0	roughness length
Z_0	mean water level above (or below) local datum
α	angle between wave crest and seabed contour
α	Phillips parameter
α_b	angle of beach slope to horizontal
α_g	change in angle of incidence due to groyning
α_i	angle between internal slope of structure and horizontal
α_0	value of wave angle α offshore
α_0	angle of wave incidence on ungroyned beach
α_B	value of wave angle α at breaking
β	ray orthogonal separation factor ($= K_R^{-1/2}$)
β	reliability index
β	angle between slope normal and direction of wave propagation or wave orthogonal.
γ	wave breaking index
γ	JONSWAP spectral peak enhancement factor
δ	declination
δ	phi mean difference (beach nourishment)
$\delta(x)$	Dirac delta function (=1 for x=0, zero otherwise)
δ_{pq}	the Kronecker delta (=1 for p=q, zero otherwise)
Δ_u	relative density of revetment system unit
ε	spectral width
ε	phi mean difference (beach nourishment)
ε_B	efficiency of bedload transport
εd	represents energy losses in the wave conservation equation
ε_m	surf similarity parameter
ε_{mc}	critical value of surf similarity parameter
ε_s	efficiency of suspended load transport
ϕ	velocity potential (2 and 3d)
ϕ	stability function
ϕ	phi scale (sediment grading)
ϕ_i	phase of ith harmonic
ϕ_p	stone arrangement packing factor
Φ	bedload transport rate factor
$\Phi(x,y)$	velocity potential (2d)
$\Phi(z)$	cumulative standard Normal distribution function
$\Gamma(x)$	Gamma function

$\Gamma_S, \Gamma_R, \Gamma$	safety factors for load, strength and combined effects respectively
η	water surface elevation above a fixed datum
κ	longshore transport coefficient
κ	von Karman's constant
λ	pipe friction factor
λ	longitude
λ_p	p'th eigenvalue
μ	viscosity
μ	phi mean value (beach nourishment)
v	kinematic viscosity
θ	latitude
θ	Shield's parameter/Densimetric Froude number
θ_{CR}	critical Shield's parameter
ρ	density of water
ρ_r	density of rock
σ	phi standard deviation (beach nourishment)
τ_0	shear force
τb	mean seabed shear stress
τ_{bx}, τ_{by}	components of bottom stress along the directions of the x- and y- axes respectively
τ_f	form drag
τ_s	skin friction
τ_t	sediment transport drag
τ_{ws}	shear stress at the bed
τ_{cr}	critical shear stress
τ	bed shear stress vector
ζ, ξ	particle displacements in x and z directions
ξ	Iribarren number
ξ_b	Iribarren number at wave breaking
ω	wave frequency ($= 2\pi/T$)
Ω_i	frequency of ith harmonic (obtained from astronomical theory)
Ω	rate of Earth's angular rotation
ψ_u	empirical stability upgrading factor

Subscripts

$x_{0, i, r, b, t}$	value of parameter x offshore, incident, reflected, at breaking, at toe of structure

Foreword

Activities and development on the coast are as old as humankind. Initially, methods and practices for design of structures in coastal areas were based purely on experience with limited possibilities for optimal design with regard to safety, economy, and the environment. However, at the beginning of the twentieth century, coastal engineering started to emerge as a discipline within civil engineering. Experimental investigations were carried out early on to understand basic processes and to arrive at empirical formulas for the design of structures. Field studies were undertaken, often as a part of solving a coastal engineering problem at a specific site. With the advent of the computer, analytical approaches to solve governing equations could be replaced by numerical techniques, allowing for the handling of more general equations applied to complex geometries with involved boundary and initial conditions.

Dr Nicholas C. Kraus identified three phases in the evolution of coastal engineering practice, namely (1) exploitation and utilisation of the coast, (2) development of protection from coastal hazards such as flooding and erosion, and (3) preserving and creating harmony between nature and coastal uses. Historically, although these phases have often occurred chronologically, in developed societies all the issues related with a particular phase need to be considered simultaneously. The migration towards the coastal areas that occurred during the last century, and which is still ongoing in many countries, has significantly increased the pressure on these areas. Thus, efficient planning and management in the coastal areas require the latest technology in coastal engineering that covers the broad range of issues at hand. Another important aspect in this context is the dynamics of the forcing at the coast, reflected in the great variability in time and space of winds, waves, currents, and water levels, which provide the basic conditions in coastal engineering design. The impact of climate change, and expected consequences, may considerably modify the forcing conditions along our coasts in the coming centuries, further emphasising the need for good coastal engineering practice.

This book is a comprehensive and coherent introduction to the theory and methods in coastal engineering that provides the reader with a solid foundation for dealing with all the above-mentioned aspects of coastal engineering practice. The approaches discussed include basic processes, theoretical formulations, analytical and numerical modelling, experimental work in the laboratory and field, and data collection, analysis, and simulation. A unique feature of the book is the aim to bridge the gap between theory and practice, which has developed in recent years as coastal research in academia has become more specialised and focused towards its own needs rather than towards serving society. This is an unfortunate development that I trust this book will help to arrest.

Professor Magnus Larson
Department of Water Resources Engineering
Lund University
Sweden

Preface to the 2nd Edition

The second edition of *Coastal Engineering: Processes, Theory and Design Practice* has been prompted by the need to update the material. The body of knowledge in this area has grown rapidly and substantially since the first edition was published. In updating the text we have adopted a conservative approach in that changes have been made only where the original material has been superseded or, further detail has been included in the light of recent developments and observations. Specific mention is made here of: the EurOtop project, completed in 2007, that provided a comprehensive collection of work on wave overtopping of sea defences; updates in the climate change forecasts by the 4th IPCC reports in 2007 and the subsequent Copenhagen Analysis (2009); the publication of the *Beach Management Manual* (2nd Edition, 2009) and the sequence of tsunami events (Indian Ocean tsunami 2004, South Pacific Islands tsunami 2009, Tohoku tsunami 2011) which have served to remind us of the destructive power of such phenomena.

We are grateful to the readers of the first edition for their comments and feedback. We have taken this opportunity to correct typographic errors and add clarifications, as well as some additional worked examples.

Finally, we would like to acknowledge the contributions of Dr Alison Raby and Dr Jose Horrillo-Caraballo (Coastal Engineering Research Group, University of Plymouth) and Dr Nigel Pontee (Halcrow Group).

Dominic Reeve
Andrew Chadwick
Chris Fleming
2011

Preface/Acknowledgements to the 1st edition

This text is based, in part, on modules in coastal processes and engineering developed over several years in the Departments of Civil Engineering at the University of Nottingham and the University of Plymouth. It is also influenced by the authors' combined experience of applying theory, mathematical and physical modelling to practical engineering design problems.

In writing this book we have assumed that prospective readers will have a good grounding in basic fluid mechanics or engineering hydraulics, and have some familiarity with elementary statistical concepts. The text is aimed at final year undergraduate and MSc postgraduate students, to bridge the gap between introductory texts and the mainstream literature of academic papers and specialist guidance manuals. As such, we hope it will be of assistance to practitioners, both those beginning their careers in coastal engineering and established professionals requiring an introduction to this rapidly growing discipline.

The motivation for this book arose because it had become apparent that although a number of good books may be available for specific parts of modules, no text provided the required depth and breadth of the subject. It was also clear that there was a gap between the theory and design equations on one hand and on the other hand the practical application of these in real life projects where constraints of time, cost and data become important factors. While engineering experience is not something that is readily taught we have included within the text a selection of real projects and studies that illustrate the application of concepts in a practical setting. Also, throughout the text we have used worked examples to amplify points and to demonstrate calculation procedures.

This book is not intended to be a research monograph nor a design manual, although we hope that researchers and practitioners will find it of interest and a useful reference source.

The book is divided into nine chapters. A full references list is given towards the end of the book and some additional sources of material are cited at the end of individual chapters. A summary of elementary statistical definitions is included in Appendix A.

Many colleagues and friends have helped in the writing of this book. We are particularly grateful to Dominic Hames of the University of East London for his many useful comments on early drafts of the text. We would also like to acknowledge the contributions of Jose Maria Horrillo (PhD student at the University of Nottingham), Professor Jothi Shankar (National University of Singapore), Dr Peter Hawkes (HR Wallingford), Kevin Burgess (Halcrow Group) and the consultants and agencies whose work has provided many of the case studies included in the book.

DER would also like to thank his father for encouraging him to start this project; his wife Audrey and family for giving support to enable him to do this; his PhD supervisor Professor Brian Hoskins who introduced him to the interesting challenges of numerical simulation of fluid flow; and Sue Muggeridge of the School of Civil Engineering at the University of Nottingham for typing much of the early drafts of Chapters 3, 4, 6 and 7.

Dominic Reeve
Andrew Chadwick
Chris Fleming
2004

Chapter 1

Introduction

1.1   The historical context

The coastline has been 'engineered' for many centuries, initially for the development of ports and maritime trade or fishing harbours to support local communities. For example, the Port of A-ur built on the Nile prior to 3000 BC and nearby on the open coast the Port of Pharos around 2000 BC. The latter had a massive breakwater more than 2.5 km long. The Romans invented an hydraulic cement and developed the practice of pile driving for cofferdam foundations, a technique that was used for the construction of concrete sea walls. Whilst these structures were no doubt built on the basis of trial and error there is no evidence that there was any real appreciation of coastal processes with respect to the siting of maritime infrastructure.

Many early sea defences comprised embankments, but when dealing with coastal erosion problems the hard edge approach dominated, at least in the United Kingdom. In particular, the Victorians were active in their desire to construct promenades in seaside resorts which were usually vertically faced. Coastal processes were not only poorly understood, but there was some confusion as to what the driving forces were. There have been several periods of development of coastal works in the UK over the past century. There was an extensive wall-building programme during the 1930s as part of the unemployment relief schemes. These were based on dock wall designs with near vertical profiles. The consequences of 'bad design' by building a hard edge structure on a shoreline were, however, appreciated at about this time. An article written by T.B. Keay in 1941, notes that 'the efforts of man to prevent erosion are sometimes the cause of its increase, either at the site of his works or elsewhere along the coast'. This he explained with an example of a sea wall built at Scarborough in 1887. In just 3 years it was necessary to add an apron and in a further 6 years an additional toe structure and timber groynes. He went on to say that an essential preliminary of all coast protection works is to study the local natural conditions.

It was not until the post-Second World War period that the theoretical models and ideas that underlie the basic processes began to be developed, save for basic wave and tidal motion. The development of the Mulberry Harbours in the Second World War led to the concept of determining wave climate, using wind data and design parameters such as wave height and wave period. Thus contemporary coastal engineering effectively began at that time witnessed by the First Conference on Coastal Engineering at Berkeley, California sponsored by The Engineering Foundation Council on Wave Research (USA). This was closely followed in 1954 with the publication and widespread acceptance of 'Shore Protection, Planning and Design – Technical Report No.4' (TR4) by the US Army Corps of Engineers, Beach Erosion Board. The 'Planning' part of the title was later dropped and it became the well-known *Shore Protection Manual*, and more recently *The Coastal Engineering Manual*.

The history books are full of accounts of major storms that caused destruction and devastation to various sections of the coast. In more recent times, one of the most significant dates in coastal engineering in England is 31 January 1953 when an extreme storm surge travelled down the North Sea coincidentally with extreme storm waves. The effect was devastating and serves as a poignant reminder as to how vulnerable the low-lying areas of the east coast are. The post-1953 period saw great activity in the construction of sea defences along that coastline at a time when sea walls and groyne systems were the norm and the overriding criterion was to provide a secure safety barrier against any such event occurring again.

The value of attempting to retain beach material, whether for sea defence, coast protection or recreational use has been recognised for some time. This is to some extent demonstrated by the extensive lengths of coastline that have been groyned in the past. However, it has been suggested that, prior to the 1970s many responsible authorities quite naturally dealt with these matters on a parochial basis with little regard for, or appreciation of, the impact of their actions on neighbouring territory.

This has allegedly led to some rather undesirable consequences in both conservation and planning terms and engineers have been criticised for being insensitive and not paying heed to these issues. There are a number of other factors that should be taken into account before reaching this conclusion. These include the constraints that have, in effect, been imposed by interpretation of Government legislation and the nature of the responsibilities that fall upon the various authorities involved in implementing coastal works. A primary aim has been to protect people and property from the effects of erosion or flooding in situations where economic justification can be established. In this regard they have generally been demonstrably successful.

It is also evident that, in the past, the planning system has not generally taken the question of long-term coastal evolution into account when in many instances planning permission has been granted for development on sites that have been well known to be vulnerable to long-term erosion. At the same time conservation issues have developed alongside our appreciation of natural processes and the complex interactions involved.

The major influences that coastal works have had on the shoreline are centred on the degree of interference that is taking place with the natural processes. Harbours and their approach channels have had a significant impact on alongshore drift as have coastal defences themselves through the use of groynes or other similar structures. It is also evident that protection of some types of coast from erosion must deprive the local and adjacent beach system of some of its natural sediment supply. Given that nature will always try to re-establish some form of dynamic equilibrium, any shortfall in sediment supply is redressed by removing material from elsewhere. Such a situation can also be exacerbated by introducing structures that, instead of absorbing energy as a natural beach does, reflect the incident waves to do more damage on the beach in front of the structure.

By the 1960s a much greater understanding of coastal processes emerged as the theoretical development coupled with physical and numerical modelling developed.

This led to a gradual re-appraisal of coastal engineering techniques in such a way that the design process began to consider studies of the coastal regime and its interaction with the proposed works. By the early 1970s this led to the application of relatively novel solutions to coastal problems such as beach nourishment, artificial headlands and offshore breakwaters. Since then, numerical modelling techniques for deep-water wave prediction, wave transformation in the coastal zone, wave/structure interaction, coastal sediment transport and coastal evolution have all developed rapidly. An excellent first source of reference to the history of coastal engineering may be found in a book published in 1997, as part of the 25th International Conference on Coastal Engineering (Kraus 1997).

In summary, the science that underpins nearshore coastal processes and hence engineering appreciation is relatively young in its development, having only emerged as a subject in its own right over the past 60 years. During that time there have been rapid advances in knowledge and understanding, thus allowing solutions to coastal problems to become very much more sophisticated with respect to harmonisation with the natural environment. There has thus been an evolution of design practice that has progressively been moving towards 'softer' engineering solutions. That is, those solutions which attempt to have a beneficial influence on coastal processes and in doing so improve the level of service provided by a sea defence or coast protection structure.

1.2   The coastal environment

1.2.1   Context

The United Nations estimate that by 2020, in excess of 75 per cent of the world's population will live within the coastal zone. These regions are therefore of critical importance to a majority of the world's citizens and effect an increasing percentage of our economic activities. The coastal zone provides important economic, transport, residential and recreational functions, all of which depend upon its physical characteristics, appealing landscape, cultural heritage, natural resources and rich marine and terrestrial biodiversity. This resource is thus the foundation for the well being and economic viability of present and future generations of coastal zone residents (European Commission 2000).

The pressure on coastal environments is being exacerbated by rapid changes in global climate, with conservative estimates of sea level rise of the order of 0.5 m over the next century. The English coastline alone spans some 3763 km, and even with sea level at its current position, 1000 km of this coastline requires protection against tidal flooding and 860 km is protected against coastal erosion, at a cost of more than half a billion pounds sterling per annum to the UK flood and coastal defence budget. The value of the coastal zone to humanity, and the enormous pressure on it, provide strong incentives for a greater scientific understanding which can ensure effective coastal engineering practice and efficient and sustainable management.

1.2.2   Beach origins

The current world's coastlines were formed as a result of the last ice age, which ended about 10 000 years ago. At that time large ice sheets covered more of the world's land masses than they do at present. As they melted there was a rapid rise of sea level (about 120 m between 20 000 and 6000 years ago). Vast quantities of sediment were carried by rivers to the sea during this period, eventually forming the precursor to our present coastlines as the rate of sea level rise rapidly reduced about 6000 years ago. Today many of our beaches are composed of the remnants of these sediments, composed predominantly of sand and gravel. These sources of beach material have subsequently been supplemented by coastal erosion of soft cliffs and the reduced but continuing supply of sediments from rivers. Material may also be derived from offshore banks left behind by relatively rapid rises of sea level after cold episodes.

1.2.3 Time and space scales

Beaches are dynamic, changing their profile and planform in both space and time in response to the natural forcing of waves and currents, sediment supply and removal, the influence of coastal geological features and the influence of coastal defences and ports and harbours. Timescales range from *micro* (for wave by wave events) through *meso* (for individual storm events) to *macro* (for beach evolution over seasons, years and decades). Similarly space scales have a range of micro (for changes at a point) through meso (for example changes of beach profile) to macro (for example changes in planform evolution over large coastal areas).

1.2.4 The action of waves on beaches

The action of waves on beaches depends on the type of wave and the beach material. For simplicity, wave types are generally categorised as storm waves or swell waves and beach materials as sand or gravel. As waves approach the shore they initially begin to feel the bottom in transitional water depths and begin to cause oscillatory motions of the seabed sediments, before breaking. Where the bed slope is small (as on sand beaches), the breaking commences well offshore. The breaking process is gradual and produces a surf zone in which the wave height decreases progressively as waves approach the shore. Where the bed slope is steeper (say roughly 1 in 10 as on gravel beaches), the width of the surf zone may be small or negligible and the waves break by plunging. For very steep slopes the waves break by surging up on to the shore. The incoming breaker will finally impact on the beach, dissipating its remaining energy in the 'uprush' of water up the beach slope. The water velocities reduce to zero and then form the 'backwash', flowing down the beach, until the next breaker arrives. This is known as the 'swash zone'.

In the surf zone, the seabed will be subject to a complex set of forces. The oscillatory motion due to the passage of each wave produces a corresponding frictional shear stress at the bed, and both incoming and reflected waves may be present. For oblique wave incidence, a current in the longshore direction will also be generated, producing an additional bed shear stress. Finally, the bed slope itself implies the existence of a component of the gravitational force along the bed. On the beach forces are produced due to bed friction and due to the impact of the breaker, which generates considerable turbulence. All of these processes are illustrated in Figure 1.1.

If the seabed and beach are of mobile material (sand or gravel), then it may be transported by the combination of forces outlined above. The 'sorting' of beach

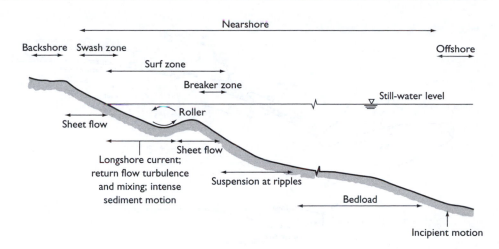

Figure 1.1 **Long and cross-shore beach hydrodynamics and sediment dynamics (reproduced by kind permission of CIRIA, from Simm *et al.* (1996)).**

material (with larger particles deposited in one position and finer particles in another) can also be explained. For convenience, coastal sediment transport is divided into two components: perpendicular to the coastline (cross-shore transport) and parallel to the coastline (longshore transport or 'littoral drift'). Whether beaches are stable or not depends on the rates of sediment transport over *meso and macro* timescales. The transport rates are a function of the wave, breakers and currents. Waves usually approach a shoreline at an oblique angle. The wave height and angle will vary with time (depending on the weather). Sediment may be transported by unbroken waves and/or currents; however, most transport takes place in the surf and swash zones. Further details of cross and longshore transport are discussed in Chapters 5 and 6.

1.2.5   Coastal features

Figure 1.2 illustrates the main types of coastal features that exist. As can be seen from this figure, these features are quite diverse and will not necessarily all exist in close proximity! Real examples of coastal features around the UK are given in Figures 1.3 to 1.11. The formation of these varying coastal features is a function of the effects and interactions of the forcing action of waves and currents, the geological and man-made features and the supply and removal of sediment.

Tombolos form due to the sheltering effect of offshore islands or breakwaters on the predominant wave directions, salients being produced where the island/breakwater is too far offshore to produce a tombolo. Spits are formed progressively from headlands which have a plentiful supply of sediment and where the predominant wave direction induces significant longshore drift into deeper water. These spits can then become 'hooked' due the action of waves from directions opposing the predominant one. Where spits form initially across a natural inlet, they may eventually form a barrier beach, which in turn may be breached by trapped water in a lagoon to form a barrier island. Pocket beaches are a relict feature, generally of small scale, formed by eroded

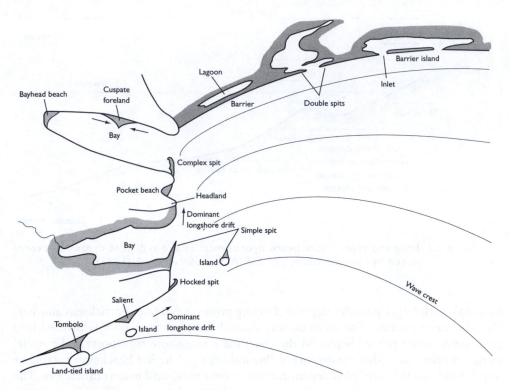

Figure 1.2 Coastal features (reproduced by kind permission of CIRIA, from Simm *et al.* (1996)).

material trapped between hard headlands. On a larger scale these may naturally tend towards a generally stable bay shape, as discussed in the next section.

1.2.6 Natural bays and coastal cells

Where an erodible coastline exists between relatively stable headlands, a bay will form (e.g. Figure 1.9). The shape of such bays is determined by the predominant wave climate and, if stable, is half heart shaped. These are called 'crenulate bays'. The reason why crenulate bays are stable is that the breaker line is parallel to the shore along the whole bay, due to refraction and diffraction of the incoming waves. Littoral drift is therefore zero. These results have several significant implications. For example, the ultimately stable shape of the foreshore, for any natural bay, may be determined by drawing the appropriate crenulate bay shape on a plan of the natural bay. If the two coincide, then the bay is stable and will not evolve further unless the wave conditions alter. If the existing bay lies seaward of the stable bay line, then either upcoast littoral drift is maintaining the bay, or the bay is receding. Also, naturally stable bays act as 'beacons' of the direction of littoral drift. Finally, the existence of crenulate bays suggests a method of coastal protection in sympathy with the natural processes, by the use of artificial headlands. This is discussed further in Section 6.3.2 and Chapter 9.

Figure 1.3 Beach types (reproduced by kind permission of Halcrow).

Dunes

Figure 1.4 Dune types (reproduced by kind permission of Halcrow).

Cliffs

Figure 1.5 Cliff types (reproduced by kind permission of Halcrow).

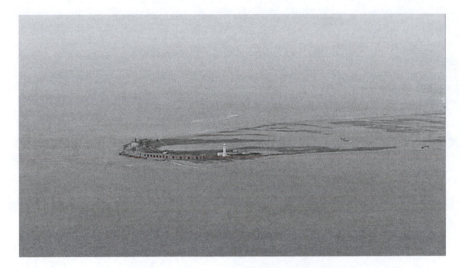

Figure 1.6 Hurst Castle Spit (reproduced by kind permission of Halcrow).

Figure 1.7 Natural tombolo, Burgh Island (reproduced by kind permission of Halcrow).

The concept of a coastal cell follows on quite naturally from the crenulate, stable bay. It is also of crucial importance to coastal zone management, allowing a rational basis for the planning and design of coastal defence schemes. The definition of a coastal cell is a frontage within which the longshore and cross transport of beach material takes place independently of that in adjacent cells. Such an idealised coastal cell is shown in Figure 1.12. Within such a cell, coastal defence schemes can be implemented without causing any effects in the adjacent cells. However, a more detailed review of

Figure 1.8 Tombolo formation, Happisburgh to Winterton Coastal Defences (reproduced by kind permission of Halcrow).

Figure 1.9 Natural Bay, Osgodby (reproduced by kind permission of Halcrow).

this concept reveals that a coastal cell is rather difficult to define precisely, depending
on both the timescale and the sediment transport mode. For meso timescales, the local longshore drift direction can be the reverse of the macro drift direction, possibly
allowing longshore transport from one cell to another. With regard to sediment transport, this may be either as bed or suspended load. Longshore transport of coarse material is predominantly by bedload across the active beach profile and largely confined
to movements within the coastal cell. Conversely, longshore transport of fine material

Figure 1.10 Erme Estuary (reproduced by kind permission of Halcrow).

Figure 1.11 Salt Marsh, Lymington (reproduced by kind permission of Halcrow).

is predominantly by suspended load which is induced by wave action but then carried by tidal as well as wave-induced currents, possibly across cell boundaries.

1.2.7 Coastal zone management principles

Despite the inherent fuzziness of the boundaries of a coastal cell, it is nevertheless a very useful concept for coastal zone management. In the UK, for example, the coastline of

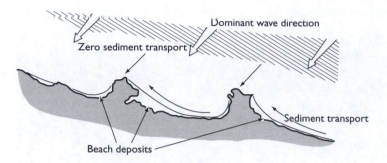

Figure 1.12 Idealised coastal cell.

England has been divided into 11 primary cells and a series of sub cells defined within each primary cell. Within the UK, the planning of new coastal defence schemes is now carried out within the context of a shoreline management plan. Many other coastal authorities throughout the world have adopted or are beginning to adopt a similar policy.

The aims of a shoreline management plan are to provide the basis for sustainable coastal defence policies within a coastal cell and to set objectives for the future management of the shoreline. To fulfil these aims four key components and their interrelationships need to be considered. These are the coastal processes, the coastal defences, land use and the human and built environment and finally the natural environment. An understanding of the interrelationships between coastal processes and coastal defence is fundamental to developing a sustainable defence policy. The need for coastal defence schemes arises from effects on land use, and the funding of such schemes relies on an economic assessment of whether the benefits of defence outweigh the costs of construction. Finally, the effects of defence schemes on the natural environment must be very carefully considered and an environmental assessment carried out. Environmental hazards and opportunities should be identified and schemes should be designed to conserve or enhance the natural environment. Where conflicts arise between the needs for defence and conservation, these must be resolved by the environmental assessment.

1.2.8   Coastal defence principles

Coastal defence is the general term used to cover all aspects of defence against coastal hazards. In the United Kingdom two specific terms are generally used to distinguish between different types of hazard. The term 'sea defence' is normally used to describe schemes which are designed to prevent flooding of coastal regions under extremes of wave and water levels. By contrast, the term 'coast protection' is normally reserved to describe schemes designed to protect an existing coastline from further erosion. The origins of these terms lie in the names of the different acts of Parliament that address these issues.

There are two approaches to the design of coastal defence schemes. The first is referred to as 'soft engineering', which aims to work in sympathy with the natural processes by mimicking natural defence mechanisms. Such an approach has the potential for achieving economies whilst minimising environmental impact and creating environmental opportunities. The second is referred to as 'hard engineering', whereby structures are constructed on the coastline to resist the energy of waves and tides. Elements of hard and soft engineering are often used together to provide an optimal coastal defence scheme, for example, the combined use of beach feeding with groynes or breakwaters. These principles and the associated forms of coastal defence techniques are discussed in detail in Chapter 9.

1.3   Understanding coastal system behaviour

1.3.1   Introduction

This section has been abstracted from the FUTURECOAST project (DEFRA 2002). This major project was commissioned by the UK government Department for the Environment Food and Rural Affairs and undertaken by Halcrow. It represents a major step forward in conceptualising the factors affecting coastal change. Analysis of coastal dynamics and evolution is difficult due to both the range of spatial and temporal scales over which coastal changes occur, and the complex interactions that result in shoreline responses of varying, non-linear and often unpredictable nature. There is also inter-dependence between different geomorphic features that make up the natural system, such that the evolution of one particular element of the coast is influenced by evolution in adjacent areas. Often these influences extend in a number of directions, thereby further complicating the task of assessing change.

Whilst there are a variety of modelling techniques for predicting coastal behaviour, many of these focus on short-term, relatively local-scale analysis based upon contemporary hydrodynamic forcing, as opposed to considering larger-scale and longer-term evolutionary behaviour. Although such modelling provides vital information, it does not necessarily provide the complete picture of influence and change.

It is also important to understand how the coastal zone functions on a wider scale both in time and space. Within the discipline of coastal engineering, there is a strong focus upon littoral processes and this approach is frequently used as a basis for analysing coastal change and assessing future policy options and impacts. Whilst the littoral cell concept is a valid approach, it is only one aspect of coastal system behaviour and other factors also need to be taken into account when assessing future shoreline evolution. Therefore, in terms of making large-scale or longer-term predictions of coastal evolution, the cell concept can have a number of shortcomings.

A 'behavioural systems' approach, such as was adopted by the FUTURECOAST project (see also Section 6.5.3), involves the identification of the different elements that make up the coastal structure and developing an understanding of how these elements interact on a range of both temporal and spatial scales. In this approach it is the interaction between the units that is central to determining the behaviour. Feedback invariably plays an important role and changes in energy/sediment inputs that affect one unit can in turn affect other units, which themselves give rise to a change in the level of energy/sediment input.

Whilst the starting point for a behavioural system is the energy and sediment pathways, it is important to identify the causative mechanism as a basis for building a robust means of predicting the response to change. This must take account of variations in sediment supply and forcing parameters, such as tide and wave energy. However, it is also important to look for situations where the system response is to switch to a different state, for example, the catastrophic failure of a spit, or the switching of channels as a consequence of episodic storm events.

1.3.2   Recognising shoreline types

Key influences upon planform shape, and evolution, are the underlying geology and coastal forcing, for example, prevailing wave activity. Large-scale shoreline evolution may be broadly considered in terms of those areas that are unlikely to alter significantly (i.e. hard rock coasts) and those areas that are susceptible to change (i.e. soft coasts).

The evolution of hard rock coasts is almost exclusively a function of the resistant nature of the geology, with the influence of prevailing coastal forcing on the orientation of these shorelines only occurring over very long timescales (millennia). Differential erosion may occur along these coastlines to create indentations or narrow pockets where there is an area of softer geology, or faulting, which has been exploited by wave activity.

The evolution of softer shorelines is more strongly influenced by coastal forcing, although geology continues to play a significant role in both influencing this forcing (e.g. diffraction of waves around headlands) and dictating the rate at which change may occur. The planform of these shorelines will, over timescales of decades to centuries, tend towards a shape whose orientation is in balance with both the sediment supply and the capacity of the forcing parameters to transport available sediment. In general soft shorelines have already undergone considerable evolution. Some shorelines may have reached their equilibrium planform in response to prevailing conditions, whilst others have not and continue to change.

Reasons why such shorelines have not reached a dynamic equilibrium, and may still be adjusting in orientation, include constraints upon the rate of change (e.g. the level of resistance of the geology) and changes in conditions (e.g. sediment availability, emergence of new controls, breakdown of older features, changes in offshore topography). It should be recognised that we are presently at a point in time when most of England and Wales is in a generally transgressive phase (i.e. a period of rising relative sea levels) and shorelines are still adjusting to this. Under rising sea levels there are two main possible responses: (1) the feature adjusts to maintain its form and position relative to mean water level (i.e. moves inland); or (2) the feature becomes over-run and is either drowned or eroded and lost.

It is the softer shorelines that are most sensitive to changes in environmental conditions, such as climate change impacts, which may alter the coastal forcing. Such changes in conditions are not necessarily instantaneous, and can take many decades or centuries to occur. Therefore, some of the changes taking place at the shoreline over the next century may be a continuation of a response to events that occurred at some time in the past.

The natural tendency for most shorelines is to become orientated to the predominant wave direction, although clearly there are many constraints and influences upon this. This concept applies equally to the shoreface, foreshore and backshore, although is perhaps best illustrated by beach behaviour; the shoreline adjusts in form because sediment is moved, giving rise to areas of erosion and deposition.

Swash-aligned, or swash-dominated, coasts build parallel to incoming wave crests, whereas drift-aligned, or drift-dominated, coasts are built parallel to the line of maximum longshore sediment transport and are generated by obliquely incident waves (but not necessarily uni-directional). In general, swash-dominated coasts are smoother in outline than those that are drift-dominated, which tend to exhibit intermittent spits and sediment accumulations, such as nesses. Due to variability in the wave climate, few beaches are entirely swash- or drift-aligned, but identification of the predominant characteristic can help in predicting likely future evolution.

Where shorelines have become adjusted to the prevailing pattern of the waves, that is they are in 'dynamic equilibrium', they reach a state of relative stability. Where changes are made to the shoreline controls, whether natural or anthropogenic, for example removal of defences, there may be a tendency towards greater drift-alignment, with increased mobility of foreshore sediments and backshore erosion. Shorelines of any form have the potential to evolve in three ways: continuation of present form; breakdown of present form; or transition to a different form. These changes could occur for various reasons, including:

- changes in the rate or volume of sediment input/output, for example, due to construction or demolition of coastal defences or exhaustion of a relict sediment source;
- changes in composition of sediment input, for example, due to the loss of sand over time through winnowing;
- changes in wave energy or approach, resulting in a change in the drift rate and/or direction;
- changes in the balance between longshore and cross-shore sediment transfer.

Hard rock coasts are resilient to significant changes in orientation over decadal to century timescales and require little further discussion. In most cases embayments within these hard strata may have been formed by submergence (sea level rise) in combination with abrasive and marine erosive processes, although in some cases they may be formed by marine erosion alone. In the latter case, the geology can have a major influence upon the coastal processes and the resultant orientation of the shoreline, in particular the formation and evolution of deeper embayments, some of which are referred to as 'zeta bays'. These form due to wave diffraction around at least one fixed point, although often between two fixed points, for example, headlands; a soft coast between two resistant points will readjust its orientation to minimise the wave-generated longshore energy. Most bay forms that have reached an equilibrium state exhibit an almost circular section behind the updrift headlands, which reflects the wave crests diffracted around the fixed points.

Notwithstanding this, differential resistance of the backing geology will also influence the position of the shoreline, which in some cases may produce secondary embayments as new headlands emerge. A further influence is the response of the different geomorphological elements that comprise the backshore. This could, for example, create floodplains or inlets that alter the hydrodynamics operating within the bay and thus the alignment tendencies at the shoreline.

All coasts are affected by tides, but only a few types of coastal environments can be considered to be tide dominated. Tide-dominated coasts generally occur in more sheltered areas where wave action is largely removed, for example, due to shoaling or by direct shelter, such as by a spit at a river mouth, and are therefore most commonly associated with estuaries, although there are parts of the open coast around England and Wales where tidal influences are most dominant upon the shore planform. The landforms reflect the change in dominant influence from waves to tides and sediments tend to be characterised by silts and muds due to the lower energy levels. These shorelines are generally low-lying and the shoreline planform arises from the deposition of fine sediments, which creates large intertidal flats.

The future planform evolution of tide-dominated shoreline is perhaps the most difficult to predict accurately due to the complex interactions within these environments. One of the key influences on the evolution of tide-dominated coasts is the change in tidal currents. This may occur for a number of reasons, but one of the key causes is due to changing tidal prisms, that is the amount of water that enters and exits an estuary every ebb-flood tidal cycle. Another influence on evolution is the configuration of ebb and flood channels, which affects the pattern of erosion, transport and deposition both across the intertidal zone and at the shoreline. In many estuaries, changes in the position of these channels have had a significant impact upon the adjacent shorelines (e.g. Morecambe Bay). Where a major channel lies close to the shoreline it allows larger, higher-energy waves to attack the marsh cliff, whereas where there are sandbanks adjacent to the marsh, wave energy is attenuated. It is often not clear what causes a channel to meander because there are a number of interacting factors involved.

1.3.3 Influences upon coastal behaviour

To understand the morphological evolution of the shoreline will, in many instances, require the identification of key controls and influences on large-scale shoreline behaviour, and the interactions taking place within coastal systems. Many changes tend to occur at scales that relate to long-term responses to past conditions. Often the underlying pressures for shoreline change are related to large-scale re-orientation of the coast, which may include the emergence of new features and/or the deterioration of existing features. Some examples include:

- changes in geological controls (e.g. emergence of headlands in eroding cliffs, changes in backshore geology);
- alteration to hydrodynamic forcing (e.g. increased or decreased wave diffraction around headlands or over offshore banks);
- changes in hydrodynamic influences (e.g. interruption of drift by newly created tidal inlets, development of tidal deltas);
- changes in sediment budget (e.g. exhaustion of relict sediment sources, shorelines switching from drift- to swash-alignment);
- human intervention (e.g. cessation of sediment supply due to cliff protection).

Appreciation of these factors enables the long-term and large-scale evolutionary tendencies to be broadly established and in particular identify where a change from past evolution may be expected. To understand the impacts of these factors on behaviour of the local-scale geomorphology, requires the following points to be considered:

- changes in foreshore response to wider-scale and local factors;
- assessment of the implications of foreshore response on backshore features;
- wider-scale geomorphological assessment of this coastal response (e.g. feedback interactions);
- identification of any potential changes in geomorphological form (e.g. breakdown of gravel barriers).

1.3.4 Generic questions

In assessing coastal and shoreline behaviour a range of factors need to be considered. To ensure that all relevant factors are addressed it is useful to have a framework of generic questions, which have been developed and are outlined below. These questions detail the main issues that need to be addressed when assessing future geomorphological behaviour and coastal evolution. To answer these questions requires an input of both data and understanding of processes and geomorphology.

Past evolution

Knowing how a feature or geomorphological system formed can assist in assessing how it will respond to future changes in the forcing parameters. This assessment may also provide information regarding sources or sinks of sediment. Key questions to be addressed therefore are:

- How and why has the feature formed, and over what timescales? Have some features disappeared and what are the possible reasons for this?
- What has been the historic behaviour of the feature at millennial, centennial and decadal timescales; for example, has the volume held within a dune system changed, or has there been a change in position?
- Are the processes that caused the features to form still occurring today, or can the features be considered relict?
- How does the evolution of a certain feature, or geomorphological element, fit into a larger-scale pattern of change? Is the contemporary landscape a product of a previous different landscape form?

Controls and influences

Key to understanding larger-scale behaviour is the understanding of the main controls on the system:

- What are the key geological controls; for example, are there predominant headlands, and are these composed of hard or soft geology? How is the geology changing over time, that is, what is the resistance to erosion and what is the main failure mechanism?
- Are there any offshore or nearshore controls, for example, banks or islands? Are these changing and/or is there potential for them to change in the future? What

control do they have on the shoreline; for example, are they providing shelter or causing wave focussing?

- What estuarine or inlet controls are present? Is there a delta that is an influence on the shoreline, for example, by providing protection? Are there spits, and how are these behaving?

Forcing

Coastal morphology changes due to the processes that act upon it. Although much of our knowledge regarding processes is contemporary, a good understanding of the coastal response to current conditions informs and improves our predictions of future coastal response. For example:

- What tidal processes operate? Are coastal processes effectively tidal-driven? Is the shoreline subject to storm surges? What has been the past response to such events?
- What wave processes operate? Are coastal processes effectively wave-driven? What are the predominant directions of wave approach? Are there differences in wave energy along the shoreline, for example, due to wave diffraction?

Linkages

The formation and maintenance/growth of geomorphological features is dependent upon a supply of sediment of an appropriate size-grade. This therefore depends upon a suitable source and a transport pathway.

- What are the key sources of sediment within the system? What sizes of sediments are released? How does this compare with the composition of the depositional features present, for example, dunes or beach ridges?
- Have previous sources of sediment now been exhausted or removed from the system, for example, due to rising sea levels?
- Are there key sinks of sediment? Can these be considered permanent or temporary stores? If temporary, are these volumes likely to be released in the future and under what processes, for example, cannibalisation of a barrier as it migrates landwards?
- What are the key mechanisms of sediment transport, for example, suspended or bedload, onshore or longshore?
- What are the interactions between features? Over what temporal and spatial scales are linkages evident? What is the relative strength/importance of these linkages?

Morphology

At the local scale, response of the geomorphological elements is key to the predictions of future coastal evolution. Understanding why the feature is where it is and its particular morphology is essential to the understanding of future behaviour.

- What are the key internal physical controls on the behaviour of the feature, for example, geology/composition, resistance to erosion, height, width, position etc.?

- What are the key external physical controls on the behaviour of the feature, for example, is it a wind-created feature, wave-dominated feature etc.?
- Does it depend on a sediment supply and if so what are the key sources? Is the source of sediment contemporary or relict?
- What are its links with neighbouring geomorphic units, for example, does is depend upon another feature for its sediment supply, or is it a source for other features?
- Does its evolution fit into a larger-scale pattern of change?

The results of the FUTURECOAST project, which addresses these generic questions for the coastline of England and Wales, are available on CD and have been widely distributed in the UK.

1.4   Scope

Coastal engineering is a relatively new and rapidly growing branch of civil engineering. It requires knowledge in a number of specialist subjects including wave mechanics, sediment transport, tide generation and numerical methods in order to understand the behaviour and interaction of coastal features. An appreciation of the power and limitations of numerical prediction methods is becoming increasingly important due to the improvements in computing power and development of computational methods for describing fluid flow and sediment transport. Indeed, some aspects of coastal engineering, such as storm surge prediction, can only be effectively handled with a numerical model. For other aspects, such as the long-term prediction of shoreline evolution, there are as yet no well-established techniques.

There is now a marked trend towards soft engineering rather than the traditional hard concrete structures that were constructed in many parts of the world in the past. In fact, in some coastal areas hard structures are now actively discouraged or prohibited by legislation. Soft engineering does not exclude hard structures but describes the more holistic approach to coast and flood defence being promoted worldwide. This encourages strategic design that takes into account the impact that construction will have on the surrounding coastal area. By necessity this requires a more detailed appreciation of the natural processes of hydrodynamics, sediment transport and morphodynamics in design than in the past.

This book is intended to provide an introduction to coastal engineering; it is not a design guide. It includes development of the theory necessary to understand the processes that are important for coastal engineering design. Much design and assessment work now makes use of mathematical or numerical models, and the use of such models is a persistent theme throughout the book. Despite the rise in popularity of numerical models, final designs are often tested in scale models in laboratories, and the important issue of how to scale full-size design to the laboratory is given in Chapter 8.

Topics covered include: linear wave theory; wave transformation in water of varying depth; a description of non-linear wave characteristics sufficient for application to beach morphology prediction and wave-induced water level changes near the shore; methods of describing the statistical characteristics of wave climate for design purposes; water level variations associated with astronomical and meteorological forces as well as long-wave activity; sediment transport; analysis, modelling and prediction of coastal morphology; design, reliability and risk; field surveying and physical modelling; design philosophy, design equations and design practice.

A key part of the engineer's repertoire is judgement, based on experience gained from design and construction projects. While it is not possible to teach design experience we have included in the book a set of case studies of real projects from across the world. These are used to illustrate how theory, modelling and design principles are drawn together in practice and used in conjunction with engineering judgement in coastal management schemes.

We hope this book will be of assistance to those university and college lecturers teaching modules covering coastal engineering and management. It is also intended to be amenable to practicing engineers, both coastal specialists and others, who require a reference source to consult on specific issues. To this end we have included a number of worked examples throughout the book to illustrate the application of design procedures and calculations.

The material presented in this book draws on the authors' many years combined experience, but is not intended to be exhaustive. Coastal engineering is an active research discipline and new ideas, measurements and techniques are becoming available all the time.

Chapter 2

Wave theory

2.1 Introduction

This chapter is concerned with the theories of periodic progressive waves and their interaction with shorelines and coastal structures. This introduction provides a descriptive overview of the generation of wind waves, their characteristics, the processes which control their movement and transformation and some of the concepts which are employed in the design process for coastal engineering studies.

Ocean waves are mainly generated by the action of wind on water. The waves are formed initially by a complex process of resonance and shearing action, in which waves of differing wave height, length and period are produced and travel in various directions. Once formed, ocean waves can travel for vast distances, spreading in area and reducing in height, but maintaining wavelength and period as shown in Figure 2.1. For example, waves produced in the gales of the 'roaring forties' have

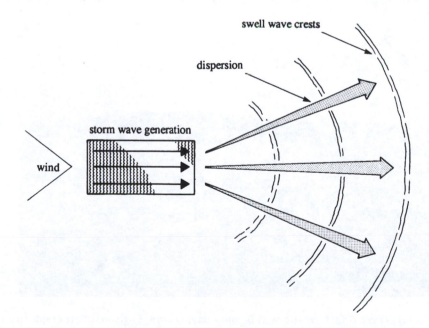

Figure 2.1 Wave generation and dispersion.

been monitored all the way north across the Pacific Ocean to the shores of Alaska (a distance of 10 000 km).

In the storm zone generation area, high-frequency wave energy (e.g. waves with small period) is both dissipated and transferred to lower frequencies. As will be shown later, waves of differing frequencies travel at different speeds, and therefore outside the storm generation area the sea state is modified as the various frequency components separate. The low-frequency waves travel more quickly than the high-frequency waves resulting in a swell sea condition as opposed to a storm sea condition. This process is known as 'dispersion'. Thus wind waves may be characterised as irregular, short crested and steep containing a large range of frequencies and directions. On the other hand swell waves may be characterised as fairly regular, long crested and not very steep containing a small range of low frequencies and directions.

As waves approach a shoreline, their height and wavelength are altered by the processes of refraction and shoaling before breaking on the shore. Once waves have broken, they enter what is termed the surf zone. Here some of the most complex transformation and attenuation processes occur, including generation of cross and longshore currents, a set-up of the mean water level and vigorous sediment transport of beach material. Some of these processes are evident in Figure 2.2(a).

Where coastal structures are present, either on the shoreline or in the nearshore zone, waves may also be diffracted and reflected resulting in additional complexities in the wave motion. Figure 2.2(b) shows a simplified concept of the main wave transformation and attenuation processes which must be considered by coastal engineers in designing coastal defence schemes.

Figure 2.2(a) Wave transformations at Bigbury Bay, Devon, England. Photograph courtesy of Dr S.M. White.

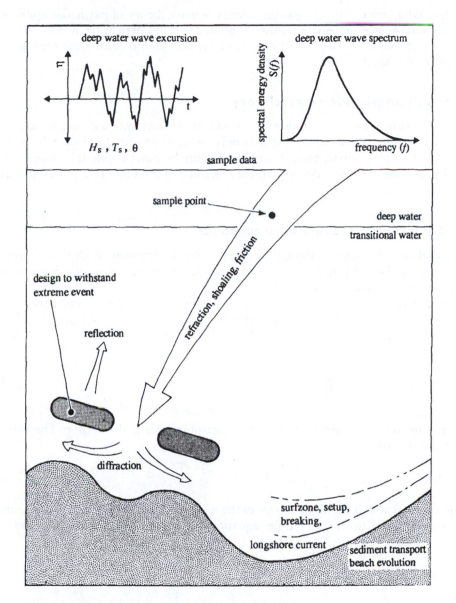

Figure 2.2(b) Wave transformations: main concepts.

Additionally, the existence of wave groups are of considerable significance as they have been shown to be responsible for the structural failure of some maritime structures designed using the traditional approach. The existence of wave groups also generates secondary wave forms of much lower frequency and amplitude called 'bound longwaves'. Inside the surf zone these waves become separated from the 'short' waves and have been shown to have a major influence on sediment transport and beach morphology producing long and cross-shore variations in the surf zone wave field.

The following sections describe some aspects of wave theory of particular application in coastal engineering. Some results are quoted without derivation, as the derivations are often long and complex. The interested reader should consult the references provided for further details.

2.2   Small-amplitude wave theory

The earliest mathematical description of periodic progressive waves is that attributed to Airy in 1845. Airy wave theory is strictly only applicable to conditions in which the wave height is small compared to the wavelength and the water depth. It is commonly referred to as linear or first-order wave theory, because of the simplifying assumptions made in its derivation.

2.2.1   Derivation of the Airy wave equations

Airy wave theory was derived using the concepts of two-dimensional ideal fluid flow. This is a reasonable starting point for ocean waves, which are not greatly influenced by viscosity, surface tension or turbulence.

Figure 2.3 shows a sinusoidal wave of wavelength L, height H and period T, propagating on water with undisturbed depth h. The variation of surface elevation with time, from the still water level, is denoted by η (referred to as excursion) and given by

$$\eta = \frac{H}{2}\cos\left\{2\pi\left(\frac{x}{L} - \frac{t}{T}\right)\right\}$$

(2.1)

where x is the distance measured along the horizontal axis and t is time. The wave celerity, c, is given by

$$c = L/T$$

(2.2)

It is the speed at which the wave moves in the x-direction. Equation (2.1) represents the surface solution to the Airy wave equations. The derivation of the Airy wave

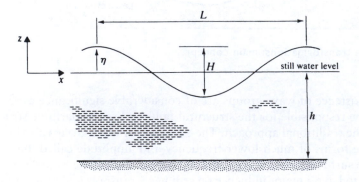

Figure 2.3 Definition sketch for a sinusoidal wave.

equations starts from the Laplace equation for irrotational flow of an ideal fluid. The Laplace equation is simply an expression of the continuity equation applied to a flow net and is given by

$$\frac{\partial u}{\partial x} + \frac{\partial w}{\partial z} = 0 = \frac{\partial^2 \phi}{\partial x^2} + \frac{\partial^2 \phi}{\partial z^2}$$

continuity   Laplace

where u is the velocity in the x-direction, w is the velocity in the z-direction and ϕ is the velocity potential and $u = \dfrac{\partial \phi}{\partial x}$, $w = \dfrac{\partial \phi}{\partial z}$

A solution for ϕ is sought which satisfies the Laplace equation throughout the body of the flow. Additionally this solution must satisfy the boundary conditions at the bed and on the surface. At the bed, assumed horizontal, the vertical velocity w must be zero. At the surface, any particle on the surface must remain on the surface, hence

$$w = \frac{\partial \eta}{\partial t} + u \frac{\partial \eta}{\partial x} \qquad \text{at } z = \eta$$

and the (unsteady) Bernoulli's energy equation must be satisfied,

$$\frac{p}{\rho} + \frac{1}{2}\left(u^2 + w^2\right) + g\eta + \frac{\partial \phi}{\partial t} = C(t) \qquad \text{at } z = \eta$$

Making the assumptions that $H \ll L$ and $H \ll h$ results in the linearised boundary conditions (in which the smaller, higher order and product terms are neglected). The resulting kinematic and dynamic boundary equations are then applied at the still water level, given by

$$w = \frac{\partial \eta}{\partial t} \qquad \text{at } z = 0$$

and

$$g\eta + \frac{\partial \phi}{\partial t} = 0$$

The resulting solution for ϕ is given by

$$\phi = -gH\left[\frac{T}{4\pi}\right] \frac{\cosh\left\{\left(\dfrac{2\pi}{L}\right)(h+z)\right\}}{\cosh\left\{\left(\dfrac{2\pi}{L}\right)h\right\}} \sin\left(\frac{2\pi x}{L} - \frac{2\pi t}{T}\right)$$

Substituting this solution for ϕ into the two linearised surface boundary conditions yields the surface profile given in Equation (2.1) and the wave celerity c given by

$$c = (gT/2\pi)\tanh(2\pi h/L) \tag{2.3a}$$

Most modern texts concerning wave theory use the quantities wave number ($k = 2\pi/L$) and wave angular frequency ($\omega = 2\pi/T$). Thus Equation (2.3a) may be more compactly stated as

$$c = (g/\omega)\tanh(kh) \tag{2.3b}$$

Substituting for c from Equation (2.2) gives

$$c = \frac{L}{T} = \frac{\omega}{k} = \left(\frac{g}{\omega}\right)\tanh(kh)$$

or

$$\omega^2 = gk\tanh(kh) \tag{2.3c}$$

Equation (2.3c) is known as the wave dispersion equation. It may be solved, iteratively, for the wave number, k, and hence wavelength and celerity given the wave period and depth. Further details of its solution and its implications are given in Section 2.3.4. Readers who wish to see a full derivation of the Airy wave equations are referred to Sorensen (1993) and Dean and Dalrymple (1991), in the first instance, for their clarity and engineering approach.

2.2.2   Water particle velocities, accelerations and paths

The equations for the horizontal, u, and vertical, w, velocities of a particle at a mean depth $-z$ below the still-water level may be determined from $\partial\phi/\partial x$ and $\partial\phi/\partial z$ respectively. The corresponding *local* accelerations, a_x and a_z, can then be found from $\partial u/\partial t$ and $\partial w/\partial t$. Finally the horizontal, ζ, and vertical, ξ, displacements can be derived by integrating the respective velocities over a wave period. The resulting equations are given by

$$\zeta = -\frac{H}{2}\left[\frac{\cosh\{k(z+h)\}}{\sinh kh}\right]\sin\left\{2\pi\left(\frac{x}{L} - \frac{t}{T}\right)\right\} \tag{2.4a}$$

$$u = \frac{\pi H}{T}\left[\frac{\cosh\{k(z+h)\}}{\sinh kh}\right]\cos\left\{2\pi\left(\frac{x}{L} - \frac{t}{T}\right)\right\} \tag{2.4b}$$

$$a_x = \frac{2\pi^2 H}{T^2}\left[\frac{\cosh\{k(z+h)\}}{\sinh kh}\right]\sin\left\{2\pi\left(\frac{x}{L} - \frac{t}{T}\right)\right\} \tag{2.4c}$$

and $\xi = \dfrac{H}{2}\left[\dfrac{\sinh\{k(z+h)\}}{\sinh kh}\right]\cos\left\{2\pi\left(\dfrac{x}{L} - \dfrac{t}{T}\right)\right\}$ $\tag{2.5a}$

$$w = \frac{\pi H}{T}\left[\frac{\sinh\{k(z+h)\}}{\sinh kh}\right]\sin\left\{2\pi\left(\frac{x}{L}-\frac{t}{T}\right)\right\} \tag{2.5b}$$

$$a_z = \frac{-2\pi^2 H}{T^2}\left[\frac{\sinh\{k(z+h)\}}{\sinh kh}\right]\cos\left\{2\pi\left(\frac{x}{L}-\frac{t}{T}\right)\right\} \tag{2.5c}$$

All the equations have three components. The first is a magnitude term, the second describes the variation with depth and is a function of relative depth and the third is a cyclic term containing the phase information. Equations (2.4a) and (2.5a) describe an ellipse, which is the path line of a particle according to linear theory. Equations (2.4b, c) and (2.5b, c) give the corresponding velocity and accelerations of the particle as it travels along its path. The vertical and horizontal excursions decrease with depth, the velocities are 90° out of phase with their respective displacements and the accelerations are 180° out of phase with the displacements. These equations are illustrated graphically in Figure 2.4.

2.2.3   Pressure variation induced by wave motion

The equation for pressure variation under a wave is derived by substituting the expression for velocity potential into the unsteady Bernoulli equation and equating the energy at the surface with the energy at any depth. After linearising the resulting equation by assuming that the velocities are small, the equation for the pressure is given by

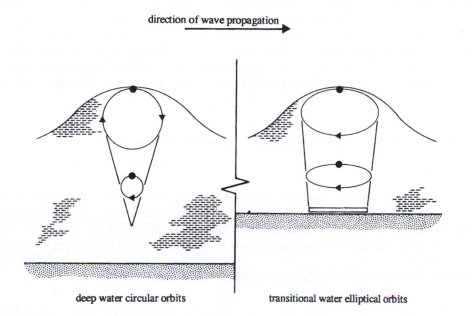

deep water circular orbits   transitional water elliptical orbits

Figure 2.4 Particle displacements for deep and transitional waves.

$$p = -\rho g z + \rho g \frac{H}{2} \cos(kx - \omega t) \frac{\cosh\{k(h+z)\}}{\cosh kh}$$

(valid at or below the still water level, $z = 0$)

or

$$p = -\rho g z + \rho g \eta K_p(z)$$

where $K_p(z)$ is known as the pressure attenuation factor, given by

$$K_p(z) = \frac{\cosh\{k(h+z)\}}{\cosh kh}$$

The pressure attenuation factor is unity at the still-water level, reducing to zero at the deep water limit (i.e. $h/L \geq 0.5$). At any depth ($-z$) under a wave crest, the pressure is a maximum and comprises the static pressure, $-\rho g z$, plus the dynamic pressure, $\rho g \frac{H}{2} K_p(z)$. The reason why it is a maximum under a wave crest is because it is at this location that the vertical particle accelerations are at a maximum and are negative. The converse applies under a wave trough.

Pressure sensors located on the seabed can therefore be used to measure the wave height, provided they are located in the transitional water depth region. The wave height can be calculated from the pressure variation by calculating $K_p(z)$ and subtracting the hydrostatic pressure (mean value of recorded pressure). This requires the solution of the wave dispersion equation for the wavelength in the particular depth, knowing the wave period. This is easily done for a simple wave train of constant period. However, in a real sea comprising a mixture of wave heights and periods, it is first necessary to determine each wave period present (by applying Fourier analysis techniques). Also, given that the pressure sensor will be located in a particular depth, it will not detect any waves whose period is small enough for them to be deep-water waves in that depth.

2.2.4 The influence of water depth on wave characteristics

Deep water

The particle displacement Equations (2.4a) and (2.5a) describe circular patterns of motion in *so-called* deep water. At a depth ($-z$) of $L/2$, the diameter is only 4 per cent of the surface value and this value of depth is normally taken as the lower limit of deep-water waves. Such waves are unaffected by depth, and have little or no influence on the seabed.

For $h/L \geq 0.5$, $\tanh(kh \cong 1)$. Hence Equation (2.3a) reduces to

$$c_0 = gT/2\pi \qquad\qquad (2.6)$$

where the subscript 0 refers to deep water. Alternatively, using Equation (2.2) to eliminate T in Equation (2.6)

$$c_0 = (gL_0/2\pi)^{\frac{1}{2}}$$

Thus, the deep-water wave celerity and wavelength are determined solely by the wave period.

Shallow water

For $h/L \leq 0.04$, $\tanh(kh) \cong 2\pi h/L$). This is normally taken as the upper limit for shallow-water waves. Hence Equation (2.3b) reduces to

$$c = gTh/L$$

and substituting this into Equation (2.2) gives $c = \sqrt{gh}$

Thus, the shallow-water wave celerity is determined by depth, and not by wave period. Hence shallow water waves are not dispersive whereas deep-water waves are.

Transitional water

This is the zone between deep water and shallow water, i.e. $0.5 > h/L > 0.04$. In this zone $\tanh(kh) < 1$, hence

$$c = \frac{gT}{2\pi} \tanh(kh) = c_0 \tanh(kh) < c_0$$

This has important consequences, exhibited in the phenomena of refraction and shoaling, which are discussed in Section 2.3. In addition, the particle displacement equations show that, at the sea bed, vertical components are suppressed so only horizontal displacements now take place (see Figure 2.4). This has important implications regarding sediment transport.

2.2.5   Group velocity and energy propagation

The energy contained within a wave is the sum of the potential, kinetic and surface tension energies of all the particles within a wavelength and it is quoted as the total energy per unit area of the sea surface. For Airy waves, the potential (E_p) and kinetic (E_K) energies are equal and $E_p = E_K = \rho g H^2 L/16$. Hence, the energy ($E$) per unit area of ocean is

$$E = \rho g H^2 / 8 \tag{2.7}$$

(ignoring surface tension energy which is negligible for ocean waves). This is a considerable amount of energy. For example, a (Beaufort) Force 8 gale blowing for 24 h will produce a wave height in excess of 5 m, giving a wave energy exceeding 30 kJ/m².

One might expect that wave power (or the rate of transmission of wave energy) would be equal to wave energy times the wave celerity. This is incorrect, and the derivation of the equation for wave power leads to an interesting result which is of considerable importance. Wave energy is transmitted by individual particles which possess potential, kinetic and pressure energy. Summing these energies and multiplying by the particle velocity in the x-direction for all particles in the wave gives the rate of transmission of wave energy or wave power (P), and leads to the result (for an Airy wave).

$$P = \frac{\rho g H^2}{8} \frac{c}{2}\left(1 + \frac{2kh}{\sinh 2kh}\right) \tag{2.8}$$

or

$$P = Ec_g$$

where c_g is the group wave celerity, given by

$$c_g = \frac{c}{2}\left(1 + \frac{2kh}{\sinh 2kh}\right) \tag{2.9}$$

In deep water ($h/L > 0.5$) the group wave velocity $c_g = c/2$, and in shallow water $c_g = c$.

Hence, in deep water wave energy is transmitted forward at only half the wave celerity. This is a difficult concept to grasp, and therefore it is useful to examine it in more detail.

Consider a wave generator in a model bay supplying a constant energy input of 128 units and assume deep-water conditions. In the time corresponding to the first wave period all of the energy supplied by the generator must be contained within one wavelength from the generator. After two wave periods, half of the energy contained within the first wavelength from the generator (64 units) will have been transmitted a further wavelength (i.e. two wavelengths in total). Also, the energy within the first wavelength will have gained another 128 units of energy from the generator and lost half of its previous energy in transmission (64 units). Hence, the energy level within the first wavelength after two wave periods will be 128 + 128 − 64 = 192 units. The process may be repeated indefinitely. Table 2.1 shows the result after eight wave

Table 2.1 Wave generation: to show group wave speed.

Number of wave periods	Wave energy within various wavelengths from generator								Total wave energy/ generated energy
	1	2	3	4	5	6	7	8	
1	128	0	0	0	0	0	0	0	1
2	192	64	0	0	0	0	0	0	2
3	224	128	32	0	0	0	0	0	3
4	240	176	80	16	0	0	0	0	4
5	248	208	128	48	8	0	0	0	5
6	252	228	168	88	28	4	0	0	6
7	254	240	198	128	58	16	2	0	7
8	255	247	219	163	93	37	9	1	8

periods. This demonstrates that although energy has been radiated to a distance of eight wavelengths, the energy level of 128 units is only propagating one wavelength in every two wave periods. Also the eventual steady wave energy at the generator corresponds to 256 units of energy in which 128 units is continuously being supplied and half of the 256 units continuously being transmitted.

The appearance of the waveform to an observer, therefore, is one in which the leading wave front moves forward but continuously disappears. If the wave generator were stopped after eight wave periods, the wave group (of eight waves) would continue to move forward but, in addition, wave energy would remain at the trailing edge in the same way as it appears at the leading edge. Thus, the wave group would appear to move forward at half the wave celerity, with individual waves appearing at the rear of the group and moving through the group to disappear again at the leading edge. Returning to our example of a Force 8 gale, a typical wave celerity is 14 m/s (for a wave period of 9 s), the group wave celerity is thus 7 m/s, giving a wave power of 210 kW/m.

2.2.6   Radiation stress (momentum flux) theory

This is defined as the *excess flow of momentum due to the presence of waves* (with units of force/unit length). It arises from the orbital motion of individual water particles in the waves. These particle motions produce a net force in the direction of propagation (S_{XX}) and a net force at right angles to the direction of propagation (S_{YY}). The original theory was developed by Longuet-Higgins and Stewart (1964). Its application to long-shore currents was subsequently developed by Longuet-Higgins (1970). The interested reader is strongly recommended to refer to these papers that are both scientifically elegant and presented in a readable style. Further details may also be found in Horikawa (1978) and Komar (1976). Here only a summary of the main results is presented.

The radiation stresses were derived from the linear wave theory equations by integrating the dynamic pressure over the total depth under a wave and over a wave period, and subtracting from this the integral static pressure below the still water depth. Thus, using the notation of Figure 2.3

$$S_{XX} = \overline{\int_{-h}^{\eta} (p + \rho u^2)\,dz} - \int_{-h}^{0} p\,dz$$

The first integral is the mean value of the integrand over a wave period where u is the horizontal component of orbital velocity in the x direction. After considerable manipulation it may be shown that

$$S_{XX} = E\left(\frac{2kh}{\sinh 2kh} + \frac{1}{2}\right) \tag{2.10}$$

Similarly

$$S_{YY} = \overline{\int_{-h}^{\eta} (p + \rho v^2)\,dz} - \int_{-h}^{0} p\,dz$$

where v is the horizontal component of orbital velocity in the y-direction.

For waves travelling in the x-direction $v = 0$ and

$$S_{YY} = E\left(\frac{2kh}{\sinh 2kh}\right) \tag{2.11}$$

In deep water

$$S_{XX} = \frac{1}{2}E \qquad S_{YY} = 0$$

and in shallow water

$$S_{XX} = \frac{3}{2}E \qquad S_{YY} = \frac{1}{2}E$$

Thus both S_{XX} and S_{YY} increase in reducing water depths.

2.3 Wave transformation and attenuation processes

As waves approach a shoreline, they enter the transitional depth region in which the wave motions are affected by the seabed. These effects include reduction of the wave celerity and wavelength, and thus alteration of the direction of the wave crests (refraction) and wave height (shoaling) with wave energy dissipated by seabed friction and finally breaking.

2.3.1 Refraction

Wave celerity and wavelength are related through two Equations (2.2) and (2.3a) to wave period (which is the only parameter which remains constant for an individual wave train. However, it should be noted that the peak energy wave period of a directional wave spectrum can change when refracting from deep to shallow water, particularly if any sheltering is involved). This can be appreciated by postulating a change in wave period (from T_1 to T_2) over an area of sea. The number of waves entering the area in a fixed time t would be t/T_1, and the number leaving would be t/T_2. Unless T_1 equals T_2 the number of waves within the region could increase or decrease indefinitely. Thus,

$$c/c_0 = \tanh(kh) \qquad \text{(from 2.3a)}$$

and

$$c/c_0 = L/L_0 \qquad \text{(from 2.2a)}$$

To find the wave celerity and wavelength at any depth h, these two equations must be solved simultaneously. The solution is always such that $c < c_0$ and $L < L_0$ for $h < h_0$ (where the subscript o refers to deep water conditions).

Consider a deep-water wave approaching the transitional depth limit ($h/L_0 = 0.5$),

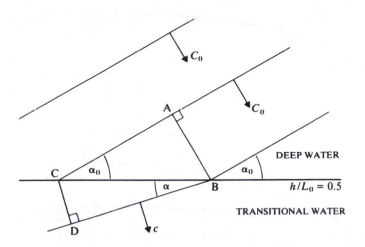

Figure 2.5 Wave refraction.

as shown in Figure 2.5. A wave travelling from A to B (in deep water) traverses a distance L_0 in one wave period T. However, the wave travelling from C to D traverses a smaller distance, L, in the same time, as it is in the transitional depth region. Hence, the new wave front is now BD, which has rotated with respect to AC.

Letting the angle α represent the angle of the wave front to the depth contour, then

$$\sin \alpha = L/BC \text{ and } \sin \alpha_0 = L_0/BC$$

Combining

$$\frac{\sin \alpha}{\sin \alpha_0} = \frac{L}{L_0}$$

Hence

$$\frac{\sin \alpha}{\sin \alpha_0} = \frac{L}{L_0} = \frac{c}{c_0} = \tanh(kh) \tag{2.12}$$

As $c < c0$ then $\alpha < \alpha_0$, which implies that as a wave approaches a shoreline from an oblique angle the wave fronts tend to align themselves with the underwater contours. Figure 2.6 shows the variation of c/c0 with h/L0, and α/α_0 with h/L_0 (the later specifically for the case of parallel contours). It should be noted that L0 is used in preference to L as the former is a fixed quantity.

In the case of non-parallel contours, individual wave rays (i.e. the orthogonals to the wave fronts) must be traced. Figure 2.6 can still be used to find α at each contour if α_0 is taken as the angle (say α_1) at one contour and α is taken as the new angle (say α_2) to the next contour. The wave ray is usually taken to change direction midway between contours. This procedure may be carried out by hand using tables or figures (see Silvester 1974) or by computer as described later in this section.

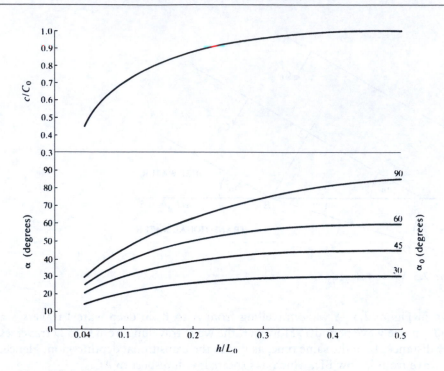

Figure 2.6 Variations of wave celerity and angle with depth.

2.3.2 Shoaling

Consider first a wave front travelling parallel to the seabed contours (i.e. no refraction is taking place). Making the assumption that wave energy is transmitted shorewards without loss due to bed friction or turbulence, then

$$\frac{P}{P_0} = 1 = \frac{Ec_g}{E_0 c_{g_0}}$$

(from Equation 2.8)

Substituting

$$E = \rho g H^2 / 8$$

(Equation 2.7)

then,

$$\frac{P}{P_0} = 1 = \left(\frac{H}{H_0}\right)^2 \frac{c_g}{c_{g_0}}$$

or

$$\frac{H}{H_0} = \left(\frac{c_{g_0}}{c_g}\right)^{1/2} = K_S$$

where K_S is the shoaling coefficient.

The shoaling coefficient can be evaluated from the equation for the group wave celerity, Equation (2.9).

$$K_S = \left(\frac{c_{g_0}}{c_g}\right)^{1/2} = \left(\frac{c_0/2}{\frac{c}{2}\left[1 + \frac{2kh}{\sinh 2kh}\right]}\right)^{1/2} \qquad (2.13)$$

The variation of K_S with d/L_0 is shown in Figure 2.7.

2.3.3   Combined refraction and shoaling

Consider next a wave front travelling obliquely to the seabed contours as shown in Figure 2.8. In this case, as the wave rays bend, they may converge or diverge as they travel shoreward. At the contour $h/L_0 = 0.5$,

$$BC = \frac{b_0}{\cos\alpha_0} = \frac{b}{\cos\alpha}$$

or

$$\frac{b}{b_0} = \frac{\cos\alpha}{\cos\alpha_0}$$

Again, assuming that the power transmitted between any two wave rays is constant (i.e. conservation of wave energy flux), then

$$\frac{P}{P_0} = 1 = \frac{Ebc_g}{E_0 b_0 c_{g_0}}$$

Substituting for E and b

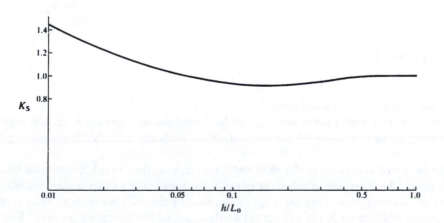

Figure 2.7 Variation of the shoaling coefficient with depth.

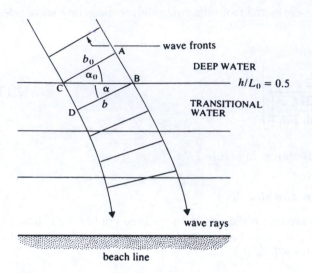

Figure 2.8 Divergence of wave rays over parallel contours.

$$\left(\frac{H}{H_0}\right)^2 \frac{\cos\alpha}{\cos\alpha_0}\frac{c_g}{c_{g_0}} = 1$$

or

$$\frac{H}{H_0} = \left(\frac{\cos\alpha_0}{\cos\alpha}\right)^{1/2}\left(\frac{c_{g_0}}{c_g}\right)^{1/2}$$

Hence

$$H/H_0 = K_R K_S \qquad\qquad (2.14)$$

where

$$K_R = \left(\frac{\cos\alpha_0}{\cos\alpha}\right)^{1/2}$$

and is called the refraction coefficient.

For the case of parallel contours, K_R can be found using Figure 2.6. In the more general case, K_R can be found from the refraction diagram directly by measuring b and b_0.

As the refracted waves enter the shallow-water region, they break before reaching the shoreline. The foregoing analysis is not strictly applicable to this region, because the wave fronts steepen and are no longer described by the Airy waveform. However, it is common practice to apply refraction analysis up to the so-called 'breaker line'. This is justified on the grounds that the inherent inaccuracies are small compared with the initial predictions for deep-water waves, and are within acceptable engineering tolerances. To find the breaker line, it is necessary to estimate the wave height as the wave progresses inshore and to compare this with the estimated breaking wave height at any particular depth. As a general guideline, waves will break when

$$h_b = 1.28 H_b \tag{2.15}$$

where the subscript b refers to breaking. The subject of wave breaking is of considerable interest both theoretically and practically. Further details are described in Section 2.6.2.

Example 2.1   Wave refraction and shoaling
A deep-water wave has a period of 8.5 s, a height of 5 m and is travelling at 45° to the shoreline. Assuming that the seabed contours are parallel, find the height, depth, celerity and angle of the wave when it breaks.

Solution
(a) Find the deep water wavelength and celerity. From (2.6)

$$c_0 = gT/2\pi = 13.27 m/s$$

From (2.2)

$$L_0 = c_0 T = 112.8m$$

(b) At the breaking point, the following conditions (from (2.15) and (2.14)) must be satisfied:

$$h_b = 1.28 H_b \text{ and } H/H_0 = K_R K_S$$

For various trial values of h/L_0, H_b/H_0 can be found using Figures 2.6 and 2.7. The correct solution is when (2.15) and (2.14) are satisfied simultaneously. This is most easily seen by preparing a table, as shown in Table 2.2. For $h/L_0 = 0.05$, $h = 5.6$ m and $H = 4.45$, requiring a depth of breaking of 5.7 m. This is sufficiently accurate for an acceptable solution, so

$$H_b = 4.45 \text{ m} \qquad c = 6.9 \text{ m/s} \qquad h_b = 5.7 \text{ m} \qquad \alpha_b = 22°$$

Table 2.2 Tabular solution for breaking waves.

h/L_0	h (m)	c/c_0	c (m/s)	K_S	α (degrees)	K_R	H/H_0	H (m)	h_b (m)
0.1	11.3	0.7	9.3	0.93	30	0.9	0.84	4.2	5.4
0.05	5.6	0.52	6.9	1.02	22	0.87	0.89	4.45	5.7

2.3.4 Numerical solution of the wave dispersion equation

In order to solve this problem from first principles it is first necessary to solve the wave dispersion equation for L in any depth h. This may be done by a variety of numerical methods. Starting from Equation (2.12)

$$\frac{L}{L_0} = \frac{c}{c_0} = \tanh(kh)$$

hence

$$L = \frac{gT^2}{2\pi} \tanh\left(\frac{2\pi h}{L}\right)$$

Given T and h an initial estimate of L (L_1) can be found by substituting L_0 into the *tanh* term. Thereafter successive estimates (say L_2) can be taken as the average of the current and previous estimates (e.g. $L_2 = (L_0 + L_1)/2$) until sufficiently accurate convergence is obtained. A much more efficient technique is described by Goda (2000), based on Newton's method, given by

$$x_2 = x_1 - \frac{(x_1 - D\coth x_1)}{\left(1 + D\left[\coth^2 x_1 - 1\right]\right)}$$

where $x = 2\pi h/L$, $D = 2\pi h/L_0$ and the best estimate for the initial value is

$$x_1 = \begin{bmatrix} D \text{ for } D \geq 1 \\ D^{\frac{1}{2}} \text{ for } D < 1 \end{bmatrix}$$

This provides an absolute error of less than 0.05 per cent after three iterations.
   A direct solution was derived by Hunt (1979), given by

$$\frac{c^2}{gh} = \left[y + \left(1 + 0.6522y + 0.4622y^2 + 0.0864y^4 + 0.0675y^5\right)^{-1}\right]^{-1}$$

Where $y = k_0 h$, which is accurate to 0.1 per cent for $0 < y < \infty$.

2.3.5 Seabed friction

In the foregoing analysis of refraction and shoaling it was assumed that there was no loss of energy as the waves were transmitted inshore. In reality, waves in transitional and shallow water depths will be attenuated by wave energy dissipation through seabed friction. Such energy losses can be estimated, using linear wave theory, in an analogous way to pipe and open channel flow frictional relationships. In contrast to the velocity profile in a steady current, the frictional effects under wave action produce an oscillatory wave boundary layer which is very small (a few millimetres or centimetres). In consequence, the velocity gradient is much larger than in an equivalent uniform current that in turn implies that the wave friction factor will be many times larger.

Firstly, the mean seabed shear stress (τ_b) may be found using

$$\tau_b = \frac{1}{2}f_w \rho u_m^2$$

where f_w is the wave friction factor and u_m is the maximum near-bed orbital velocity; f_w is a function of a local Reynolds' number (Re_w) defined in terms of u_m (for velocity) and either a_b, wave amplitude at the bed or the seabed grain size k_s (for the characteristic length). A diagram relating f_w to Re_w for various ratios of a_b/k_s, due to Jonsson, is given in Dyer (1986). This diagram is analogous to the Moody diagram for pipe friction factor (λ). Values of f_w range from about 0.5×10^{-3} to 5. Hardisty (1990) summarises field measurements of f_w (from Sleath) and notes that a typical field value is about 0.1. Soulsby (1997) provides details of several equations which may be used to calculate the wave friction factor. For rough turbulent flow in the wave boundary layer, he derived a new formula which best fitted the available data, given by

$$f_w = 0.237 r^{-0.52}$$

where

$$r = A/k_s$$

$$A = u_m T/2\pi$$

Using linear wave theory u_m is given by

$$u_m = \frac{\pi H}{T \sinh kh}$$

The rate of energy dissipation may then be found by combining the expression for τ_b with linear wave theory to obtain

$$\frac{dH}{dx} = -\frac{4f_w k^2 H^2}{3\pi \sinh(kh)(\sinh(2kh) + 2kh)}$$

The wave height attenuation due to seabed friction is of course a function of the distance travelled by the wave as well as the depth, wavelength and wave height. Thus the total loss of wave height (ΔH_f) due to friction may be found by integrating over the path of the wave ray.

   BS6349 presents a chart from which a wave height reduction factor may be obtained. Except for large waves in shallow water, seabed friction is of relatively little significance. Hence, for the design of maritime structures in depths of 10 m or more, seabed friction is often ignored. However, in determining the wave climate along the shore, seabed friction is now normally included in numerical models, although an appropriate value for the wave friction factor remains uncertain and is subject to change with wave-induced bed forms. Furthermore, wave energy losses due to other physical processes such as breaking can be more significant.

2.3.6 Wave-current Interaction

So far, consideration of wave properties has been limited to the case of waves generated and travelling on quiescent water. In general, however, ocean waves are normally travelling on currents generated by tides and other means. These currents will also, in general, vary in both space and time. Hence, two distinct cases need to be considered here. The first is that of waves travelling *on* a current and the second when waves generated in quiescent water encounter a current (or travel over a varying current field).

For waves travelling on a current, two frames of reference need to be considered. The first is a moving or *relative* frame of reference, travelling at the current speed. In this frame of reference, all the wave equations derived so far still apply. The second frame of reference is the stationary or *absolute* frame. The concept which provides the key to understanding this situation is that the wavelength is the *same* in both frames of reference. This is because the wavelength in the relative frame is determined by the dispersion equation and this wave is simply moved at a different speed in the absolute frame. In consequence, the absolute and relative wave periods are different.

Consider the case of a current with magnitude (u) following a wave with wave celerity (c), the wave speed with respect to the seabed (c_a) becomes $c + u$. As the wavelength is the same in both reference frames, the absolute wave period will be less than the relative wave period. Consequently, if waves on a current are measured at a fixed location (e.g. in the absolute frame), then it is the absolute period (T_a) which is measured. The current magnitude must, therefore, also be known in order to determine the wavelength. This can be shown as follows:

Starting from the dispersion Equation (2.3a) and noting that $c=L/T_r$ leads to

$$c = \left(\frac{gL}{2\pi} \tanh \frac{2\pi h}{L} \right)^{\!\frac{1}{2}}$$

As $c_a = c + u$ and $c_a = L/T_a$, then

$$L = \left[\left(\frac{gL}{2\pi} \tanh \frac{2\pi h}{L} \right)^{\!\frac{1}{2}} + u \right] T_a$$

This equation thus provides an implicit solution for the wavelength in the presence of a current when the absolute wave period has been measured.

Conversely, when waves travelling in quiescent water encounter a current, changes in wave height and wavelength will occur. This is because as waves travel from one region to the other requires that the absolute wave period remains constant for waves to be conserved. Consider the case of an opposing current, the wave speed relative to the seabed is reduced and therefore the wavelength will also decrease. Thus wave height and steepness will increase. In the limit the waves will break when they reach limiting steepness. In addition, as wave energy is transmitted at the group wave speed, waves cannot penetrate a current whose magnitude equals or exceeds the group wave speed and thus wave breaking and diffraction will occur under these circumstances. Such conditions can occur in the entrance channels to estuaries when strong ebb tides are running, creating a region of high, steep and breaking waves.

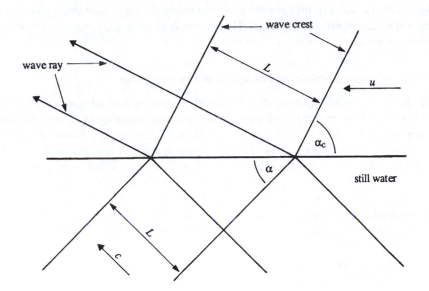

Figure 2.9 Deep water wave refraction by a current.

Another example of wave-current interaction is that of current refraction. This occurs when a wave obliquely crosses from a region of still water to a region in which a current exits or in a changing current field. The simplest case is illustrated in Figure 2.9 showing deep-water wave refraction by a current.

In an analogous manner to refraction caused by depth changes, Jonsson showed that in the case of current refraction

$$\sin \alpha_c = \frac{\sin \alpha}{\left(1 - \dfrac{u}{c} \sin \alpha\right)^2}$$

The wave height is also affected and will decrease if the wave orthogonals diverge (as shown) or increase if the wave orthogonals converge.

For further details of wave-current interactions, the reader is referred to Hedges (1987) in the first instance.

2.3.7   The generalised refraction equations for numerical solution techniques

The foregoing equations for refraction and shoaling may be generalised for application to irregular bathymetry and then solved using a suitable numerical scheme. Two approaches have been developed. The first is the numerical equivalent of the ray (i.e. wave orthogonal) tracing technique and allows determination of individual ray paths, giving a clear picture of wave refraction patterns for any bathymetry. The wave height at any location, however, has to be calculated separately using the local ray spacing (b) to find the refraction coefficient (K_R). The second method computes the local wave height and direction at each point on a regular grid using the wave and energy conservation equation in Cartesian coordinates. This is much more useful as input to other models (for example for wave-induced currents).

2.3.8 The wave conservation equation in wave ray form

Figure 2.10 shows a pair of wave crests and a corresponding pair of wave rays. The wave rays are everywhere at right angles to the wave crests resulting in an orthogonal grid. This implies that only wave refraction and shoaling can occur. Wave energy is therefore conserved between wave rays. The wave ray at point A is at an angle θ with the x-axis and is travelling at speed c. The wave ray at B is a small distance δb from A and is travelling at a speed $c + \delta c$, as it is in slightly deeper water than point A.

In a small time δt, the wave ray at A moves to E at a speed c and the wave ray at B moves to D at speed $c + \delta c$. Thus the wave orthogonal rotates through $\delta \theta$. Let point M be the centre of rotation at distance R from A and E. Using similar triangles

$$\frac{c\delta t}{R} = \frac{(c + \delta t)\delta t - c\delta t}{\delta b}$$

Simplifying and rearranging

$$\frac{\delta b}{R} = \frac{\delta c}{c} \qquad\qquad (2.16)$$

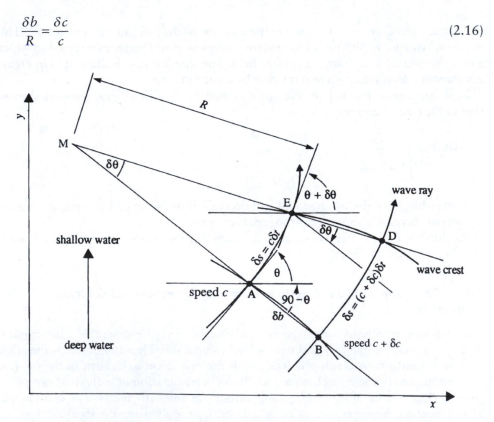

Figure 2.10 Derivation of the wave conservation equation in wave ray form.

also

$$\frac{\delta s}{R} = -\delta\theta$$

The negative sign is introduced to ensure that the orthogonal bends in the direction of reducing c. Rearranging

$$-\frac{\delta\theta}{\delta s} = \frac{1}{R} \qquad (2.17)$$

combining Equations (2.16) and (2.17) and in the limit

$$\frac{d\theta}{ds} = -\frac{1}{c}\frac{dc}{db} \qquad (2.18)$$

Considering a ray path, by trigonometry

$$\delta x = \delta s \cos\theta$$

$$\delta y = \delta s \sin\theta$$

and as

$$\delta s = c\delta t$$

then in the limit

$$\frac{dx}{dt} = c\cos\theta \qquad (2.19)$$

$$\frac{dy}{dt} = c\sin\theta \qquad (2.20)$$

Returning to Equation (2.18) and given that

$$c = f(x,y) \text{ and } x,y = f(b)$$

then applying the chain rule

$$\frac{dc}{db} = \frac{\partial c}{\partial x}\frac{\partial x}{\partial b} + \frac{\partial c}{\partial y}\frac{\partial y}{\partial c} \qquad (2.21)$$

Along a wave crest

$$\frac{\partial x}{\partial b} = \cos(90 - \theta) = -\sin\theta \tag{2.22}$$

$$\frac{\partial y}{\partial b} = \sin(90 - \theta) = \cos\theta \tag{2.23}$$

Substituting Equations (2.21), (2.22) and (2.23) into Equation (2.18) yields

$$\frac{d\theta}{ds} = \frac{1}{c}\left(\frac{\partial c}{\partial x}\sin\theta - \frac{\partial c}{\partial y}\cos\theta\right) \tag{2.24}$$

Finally, we note that $\partial s = c\partial t$. Substituting this into Equation (2.24) gives

$$\frac{d\theta}{dt} = \frac{\partial c}{\partial x}\sin\theta - \frac{\partial c}{\partial y}\cos\theta \tag{2.25}$$

Equations (2.19), (2.20) and (2.25) may be solved numerically along a ray path sequentially through time. Koutitas (1988) gives a worked example of such a scheme. If two closely spaced ray paths are calculated, the local refraction coefficient may then be found and hence the wave heights along the ray path determined. However, a more convenient method to achieve this was developed by Arthur *et al.* (1952). They derived an expression for the orthogonal separation factor $\beta = b/b_c = K_R^{-\frac{1}{2}}$ given by

$$\frac{d^2\beta}{ds^2} + p\frac{d\beta}{ds} + q\beta = 0$$

where

$$p = \frac{\cos\theta}{c}\frac{\partial c}{\partial x} - \frac{\sin\theta}{c}\frac{\partial c}{\partial y}$$

and

$$q = \frac{\sin^2\theta}{c}\frac{\partial^2 c}{\partial x^2} - 2\frac{\sin\theta\cos\theta}{c}\frac{\partial^2 c}{\partial x\partial y} + \frac{\cos^2\theta}{c}\frac{\partial^2 c}{\partial y^2}$$

The derivation of these equations may be found in Dean and Dalrymple (1991) together with some references to the numerical solution techniques.

2.3.9   Wave conservation equation and wave energy conservation equation in Cartesian coordinates

The wave conservation Equation (2.18) may be reformulated in Cartesian coordinates by transformation of the axes. The result, in terms of the wave number ($k = 2\pi/L = \omega/c$) is given by

$$\frac{\partial(k\sin\theta)}{\partial x} - \frac{\partial(k\cos\theta)}{\partial y} = 0 \tag{2.26}$$

The proof that Equation (2.26) is equivalent to (2.18) is given in Dean and Dalrymple (1991). The wave energy conservation equation is given

$$\frac{\partial(EC_g\cos\theta)}{\partial x} - \frac{\partial(EC_g\sin\theta)}{\partial y} = -\varepsilon_d \tag{2.27}$$

where ε_d represents energy losses (due to seabed friction, cf. Section 2.3.5). Again, Koutitas gives a worked example of a numerical solution to Equations (2.26) and (2.27).

2.3.10   Wave reflection

Waves normally incident on solid vertical boundaries (e.g. harbour walls and sea walls) are reflected such that the reflected wave has the same phase but opposite direction and substantially the same amplitude as the incident wave. This fulfils the necessary boundary condition that the horizontal velocity is always zero. The resulting wave pattern set up is called a 'standing wave', as shown in Figure 2.11. Reflection can also occur when waves enter a harbour or estuary. This can lead to 'resonance' where the waves are amplified (see Section 4.8.3).

The equation of the standing wave (subscript s) may be found by adding the two waveforms of the incident (subscript i) and reflected (subscript r) waves. Thus,

$$\eta_i = \frac{H_i}{2}\cos\left\{2\pi\left(\frac{x}{L} - \frac{t}{T}\right)\right\}$$

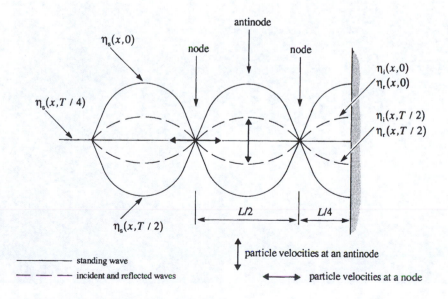

Figure 2.11(a) Standing waves, idealised.

$$\eta_r = \frac{H_r}{2}\cos\left\{2\pi\left(\frac{x}{L}+\frac{t}{T}\right)\right\}$$

$$\eta_s = \eta_i + \eta_r$$

Taking

$$H_r = H_i = H_s/2$$

then

$$\eta_s = H_s\cos(2\pi x/L)\cos(2\pi t/T) \tag{2.28}$$

At the nodal points there is no vertical movement with time. By contrast, at the anti-nodes, crests and troughs appear alternately. For the case of large waves in shallow water and if the reflected wave has a similar amplitude to the incident wave, then the advancing and receding crests collide in a spectacular manner, forming a plume known as a 'clapotis' (see Figure 2.11(b)). This is commonly observed at sea walls. Standing waves can cause considerable damage to maritime structures, and bring about substantial erosion.

Clapotis gaufre

When the incident wave is at an angle α to the normal from a vertical boundary, then the reflected wave will be in a direction α on the opposite side of the normal. This is

Figure 2.11(b) Standing waves, observed clapotis.

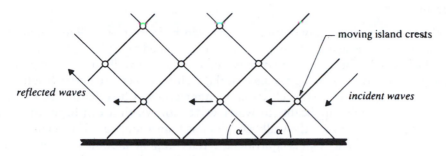

Figure 2.12 Plan view of oblique wave reflection.

illustrated in Figures 2.12 and 2.13. The resulting wave motion (the 'clapotis gaufre') is complex, but essentially consists of a diamond pattern of island crests which move parallel to the boundary. It is sometimes referred to as a short-crested system. The crests form at the intersection of the incident and reflected wave fronts. The resulting particle displacements are also complex, but include the generation of a pattern of moving vortices. A detailed description of these motions may be found in Silvester (1974). The consequences of this in terms of sediment transport may be severe. Very substantial erosion and longshore transport may take place. Considering that oblique wave attack to sea walls is the norm rather than the exception, the existence of the clapotis gaufre has a profound influence on the long-term stability and effectiveness of coastal defence works. This does not seem to have been fully understood in traditional designs of sea walls, with the result that collapsed sea walls and eroded coastlines have occurred.

Figure 2.13 Wave impact and reflection during a storm.

Mach stem

When periodic or solitary waves approach a steep barrier at an oblique angle, the amplitude of the wave against the barrier may be magnified by a phenomenon known as the 'Mach stem'. The crest immediately adjacent to the wall alters its alignment to create a wave travelling along the face of the wall with increased crest height and this is the Mach stem wave. This reflection phenomenon was first discovered in aerodynamics, but is equally applicable to water waves for which it can begin to occur when the angle of obliquity to the wall becomes less than about 45°. The height of the crest gives rise to a velocity that is equivalent to the component of the incident wave's celerity in the direction of the alignment of the wall. Since the waves do not strike the wall due to a growing slipstream zone, reflection is much reduced until for angles of obliquity less than about 20° reflection becomes non-existent.

Miles (1980) demonstrated theoretically that the Mach stem wave could be amplified by as much as four times the incoming waves, two times greater than a linear superposition of incident and reflected wave. However, Melville (1980) was unable to reproduce such large amplification factors in the laboratory. More recently Yoon and Liu (1989) employed parabolic approximations to study the stem waves induced by an oblique cnoidal wave train in front of a vertical barrier and Honda and Mase (2007) have applied a non-linear frequency-domain wave model to Mach stem evolution and wave transformation on a reef structure. In practical design terms it is important to recognise the potential for Mach stem waves to exist as the enhanced wave height and high velocities running along a breakwater can result in increased overtopping, armour instability, toe scour problems and beach erosion at the root. This is discussed further in Chapter 9.

Wave reflection coefficients

Defining a reflection coefficient $K_r = H_r/H_i$ then typical values are as follows

Reflection barrier	K_r
Concrete sea walls	0.7–1.0
Rock breakwaters	0.4–0.7
Beaches	0.05–0.2

It should be noted that the reflected wave energy is equal to K_r^2 as energy is proportional to H^2.

Predictive equations for wave reflection from rock slopes

The Rock Manual (CIRIA/CUR manual 1991 and CIRIA/CUR/CETMEF 2007) gives an excellent summary of the development of wave reflection equations based on laboratory data of reflection from rock breakwaters. This work clearly demonstrates that rock slopes considerably reduce reflection compared to smooth impermeable slopes. Based on this data, the best fit equation was found to be

$$K_r = 0.125\xi_p^{0.73}$$

where ξ_p is the Iribarren Number $= \beta/\sqrt{H/L_p}$ and p refers to peak frequency.

Davidson *et al.* (1996) subsequently carried out an extensive field measurement programme of wave reflection at prototype scale at the Elmer breakwaters (Sussex, UK) and after subsequent analysis proposed a new dimensionless reflection parameter given by

$$R = \frac{d_t \lambda_0^2}{H_i D^2 \cot \beta} \qquad (2.29)$$

where d_t (m) is water depth at the toe of structure, λ_0 is deep water wavelength at peak frequency, H_i is significant incident wave height, D is characteristic diameter of rock armour (= W_{50}/ρ median mass/density) and tan β is the structure gradient. R was found to be a better parameter than ξ in predicting wave reflection. The reflection coefficient is then given by

$$K_r = 0.151 R^{0.11} \qquad (2.30)$$

or, alternatively

$$K_r = \frac{0.635 R^{0.5}}{41.2 + R^{0.5}} \qquad (2.31)$$

Wave reflection due to refraction

Wave reflection due to refraction alone can also occur due to very rapid changes in the seabed. In particular, when waves approach a deep dredged channel with the direction or propagation at a sufficiently acute angle to the dredged side slope and there is a sufficiently large change in water depth which, in turn, results in a large and rapid change in wave speed, the wave may reflect off the side of the channel. An analogous example of this phenomenon is the internal reflection of light rays in a glass prism due to changes in wave speed between the glass (shallow water) and air (deep water), the essential difference being that, as wave speed is a function of water depth, it is not a constant on the wave approach or on the channel side slope. This is a very real phenomenon and, if not recognised, can result in wave energy inadvertently being reflected into a port area. The converse also applies as this process can also be used to advantage to reflect wave energy away from a harbour entrance. It should also be appreciated that longer period waves will also be more susceptible to this phenomenon due to their relatively greater speed in deeper water. When it comes to wave modelling, as described in Section 3.9, it follows that any numerical grid used in a wave model must be fine enough to capture the detail of the dredged channel in order to properly reproduce this effect.

2.3.11   Wave diffraction

This is the process whereby waves bend round obstructions by radiation of the wave energy. Figure 2.14a shows an oblique wave train incident on the tip of a breakwater. There are three distinct regions:

1   the shadow region in which diffraction takes place;
2   the short-crested region in which incident and reflected waves form a clapotis gaufre;
3   an undisturbed region of incident waves.

In region (1), the waves diffract with the wave fronts forming circular arcs centred on the point of the breakwater. When the waves diffract, the wave heights diminish as the energy of the incident wave spreads over the region. The real situation is, however, more complicated than that presented in Figure 2.14a. The reflected waves in region (2) will diffract into region (3) and hence extend the short crested system into region (3).

Mathematical formulation of wave diffraction

Mathematical solutions for wave diffraction have been developed for the case of constant water depth using linear wave theory. The basic differential equation for wave diffraction is known as the Helmholtz equation. This can be derived from the Laplace equation (refer to Section 2.2) in three dimensions

$$\frac{\partial^2 \phi}{\partial x^2} + \frac{\partial^2 \phi}{\partial y^2} + \frac{\partial^2 \phi}{\partial z^2} = 0$$

Now, let

$$\phi(x, y, z) = Z(z)F(x, y)e^{i\omega t}$$

(i.e. ϕ is a function of depth and horizontal coordinates and is periodic and i is the imaginary number = $\sqrt{-1}$)

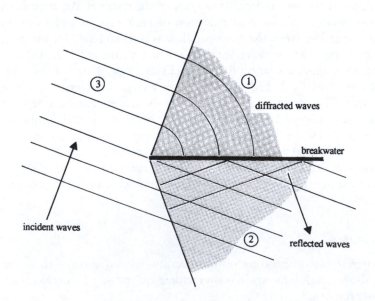

Figure 2.14(a) Idealised wave diffraction around an impermeable breakwater.

Figure 2.14(b) Photograph of real wave diffraction at the Elmer breakwater scheme, Sussex, England.

Figure 2.14 (c) Physical model study of (b) in the UK Coastal Research Facility at HR Wallingford.

Figure 2.14 (d) Aerial photograph of wave diffraction at the Happisburgh to Winterton scheme, Norfolk, England, (courtesy of Mike Page).

For uniform depth an expression for $Z(z)$ satisfying the no-flow bottom boundary condition is

$$Z(z) = \cosh \{k(h+z)\}$$

Substituting for ϕ and Z in the Laplace equation leads (after further manipulation) to the Helmholtz equation

$$\frac{\partial^2 F}{\partial x^2} + \frac{\partial^2 F}{\partial y^2} + K^2 F(x,y) = 0 \tag{2.32}$$

Solutions to the Helmholtz equation

A solution to the Helmholtz equation was first found by Sommerfeld in 1896 who applied it to the diffraction of light (details may be found in Dean and Dalrymple (1991)). Somewhat later, Penney and Price (1952) showed that the same solution applied to water waves and presented solutions for incident waves from different directions passing a semi-infinite barrier and for normally incident waves passing through a barrier gap. For the case of normal incidence on a semi-infinite barrier, it may be noted that, for a monochromatic wave, the diffraction coefficient K_d (= H_d/H_i) is approximately 0.5 at the edge of the shadow region and that K_d exceeds 1.0 in the 'undisturbed' region due to diffraction of the reflected waves caused by the (perfectly) reflecting barrier. Their solution for the case of a barrier gap is essentially the superposition of the results from two mirror image semi-infinite barriers.

Their diagrams apply for a range of gap width to wavelength (b/L) from 1 to 5. When b/L exceeds 5 the diffraction patterns from each barrier do not overlap and hence the semi-infinite barrier solution applies. For b/L less than one the gap acts as a point source and wave energy is radiated as if it were coming from a single point at the centre of the gap. It is important to note here that these diagrams should not be used for design. This is because of the necessity of considering directional wave spectra, which are discussed in Chapter 3.

2.3.12   Combined refraction and diffraction

Refraction and diffraction often occur together. For example, the use of a wave ray model over irregular bathymetry may produce a caustic (i.e. a region where wave rays cross). Here diffraction will occur spreading wave energy away from regions of large wave heights. Another example is around offshore breakwaters; here diffraction is often predominant close to the structure with refraction becoming more important further away from the structure. A solution to the Laplace equation over irregular bathymetry is required, which allows diffraction as well as refraction. Such a solution was first derived in 1972 by Berkhoff. This is generally known as the mild slope equation because the solution is restricted to bathymetry that varies slowly relative to the wavelength.

It may be written as

$$\frac{\partial}{\partial x}\left(cC_g\frac{\partial\phi}{\partial x}\right)+\frac{\partial}{\partial y}\left(cC_g\frac{\partial\phi}{\partial y}\right)+\omega^2\frac{C_g}{c}\phi=0 \tag{2.33}$$

where $\phi\,(x,\,y)$ is a complex wave potential function. The solution of this equation is highly complex and beyond the scope of this text. However, the interested reader is directed to Dingemans (1997) for a review of the subject. This type of model is also discussed further in Section 3. 9. One of the more recent developments in solving the mild slope equation is that due to Li (1994a). This version of the mild slope equation allows the simultaneous solution of refraction, diffraction *and* reflection. It has also been the subject of a field validation study. Initial results may be found in Ilic and Chadwick (1995). They tested this model at the site of the Elmer offshore breakwater scheme (shown in Figure 2.14b) where refraction and reflection are the main processes seaward of the breakwaters with diffraction and refraction taking place shoreward of the breakwaters, and in a physical model (shown in Figure 2.14c).

2.4   Finite amplitude waves

It has already been noted that the Airy wave equations only strictly apply to waves of relatively small height in comparison to their wavelength and water depth. For steep waves and shallow-water waves the profile becomes asymmetric with high crests and shallow troughs. For such waves, celerity and wavelength are affected by wave height and are better described by other wave theories. To categorise finite amplitude waves, three parameters are required. These are the wave height (H), the water depth (h) and wavelength (L). Using these parameters various non-dimensional parameters can be defined, namely relative depth (h/L), wave steepness (H/L) and wave height to water depth ratio (H/h). Another useful non-dimensional parameter is the Ursell number $(U_r = HL^2/h^3)$, first introduced in 1953.

The first finite amplitude wave theory was developed by Stokes in 1847. It is applicable to steep waves in deep and transitional water depths. Following Stokes, Korteweg and de Vries developed a shallow-water finite amplitude wave theory in 1895. They termed this Cnoidal theory, analogous to the sinusoidal Airy wave theory. Both of these theories relax the assumptions made in Airy theory which, as previously described, linearises the kinematic and dynamic surface boundary conditions. In Stokes' wave theory H/L is assumed small and h/L is allowed to assume a wide range of values. The kinematic free surface boundary condition is then expressed as a power series in terms of H/L, and solutions up to and including the nth order of this power series are sought. Stokes derived the second-order solution. In Cnoidal theory, H/h is assumed small and U_r of the order of unity. Korteweg and de Vries derived a first-order solution. Much more recently (1960s to 1980s), these two theories have been extended to higher orders (third and fifth). The mathematics is complex and subsequently other researchers developed new methods whereby solutions could be obtained to any arbitrary order by numerical solution.

Stokes' solution for the surface profile is given by:

$$\eta = \frac{H}{2}\cos\left\{2\pi\left(\frac{x}{L} - \frac{t}{T}\right)\right\} + \frac{\pi H}{8}\left(\frac{H}{L}\right)\frac{\cosh\{kh(2 + \cosh 2kh)\}}{\sinh^3 kh}\cos\left\{4\pi\left(\frac{x}{L} - \frac{t}{T}\right)\right\}$$

This equation differs from the linear solution by the addition of the second-order term. Its frequency is twice that of the first-order term, which therefore increases the crest height, decreases the trough depth and thus increases the wave steepness. To second order, the wave celerity remains the same as linear theory. However, to third order the wave celerity increases with wave steepness and is approximately 20 per cent higher than given by linear theory in deep water at the limiting steepness (1/7).

A full mathematical description of all these theories is beyond the scope of this book and the reader is referred to Dean and Dalrymple (1991) and Sorenson (1993) for further details. However, it is useful here to provide some information on the circumstances under which these finite amplitude wave theories can be applied. Figure 2.15, taken from Hedges (1995), provides useful guidance. It may be noted that the range of validity of linear theory is reassuringly wide, covering all of the transitional water depths for most wave steepnesses encountered in practice. For engineering design purposes, the main implication of using linear theory outside its range of validity is that wave celerity and wavelength are not strictly correct, leading to (some) inaccuracies in refraction and shoaling analysis. In addition, the presence of asymmetrical wave forms will produce harmonics in the Fourier analysis of recorded wave traces which could be incorrectly interpreted as free waves of higher frequency.

2.5 Wave forces

Wave forces on coastal structures are highly variable and depend on both the wave conditions and the type of structure being considered. Three cases of wave conditions

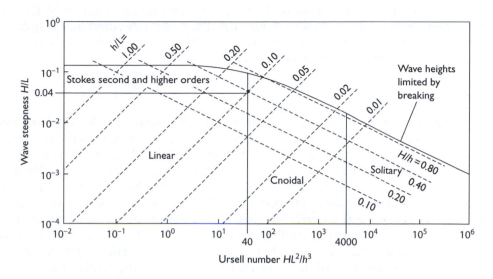

Figure 2.15 Approximate regions of validity of analytical wave theories.

need to be considered, comprising unbroken, breaking and broken waves. Coastal structures may also be considered as belonging to one of three types, vertical walls (e.g. sea walls, caisson breakwaters), rubble mound structures (e.g. rock breakwaters, concrete armoured breakwaters) and individual piles (e.g. for jetty construction). Here consideration is limited to outlining some of the concepts and mentioning some of the design equations that have been developed. Specific equations recommended for design purposes may be found in Chapter 9.

2.5.1 Vertical walls

The forces exerted on a vertical wall by wave action can be considered to be composed of three parts: the static pressure forces, the dynamic pressure forces and the impulsive forces. When the structure is placed such that the incident waves are unbroken, then under normal incidence a standing wave will exist seaward of the wall and only the static and dynamic forces will exist. These can be readily determined from linear wave theory. As a standing wave comprises two superposed progressive waves, travelling in opposite directions, the resulting equation for pressure under a standing wave is of the same form as that for a progressive wave. The standing wave height must be used in the equation, rather than the incident wave height (see Section 2.3.10 for the equation and Dean and Dalrymple (1991) for further details). However, more commonly the structure will need to resist the forces produced by breaking or broken waves. The most widely used formulae for estimating the quasi static pulsating forces for either broken or unbroken waves are due to Goda (1974, 2000).

Additionally, very high localised impulsive forces may also arise due to breaking waves. These trap pockets of air, which are rapidly compressed, resulting in highly variable impulse forces(between 10 to 50 times higher than the pulsating forces). The study of this phenomenon is an ongoing area of research and currently there are no widely accepted formulae for the prediction of such forces (see Cuomo *et al.* (2010a) for recent results). Impact pressure forces are of very short duration (of the order of tenths of a second) and consequently typically affect the dynamic response of the structure rather than its static equilibrium.

2.5.2 Rubble mounds

In the case of rubble mound structures, the waves will generally break on the structure and their energy is partly dissipated by turbulence and friction, with the remaining energy being reflected and possibly transmitted. Many breakwaters are constructed using large blocks of rock (the 'armour units') placed randomly over suitable filter layers. More recently, rock has been replaced by numerous shapes of massive concrete blocks (for example, Dolos, Tetrapod and Cob). The necessary size of the armour units depends on several interrelated factors (wave height, amour unit type and density, structure slope and permeability). Traditionally, the Hudson formula has been used. This was derived from an analysis of a comprehensive series of physical model tests on breakwaters with relatively permeable cores and using regular waves. More recently (1985–1993) these equations have been superseded by Van der Meer's equations for rock breakwaters. These equations were also developed from an extensive series of physical model tests. In these tests random waves were used and the influence of wave period and number of storm waves were also considered. A new damage criterion and a notional core permeability factor were developed. The equations are for use where the structure is placed in deep water with the waves either breaking on the structure or causing surging.

2.5.3 Vertical piles

Finally for the case of unbroken wave forces on piles, the Morrison equation is an option that is used for design. This equation presumes that there are two forces acting. These are a drag force (F_D) induced by flow separation around the pile and an inertia force (F_I) due to the flow acceleration. For the case of a vertical pile, only the horizontal velocities (u) and accelerations (a_x) need be considered (see Equations 2.4b, 2.4c). The drag and inertia forces per unit length of pile of diameter D are given by:

$$F_D = 1/2 C_D \rho D u |u|$$

where C_D is a drag coefficient
and

$$F_I = \rho C_M (\pi D^2 / 4) a_x$$

where C_M is an inertia coefficient, so that the total 'in-line' force is given by:

$$F = \frac{C_D \rho \, |u| \, u D}{2} + \frac{C_M \rho \pi D^2}{4} \frac{du}{dt} \qquad (2.34)$$

Morison's Equation (Equation 2.34) is derived from a combination of theoretical considerations and empirical evidence, not from first principles. The equation does not include lift and slam forces and is most appropriately applied to slender circular piles or pipes subject to unbroken waves. Considering linear waves the velocity u and corresponding component of acceleration are given by Equation 2.4b and 2.4c respectively. Writing

$$u = \frac{\omega A}{2} \cos\left\{ 2\pi \left(\frac{x}{L} - \frac{t}{T} \right) \right\}$$

where

$$A = \frac{H \cosh(k(z+h))}{\sinh(kh)}$$

and

$$\frac{du}{dt} = \frac{A\omega^2}{2} \sin(kx - \omega t)$$

allows the drag term due to linear waves to be written as

$$Drag\ term = \frac{C_d \rho D A^2 \omega^2}{(2)(2^2)} \cos(kx - \omega t)|\cos(kx - \omega t)|$$

and the inertia term as

$$Inertia\ term = \frac{C_M \rho \pi D^2}{4} \frac{A\omega^2}{2} \sin(kx - \omega t)$$

Both these forces reduce with increasing depth and are 90° out of phase. The total force acting on a vertical pile must be found as their sum, integrated over the length of the pile. The ratio of the maximum of the inertia and drag forces is found by setting the trigonometrical terms in each expression equal to one:

$$\frac{Maximum\ inertia}{Maximum\ drag} = \frac{\pi C_M}{C_d} \frac{D}{A} \approx \frac{2\pi D}{A}$$

Typical values of C_D and C_M for cylinders are 1 and 2 respectively, leading to the approximation in the above equation. The number A/D has special significance and is known as the 'Keulegan–Carpenter number'. Accurate values of C_D and C_M are difficult to establish from field measurements, but recommended values have been published (see the *Shore Protection Manual* (1984) and BS6349 (1984). Both C_D and C_M are functions of Reynolds' number and Keulegan–Carpenter number. Results from many laboratory and field experiments have been compiled by the US Army Coastal Engineering Centre who have recommended design values for C_D and C_M shown in the tables below.

Reynolds Number	C_M
$R < 2.5 \times 10^5$	2.0
$2.5 \times 10^5 < R < 5 \times 10^5$	$C_M = 2.5 - R/(5 \times 10^5)$
$R > 5 \times 10^5$	$C_M = 1.5$

Reynolds Number	C_D
$R < 10^5$	1.2
$10^5 < R < 4 \times 10^5$	$1.2 < C_D < 0.6$
$R > 4 \times 10^5$	$C_D = 0.6–0.7$

If these tables are used then the Reynolds number must be calculated using the maximum velocity associated with the wave.

Example 2.2   Wave forces on a pile

A vertical cylindrical pile having a diameter of 0.4 m is installed in water that is 10 m deep. For an incident wave having a height of 2 m and a period of 8 s, determine the horizontal force experienced by the pile, per unit length, at mean water level at

(a) the peak of the waves;
(b) the trough of the waves.

Also, determine the horizontal force experienced by the pile, per unit length, at the seabed at the peak of the wave.

The kinematic viscosity of seawater may be taken as $1.5 \times 10^{-6}\ m^2 s^{-1}$ and the density as 1028 kg/m³.

Solution

First, the structure of the solution strategy is as follows:

1   Calculate L
2   Calculate maximum u
3   Calculate Reynolds number
4   Find values of C_D and C_M
5   Calculate force from Morison's Equation for the different cases

For $T = 8$ s, $L_0 = gT^2/2\pi = 9.81\ (8)^2/2\pi = 99.9$ m

The calculation is illustrated with two methods. First, using an iterative technique and secondly using Hunt's expansion. By trial and error, working across each line in turn, a solution for L (the wavelength) at a water depth of 10 m is found iteratively:

L (guess)	L/L_0	$tanh(2\pi h/L)$	$tanh(2\pi h/L) < L/L_0$?	Comment
80	0.801	0.656	yes	Reduce L
70	0.701	0.715	no	Increase L
72	0.721	0.703	yes	Reduce L
71	0.7107	0.7089	yes	Reduce L
70.8	0.7087	0.7101	no	Increase L
70.9	0.7097	0.7095	Slightly	Close enough!

giving $L = 70.9$ m and $k = 2\pi/L = 0.0886$. Also, $\omega = 2\pi/T = 0.7854$.

Alternatively, using Hunt's approximation (see Section 2.3.4), $y = k_0 h$. Now, $h = 10$ m and $k_0 = 2\pi/L_0$. But $L_0 = 99.9$ m. Hence, $k_0 = 2\pi/99.9 = 0.0629$, and so $y = 0.629$. Substituting this value of y into Hunt's formula gives $c^2/gh = 78.55$. From this, $c = \sqrt{(gh \times 78.55)} = 8.86$ m/s. As $c = L/T$, this means that $L = cT = 8.86 \times 8 = 70.88$ m.

Now calculate the maximum horizontal velocity for mean water level and at the seabed:

$$u = \frac{\omega A}{2} \cos\left\{ 2\pi \left(\frac{x}{L} - \frac{t}{T} \right) \right\}$$

so

$$u_{max} = \frac{\omega}{2} \frac{H \cosh(k(z+h))}{\sinh(kh)}$$

Substituting in the values for ω, H, k and h for values of $z = 0$ (mean water level) and $z = -10$ m (seabed) gives the maximum values of u in the two cases as 1.11 m/s and 0.780 m/s, respectively.

The Reynolds number, R, in each case is uD/v or $1.11 \times 0.4/1.5 \times 10^{-6} = 296\,000$ and $0.78 \times 0.4/1.5 \times 10^{-6} = 208\,000$, respectively. Using the tables above gives:

For mean water level, $C_M = 2.5 - R/(5 \times 10^5) = 1.91$ and $1.2 < C_D < 0.6$. Linear interpolation could be employed to calculate the value of C_D but here the upper limit is taken as being representative of a worst case, i.e. $C_D = 1.2$.

For the seabed level, $C_M = 2.0$ and $1.2 < C_D < 0.6$. Again, the upper limit for C_D is used.

(a) The force at mean sea level at the peak of the waves is found from

$$Force = \frac{C_d \rho D A^2 \omega^2}{(2)(2^2)} \cos(kx - \omega t)|\cos(kx - \omega t)| + \frac{C_M \rho \pi D^2}{4} \frac{A \omega^2}{2} \sin(kx - \omega t)$$

by setting $x = 0$ (take the pile to be situated at $x = 0$), $t = 0$ (corresponding to the peak of the wave) and $z = 0$. This gives

$$Force = \frac{1.2 \times 1028 \times 0.4 \times A^2 \times \omega^2}{8} + 0$$

$$= 61.68 \times \left(\frac{2 \times \cosh(0.0886 \times 10)}{\sinh(0.0886 \times 10)} \right)^2 \times 0.7854^2$$

$$= 302 N\,/\,m$$

(b) The force at mean sea level at the trough of the waves is found in exactly the same way but setting $x = 0$, $t = \pi/\omega$ (corresponding to the trough of the wave) and $z = 0$. This gives

$$Force = 0 + \frac{1.91 \times 1028 \times \pi \times (0.4)^2 \times A \times \omega^2}{8}$$

$$= 123.4 \times \left(\frac{2 \times \cosh(0.0886 \times 10)}{\sinh(0.0886 \times 10)} \right) \times 0.7854^2$$

$$= 215 N\,/\,m$$

(c) The force at the seabed level at the peak of the waves is found as in (a) but setting $z = -10$. This gives

$$Force = \frac{1.2 \times 1028 \times 0.4 \times A^2 \times \omega^2}{8} + 0$$

$$= 61.68 \times \left(\frac{2 \times \cosh(0)}{\sinh(0.0886 \times 10)} \right)^2 \times 0.7854^2$$

$$= 150 N / m$$

2.6 Surf zone processes

2.6.1 A general description of the surf zone

For simplicity, consider the case of a coast with the seabed and beach consisting of sand. The bed slope will usually be fairly shallow (say $0.01 < \beta < 0.03$). Waves will therefore tend to start to break at some distance offshore of the beach or shoreline (i.e. the beach contour line which corresponds to the still-water level, see Figure 2.16a). At this initial break point the wave will be of height H_b and at angle α_β to the beach line. The region between this initial point and the beach is known as the surf zone. In this region, the height of an individual wave is largely controlled by the water depth. The wave height will progressively attenuate as it advances towards the beach, and the characteristic foam or surf formation will be visible on the wave front (see Figure 2.16b for a real example). The mechanics of this progressive breaking are very complex. A brief summary is as follows:

1 Turbulence and aeration are produced.
2 Significant rates of change are induced in the momentum of the elements of fluid which constitute the wave. This produces a momentum force which may be resolved into two components (Figure 2.16a). The component which lies parallel to the shoreline is the cause of a corresponding 'longshore current'. The component which is perpendicular to the shoreline produces an increase in the depth of water above the still-water level, and this is usually called the 'set-up'.
3 Energy is lost due to bed friction and due to the production of turbulence. The frictional losses are produced both by the oscillatory motion at the seabed due to the wave and by the unidirectional motion of the longshore current. The two motions are not completely independent, and their interaction has significant effects on the bed friction.

2.6.2 Wave breaking

There are two criteria which determine when a wave will break. The first is a limit to wave steepness and the second is a limit on the wave height to water depth ratio. Theoretical limits have been derived from solitary wave theory, which is a single wave with a crest and no trough. Such a wave was first observed by Russell in 1840, being produced by a barge on the Forth and Clyde canal. The two criteria are given by:

1 Steepness H/L < 1/7. This normally limits the height of deep-water waves
2 Ratio of height to depth: the breaking index

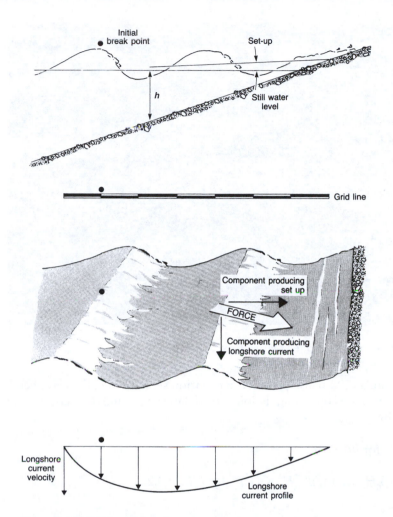

Figure 2.16(a) The surf zone, conceptual.

$$\gamma = H/h = 0.78 \tag{2.35}$$

In practice γ can vary from about 0.4 to 1.2 depending on beach slope and breaker type.

Goda (2000) provides a design diagram for the limiting breaker height of regular waves, which is based on a compilation of a number of laboratory results. He also presents an equation, which is an approximation to the design diagram, given by:

$$\frac{H_b}{L_0} = 0.17\left\{1 - \exp\left[-\frac{1.5\pi h}{L_0}\left(1 + 15\tan^{4/3}\beta\right)\right]\right\}$$

where $\tan(\beta)$ is the beach slope. For the case of random waves (see Chapter 3 for a

Figure 2.16(b) A real surf zone at Hope Cove, Devon, England.

full discussion), Goda (2000) also presents an equation set to predict the wave heights within the surf zone, based on a compilation of field, laboratory and theoretical results. These are given by:

$$H_{1/3} = K_S H_0' \quad \text{for } h/L_0 \geq 0.2$$

$$H_{1/3} = \min \{(\beta_0 H_0' + \beta_1 h), \beta_{max} H_0', K_S H_0'\} \quad \text{for } h/L_0 < 0.2$$

where

$$H_0' = K_d K_R (H_{1/3})_0$$

and

$$\beta_0 = 0.028 (H_0' / L_0)^{-0.38} \exp\left[20 \tan^{1.5} \beta\right]$$

$$\beta_1 = 0.52 \exp\left[4.2 \tan \beta\right]$$

$$\beta_{max} = \max\left\{0.92, 0.32 (H_0' / L_0)^{-0.29} \exp\left[2.4 \tan \beta\right]\right\}$$

Breaker types

Breaking waves may be classified as one of three types as shown in Figure 2.17. The type can be approximately determined by the value of the surf similarity parameter (or Iribarren Number)

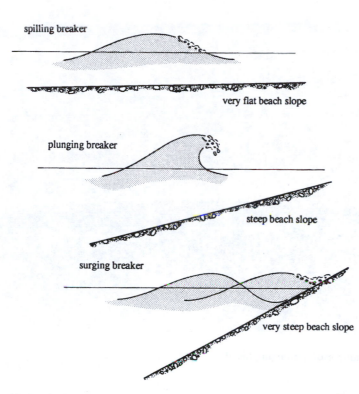

spilling breaker

very flat beach slope

plunging breaker

steep beach slope

surging breaker

very steep beach slope

Figure 2.17(a) Principal types of breaking waves.

Figure 2.17(b) Example of a spilling breaker.

Figure 2.17(c) Example of a plunging breaker.

$$\xi_b = \tan \beta \, / \, \sqrt{H_b / L_b} \tag{2.36}$$

where $\tan \beta$ = beach slope, and for

Spilling breakers	$\xi_b < 0.4$
Plunging breakers	$0.4 \leq \xi_b \leq 2.0$
Surging breakers	$\xi_b > 2.0$

Battjes found from real data that

$$\gamma \cong \xi_0^{0.17} + 0.08 \quad \text{for } 0.05 < \xi < 2 \tag{2.37}$$

Further details may be found in Horikawa (1988) and Fredsoe and Deigaard (1992).

2.6.3   Wave set-down and set-up

The onshore momentum flux (i.e. force) S_{xx}, defined in Section 2.2.6, must be balanced by an equal and opposite force for equilibrium. This manifests itself as a *slope* in the mean still water level (given by $d\eta/dx$).

Consider the control volume shown in Figure 2.18 in which a set-up $\bar{\eta}$ on the still-water level exists induced by wave action. The forces acting are the pressure forces, the reaction force on the bottom and the radiation stresses (all forces are wave period averaged). For equilibrium the net force in the x direction is zero. Hence

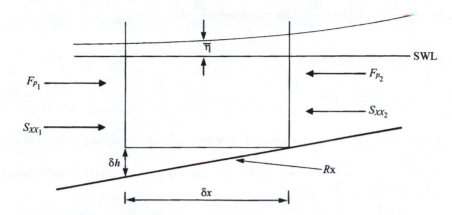

Figure 2.18 Diagram for derivation of wave set-down/up.

$$(F_{p_1} - F_{p_2}) + (S_{XX_1} - S_{XX_2}) - R_x = 0 \qquad\qquad (2.38)$$

as

$$F_{p_2} = F_{p_1} + \frac{dF_p}{dx}\delta x$$

$$S_{XX_2} = S_{XX_1} + \frac{dS_{XX}}{dx}\delta x$$

then by substitution into Equation (2.38)

$$\frac{dF_p}{dx}\delta x + \frac{dS_{XX}}{dx}\delta x = R_x \qquad\qquad (2.39)$$

As

$$F_p = \frac{1}{2}\rho g(h + \bar{\eta})^2 \quad \text{(i.e. the hydrostatic pressure force)}$$

then

$$\frac{dF_p}{dx} = \frac{1}{2}\rho g \frac{d}{dx}(h + \bar{\eta})^2 = \rho g(h + \bar{\eta})\left(\frac{dh}{dx} + \frac{d\bar{\eta}}{dx}\right)$$

and as R_x for a mildly sloping bottom is due to bottom pressure,

$$R_x = \bar{p}\delta h = \bar{p}\frac{dh}{dx}\delta x = \rho g(h + \bar{\eta})\frac{dh}{dx}\delta x$$

then substituting for F_p and R_x in Equation(2.38)

$$\rho g(h+\overline{\eta})\left(\frac{dh}{dx}+\frac{d\overline{\eta}}{dx}\right)\delta x+\frac{dS_{XX}}{dx}\delta x=\rho g(h+\overline{\eta})\frac{dh}{dx}\delta x$$

and finally, simplifying we obtain

$$\frac{dS_{XX}}{dx}+\rho g(h+\overline{\eta})\frac{d\overline{\eta}}{dx}=0 \qquad (2.40)$$

where $\overline{\eta}$ is the difference between the still-water level and the mean water level in the presence of waves.

Outside the breaker zone Equation (2.40) (in which Equation (2.10) is substituted for S_{XX}), may be integrated to obtain

$$\overline{\eta_d}=-\frac{1}{8}\frac{kH_b^2}{\sinh(2kh)} \qquad (2.41)$$

this is referred to as the *set-down* ($\overline{\eta_d}$) and demonstrates that the mean water level decreases in shallower water. Inside the breaker zone the momentum flux rapidly reduces as the wave height decreases. This causes a *set-up* ($\overline{\eta_u}$) of the mean still-water level. Making the assumption that inside the surf zone the broken wave height is controlled by depth such that

$$H=\gamma(\overline{\eta}+h) \qquad (2.42)$$

where $\gamma\approx 0.8$ then combining Equations (2.10), (2.40) and (2.42) leads to the result

$$\frac{d\overline{\eta}}{dx}=\left(\frac{1}{1+\dfrac{8}{3\gamma^2}}\right)\tan\beta$$

where β is the beach slope angle. Thus for a uniform beach slope it may be shown that

$$\overline{\eta_u}=\left(\frac{1}{8/3\gamma^2}\right)(h_b-h)+\overline{\eta_{d_b}} \qquad (2.43)$$

demonstrating that inside the surf zone there is a rapid increase in the mean water level.

Thus it may be appreciated that set-down is quite small and the set-up much larger (see Example 2.3). In general wave set-down is less than 5 per cent of the breaking depth and wave set-up is about 20–30 per cent of the breaking depth. It may also be noted that for a real sea, composed of varying wave heights and periods, the wave set-up will vary along a shoreline at any moment. This can produce the phenomenon referred to as surf beats (refer to Section 2.6.6 for further details). Wave set-up also contributes to the overtopping of sea defence structures, during storm conditions, and may thus be a contributory factor in coastal flooding.

2.6.4 Radiation stress components for oblique waves

The radiation stresses S_{XX}, S_{YY} are, in fact, principal stresses. Utilising the theory of principal stresses, shear stresses will also act on any plane at an angle to the principal axes. This is illustrated in Figure 2.19 for the case of oblique wave incidence to a coastline. The relationships between the principal radiation stresses and the direct and shear components in the x-, y-directions are

$$S_{xx} = S_{XX}\cos^2\theta + S_{YY}\sin^2\theta = \frac{1}{2}E\left[(1+G)\cos^2\theta + G\right]$$

$$S_{yy} = S_{XX}\sin^2\theta + S_{YY}\cos^2\theta = \frac{1}{2}E\left[(1+G)\sin^2\theta + G\right]$$

$$S_{xy} = S_{XX}\sin\theta\cos\theta + S_{YY}\sin\theta\cos\theta = \frac{1}{2}E\left[(1+G)\sin\theta\cos\theta\right]$$

where $G = 2kh / \sinh(2kh)$

2.6.5 Longshore currents

Radiation stress theory has been successfully used to explain the presence of longshore currents. The original theory is eloquently explained by Longuet-Higgins (1970). Subsequently Komar (1976), as a result of his own theoretical and field investigations, developed the theory further and presented revised equations. All of the foregoing is succinctly summarised in Hardisty (1990). Here a summary of the main principles is given together with a statement of the main equations.

An expression for the mean wave *period-averaged* longshore velocity ($\overline{v_l}$) was derived from the following considerations. Firstly outside the surf zone the energy flux towards the coast (P_x) of a wave travelling at an oblique angle (α) is constant and given by

$$P_x = Ec_g\cos\alpha \qquad \text{(cf. Equation (2.8))} \qquad (2.44)$$

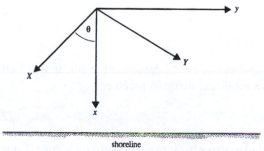

Figure 2.19 Relationships between principal axes and shoreline axes.

Secondly, the radiation stress (S_{xy}) which constitutes the flux of y momentum parallel to the shoreline across a plane x = constant is given by

$$S_{xy} = S_{XX} \sin\alpha \cos\alpha - S_{YY} \sin\alpha \cos\alpha$$

$$= E\left(\frac{1}{2} + \frac{kh}{\sinh 2kh}\right)\cos\alpha \sin\alpha$$

$$= E\left(\frac{c_g}{c}\right)\cos\alpha \sin\alpha \tag{2.45}$$

hence combining Equations (2.44) and (2.45).

$$S_{xy} = P_x\left(\frac{\sin\alpha}{c}\right)$$

outside the surf zone. S_{xy} is therefore constant, as $\sin\alpha/c$ is also constant (cf. Equation 2.12). However, inside the surf zone this is no longer the case as wave energy flux is rapidly dissipated. The net thrust (F_y) per unit area exerted by the waves is given by

$$F_y = \frac{-\partial S_{xy}}{\partial x} \tag{2.46}$$

Substituting for S_{xy} from Equation (2.46) and taking conditions at the wave break point (at which $c_g = c = \sqrt{gh_b}$, $H_b/h_b = \gamma$, $u_m = \gamma/2\sqrt{gh_b}$), Longuet-Higgins derived an expression for F_y given by

$$F_y = \frac{5}{4}\rho u_{mb}^2 \tan\beta \sin\alpha \tag{2.47}$$

Finally, by assuming that this thrust was balanced by frictional resistance in the longshore (y) direction he derived an expression for the mean longshore velocity $(\overline{v_l})$ given by

$$\overline{v_l} = \frac{5\pi}{8C} u_{mb} \tan\beta \sin\alpha_b \tag{2.48}$$

where C was a friction coefficient.
Subsequently, Komar found from an analysis of field data that $\tan\beta/C$ was effectively constant and he therefore proposed a modified formula given by

$$\overline{v_l} = 2.7 u_{mb} \sin\alpha_b \cos\alpha_b \tag{2.49}$$

in which the $\cos\alpha_b$ term has been added to cater for larger angles of incidence (Longuet-Higgins assumed α small and therefore $\cos\alpha \to 1$).

The distribution of longshore currents within the surf zone was also studied by both Longuet-Higgins and Komar. The distribution depends upon the assumptions made concerning the horizontal eddy coefficient, which has the effect of transferring horizontal momentum across the surf zone. Komar (1976) presents a set of equations to predict the distribution.

Example 2.3   Wave set-down, set-up and longshore velocity
(a) A deep-water wave of period 8.5 s and height 5 m is approaching the shoreline normally. Assuming the seabed contours are parallel, estimate the wave set-down at the breakpoint and the wave set-up at the shoreline.
(b) If the same wave has a deep-water approach angle of 45°, estimate the mean longshore current in the surf zone.

Solution
(a) The first stage of the solution is analogous to Example (2.1), except that no refraction occurs, thus at the break point we obtain

h/L_0	h (m)	c/C_0	c(m/s)	K_S	$\alpha(°)$	K_R	H/H_0	H(m)	h_B(m)
0.06	6.4	0.56	7.5	1.00	0	1.00	1.0	5.0	6.4

The set-down may now be calculated from Equation (2.41), that is:

$$\overline{\eta_d} = -\frac{1}{8}\frac{kH_b^2}{\sinh(2kh)}$$

As

$$k = 2\pi/L = 2\pi/(112.8x0.56) = 0.099$$

and

$$2kh = 2x0.099x6.4 = 1.267$$

$$\overline{\eta_d} = -\frac{1}{8}\frac{5^2 x0.099}{\sinh(1.267)}$$

then

$$\overline{\eta_d} = -0.19$$

The set-up may be calculated from Equation (2.43)

$$\overline{\eta_u} = \left(\frac{1}{8/3\gamma^2}\right)(h_b - h) + \overline{\eta_{d_b}}$$

i.e. at the shoreline $h = 0$ and taking $\gamma = 0.78$ then

$$\overline{\eta_u} = \left(\frac{1}{8/3x0.78^2}\right)(6.4) - 0.19$$
$$= 1.27 \text{ m}$$

(b) Here the same wave as in Example 2.1 has been used. Recalling that at the wave breakpoint $\alpha_b = 22°$ and $h_b = 5.7$ m then Equation (2.49) may be used to estimate $\overline{v_l}$:

$$\overline{v_l} = 2.7u_{mb}\sin\alpha_b\cos\alpha_b$$

Recalling that

$$u_{mb} = \frac{\gamma}{2}\sqrt{gh_b}$$

then

$$u_{mb} = \frac{0.78}{2}\sqrt{9.81x5.7}$$
$$= 2.92 \text{ m/s}$$

Hence

$$\overline{v_l} = 2.7 \times 2.92\sin(22)\cos(22)$$
$$= 2.74 \text{ m/s}$$

2.6.6 Infragravity waves

Waves often travel in groups as shown in Figure 2.20, hence under large waves the set-down is larger than under small waves, this results in a second order wave – the bound long wave. The bound long waves travel with the wave groups with a celerity corresponding to the group celerity of the short waves and thus are refracted with the short waves. In shallow water the height of the bound long waves will increase quite dramatically due to shoaling.

In the surf zone the short waves lose height and energy and can no longer balance the bound long waves *which are therefore released as free long waves*. The free long waves are substantially *reflected* from the beach and either progress back out to sea (for normally incident short waves), termed the 'leaky mode' or *refract and turn back to the shore* to be re-reflected, termed the 'trapped mode'. The trapped free long waves then form *3D edge waves* with a wave height which decreases with distance from the shore.

Another mechanism for generating long waves in the surf zone is *variation in set-up caused by breaking wave groups*. Surf beat is the variation of set-up on the shoreline and may be caused by a combination of free long waves in the surf zone, generated at sea as bound long waves and free long waves generated in the surf zone due to variations in set-up.

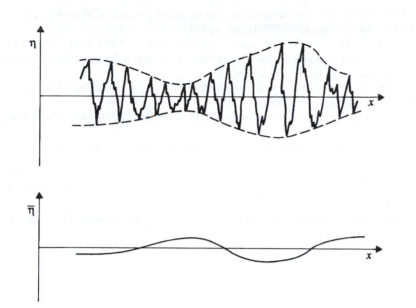

Figure 2.20 The wave groups and the associated mean water level.

'Cell circulation' is the term used to describe currents within the surf zone that are not parallel to the shore. The existence of cell circulations is evidenced by rip currents (a common hazard for swimmers!). Rip currents are a seaward return flow of water concentrated at points along the beach. They are caused by a longshore variation of wave height and hence set-up which provides the necessary hydraulic head to drive them. The longshore variation of wave height can be caused either by refraction effects or by the presence of edge waves. Under the latter circumstances a regular pattern of cell circulations and rip currents will exist and beach cusps may be formed. The interested reader is referred to Komar (1976) and Huntley *et al.* (1993) for further details.

Further reading

Battjes, J. A., 1968. Refraction of water waves. *J. Waterways and Harbours Div.*, WW4, pp. 437–457.

Berkhoff, J.C.W., 1972. Computation of combined refraction-diffraction, *Proc. 13th Int. Conf. Coastal Eng.*, Lisbon, pp. 55–69.

Briggs, M.J., Thompson, E. F. and Vincent, C. L., 1995. Wave diffraction around breakwater. *Journal of Waterways, Port, Coastal and Ocean Engineering*, 121, pp. 23–35.

Honda, K. and Mase, H., 2007. Application of non-linear frequency-domain wave model to mach stem evolution and wave transformation on reef. *Proceedings of the 5th Coastal Structures International Conference*, ASCE, Venice, Italy. 2–4 July.

Horikawa, K. (ed.), 1988. *Nearshore Dynamics and Coastal Processes, Theory Measurement and Predictive Models*, University of Tokyo Press, Tokyo.

Hughes, S.A., 1984. *The TMA Shallow Water Spectrum Descriptions and Applications*, Technical Report 84–87, US Army Corps of Engineers.

Hunt, J. N., 1979. Direction solution of wave dispersion equation. *Journal of Waterways, Port, Coastal and Ocean Engineering,* 105 (WW4), pp. 457–459.

Huntley, D. A., Davidson, M., Russell, P., Foote, Y. and Hardisty, J., 1993. Long waves and sediment movement on beaches: recent observations and implications for modelling. *Journal of Coastal Research Special Issue,* 15, pp. 215–229.

MAFF, 1993. *Coastal Defence and the Environment: A Guide to Good Practice,* Ministry of Agriculture, Fisheries and Food, London.

Melville,W.K., 1980. On the mach reflection of a solitary wave. *Journal of Fluid Mechanics,* 98, pp. 285–297.

Miles J., 1980, Solitary Waves. *Annual Review of Fluid Mechanics,* Vol. 12, pp. 11–43, January.

Morison, J.R., Johnson, J.W., O'Brien, M.P. and Schaaf, S.A., 1950. The forces exerted by surface waves on piles, *Petroleum Transactions, American Institute of Mining Engineers,* 189, pp. 145–154.

Open University, 1989. *Waves, Tides and Shallow Water Processes,* Pergamon Press, Oxford.

Yoon, S.B. and Liu, P.L.-F., 1989. Stem waves along breakwater. *Journal of Waterways, Port, Coastal and Ocean Engineering,* 115, pp. 635–648.

Chapter 3

Design wave specification

3.1 Introduction

This chapter covers the description of wave climate for design purposes. As waves are generally random, being driven by the near-surface winds, a statistical approach is often taken to define design conditions. Thus, the wave climate is described in terms of representative measures of wave height, period and direction. Formulae for estimating these quantities from first principles and from empirical equations are used in design. Some of the more widely used methods are introduced in this chapter, as well as a discussion of their relative advantages.

Wind-generated ocean waves are complex, incorporating many superimposed components of wave periods, heights and directions. If the sea state is recorded in a storm zone, then the resulting wave trace appears to consist of random periodic fluctuations. To find order in this apparent chaos, considerable research and measurement has been, and is being, undertaken.

Wave records are available for certain locations. These are normally gathered by either ship-borne wave recorders (for fixed locations) or wave rider buoys (which may be placed at specific sites of interest). These records generally consist of a wave trace for a short period (typically 20 min) recorded at fixed intervals (normally 3 h) and sampled at 2 readings per second (2 *Hz*). In this way, the typical sea state may be inferred without the necessity for continuous monitoring. An example wave trace is shown in Figure 3.1 (a). (Note: this was recorded in shallow water.)

3.2 Short-term wave statistics

Two types of analysis may be performed with such wave trace records. The first type is referred to as *time domain analysis* and the second *frequency domain analysis*. Both methods assume a state of stationarity (i.e. the statistics of the sea state do not vary with time).

3.2.1 Time domain analysis

For a given wave record (e.g. a 20 min record representing a 3 h period) the following parameters may be directly derived in the time domain (refer to Figure 3.2) using either up-crossing or down-crossing analysis.

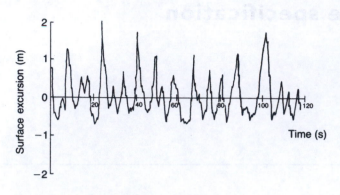

Figure 3.1(a) Recorded wave trace.

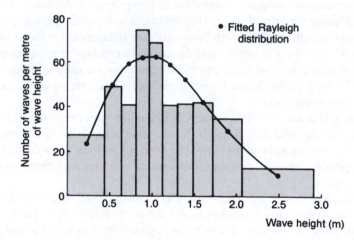

Figure 3.1(b) Histogram of wave heights.

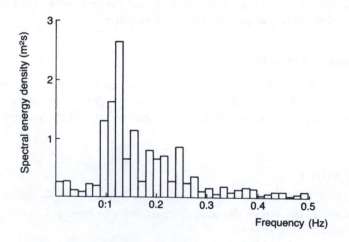

Figure 3.1(c) Histogram of spectral energy density.

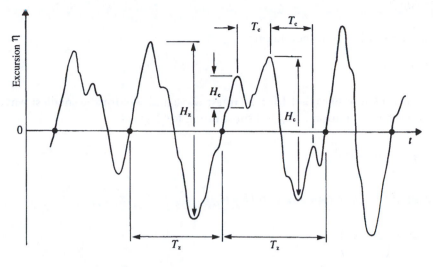

● up-crossings of the mean level

Figure 3.2 Time domain analysis.

1 H_z (mean height between zero upward crossing);
2 T_z (mean period between zero upward (or downward) crossings);
3 H_c (mean height between wave crests);
4 T_c (mean period between wave crests);
5 H_{max} (maximum difference between adjacent crest and trough);
6 H_{rms} (root-mean-square wave height);
7 $H_{1/3}$ or H_s (mean height of the highest one-third of the waves);
8 $H_{1/10}$ (mean height of the highest one-tenth of the waves).

It should also be noted that T_{max} and $T_{1/10}$ are the periods for the *corresponding* wave heights.

Based on previous experience, wave record analysis is greatly simplified if some assumptions regarding the probability distributions of wave heights are made. The distribution of wave heights is often assumed to follow the Rayleigh distribution, thus

$$P(h < H) = 1 - \exp[-2(H/H_s)^2]$$
or equivalently,
$$P(h \geq H) = \exp[-2(H/H_s)^2]$$
(3.1)

where $P(h \geq H)$ is the probability that the wave height (h) will equal or exceed the given value (H).

The corresponding probability density function $f(h)$ is given by

$$f(h) = (2h/H_{rms}^2)\exp[-(h/H_{rms})^2]$$
(3.2)

where $\int_0^\infty f(h)dh = 1$

For a wave record of N waves, taking

$$P(h \geq H) = i / N$$

where i is the rank number ($i = 1$ for the largest wave and $= N$ for the smallest wave) then rearranging Equation (3.1) and substituting for P gives

$$H = H_s \left[\frac{1}{2} \ln (N / i) \right]^{\frac{1}{2}}$$

For the case of $i = 1$, corresponding to $H = H_{max}$,

$$H_{max} = H_s \left(\frac{1}{2} \ln N \right)^{\frac{1}{2}}$$

However, H_{max} itself varies quite dramatically from one wave record to the next and needs to be treated as a statistical quantity. Hence it may be shown that the most probable value of H_{max} is

$$H_{max} = H_s \left(\frac{1}{2} \ln N \right)^{\frac{1}{2}} \qquad \text{as above} \qquad (3.3a)$$

and the mean value of H_{max} is

$$H_{max} = \frac{H_s}{\sqrt{2}} \left[\left(\ln N^{\frac{1}{2}} \right) + 0.2886 \left(\ln N \right)^{\frac{1}{2}} \right]. \qquad (3.3b)$$

Other useful results which have been derived include:

$$H_{rms} \cong 1.13 H_z$$

$$H_s \cong \sqrt{2} H_{rms} = 1.414 H_{rms}$$

$$H_{1/10} \cong 1.27 H_s \cong 1.8 H_{rms}$$

$$H_{1/100} \cong 1.67 H_s \cong 2.36 H_{rms}$$

$$H_{max} \cong 1.6 H_s \cong 2.26 H_{rms} \quad \text{(for a typical 20 min wave trace)}$$

Thus, if the value of H_{rms} is calculated from the record, the values of H_s etc. may easily be estimated.

The Rayleigh distribution was originally derived by Lord Rayleigh in the late nineteenth century for sound waves. It is commonly assumed to apply to wind waves and swell mixtures and gives a good approximation to most sea states. However, the Rayleigh distribution is theoretically only a good fit to sine waves with a small range of periods with varying amplitudes and phases. This is more characteristic of swell waves than storm waves. To determine what type of distribution is applicable, the parameter ε, known as the 'spectral width', may be calculated:

$$\varepsilon^2 = 1 - \left(\frac{T_c}{T_z}\right)^2$$

For the Rayleigh distribution (i.e. a small range of periods) $T_c \cong T_z$, hence $\varepsilon \to 0$. For a typical storm sea (containing many frequencies) the period between adjacent crests is much smaller than the period between zero upward crossings, hence $\varepsilon \to 1$.

Actual measurements of swell and storm waves as given by Silvester (1974) are as follows:

	Swell waves	Storm waves
ε	$\cong 0.3$	$\cong 0.6$ to 0.8
Hs	$\cong 1.42\ H_{rms}$	$\cong 1.48\ H_{rms}$
$H_{1/10}$	$\cong 1.8\ H_{rms}$	$\cong 2.0\ H_{rms}$

Figure 3.3 illustrates the histogram of wave heights (derived from the wave trace shown in Figure 3.2) and shows the fitted Rayleigh distribution. As a matter of

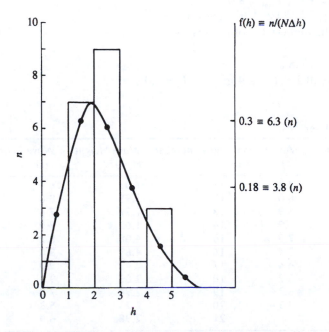

Figure 3.3 Histogram and superimposed Rayleigh pdf.

interest, these data were recorded in very shallow water, for which the Rayleigh distribution was not expected to be a good fit. Applying a statistical goodness of fit criterion, this proved not to be the case. Further details may be found in Chadwick (1989a).

Time domain analysis has traditionally been carried out using analogue data. A rapid method was developed by Tucker to find H_{rms} from which other wave parameters can be derived by assuming a Rayleigh distribution. More recently, digital data has become available and Goda (2000) gives details of how to derive time domain parameters directly.

Example 3.1
Using the time series data given in Table 3.1

(a) determine H_{max}, T_{max}, H_s, T_s, H_z, T_z;
(b) plot a histogram of the wave heights using a class interval of 1 m;
(c) determine H_{max}, H_s and H_{rms} from H_z, assuming a Rayleigh distribution;
(d) calculate the value of $f(h)$ at the centre of each class interval and hence superimpose the p d f on the histogram (note: assume that the scale equivalence is) $f(h) \equiv n/N\Delta h$);
(e) suggest reasons for the anomalies between the results in (a) and (c).

Solution
(a) From Table 3.1
     16th wave gives H_{max} = 4.89, T_{max} = 8.0
     For H_s 16th, 3rd, 15th, 5th, 21st, 19th, 18th waves
     (21/3 = 7 waves) are the highest 1/3 of the waves
     Average to obtain
     H_s = 3.6 m T_s = 7.8 s
     For H_z, T_z average all 21 H_z = 2.4 m T_z = 7.0 s

Table 3.1 Wave heights and periods.

Wave number	Wave Height H(m)	Wave Period T(s)	Wave number	Wave Height H(m)	Wave Period T(s)
1	0.54	4.2	11	1.03	6.1
2	2.05	8.0	12	1.95	8.0
3	4.52	6.9	13	1.97	7.6
4	2.58	11.9	14	1.62	7.0
5	3.20	7.3	15	4.08	8.2
6	1.87	5.4	16	4.89	8.0
7	1.90	4.4	17	2.43	9.0
8	1.00	5.2	18	2.83	9.2
9	2.05	6.3	19	2.94	7.9
10	2.37	4.3	20	2.23	5.3
			21	2.98	6.9

(b)

Class interval of wave height (m)	No. of waves
0–1	1
1–2	7
2–3	9
3–4	1
4–5	3

The histogram is shown in Figure 3.3.

(c) from part (a)

H_z = 2.4m

$\therefore H_{rms}$ = 1.13 × 2.4 = 2.71 m

$\therefore H_s$ = 1.414 H_{rms} = 3.83 m

$\therefore H_{max}$ = H_s (½ lnN)½

= 3.83 (½ ln21)½

= 4.73 m

(d) using (3.2)

$$f(h) = (2h/H^2_{rms})\exp - (h/H_{rms})^2$$

(h)	f(h)	n = f(h) NΔh
0.5	0.13	2.8
1.5	0.3	6.3
2.5	0.29	6.1
3.5	0.18	3.8
4.5	0.078	1.6
5.5	0.024	0.5

These results are also plotted in Figure 3.3.

(e) Visually, the Rayleigh distribution is apparently not a good fit. However, this should be checked by undertaking a statistical goodness of fit test (see Chadwick (1989a) for details).

3.2.2 Frequency domain analysis

The wave trace shown in Figure 3.2 can also be analysed in the frequency domain. This is made possible by application of the Fourier series representation. In essence, any uni-directional sea state can be described mathematically as being composed of an infinite series of sine waves of varying amplitude and frequency. Thus, the surface excursion at any time $\eta(t)$ (defined in Figure 2.3) may be represented as

$$\eta(t) = \sum_{n=1}^{\infty} [a_n \cos \omega nt + b_n \sin \omega nt] \tag{3.4}$$

where ω is the angular frequency $(2\pi/T)$ and $t = 0$ to $t = T$; a_n and b_n are amplitudes.

Equation (3.4) may be equivalently written as

$$\eta(t) = \sum_{n=1}^{\infty} c_n \cos(\omega nt + \phi_n) \tag{3.5}$$

where

$$c_n^2 = a_n^2 + b_n^2$$

$$\tan \phi_n = -b_n / a_n$$

(This is shown graphically in Figure 3.4.)

Noting that the equation for wave energy is $E = \rho g H^2/8$ then wave energy is proportional to (amplitude)^2/2 (with units of m^2). Thus the spectral energy density curve $S(f)$ (with units of m^{2}s) may be found from

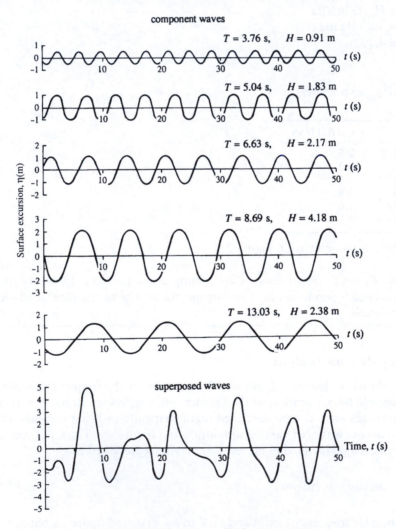

Figure 3.4 Graphical representation of a Fourier series.

$$S(f)\Delta f = \sum_{f}^{f+\Delta f} \frac{1}{2}c_n^2 \qquad (3.6)$$

To accomplish this, values of c_n must be found from Equation (3.5). The technique commonly used for doing this is termed the Fast Fourier Transform (FFT). A description of the FFT techniques is beyond the scope of this chapter, but the reader is directed to Carter *et al.* (1986) for a description of its application to sea waves and Broch (1981) for details of the principles of digital frequency analysis.

Suffice to say here, that a given wave trace record may be analysed using FFT techniques to produce the spectral density histogram. An example is shown in Figure 3.1(c). Having obtained the spectral density histogram then the *frequency domain wave parameters* may be found from the following equations:

$$H_{m0} = 4(m_0)^{0.5}$$

$$T_{m01} = m_0 / m_1$$

$$T_{m02} = (m_0 / m_2)^{0.5}$$

$$T_p = 1/F_p$$

where f_p=frequency at the maximum value of $S(f)$

$$Q_p = \left(\frac{2}{m_0^2} \right) \int_0^\infty f S(f)^2 df \quad \text{(spectral peakedness)}$$

$$\varepsilon = \left[1 - \frac{m_2^2 m_4}{m_0} \right]^{0.5} \quad \text{(spectral width)}$$

$$\sigma^2 = \int_0^\infty S(f) df = m_0 \quad \text{(spectral variance)}$$

where

$$m_n = \int_0^\infty S(f) f^n df \quad (n\text{th spectral moment})$$

Frequency domain wave parameters do not have direct equivalent parameters in the time domain. However, as a useful guide, the following parameters have been found to be roughly equivalent:

Time domain parameter	Equivalent frequency domain parameter
H_s	H_{m0} (approximate)
η_{rms}	$m_0^{0.5}$ (exact)
T_z	T_{m02} (approximate)
T_s	$0.95T_p$ (approximate)

Due to the proliferation of wave parameters in both the time and frequency domain, there exists confusion in the literature as to the precise definition of some of those parameters. For example H_{mo} and H_s are often confused. For this reason (and others) a standard set of sea state parameters was proposed by the International Association for Hydraulic Research. Details may be found in Darras (1987).

3.3   Directional wave spectra

The sea state observed at any particular point consists not only of component waves of various heights and periods but also from different directions. Therefore, a complete description of the sea state needs to include directional information. Mathematically this may be expressed as

$$\eta(x,y,t) = \sum_{n=1}^{\infty} \sum_{m=0}^{2\pi} a_{n,m} \cos(k_n x \cos\theta_m + k_n y \sin\theta_m - 2\pi f_n t + \phi_{n,m}) \tag{3.7}$$

where a is amplitude, k is wave number $= 2\pi/L$, f is frequency, θ is wave direction, ϕ is phase angle, n is frequency counter and m is direction counter.

Equivalently, extending the concept of spectral density $S(f)$ to include direction, the directional spectral density $S(f,\theta)$ can be defined as

$$S(f,\theta) = S(f)G(f,\theta) \tag{3.8}$$

where $\displaystyle\sum_{f}^{f+\Delta f} \sum_{\theta}^{\theta+\Delta\theta} \frac{1}{2} a_n^2 = S(f,\theta)\Delta f \Delta\theta$

and $G(f,\theta)$ is the directional spreading function, where $\int_{-\pi}^{\pi} G(f,\theta)d\theta = 1$

An idealised directional spectrum is shown in Figure 3.5(a),(b) and a measured one, which contains both incident and reflected waves, is shown in Figure 3.5(c).

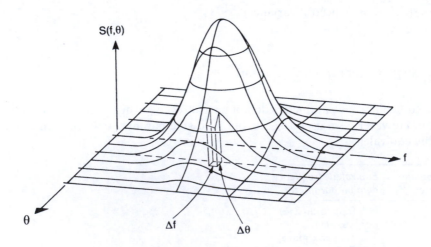

Figure 3.5(a) Idealised directional spectral density.

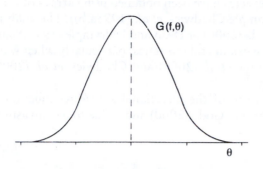

Figure 3.5(b) Idealised directional spreading function.

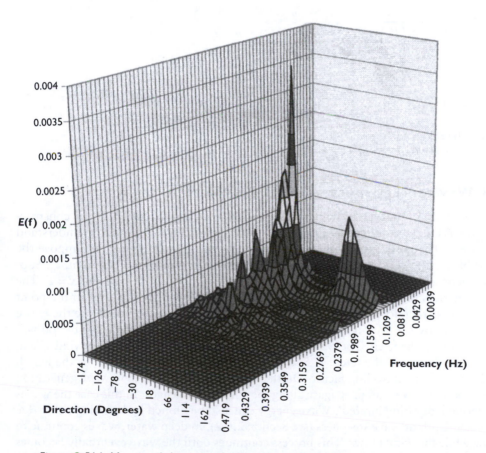

Figure 3.5(c) Measured directional energy spectrum.

The notation $E(f,\theta)$ is sometimes used in place of $S(f, \theta)$ in Equation (3.8). In what follows we use $S(f,\theta)$ to denote the directional spectral density and $S(f)$ to denote its integral over all directions.

Direct measurements of directional spectra have been obtained using arrays of wave recorders of various forms (see for example Chadwick *et al.* (1995a,b)). The analysis of such records is complex (refer to Goda (2000) or Dean and Dalrymple (1991)), and the analysis techniques do not always work in real sea states, particularly when wave reflections are present. The papers by Ilic *et al.* (2000) and Chadwick *et al.* (2000) provide interesting reading in this regard.

For design purposes a parametric form of the directional spreading function is required. Such a parametric form is given in Goda (2000) and is due to Mitsuyasu:

$$G(f,\theta) = N\cos^{2s}\left(\frac{\theta}{2}\right) \tag{3.9}$$

where N is a normalising factor, given by

$$N = \frac{1}{\int_{-\pi}^{\pi}\cos^{2s}\left(\frac{\theta}{2}\right)}$$

and

$$s = s_m(f/f_p)^{\mu}$$

s_m may be taken equal to: 10 for wind waves; 25–75 for swell waves
$\mu = -2.5$ for $f \geq f_p$, 5 for $f < f_p$

3.4 Wave energy spectra, the JONSWAP spectrum

The generation of waves by wind was discussed briefly in Section 2.1. It is constructive to consider a simplified situation, shown in Figure 3.6(a), in order to understand the effect of wind duration and limited fetch on the growth of waves. Suppose the wind blows at a constant speed, U, along the positive x-axis for a duration, t_D. Suppose there is land where $x < 0$ and we require wave conditions at a point $x = F$. The fetch length is the distance along which the wind is blowing over water to the point $x = F$, and is therefore equal to F. Denote the time for waves to propagate the entire fetch length by $t_F = F/c_g$. If t_D is greater than t_F then both wave height and period will increase along the fetch. The wave conditions at $x = F$ will depend on the wind speed, U, and the length of the fetch, F. Such waves are termed 'fetch-limited' as the length of the fetch is a controlling factor. On the other hand, if $t_D < t_F$ then wave generation ceases before waves can propagate along the entire fetch length. In this case the waves are termed 'duration limited'. Wave energy can also be limited by breaking either due to water depth or wave steepness (see Section 2.6.2). In deep water, waves continue to grow while the wind blows. This process continues until the wave eventually becomes unstable and breaks. Figure 3.6(b) illustrates the growth and limitation of wave height and period due to these influences.

Engineers require estimates of sea conditions when designing coastal and offshore structures. These conditions are often described in terms of statistical measures of wave height and period, such as those described in Section 3.2. The values of these parameters can be derived from simple predictive formulae. However, sometimes a

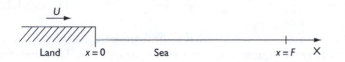

Figure 3.6(a) Schematic diagram of wind blowing along a fetch.

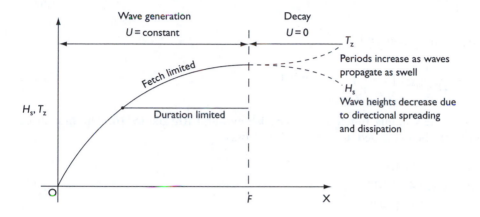

Figure 3.6(b) Idealised wave growth and decay for constant wind speed along a fixed fetch.

more detailed description of the sea surface is required. For instance, the engineer may need to evaluate the impact on a structure from all wave components that would be present in a directional wave spectrum.

As noted in Section 3.2, the distribution of wave energy across frequency and direction is described by the directional wave spectrum, $S(f,\theta)$, where f is wave frequency (in Hertz) and θ is wave direction (in radians). The directional spreading function is scaled so that its integral over all directions is equal to 1 (Equation 3.9), and so the frequency spectrum satisfies

$$S(f) = \int_{0}^{2\pi} S(f,\theta)d\theta \qquad (3.10)$$

Various analytical forms have been proposed for $S(f)$ on the basis of theoretical and observational considerations. The exact shape and scale of the spectrum will depend on the generating factors such as wind speed, duration, fetch, propagation and dissipation. Phillips (1958) considered the shape of the high-frequency side of the frequency spectrum. Under the assumption that waves in this portion of the spectrum are controlled by gravity he found

$$S(f) \propto g^2 f^{-5} \qquad (3.11)$$

The low-frequency part of the spectrum can be modelled fairly well with an exponential form, leading to a general form of spectrum:

$$S(f) = \frac{A}{f^5} \exp\left[-\frac{B}{f^4}\right] \tag{3.12}$$

The functional form in Equation (3.12) can be used to find analytical expressions for the spectral moments, and hence frequency domain wave parameters. The peak frequency is given from Equation (3.12) by

$$f_p = \left(\frac{4B}{5}\right)^{1/4} \tag{3.13}$$

The nth moment is given by

$$m_n = \frac{A}{4B^{(1-n/4)}} \Gamma\left(1 - \frac{n}{4}\right) \qquad n < 4 \tag{3.14}$$

where Γ is the Gamma function (see Gradshteyn and Rhyzik 1980). The first three moments can be expressed in terms of A and B as

$$m_0 = A / 4B$$
$$m_1 \approx 1.2254A / 4B^{3/4} \tag{3.15}$$
$$m_2 \approx 1.77245A / 4B^{1/2}$$

and hence

$$T_{m01} \approx 0.816B^{-1/4}$$
$$T_z \approx 0.751B^{-1/4} \tag{3.16}$$
$$T_p \approx 1.057B^{-1/4}$$

Some of the more widely used forms developed for deep water waves are the Bretschneider, Pierson–Moskowitz and JONSWAP spectra. These spectra are described below, and provide a historical perspective of the development of our understanding of wave spectra.

3.4.1 Bretschneider spectrum (Bretschneider 1959)

This has the form given in Equation (3.12) where the variables A and B can be specified by wave height and period. B is determined by the period through Equation (3.16) and A is found from Equation (3.14) and Equation (3.15). In terms of significant wave height, H_s, and period T_z, the spectrum becomes (Carter 1982):

$$S(f) = 0.08 \frac{H_s T_z}{(T_z f)^5} \exp\left[-0.318\left(\frac{1}{T_z f}\right)^4\right] \tag{3.17}$$

Bretschneider (1959) developed an empirical relationship between wave height and period and the wind speed, the fetch and the wind duration to develop prediction formulae for the average wave height and wave period, and thence the wave spectrum.

Bretschneider (1977) defined the significant wave period, $T_s = (0.8)^{1/4}/f_p = B^{-1/4}$, and so for this spectrum $T_s \approx 0.946T_p \approx 1.23T_{m01} \approx 1.33T_z$.

3.4.2   Pierson–Moskowitz spectrum (Pierson and Moskowitz 1964)

This spectrum was developed from an analysis of wind and wave records from British weather ships positioned in the North Atlantic. Only those records representing fully developed seas (for wind speeds between 20 and 40 knots) were used. The spectrum has the form

$$S(f) = \frac{\alpha g^2}{(2\pi)^4 f^5} \exp\left[-1.25\left(\frac{f_p}{f}\right)^4\right] \tag{3.18}$$

The constant α has a value of 8.1×10^{-3}. Ochi (1982) developed the following predictive formulae for wave height and frequency based on the Pierson-Moskowitz spectrum:

$$H_{m0} = \frac{0.21U_{19.5}^2}{g} \qquad\qquad f_p = \frac{0.87g}{2\pi U_{19.5}} \tag{3.19}$$

This spectrum is of the same form as Equation (3.12), with $A = \alpha g^2/(2\pi)^4$ and $B = 5f_p^4/4$. Hence, from Equations (3.14) and (3.19),

$$
\begin{aligned}
T_{m01} &= 3.86\sqrt{H_s} \\
T_z &= 3.55\sqrt{H_s} \\
T_p &= 5.00\sqrt{H_s}
\end{aligned}
\tag{3.20}
$$

Note that as A is a constant the Pierson–Moskowitz spectrum can be fully specified by either wave height or wave period. Note also that the wind speed is measured at an elevation of 19.5 m and is typically 5–10 per cent greater than U_{10}, a commonly quoted value. The use of a 1/7th power law vertical wind profile is suggested by CIRIA/CUR (1991). Thus the wind speed at height h is determined from $U_h/U_{10} = (h/10)^{1/7}$.

3.4.3   JONSWAP spectrum (Hasselmann et al. 1973)

$$S(f) = \frac{\alpha g^2}{(2\pi)^4 f^5} \exp\left[-1.25\left(\frac{(f_p)}{f}\right)^4\right]\gamma^q \tag{3.21}$$

where

$$\alpha = 0.076\left(\frac{gF}{U_{10}^2}\right)^{-0.22} \qquad\qquad f_p = \frac{3.5g}{U_{10}}\left(\frac{gF}{U_{10}^2}\right)^{-0.33} \tag{3.22}$$

where F = the fetch length and

$$q = \exp\left(-\frac{(f - f_p)^2}{2\sigma^2 f_p^2}\right)$$

with

$$\sigma = \begin{cases} 0.07 & f \leq f_p \\ 0.09 & f > f_p \end{cases}$$

The frequency at which the spectrum attains its maximum value is denoted by f_p. The magnitude of the peak enhancement parameter, γ, lies between 1 and 7 with an average value of 3.3. Figure 3.7 shows a schematic comparison of the JONSWAP and Pierson-Moskowitz spectra, and the significance of γ.

The moments of this spectrum cannot be calculated analytically but may be estimated by numerical integration. For $\gamma = 3.3$, the following approximate relationships hold: $T_{m01} = 0.8345T_p$; $T_z = 0.7775T_p$.

When $\gamma = 1.0$ the JONSWAP spectrum simplifies to the Pierson–Moskowitz spectrum. The JONSWAP spectrum has become one of the most widely used spectra, both in laboratory experiments and for design.

It should be borne in mind that the above three spectra have important limitations:

- They are not applicable to intermediate or shallow-water conditions.
- The Pierson–Moskowitz spectrum is for fully developed seas only.
- The JONSWAP spectrum was developed under fetch-limited conditions.
- The Bretschneider spectrum accounts for duration and fetch limitation in an empirical manner.
- They are all single-peaked spectra. (Ochi and Hubble (1976) describe a method of modelling double-peaked spectra.)

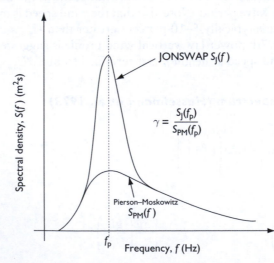

Figure 3.7 Illustrative plot of the Pierson–Moskowitz and JONSWAP frequency spectra.

3.5 Swell waves

The sea state at any deep-water site comprises a combination of waves generated locally by the wind and waves that have propagated to the site from outside the immediate area. The former are often termed 'wind-sea' while the latter are termed 'swell'. As noted in Section 2.1 'swell' refers to waves that have propagated away from the region in which they were generated and can therefore not generally be predicted from a knowledge of local conditions. As swell waves propagate, their average height decays very slightly due to air resistance and friction. More importantly, angular spreading will cause a reduction in wave height as the swell propagates. In addition frequency dispersion occurs, so that longer waves travel faster than shorter waves. However, in cases where the swell has only a narrow directional and frequency spread it can travel extremely long distances without much apparent decay. Snodgrass *et al.* (1966) observed that swell generated to the south of New Zealand could propagate across the Pacific Ocean without any discernible decay in height.

A precise definition of 'swell', in terms of its contribution to the wave spectrum, does not exist. However, it may generally be characterised by long periods (above 8 s) and relatively low wave heights. An estimate of the lowest frequency generated by the wind field can be obtained by noting that waves propagating with a phase speed greater than the wind speed, U, cannot receive energy from the wind. In deep water the phase speed of Airy waves is given by $g/2\pi f$ and so the lowest frequency affected by the wind is $g/2\pi U$. This can be used to define a 'separation' frequency such that all energy in the spectrum below this frequency can be attributed to swell and all energy above the frequency defined as wind-sea (see Figure 3.8). In practice, the wind speed is not a constant, and swell may have energy at all frequencies. A detailed discussion of the swell climate near the United Kingdom is provided by Hawkes *et al.* (1997).

At first glance it might appear that wind waves are more important for the design of coastal works as they are associated with the highest waves. However, the longer periods of swell waves mean that they can be a significant concern in coastal and harbour studies where sediment mobility, armour stability, harbour resonance and overtopping are important. Wind-sea and swell provide alternative extreme wave conditions for design, and in situations where both can occur together the design should take this into account. One important consequence arising from the presence of swell is that the full wave spectrum may no longer have a single peak.

3.6 Prediction of deep-water waves

The accurate prediction of deep-water waves requires knowledge of both local conditions, to estimate the wind-sea, and more distant wave conditions and associated transformations to estimate the swell. Numerical models that can predict average wave conditions over large areas have been developed since the 1960s. They are often termed 'phase-averaged' models because rather than resolving individual waves the models predict average wave quantities. In their simplest form such models predict integrated wave parameters such as H_s and T_z. 'Phase-resolving' models are discussed in Section 3.9. Modern phase-averaging models solve a single equation describing the evolution of the wave energy spectrum with time over an area. A simple deep water form of this equation is:

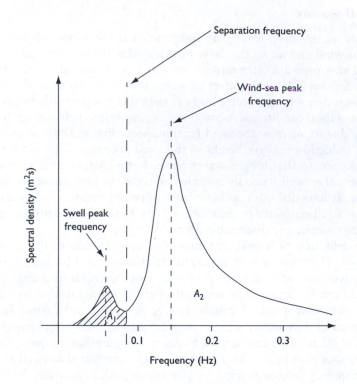

Figure 3.8 Schematic bi-modal spectrum with swell and wind-sea components. The significant wave heights of the swell and wind-sea components are given approximate-
ly by $4\sqrt{A_1}$ and $4\sqrt{A_2}$ respectively, where A_1 and A_2 are the areas shown above.

$$\frac{\partial E}{\partial t} + \mathbf{c_g}.\nabla E = S_{in} + S_{ds} + S_{nl} + S_{bf} \qquad (3.23)$$

where $E = E(f, \theta, x, y, t)$ is the directional energy spectrum and c_g is the group velocity
vector. The terms on the right-hand side of Equation (3.23) represent source terms:
S_{in} is the energy input from wind stress; S_{ds} is the energy loss from white capping; S_{nl}
represents non-linear interactions that redistribute energy within the spectrum; S_{bf} is
the energy loss due to bottom friction. In shallow water the effects of refraction may
be included through an additional term on the left-hand side of Equation (3.23). Further details of the source terms may be found in Young (1999). Equation (3.23) has
the form of an advective transport equation. Its solution requires suitable initial and
boundary conditions to be specified and an appropriate numerical scheme to calculate the transport of energy within the computational domain. Global phase-averaged models are run by meteorological or oceanographic organisations throughout
the world, in much the same way that atmospheric models are run to provide weather
forecasts. In fact, many global wave models use a mixture of observations and atmospheric models to specify the term S_{in}.

As the understanding of the source terms has improved so the sophistication with
which the source terms have been represented in Equation (3.23) has increased. Three

different stages of development of the model are now recognised and are labelled first-, second- and third-generation models. First-generation models, such as those of Gelci *et al.* (1957), Pierson *et al.* (1966) and Cavaleri and Rizzoli (1981), include only the first two terms on the right-hand side of Equation (3.23). The wind stress input term in these models was characterised by the sum of a linear (Phillips) term and an exponential (Miles) term. The dissipation term was effectively a numerical convenience that prevented the spectrum exceeding a pre-determined saturation limit (Phillips 1958). This had the effect of constraining the spectrum shape at the high frequency range. These models typically performed well in the regions for which they were developed. However, they could not be guaranteed to be accurate for other regions without extensive 'recalibration' of the model parameters. Second-generation models included the first three terms on the right-hand side of Equation (3.23). The JONSWAP study (Hasselmann *et al.* 1973) demonstrated the importance of wave–wave interactions in determining the evolution of the spectrum. However, this term required a large amount of computational effort to evaluate and so was included in a variety of approximate forms (e.g. Barnett 1968; Ewing 1971; Sobey and Young 1986). In many second-generation models the dissipation term was altered to allow a variable saturation limit to the spectrum. The 1980s saw a number of first- and second-generation models enter into operational service. The UK Meteorological Office wave model, described by Golding (1983), is one example. The developers of third-generation models sought to relax the approximations and parameterisations made in the previous generations of models. An international collaboration resulted in the WAM Model (WAMDI 1988; Hasselman and Hassellman 1985). A major difference with earlier models is the relaxation of the saturation to a condition requiring a zero balance between the source terms at high frequencies. Figure 3.9 shows an illustrative diagram of the contributions of the different source terms to the frequency spectrum.

The numerical solution of Equation (3.23) presents formidable difficulties. The directional wave spectrum will be 'discretised' and represented as a finite set of frequencies and directions. The resulting discrete spectrum has then to be predicted

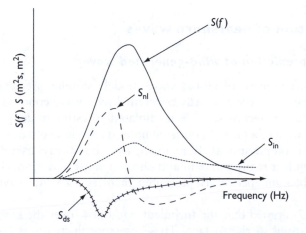

Figure 3.9 Source terms for third-generation wave models, and corresponding frequency spectrum.

over a two-dimensional spatial grid. Advective transport equations are notorious for problems associated with artificial numerical dispersion. These problems are often addressed by using a high-order scheme (which has smaller numerical errors associated with it) and accepting the additional computational expense this involves. As a result, most phase-averaged models use the finite-difference technique, for example WAM (WAMDI, 1988), SWAN (Booij *et al.* 1996). The combination of the requirements for an accurate numerical scheme, associated computational effort and global coverage necessarily leads to a compromise on spatial resolution. As an example, the European Centre for Medium Range Forecasting (ECMWF) originally operated its global wave model (a version of the WAM model), at a spatial resolution of 3°, a time step of 20 min and the spectrum being represented by 25 frequencies and 12 directions. The UK Meteorological Office has run a global wave model on an operational basis for many years and uses this to specify boundary conditions for regional wave models covering, for example, Europe and the Middle East. Details of the model physics as used operationally up to April 1987 may be found in Golding (1983). The European Wave Model has a grid resolution of approximately 30 km, which is not fine enough to fully resolve the coastal geometry. Therefore, output from such models still requires further transformation calculations to determine design conditions at or close to the coast. More recent developments have been encapsulated in the WAVEWATCH sequence of models. The third in the sequence, WAVEWATCH III (Tolman 1997, 1999, 2009) is a third generation wave model developed at NOAA/NCEP following the conceptual development of the WAM model. WAVEWATCH III represents a signal change in relation to its predecessors as it includes new governing equations, model structure, numerical methods and physical parameterisations. WAVEWATCH III solves the random phase spectral action density balance equation for wave number-direction spectra. This is appropriate for predicting wave evolution where variations in water depth, currents and wave height are slow in comparison to the spatial and temporal scales of a single wave. In the most recent releases new options for extremely shallow water (surf zone) have been included, although these remain at a relatively rudimentary level.

3.7   Prediction of nearshore waves

3.7.1   Point prediction of wind-generated waves

Prior to the development of global wave models simpler prediction methods had evolved. As engineers were primarily concerned with wave conditions at a particular site, these prediction techniques are formulated to estimate design conditions at a point, rather than over a grid of points. Simpler methods also require much less input information and computational effort. In any case, it is always useful to have a second method to provide a ready estimate and check against wave model output. In 1957 Phillips and Miles independently presented theories on wind wave generation and growth.

Phillips (1957) argued that the turbulent wind flow over the sea surface produces pressure fluctuations at the surface. These pressure fluctuations contain a range of frequencies and directions which can generate infinitesimal waves. The wind pressure fluctuations propagate according to the variations in the wind field. On the other hand, an infinitesimal wave generated on the sea surface will propagate according to Airy theory. When these two velocities are different the wave will be damped; when the velocities are similar the wave will amplify. This theory provides an explanation of wave generation at its initial stage, and is sometimes referred to as the 'resonance model'.

Miles (1957) theory explained the continued growth of waves, rather than the initiation of waves. The theory relies on variations in wind velocity in the vertical, and is hence known as the 'shear flow model'. When waves appear on the sea surface, the vertical wind velocity profile near the surface is modified by the waves. As the wind blows over a forward-moving wave a secondary circulation is set up in the down-wind edge of the wave. The net result of this complex air circulation is positive pressure anomaly on the up-wind edge of the wave and a negative pressure anomaly on the down-wind side. This pressure imbalance supplies a component of force on the wave in the direction of the primary wind flow. As this force does work, energy is supplied to the wave, increasing its amplitude. The greater the wave height the greater the pressure imbalance, and the greater the wave energy growth. This theory describes wave growth from the initial stage to a developed stage where linear theory is no longer valid. Then non-linear processes such as wave breaking and wave–wave interactions must be considered.

These theories were pre-dated by a semi-empirical ocean wave forecasting method developed during the Second World War by Sverdrup and Munk which became a standard procedure for many years. This method and some subsequent modifications are presented in the following sections. Although these formulae are predictive, in practice they are often used to reconstruct wave conditions from wind records, and are termed 'hindcasting' equations.

3.7.2   The SMB method

Horikawa (1978) provides further historical background to the development of the Sverdrup and Munk formulae. These formulae were subsequently refined using additional data by Bretschneider (1952, 1958) and the method became known as the SMB wave prediction method after the three authors. The prediction formulae are often presented in terms of nomograms or charts (Shore Protection Manual 1984). The SMB curves for deep-water fetch-limited wave height and period are based on the formulae:

$$\hat{H}_s = 0.283 \tanh(0.0125 \hat{F}^{0.42}) \tag{3.24}$$

$$\hat{T}_z = 7.54 \tanh(0.077 \hat{F}^{0.25}) \tag{3.25}$$

where the dimensionless parameters denoted by the caret (^) are defined by

$$\hat{H}_s = H_s \frac{g}{U_{10}^2}$$

$$\hat{T}_z = T_z \frac{g}{U_{10}} \tag{3.26}$$

$$\hat{F} = F \frac{g}{U_{10}^2}$$

Updated forms of these equations have been published in SPM (1984) and are based on an intermediate calculation that replaces U_{10} by $U_a = 0.71 U_{10}^{1.23}$. However, the reliability of this intermediate calculation for all cases has been questioned in, for example, CIRIA (1996).

3.7.3 The JONSWAP method

In SPM (1984) a parametric method, based on the JONSWAP spectrum is recommended for deep water, replacing the SMB parametric prediction formulae. These take the form:

$$\hat{H}_s = a\hat{F}^{0.5} \tag{3.27}$$

$$\hat{T}_p = b\hat{F}^{0.33} \tag{3.28}$$

$$\hat{T}_p = T_p \frac{g}{U_a} \tag{3.29}$$

where
  The constants a and b take the values 0.0016 and 0.2857, respectively. The original form, using U_{10} instead of U_a, is quoted in CIRIA (1996) with the values of a and b being 0.00178 and 0.352, respectively.

Example 3.2
Calculate the significant wave height and zero up-crossing period using the SMB method (with and without the SPM modification) and the JONSWAP method (using the SPM and CIRIA formulae) for a fetch length of 5 km and a wind speed of $U_{10} = 10$ m/s. In all cases the first step is to calculate the non-dimensional fetch length.

SMB METHOD (ORIGINAL VERSION)

$$\hat{F} = \frac{5000 x 9.81}{100} = 490.5$$

SO

$$H_s = \frac{100 x 0.283}{9.81} \tanh\{0.0125 F^{0.42}\} = 0.5m$$

$$T_z = \frac{10 x 7.54}{9.81} \tanh\{0.077 F^{0.25}\} = 2.7s$$

SMB METHOD (MODIFIED VERSION)

First calculate U_a: $U_a = 0.71 U_{10}^{1.23} = 12.06$ m/s

$$\hat{F} = \frac{5000 x 9.81}{12.06^2} = 337.2$$

$$H_s = \frac{12.06^2 x 0.283}{9.81} \tanh\{0.0125 F^{0.42}\} = 0.6m$$

$$T_z = \frac{12.06 x 7.54}{9.81} \tanh\{0.077 F^{0.25}\} = 3.0s$$

JONSWAP METHOD (ORIGINAL VERSION)

We have $U_{10} = 10$ m/s and non-dimensional fetch = 490.5. These give:

$$H_s = \frac{10.0^2 x 0.00178}{9.81} F^{0.5} = 0.4m$$

$$T_p = \frac{10.0 x 0.352}{9.81} F^{0.33} = 2.8s$$

But for a JONSWAP spectrum with an average value of the peak enhancement factor we have $T_z \approx 0.7775 T_p = 2.2$ s.

JONSWAP METHOD (MODIFIED VERSION)

As above, we have $U_a = 12.06$ m/s and non-dimensional fetch = 337.2. These give:

$$H_s = \frac{12.06^2 x 0.0016}{9.81} F^{0.5} = 0.4m$$

$$T_p = \frac{12.06 x 0.2857}{9.81} F^{0.33} = 2.4s$$

But for a JONSWAP spectrum with an average value of the peak enhancement factor we have $T_z \approx 0.7775 T_p = 1.9$ s.

3.7.4 Further modifications and automated methods

Where fetches may be of restricted width (such as in estuaries, bays and lochs), some modification to the hindcasting equations is necessary. Several methods have been proposed; however, measurements obtained by Owen (1988) on UK reservoirs showed that none of the methods provided consistently good predictions over a range of conditions.

Saville's method

Saville *et al.* (1962) proposed the concept of 'effective fetch', which assumes that: waves are generated over a 45° range either side of the wind direction and energy transfer from wind to waves is proportional to the cosine of the angle between the wind and wave directions; wave growth is proportional to fetch length. Hence,

$$F_{eff} = \frac{\sum_i F_i \cos^2(\theta_i)}{\sum_i \cos(\theta_i)} \tag{3.30}$$

where F_{eff} is the effective fetch and is the fetch length to be used in the SMB formula for open seas. F_i and θ_i are fetch lengths and angles measured at 6° intervals.

Seymour's method

Seymour (1977) adopted a different concept of effective fetch, using the JONSWAP formulae. The method assumes that the waves have a cosine-squared directional spreading function over a 180° arc. Further, it assumes that the energy along each direction E_i is given by the JONSWAP formulae and that the waves generated are Airy waves (with energy proportional to $H^2/8$), and so the total wave energy, E, is given by

$$E = \frac{2}{\pi} \sum_i E_i \cos^2(\theta_i - \theta_w)\Delta\theta \tag{3.31}$$

Donelan–JONSWAP method

Donelan (1980) proposed a method based on the JONSWAP formulae but used the notion that the fetch length be measured along the wave direction rather than the wind direction. For fetches of general shape, the assumption is made that the predominant wave direction is that which produces the maximum value of wave period (for the given wind speed). To determine the maximum wave period requires repeated calculations and a process of trial and error. Once the wave direction has been determined the wave height and period are determined from the JONSWAP formulae with U_{10} replaced by $U_{10}(\cos(\theta_i - \theta_w))$.

SMB shallow-water formulae

Versions of the SMB formulae, modified for use in shallow water, are given in SPM (1984). These are acknowledged to be approximate but provide a means of computing wave conditions at a point using wind information, accounting for finite depth effects.

Automated methods

A deep-water wave climate can be derived with the aid of hindcasting models. Given an historical wind record such as standard hourly wind measurements and details of the sea area, several computational methods are available for prediction of the resulting sea state. The 'predictions' are termed 'hindcasts' because past rather than future wave conditions are computed. One such model is described in Fleming *et al.* (1986). This calculates wave heights, periods and directions according to wind duration, strength and fetch length. Input data consists of wind speed and direction at sequential time intervals. Calculations are performed for every combination of wind speed and duration preceding the current time in order to select the maximum wave height that could have arisen from the conditions. Effective fetches are calculated within the programme to allow for the directional distribution of the wave spectrum. Allowances are made for wave decay when winds suddenly drop or veer to directions where there is little or no fetch. Figure 3.10 shows measured and predicted wave climate for the Canadian Coastal Sediments Study (Fleming *et al.* 1986).

3.8   The TMA spectrum

The wave prediction schemes described in Section 3.7 provide some simple means of estimating wind waves at a given location. For many coastal locations wind or wave conditions are known some distance from shore and so must be transformed inshore

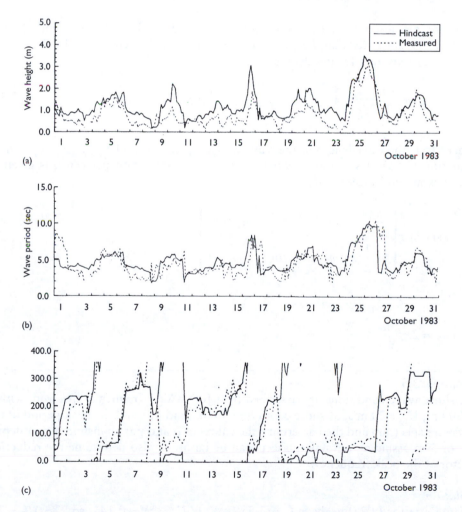

Figure 3.10 Comparison of predicted (hindcast) and measured waves (from Fleming *et al.* 1986). (a) Significant wave height; (b) Peak wave period; (c) Mean direction MA75.

to determine design conditions. The wave frequency spectra discussed in Section 3.4 describe deep-water conditions. As waves propagate into intermediate and shallow water, changes in the shape of the spectrum occur. In general, these changes are frequency dependent, since shoaling, refraction and diffraction are frequency dependent. One widely used spectrum that has been developed for shallow water conditions is the TMA spectrum (Bouws *et al.* 1985). The TMA data comprised of measurements made at three coastal sites, Texel (Dutch North Sea), Marsen (German Bight) and Arsloe (US East Coast). This spectrum is based on the assumption that waves generated in deep water propagate into intermediate or shallow water without refracting. The TMA spectrum is a form of the JONSWAP spectrum with modified α and γ coefficients and a multiplicative depth- and frequency-dependent reduction factor,

$$S_{TMA}(f,h) = S_{JONSWAP}(f)\Theta(\omega,h) \tag{3.32}$$

where the subscripts have the obvious reference and h is the still water depth. Bouws *et al.* (1985) suggested that the α and γ coefficients could be determined by the following expressions for all water depths:

$$\alpha = 0.0078\left(\frac{2\pi U_{10}^2}{gL_p}\right)^{0.49} \qquad \gamma = 2.47\left(\frac{2\pi U_{10}^2}{gL_p}\right)^{0.39} \tag{3.33}$$

where L_p is the wave length of the wave with frequency f_p and $S_{JONSWAP}(f)$ is given by Equation (3.21). A good approximation for Θ (accurate to 4 per cent), is given in Thompson and Vincent (1983) as

$$\Theta(\omega,h) = \begin{cases} \dfrac{\omega_h^2}{2} & \text{for} \quad \omega_h \leq 1 \\ 1 - \dfrac{1}{2}(2-\omega_h)^2 & \text{for} \quad \omega_h > 1 \end{cases} \tag{3.34}$$

where

$$\omega_h = 2\pi f\sqrt{\frac{h}{g}}$$

Example 3.3

Offshore wave conditions are described by a JONSWAP frequency spectrum with $F = 80$ km, $U_{10} = 20$ m/s, offshore depth $h_o = 50$ m. Find α, γ and the peak period of the offshore spectrum and then determine the values of α and γ at an inshore water depth, h_i, of 5 m assuming refraction effects can be ignored. Also determine the reduction factor at the peak frequency.

Solution

From Equation (3.22) we have $f_p = (3.5g/U_{10})(gF/U_{10}^2)^{-0.33} = 0.141$, so $T_p = 1/f_p = 7.1$ secs. In deep water the wave length of the wave with peak frequency may be calculated from $L_{po} = gT_p^2/2\pi = 78.7$ m. [As a check note that $h_o/L_{po} = 0.635 > 0.5$]. We also calculate, from Equation (3.22), $\alpha = 0.076(gF/U_{10}^2)^{-0.22} = 0.076(9.81 \times 80000/400)^{-0.22}$ = 0.014. You obtain the same value for α if you use Equation (3.33) instead, which also gives $\gamma = 2.47(2\pi U_{10}^2/gL_{po})^{0.39} = 3.9$.

Now, Equation (3.33) is valid for all water depths. Denoting the offshore and inshore values of α and γ by the subscripts 'o' and 'i' respectively we find:

$$\alpha_o = 0.0078\left(\frac{2\pi U_{10}^2}{gL_{po}}\right)^{0.49} \qquad \alpha_i = 0.0078\left(\frac{2\pi U_{10}^2}{gL_{pi}}\right)^{0.49}$$

and similarly for γ. Taking U_{10} to be the same offshore and inshore, and identifying 'offshore' with 'JONSWAP' and 'inshore' with 'TMA' gives

$$\alpha_{TMA} = \alpha_{JONSWAP}\left(\frac{L_{po}}{L_{pi}}\right)^{0.49} \qquad \gamma_{TMA} = \gamma_{JONSWAP}\left(\frac{L_{po}}{L_{pi}}\right)^{0.39}$$

L_{pi} can be determined using the methods described in Section 2.3.4 to be 46.4m. [Using a shallow-water approximation, $L_{pi} \approx \sqrt{(gh_i)}T_p$, gives L_{pi} = 49.7 m]. Hence α_{TMA} = 0.014(78.7/46.4)$^{0.49}$ = 0.018, and γ_{TMA} = 4.06. For the peak frequency $\omega_h = 2\pi f_p\sqrt{(h/g)}$ = $(2\pi/7.1)\sqrt{(5/9.81)}$ = 0.6318 < 1.0. Thus $\Theta = \omega_h^2/2$ = 0.20, and the corresponding TMA spectrum calculated. Figure 3.11 shows the JONSWAP and TMA spectra for this case.

3.9 Numerical transformation of deep-water wave spectra

An introduction to wave transformation modelling was provided in Sections 2.3.11 and 2.3.12. With recent advances in computer performance, surface wave propagation

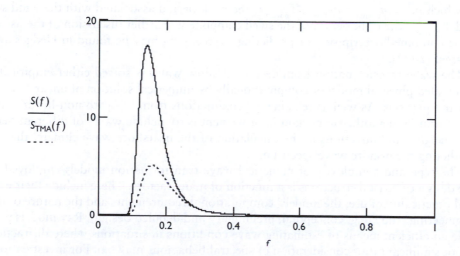

Figure 3.11 Showing JONSWAP and TMA spectra of above example. The units for the frequency spectra are m²s and frequency has units of hz.

models have become a standard tool for coastal engineers. Estimation of refraction and diffraction of waves passing over a complicated seabed surface (or bathymetry) is extremely important for coastal engineering. Although the complexity of the natural sea state is recognised, many engineering analyses of wave propagation over irregular bathymetry have been based on characterising the sea state by a representative wave of a particular height, period and direction. For wave height, H_s is often used for estimating overtopping, but H_{max} provides greater conservatism for critical parts of a structure such as the end of a breakwater. Wave periods are often characterised by T_p or T_z. Most design formulae use one or other of these measures. For wave direction, it is common to take the mean wave direction, as defined by the spectrum. However, by analysing the wave statistics (wave rose or frequency tables), it will become evident whether there is a predominant wave direction or whether there are seasonal effects resulting in preferred directions in different seasons. If seasonal effects are notable, calculations and design must take this into account. Difficulties can arise if the frequency wave spectrum is very broad, containing a wide range of periods, or if it is bi-modal. In this case it is often advisable to repeat calculations for representative wind-sea and swell-wave components, taking the wave condition that gives the worst-case result for design. If a combined swell-wind sea wave condition is required then the corresponding equivalent wave height can be calculated from:

$$H_{s(eq)} = \sqrt{H_{s\,swell}^2 + H_{s\,wind\,sea}^2}$$

The choice of a characteristic wave period is more problematic. One option is the equivalent peak period, $T_{p(eq)}$, described by van der Meer and Janssen (1995), and defined as:

$$T_{p(eq)} = \sqrt[4]{\frac{m_{0\,wind}}{m_0}T_{p\,wind}^4 + \frac{m_{0\,swell}}{m_0}T_{p\,swell}^4}$$

In which $m_0 = m_{0\,wind} + m_{0\,swell}$, $T_{p\,wind}$ is the peak period associated with the wind sea and $T_{p\,swell}$ is the peak period of the swell component. Further discussion of the use of these combined descriptors for predicting overtopping may be found in Hedges and Shareef (2002).

The wave transformation from deep to shallow water is solved either empirically in a scaled physical model or computationally by numerical solution of linear or non-linear equations. As well as relaxing approximations from linear to non-linear wave transformation another important improvement is to include waves of different periods, heights and directions in the calculation of the nearshore wave climate, that is, predicting the inshore wave spectrum.

The type and complexity of numerical wave transformation models employed in the design of coastal structures is a function of many factors. These include their ease and practicality of use, the models' computational requirements and the nature of the engineering study (e.g. conceptual, preliminary and detailed design). Ray models provide an efficient means of estimating wave conditions in situations where diffraction is not significant (but consideration of spectral behaviour may be). For investigations of wave penetration around breakwaters and into harbours, models based on a wave function description are required. For situations where wave-structure interactions are important then models based on non-linear Boussinesq or shallow-water equations have been employed. These have the advantage of being computationally efficient but have drawbacks. In the case of Boussinesq models the equations provide an approximate description of dispersion. Shallow-water equation models deal strictly with situations of shallow water where there is negligible variation in the flow in the vertical direction. To obtain a full description of the flow it is necessary to solve the Navier–Stokes equations, usually with a turbulence model. One example is the Reynolds-Averaged Navier–Stokes solution developed by Lin and Liu (1998). While giving a more complete description of the physics these models are computationally demanding and, without access to high-performance computing, it is only feasible to simulate wave propagation across the surf zone. Only in the last few years have fully three-dimensional numerical wave models been developed for application to coastal engineering problems. Some of these models that are in widespread use are discussed briefly below.

3.9.1 Spectral ray models

During the 1970s numerical ray models were developed (e.g. Abernethy and Gilbert 1975) that allowed the inshore transformation of deep-water wave spectra, accounting for refraction and shoaling. The offshore wave spectrum was discretised in both direction and frequency. A refraction and shoaling analysis was performed for each direction–frequency combination and then the resulting inshore energies were summed to assemble an inshore directional spectrum. The engineering design wave parameters could then be computed from the definitions given in Section 3.2. In short, the inshore spectrum was computed from

$$E_i(f,\theta_i) = E_o(f,\theta_o)\frac{cc_{go}}{cc_g}$$

(3.35)

where subscripts o and i refer to offshore and inshore, respectively. The offshore spectrum was considered to be known and the shoaling and angle changes were determined from numerical ray tracing over a digital representation of the seabed.

With the development of the TMA shallow-water spectrum (Bouws et al. 1985), which provided an upper bound on the energy content of the frequency spectrum, the ray-tracing models could be extended to incorporate wave breaking and other surf-zone processes in an empirical manner. This was done by reducing the energy content of the computed inshore frequency spectrum to the value predicted by the TMA spectrum for the given water depth.

Thus a spectral description of the nearshore wave climate at a point could be obtained that accounted for refraction and shoaling, together with an empirical treatment of non-linear wave processes such as wave breaking. However, this approach could not account for diffraction, and was limited to describing conditions at a selected position. Nevertheless, ray models remain in current use because of their modest computational requirements and because they can provide a spectral description of nearshore wave conditions. The effectiveness of this relatively straightforward approach for coastal flood warning was demonstrated by Reeve et al. (1996). They compared in situ measurements against transformed deep-water predictions of wave climate from the UK

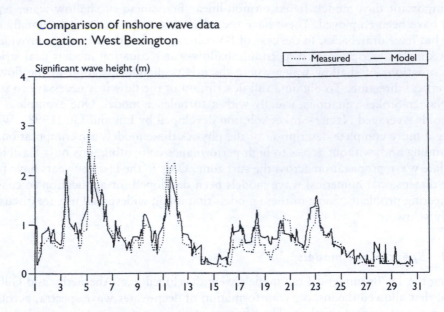

Comparison of inshore wave data
Location: West Bexington

Figure 3.12 Comparison of measured and transformed offshore wave model predictions of significant wave height for April 1993.

Meteorological Office European Wave Model at three sites around the UK. Figure 3.12 shows a short record of predicted and recorded wave heights at West Bexington, Lyme Bay, on the south coast of the UK.

This approach would not be expected to be so effective where diffraction effects are significant, as it does not account for wave diffraction.

The need to fully consider the implications of directional wave spectra as described in Section 3.3 with respect to the design of areas protected by marine structures should be readily appreciated. In real seas waves approach from a range of directions centred around the mean wave direction. The amount of spreading is dependent on the characteristics of the sea state such as locally generated wind waves or swell waves from a more distant source as demonstrated schematically in Figure 3.5(a). From Figure 2.1.4(a) it can be appreciated that the wave height reduction in the lee of a breakwater is relatively rapid in the shadow zone. Even quite small changes in the angle of approach will result in differences in the diffracted wave height which might be quite small, but not necessarily insignificant in the context of vessel motions at a berth, particularly for smaller working or recreational craft. The outcome of properly modelling directional wave spectra as opposed to a unidirectional sea state is that, whilst the total energy entering a protected area such as a port or marina might be the same, its distribution around the perimeter of the area will be different. In general it is likely that berths with the highest wave activity under unidirectional conditions will experience a reduced condition and vice versa – berths considered to be well sheltered might be subject to increased wave activity. This is not a hard and fast rule as there may be further reflection and diffraction within the protected area. However, as a general principle directional wave spectra should always be used for modelling wave activity within ports and marinas or any protected area for which potential downtime for loading/unloading or simply berthing might be a critical issue in the design. Indeed, it is rarely that this is not the case as a competent design will generally seek to optimise breakwater lengths to the minimum consistent with operational requirements.

3.9.2   Mild-slope equation

A significant step in the development was the introduction of the mild-slope equation, first derived by Berkoff (1972). The mild-slope equation is derived from the linearised governing equations of irrotational flow in three dimensions under the assumption that the bottom varies slowly over the scale of a wavelength. The mild-slope equation has been used widely to date to predict wave properties in coastal regions. The equation, which can deal with generally complex wave fields with satisfactory accuracy accounts for refraction, shoaling, diffraction (and in some forms reflection as well). The mild-slope equation may be written as

$$\nabla \cdot (cc_g \nabla \Phi) + \frac{\omega^2 c_g \Phi}{c} = 0 \tag{3.36}$$

for the complex two-dimensional potential function Φ. In a three-dimensional Cartesian coordinate system, Φ is related to the water wave velocity potential of linear periodic waves $\phi(x,y,z,t)$ by

$$\Phi(x,y) = \phi(x,y,z,t) \frac{\cosh(\kappa h)}{\cosh(\kappa(h+z))} e^{-i\omega t} \tag{3.37}$$

where the frequency ω is a function of the wavenumber $\underline{k} = (k,l)$ with $\kappa = |\underline{k}|$ by virtue of the dispersion relationship

$$\omega^2 = g\kappa \tanh(\kappa h) \tag{3.38}$$

The local water depth is $h(x,y)$, the local phase speed $c = \omega/\kappa$ and the local group velocity $c_g = (\partial \omega / \partial k, \partial \omega / \partial l)$. Writing $\psi = \Phi \sqrt{(cc_g)}$ allows the mild-slope equation to be cast into the form of a Helmholtz equation. Under the assumptions of slowly varying depth and small bottom slope, Radder (1979) showed that the equation for ψ might be approximated as the following elliptic equation:

$$\nabla^2 \psi + \kappa^2 \psi = 0 \tag{3.39}$$

Several numerical models are available that solve the elliptic form of the mild-slope equation by finite elements (e.g. Liu and Tsay 1984). However, a finite difference discretisation is generally easier to implement. This approach produces reasonably good results provided a minimum of between 8 to 10 grid nodes are used per wavelength. This requirement precluded the application of this equation from modelling large coastal areas (i.e. with dimensions greater than a few wavelengths) due to the high computational cost. As a result, a number of authors have proposed models based on different forms of the original equation.

Copeland (1985) has transformed the equation into a hyperbolic form. This class of model is based on the solution to a time-dependent form of the mild-slope equation, and involves the simultaneous solution of a set of first-order partial differential equations. In practical applications numerical convergence can be difficult to achieve with this approach (Madsen and Larsen 1987).

An alternative simplification was proposed by Radder (1979). This involved a parabolic approximation that relied on there being only small variations in wave direction. Consider an initially plane wave of unit amplitude approaching from $x = -\infty$. The rapidly varying component of the wave field is isolated by writing $\phi = \psi\exp(-ik_0x)$, where k_0 is a reference wavenumber corresponding to the positive root of Equation (3.38) with $h = h_0$, a reference depth. Equation (3.39) then becomes

$$\frac{\partial^2\phi}{\partial x^2} + \frac{\partial^2\phi}{\partial y^2} - 2ik_0\frac{\partial\phi}{\partial x} + k_0^2(n^2 - 1)\phi = 0 \qquad (3.40)$$

where the refractive index is defined as $n = k/k_0$. The parabolic approximation neglects the second term on the left-hand side of Equation (3.40) in comparison to the third term (Tappert 1977). In effect, the assumption is that the incoming wave will only deviate from its initial direction by a small amount. The advantage of such an approach is that a very computationally efficient time-stepping algorithm can be adopted and this allows solutions to be obtained over large areas. The parabolic model framework was used by Reeve (1992b) to demonstrate that random variations in the level of the seabed would lead to the development of a directional spreading function. The form of the directional spreading function obtained is very similar to that proposed by Goda (2000). The numerical solution steps forward from the seaward boundary with the given seaward boundary condition and appropriate lateral boundary conditions. In contrast, the elliptic problem given by Equation (3.40) must be solved simultaneously over the whole computational domain, subject to the boundary conditions along the sides of the domain. The disadvantages include the neglect of reflections, neglect of diffraction effects in the direction of wave propagation, and the constraint of small angular deviation from the initial direction of propagation. The last constraint can be relaxed somewhat through improvements on the approximation that allow larger angular deviations (e.g. McDaniel 1975; Kirby 1986, Dalrymple and Kirby 1988).

More recently, procedures that are both computationally efficient and stable have been developed for solution of the elliptic form of mild-slope equation (e.g. Li and Anastasiou 1992; Li 1994a). This has obviated the need to make approximations regarding wave angles, and as a result models based on the parabolic and hyperbolic forms of equation are being used less. Elliptic models have been extended to account for irregular waves (i.e. a wave spectrum) by Al-Mashouk et al. (1992) and Li et al. (1993) using the model to compute solutions for individual direction–frequency pairs. The results are then combined, following Goda (2000), as a weighted integral to provide a combined refraction/diffraction/shoaling coefficient. Thus

$$S_i(f,\theta_i) = S_o(f,\theta_o)K_R^2(f,\theta_o)K_S^2(f,\theta_o)K_d^2(f,\theta_o) \qquad (3.41)$$

which is the natural extension of Equation (3.35) where K_D is a diffraction coefficient. The numerical model is used to determine the coefficients at each point in a regular computational grid for a range of frequencies and offshore wave directions. The results can then be used in Equation (3.41) to estimate the spectrum at any point in the grid when the offshore spectrum is specified. The wave parameters H_s and T_z can then be obtained by integrating the inshore wave spectrum. This approach makes the assumption that the waves are small and so the principle of linear superposition is valid. Li *et al.* (1993) suggest that wave breaking can be accounted for in a simplified manner by applying a simple breaking criterion (e.g. Equation 2.34) to the resultant wave heights.

3.9.3 Non-linear models

The disadvantage of the models mentioned above is that they do not explicitly account for non-linear processes such as wave breaking, harmonic generation or wave–wave interaction. A class of models known as Boussinesq models is able to describe some aspects of non-linear wave behaviour. Boussinesq models are 'phase-resolving' in that they describe both the amplitude and the phase of individual waves. Equations of motion describing relatively long, small-amplitude waves propagating in water of varying depth were derived by Peregrine (1967):

$$\mathbf{u}_t + (\mathbf{u}\cdot\nabla)\mathbf{u} + g\nabla\eta = \frac{h}{2}\frac{\partial}{\partial t}\nabla[\nabla.(h\mathbf{u})] - \frac{h^2}{6}\frac{\partial}{\partial t}\nabla(\nabla.\mathbf{u}) \qquad (3.42)$$

$$\frac{\partial\eta}{\partial t} + \nabla.[(h+\eta)\mathbf{u}] = 0 \qquad (3.43)$$

where η is the surface displacement, u is the depth-averaged horizontal velocity and h is the undisturbed water depth. An additional assumption of slowly varying seabed variation is often made to simplify the terms on the right-hand side of Equation (3.42). This type of model is discussed by Beji and Battjes (1994) and Madsen *et al.* 1997) who used a Boussinesq model to simulate non-linear wave–wave interaction due to waves propagating over a submerged bar. Boussinesq models are non-hydrostatic and dispersive, and while they can describe solitary wave propagation do not describe breaking. The non-linear shallow-water equations (see Chapter 4), have also been used to predict wave propagation near the shore (see e.g. Dodd 1998). Although the equations are hydrostatic and non-dispersive these models can represent the propagation of bores and have better wave dispersion properties than Boussinesq models in shallow water. An alternative is to solve the full equations governing the fluid flow (the Navier–Stokes equations). This is much more computationally demanding than either the Boussinesq or non-linear shallow-water equations. Lin and Liu (1998) describe such a model that has been demonstrated to simulate wave breaking and wave run-up in good agreement with observations. Figure 3.13 shows the output of this type of model when applied to the situation in which random waves approach and break on a sloping seawall (see e.g. Soliman and Reeve 2003).

A more recent alternative is the Discrete Particle Method (see e.g. Koshizuka *et al.* 1995 and Gotoh *et al.* 2003), in which the fluid is represented by a large number of small particles (typically at least 10 000). The motion of each of the particles is governed by what are effectively Newton's Laws of Motion, together with rules governing what happens when two or more particles collide. Each particle is tracked in the

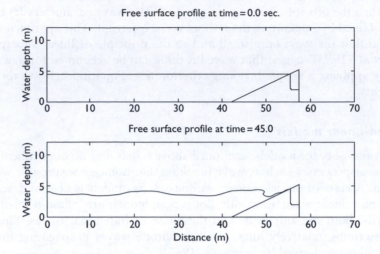

Figure 3.13 Simulation of random wave propagation, breaking and run-up on a sloping sea-wall. Top pane shows the initial condition when the water is at rest. Bottom pane shows random waves propagating towards the seawall.

numerical simulation to determine the movement of the water body as a whole.

3.10 Choosing and defining design conditions

In the past, when deep-water wave conditions were extremely scarce and wave trans-formation models were not available, it was common to define the extreme condition using deep-water wave conditions. These would have typically been annual maximum H_s and frequency tables. Extremes analysis would be performed on the wave heights to derive the heights corresponding to the required return period. The wave period would be estimated on the assumption that extreme waves have the same steepness as other waves, and the predominant steepness could be estimated directly from plotting lines of constant steepness on wave height–period frequency tables. Depending on the amount of data available, extreme conditions would be calculated either as 'omni-directional extremes' using wave heights irrespective of wave direction or extremes would be calculated for individual direction sectors (typically 15–30° wide). Wave transformation from deep water to shallow water would have been performed using refraction/shoaling nomograms or equations.

Extremes analysis involves selecting the largest recorded wave height in each year. These are then used to define the best fit for an extreme value distribution. These are similar to more everyday distributions like the normal, Rayleigh or uniform distri-butions, but they arise in the theory of the distribution of extreme values. The most widely used extreme value distributions are the Gumbel and Weibull distributions. These are defined later in Section 7.1.3.

The key point for the present discussion is that these formulae have parameters which define the central position and width of the peak in the distribution. Extreme value analysis is a procedure for fitting a chosen extreme distribution to the annual maxima. This defines the values of the parameters of the distribution. Knowing the distribution the values of wave height that correspond to specific values of the probability of occurrence can be found. Design of coastal structures use the concept of return period. This is discussed, along with extreme value methods, in Chapter 7. For now, consider the case where there is 100 years of wave height data from which 100 values are extracted – the largest wave height in each year. A histogram of number of occurrences against wave height can be constructed with this information. The numerical curve-fitting process of extreme value analysis effectively fits the equation *that you have chosen* to the data. In a collection of data containing 100 values one can be reasonably secure that values that occur ten times will correspond to a condition that is experienced once every 10 years on average. That is, the condition corresponds to the 1 in 10 year extreme. For more extreme events, say a wave height so large that it occurs only once in the record, one might feel less comfortable in ascribing a return period of 100 years. It might be close to the 1 in 100 year event, but equally it might be a more extreme event that just happened to occur during the duration of the measurements; or, it might be less severe than 1 in 100 years because the data you have happened to be taken was from a relatively quiescent period. There is no way of telling! Ideally, to estimate the 'N year' return value a sequence containing many times N years of records are required. Extreme value analysis formalises the process of converting observations into a probability distribution, and then extrapolating to determine the wave heights corresponding to particular return periods. Dealing with annual maxima gives this a particularly straightforward interpretation. For design purposes what is important is whether a particular severity of condition is exceeded or not. If the condition is exceeded the amount that it is exceeded by is of secondary importance. In terms of probability what matters is the chance of wave heights being greater than a particular value. The probability of wave heights being greater than a given value is 1 minus the probability of the wave heights being less than or equal to the given value. But the probability of the wave heights being less than a given value is just the (cumulative) probability distribution of wave heights. As an example, consider the point where the extreme value distribution is equal to 0.95; this corresponds to a condition that is exceeded $1 - 0.95 = 0.05 = 1/20$ of the time. As the data are annual maxima this corresponds to an occurrence of once in 20 years – a 20-year return period.

The above procedure, bar minor alterations, is the process through which design wave conditions at a structure have been estimated. There are of course many aspects to consider, such as: what extreme distribution to choose; the method employed to fit the distribution curve to your data; omnidirectional versus sector analysis and so on. However, several important developments have taken place to change what is considered best practice. The first is the creation of global and regional wave prediction models that are run on an operational basis. Archives of these models' output provide a useful source of deep-water wave information for design. Most of these models have a grid resolution that is coarse in comparison to the scale of variation in the coastal zone so do not provide good estimates of very nearshore conditions. Secondly, accurate and efficient wave transformation models have been written and widely distributed. These, together with the rapid increase in computing power that is now available, mean that it is possible to transform the equivalent of many years worth of wave data from deep water to a coastal site. It is now considered better practice to perform the extremes analysis on the nearshore wave conditions that have been transformed from deep water. As the water depth is an important controlling factor on wave height it is necessary to include the effects of tide and surge in the wave transformation calculations. Adding this extra variable complicates the statistical calculations and has provided an impetus for developing the theory of joint extremes. That is the probability of the joint occurrence of high large waves with high water levels. This and other topics are discussed further in Chapter 7.

3.11   Long-term wave climate changes

Long-term changes in wave climate, such as variations in wave heights, periods or directions, are clearly of importance in designing coastal works. At the coast, increases in wave heights will be restricted to the depth-limiting value and long-term changes will be controlled by long-term sea level rise and beach level trends. Changes in wave period and direction may be of more significance in this regard as they can result in large variations in overtopping of sea defences and longshore transport of beach material.

Wave archives of sufficient length to investigate long-term changes in wave climate are extremely scarce. However, the advent of Earth-orbiting satellites equipped with instruments that allow wave heights and periods to be determined and recorded now provide global coverage for a period covering approximately a decade. Young and Holland (1998) have presented global statistics of significant wave height for a dataset covering 9 years. Recent computational studies of the effect of atmospheric climate change on wave climate have not suggested a strong link between the two (Brampton 1999). However, Cotton et al. (1999) used satellite data to analyse changes in the mean wave climate in the North Atlantic, and found evidence of increases of up to 20 per cent in mean winter significant wave height from the period 1985–1989 to 1991–1996 (Figure 3.14).

Li et al. (2002) found reasonable agreement between design wave conditions derived from in situ measurements and those derived from transforming deep-water waves estimated from satellite altimeter measurements to the shore with a spectral refraction

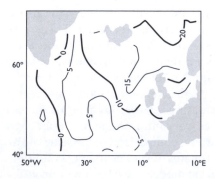

Figure 3.14 The percentage increase in mean winter significant wave height, 1985–1989 to 1991–1996 (with permission from Satellite Observing Systems Ltd).

model. As the duration of satellite records increase so they should provide an increasingly useful source of information for coastal engineers.

In addition to analysing observations of the recent past, interest in the effects of changes in our climate has prompted politicians, administrators and scientists to formulate a range of potential future scenarios. These scenarios are driven ostensibly by different rates of emissions of carbon dioxide and other gases created when fossil fuels are burnt for energy. The scenarios have been developed by the Intergovernmental Panel on Climate Change (IPCC). Many subsidiary studies investigating the impact of the scenarios on our climate, weather and much more have been supported or inspired by the IPCC initiative.

One of the possible impacts of climate change is changes to the hydrodynamic processes experienced along our coasts. This could have serious implications for existing flood and coast protection schemes. Potential climatic changes in wave climate, storm frequency and intensity, surges, and precipitation are all important in determining future shoreline trends and could enhance or counter the effects of SLR (e.g. Douglas *et al.* 2001; Stive *et al.* 2002; Walsh *et al.* 2004).

However, there are a growing number of studies on the impact of global warming in future wind, storm and wave climates, using either historical trends (e.g. Günther *et al.* 1998; Gulev and Hasse 1999; Cotton *et al.* 1999; Alexander *et al.* 2005) or climate model output (e.g. Kaas *et al.* 2001; Hulme *et al.* 2002; Debernard and Red 2008). Taken together, these suggest that future changes in wave climate are very likely. Changes in wave heights and directions in particular are likely to be significant because these characteristics are the main regulator of longshore sediment transport rates.

The studies by Ruggiero *et al.* (2006) and Dickson *et al.* (2007) explicitly forecast future shoreline shapes arising in response to potential changes in future wave climate. In both cases a long record of deep-water wave conditions were transformed to the shore using spectral wave transformation models. The resulting wave conditions were used to drive relatively simple models of shoreline and/or cliff response. Both studies concluded that future shoreline changes are more sensitive to changes in wave direction than in wave height. Hosking and McInnes (2002) gave an example of a site on the south coast of the UK which would suffer littoral drift reversal should the predominant wave direction change by the order of 1^0. They used wind data output from a Regional Climate Model (RCM) at 50 km horizontal resolution, covering a 10-year period representing current conditions and a 10-year future period representing the 2080s for the medium-high UKCIP98 (Hulme and Jenkins 1998) climate change greenhouse-gas emission scenario. Sutherland and Gouldby (2002) used global climate model outputs representing the IS92A emission scenario of the IPCC 1992 first assessment (Leggett *et al.* 1992), to analyse future drift rates at five coastal sites around the UK. They concluded that future changes are unlikely to be greater than current levels of uncertainty. The recent study by Zacharioudaki and Reeve (2011) used wind fields from several different climate change models to investigate the statistical significance of differences in predicted beach response. They found that there were statistically significant changes in some climate change scenarios, although these were limited to individual seasons. Further, there was widespread disagreement between different global climate change models on the statistical significance of a change, although all experiments agreed on future seasonal trends. In agreement with earlier studies they also found that material shoreline changes were generally linked to significant changes in future wave direction rather than wave height.

When using such studies it should be borne in mind that:

1   Scenarios are exactly that and may be unrealistic in terms of being a forecast for actual future conditions (IPCC WG1 2001; IPCC WG1 2007).
2   Coarse-resolution climate simulations are unable to capture changes in short-range variability or extremes and can be very different to average changes in climate (Feyen *et al.* 2006).
3   Climate change scenarios vary. They involve different climate models, resolutions and greenhouse-gas emission scenarios aiming to address uncertainty. Use of a single experiment provides no uncertainty estimates.
4   Assessment of the significance of any changes requires rigorous statistical analysis.

Further reading

Arnell N.W., Livermore M.T.J., Kovats R.S., Levy P.E., Nicholls R. J., Parry M.L. and Gaffin S.R., 2004. Climate and socio-economic scenarios for global-scale climate change impact assessments: characterising the SRES storylines. *Global Environmental Change*, 14, pp. 3–20.

Department of Energy, 1990. *Metocean Parameters – Wave Parameters,* Offshore Technology Report 893000, HMSO, London.

Horikawa, K., 1978. *Coastal Engineering*, University of Tokyo Press, Tokyo.

Ilic, S. and Chadwick, A. J., 1995. Evaluation and validation of the mild slope evolution equation model for combined refraction-diffraction using field data, *Coastal Dynamics 95,* Gdansk, Poland, pp. 149–160.

IPCC website, www.ipcc.ch

Plimer, I., 2009. *Heaven and Earth*, Quartet, London, 504 pp. (Provides a contrarian argument against the hypothesis that humans have caused climate change.)

Richards J., Mokrech M., Berry P.M. and Nicholls R. J., 2008. Regional assessment of climate change impacts on coastal and fluvial ecosystems and the scope for adaptation. *Climatic Change*, 90, pp. 141–167.

Chapter 4

Coastal water level variations

4.1 Introduction

This chapter is concerned with coastal water level variations caused by factors other than wind-generated waves. These variations typically, but not exclusively, take the form of long-period waves. Such water level fluctuations can be classified as:

- *Astronomical tide* – periodic variations due to the tide-generating forces. These are well understood and can be predicted to good accuracy many years in advance.
- *Storm surge* – variations in water level due to the passage of atmospheric weather systems across the surface of the sea. Storm systems are significant because of their frequency and potential for causing large water level variations in conjunction with large wind waves.
- *Basin oscillations* – resonant responses of partially enclosed water bodies to external forcing.
- *Tsunamis* – surface waves associated primarily with sub-sea seismic disturbances. These waves can travel huge distances across an ocean, with speeds sometimes in excess of 800 km/h.
- *Climatological effects* – such as long-term sea level changes.

Figure 4.1 is a schematic representation of the energy spectrum of variations in the ocean surface elevation. The spectrum at a particular point will vary with time. Wind-generated waves fall approximately in the frequency band 1–0.03 Hz (or periods of 1 to 30 s). These are the waves that one might see when visiting the beach. At periods of less than 1 s the spectrum is dominated by capillary waves which are of little significance to coastal engineering design. For periods between 30 s and approximately 5 min the spectrum is dominated by surf-beat. This is a wavelike variation in water level arising from variations in the set-up due to the incoming waves. Tsunami waves tend to dominate the spectrum from 5 min to approximately an hour. The dominant astronomical tide variations have periods close to 12 and 24 h. The period of basin oscillations is highly dependent on the geometry and depth of the basin.

The wavelike water level variations considered in this chapter have relatively long periods. Their depth to wavelength ratio is therefore low and they may be treated as shallow-water waves to a reasonable degree of accuracy, even in the deeper ocean. We may thus use the small amplitude Airy wave theory (described in Chapter 2) to compute wave speeds and particle velocities. An alternative approach, which is rather

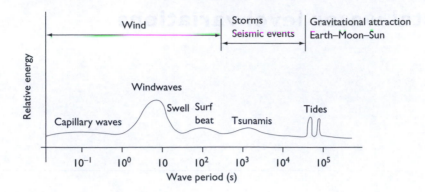

Figure 4.1 Illustrative ocean wave energy spectrum.

more flexible in general application, is based on deriving approximate forms of the equations of motion and mass conservation appropriate for describing long wave motion. This is covered in Section 4.6.

4.2   Astronomical tide generation

Tides have been studied from the earliest times. Indeed, it is documented that Aristotle spent the final part of his life on the island of Euboea, where he studied the tidal flows. It has been suggested that his failure to explain the tidal variations drove him to hurl himself into the strait where he drowned (Deacon 1971).

The modern theory of the tides is based on the equations of fluid motion developed by Euler and on Newton's theory of gravitation. Together, these provide the means of predicting the forces acting on the sea and their response to these forces. In the latter part of the eighteenth century, Laplace (1778/79) established a mathematical theory of the tides, which serves as the basis for modern tidal theory. Not only did he publish the equations for fluid motion on a rotating sphere but he also determined the *tide-generating forces*. The tides are a result of the simultaneous action of the Sun's, Moon's and Earth's gravitational forces, together with the orbital motion of the Earth and Moon and the Earth and Sun. As an aside, it should be noted that tidal movements also occur in the atmosphere and the solid Earth as well as the sea, but for coastal engineers the ocean tides are the ones of direct interest.

Consider for a moment the Earth–Moon system, ignoring the gravitational effects of the Sun. At first glance one would expect the gravitational force to result in one thing, the mutual attraction and eventual collision between the Earth and Moon. This is due to the direct gravitational attraction between the two bodies, according to Newton's Law of Gravitation. That the Earth and Moon do not collide is a consequence of the fact that the Moon is in orbit around the Earth. It is this motion that gives rise to the tides. The tide-generating force is most simply defined as the attractive force that does not affect the motion of the Earth as a whole. The Moon does not orbit the centre of the Earth but about the centre of mass of the Earth and Moon. (This point is actually inside the Earth, but not at its centre). A common picture of this situation has the Moon and Earth likened to a dumbbell with the two 'joined' by a fixed rod.

This picture is misleading because the Earth and Moon are free to rotate whereas they would not be if attached by a fixed rod. As a result the net effect is that all points on the Earth rotate in a circle of the same radius *relative to the centre of mass*. This rotation gives rise to a centripetal acceleration that must be balanced by another force in order for equilibrium to be maintained. On average this balancing force is provided by the gravitation of the Moon. However, because the Earth is a large body there is an appreciable change in the gravitational force experienced by points on the Earth closer to the Moon and those points on the side of the Earth furthest from the Moon. The tide-generating forces arise because the resultant attractive force is not uniform over the surface of the Earth. The effect is symmetrical about the line joining the Earth to the Moon, tending to pull the ocean into an ellipsoidal shape with 'bulges' at points on the Earth nearest and furthest from the Moon. The ocean does not take on a perfectly ellipsoidal shape because the oceans have inertia which means it takes time for the water to move into equilibrium with the exerted forces; and there are continents that constrain how the water can flow.

Before discussing the tidal theories several important characteristics of observed tidal signals are mentioned. Looking at a tide gauge record that shows the water level over a period time, the primary characteristic is a wave-like behaviour. In many cases one will be able to pick out an oscillation that has approximately one or two high waters each day (see Figure 4.7). Tides that have two high waters and low waters per day are called semi-diurnal and those that have one high and low water per day are known as diurnal. Looking at a longer stretch of record, say one month, it is not unusual to notice a low period modulation of the underlying tidal oscillation. This is known as the spring-neap cycle and has a period of approximately 14 days. The spring-neap cycle is a consequence of the combination of the Sun's and Moon's tide-generating forces on the Earth. When the Moon, Earth and Sun are in alignment the combined tide generating force is largest and corresponds to the large spring tides. When the Earth, Moon and Sun are in quadrature the combined tide generating force is a minimum, corresponding to neap tides. Looking at adjacent tides it is also often seen that successive high (or low) tides are of different heights. This difference is termed the 'diurnal inequality'. The diurnal inequality occurs because the Earth's axis is inclined relative to the plane of the orbit of the Moon. Again, a close inspection of tide gauge records will show that the time of high tide is later each day. This can be explained as follows. The Earth spins on its own axis every 24 h and the Moon rotates around the Earth once every 28 days. Now imagine that you are on the Earth and that the Moon is right above you. In 24 h time you will be back at the same location as the Earth spins on its axis. However, the Moon will no longer be right above you as it will have moved in its orbit – in fact it will have covered 1/28 of its monthly journey. It will take you about another 50 min before the Moon is directly above you ($1/28 \times 24 = 51.53$ min to be precise).

There are two forms of tidal theory: the equilibrium theory put forward by Newton and the dynamical theory developed by Laplace. The equilibrium theory provides an explanation for the diurnal inequality, the spring-neap cycle and the generally semi-diurnal nature of observed tides. Some notable failings of this theory are the under-prediction of observed tidal range and inaccuracies regarding the timing of high (and low) waters. The wave nature of tidal motion is more explicit in the dynamical theory where the periodic tide generating forces are used explicitly to drive the equations of fluid flow. It thus takes into account the wave travel time (inertia effects) and can predict both the amplitudes and timings of tidal variations better. However, numerical solution of the equations is required in order to account for the effects of irregular continental land masses and seabed variations.

Here, we provide a simplified account of modern equilibrium tidal theory, sometimes termed the 'equilibrium theory' of tides. In this theory it is assumed that:

1   water covers the whole of the Earth, initially at a constant depth;
2   water has no inertia (i.e. responds instantaneously);
3   water is in equilibrium, so that the water surface is normal to the imposed force.

These simplifying assumptions make it possible to derive analytical formulae for the shape of the free surface. The interested reader is referred to Godin (1972), Defant (1961) and Hendershott and Munk (1970) for further details. An outline of the theory and results is given below.

According to Newton's law of universal gravitation there is an attractive force between every pair of bodies, proportional to their masses. If the masses of the bodies are m and M, and the position vectors of their centres of mass are r_m and r_M respectively, then the attractive (gravitational) force, F, is given by

$$\mathbf{F} = -\frac{GmM}{|\mathbf{r}_m - \mathbf{r}_M|^2} \tag{4.1}$$

where G is the gravitational constant ($G = 6.67 \times 10^{-11}$ Nm^{2}kg^{-2}). This force may be written in terms of a gravitational potential, Ξ, as:

$$\mathbf{F} = -\nabla\left(\frac{GmM}{|\mathbf{r}_m - \mathbf{r}_M|}\right) = -\nabla\left(\frac{GmM}{|\mathbf{r}|}\right) \equiv -m\nabla\Xi(\mathbf{r}) \tag{4.2}$$

We shall apply this to the Earth–Moon system to determine the gravitational tide-generating forces. The two bodies are now the Earth (with mass m) and the Moon (with mass M). The positions of their centres of mass relative to an origin are r_m and r_M, and r is the position of the moon relative to the Earth's centre.

Consider a point s on the Earth; the potential at this point is (see Figure 4.2),

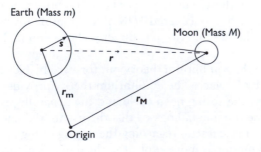

Figure 4.2 Definition of terms for Earth–Moon system.

$$\Xi(s) = -\frac{GM}{|s-r|} \qquad (4.3)$$

Writing $s = |s|$ and $r = |r|$, recall that if θ is the angle between the two position vectors r and s, then

$$|s-r|^2 = s^2 - 2sr\cos(\theta) + r^2 \qquad (4.4)$$

We shall assume that $s \ll r$, so that we may expand using the binomial theorem. Thus

$$\frac{1}{|s-r|} = \frac{1}{r}\left(1 - 2\frac{s}{r}\cos(\theta) + \frac{s^2}{r^2}\right)^{-1/2} = \frac{1}{r} + \frac{s\cos(\theta)}{r^2} + \frac{s^2}{r^3}\frac{(3\cos^2(\theta)-1)}{2} + ... \qquad (4.5)$$

and hence the potential may be written as:

$$\Xi(s) = -GM\left[\frac{1}{r} + \frac{s\cos(\theta)}{r^2} + \frac{s^2}{r^3}\left(\frac{3\cos^2(\theta)-1}{2}\right) + ...\right]$$

$$\equiv -\frac{GM}{r}\left[1 + \frac{s}{r}P_1 + \frac{s^2}{r^2}P_2 + ...\right] \qquad (4.6)$$

where P_n are Legendre polynomials with argument $\cos(\theta)$. We now substitute this expression for Ξ into Equation (4.2). The first term is a constant and so does not yield a force when we take the gradient. The second term gives a uniform acceleration GM/r^2 directed towards the Moon. This corresponds to the major effect of the Moon's gravitational force, namely to accelerate the Earth as a whole, and is not a tide-generating force. The third term gives a gravitational field

$$F = \frac{GMs}{r^3}(3\cos^2(\theta) - 1) \qquad (4.7)$$

Note that the force given by Equation (4.7) is symmetrical in the central plane, and suggests a predominantly semi-diurnal tidal variation (i.e. two high waters and low waters per day). The additional higher-order terms in the series expansion correspond to higher harmonics. An analogous argument applies when we consider the Earth–Sun system. However, in this case the distances and masses are such that the tide-generating effect of the Sun is approximately half that of the Moon.

Example 4.1
Calculate tide generating force on the Earth due to the Earth–Moon and Earth–Sun systems, given that the mass of the Earth (m) is = 5.98×10^{24} kg, the mass of the moon (M) = 7.35×10^{22} kg, the semi-major axis of the lunar orbit around the Earth (r) is 3.84×10^8 m and the mean radius of the Earth (s) is 6.37×10^6 m.

We may estimate the magnitude of the effect of the tide-generating forces by assuming that the water on the Earth's surface is in equilibrium. (This is equivalent to assuming that the natural periods of tidal oscillation are small in comparison to the rotation period of the Earth.) In this case the surface of the water takes a shape on which the gravitational potential has a single value. Let the height through which the sea surface is raised be $\eta(\theta)$. Then, under the assumption that $\eta(\theta)$ is small the change in the Earth's gravitational potential due to a rise in the sea surface of $\eta(\theta)$ is approximately $g\eta(\theta)$. This change must be balanced by the potential due to the Moon to maintain equilibrium. Thus,

$$g\eta(\theta) = \frac{GMs^2}{r^3}\left(\frac{3\cos^2(\theta)-1}{2}\right)$$

(4.8)

Thus, from Equation (4.1), $g = Gm/s^2$. Substituting this into Equation (4.8) gives

$$\eta(\theta) = \frac{Ms^4}{mr^3}\left(\frac{3\cos^2(\theta)-1}{2}\right)$$

(4.9)

Equation (4.9) gives the maximum tidal elevation for the Moon as 36 cm and for the Sun as 16 cm. Equation (4.9) describes a prolate spheroid, and indicates a lowering of the sea surface away from the equator.

Some additional characteristics of tidal variations can be inferred as follows. The principal astronomical factors are the Earth's orbit around the Sun, the Moon's orbit around the Earth, the Earth's rotation, the inclination of the Earth's equator to the ecliptic plane, and the transit of the Earth–Moon system around the Sun, see Figure 4.3.

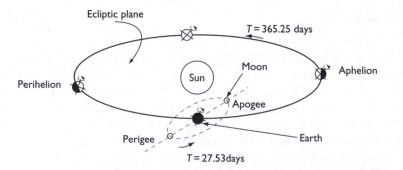

Figure 4.3 Principal astronomical definitions: (i) The plane defined by the orbit of the Earth around the Sun is called the ecliptic plane; (ii) The Earth's equator is inclined 23°27′ to the ecliptic plane; (iii) The plane defined by the orbit of the Moon around the Earth is inclined 5° to the ecliptic plane; (iv) The Moon's declination $\approx \pm$ 28°30′; (v) Aphelion occurs when the Earth is furthest from the sun and perihelion occurs when the Earth is closest to the Sun; (vi) Similarly apogee and perigee occur when the Moon is furthest and closest to the Earth respectively.

4.2.1 Diurnal inequality

As the Earth spins at a non-zero declination to the ecliptic plane an observer on the Earth's surface at latitude θ will be moved relative to the prolate spheroid, and will observe the height of the free surface to be given by

$$\eta = \frac{r_e}{2}\left(\frac{M}{m}\right)\left(\frac{r_e}{r}\right)^3 [(3\sin^2\theta\sin^2\delta-1)+\frac{3}{2}\sin2\theta\sin2\delta\cos\lambda+ \\ 3\cos^2\theta\cos^2\delta\cos^2\lambda+...]$$

(4.10)

where δ = declination, λ= Earth's angular displacement (0 → 360° in 24 h) and θ = latitude.

From the above equation tidal curves can be drawn for specific values of declination and latitude, as a function of angular displacement. Illustrative plots for the case of maximum declination, $\delta = 28°$ 30', are shown in Figure 4.4.

These exhibit the well-known phenomenon of semi-diurnal inequality, i.e. that the height of successive tides is different. The inequality becomes more pronounced at higher latitudes, and in some cases the tide may even become diurnal.

4.2.2 Tidal species

The expression for $\eta(\theta)$ (Equation 4.10), may be rewritten in terms of a series involving $cos(n\lambda)$ using standard trigonometrical relationships:

$$\eta_m = K_0 + K_1\cos\lambda + K_2\cos2\lambda +...$$

(4.11)

This form makes explicit that the tidal water level changes may be considered as the superposition of tidal harmonics that have distinct groups of periods related to the day length. In Equation (4.11)

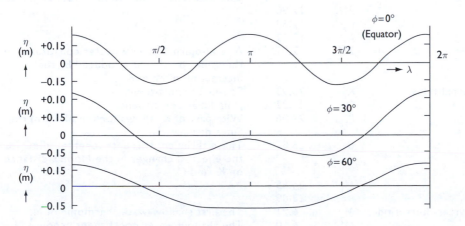

Figure 4.4 Tides at three latitudes (0°, 30° and 60°N) for Moon declination of 28°30'. Longitude, λ, is shown in radians.

K_0 = long period tides, which are generated by the monthly variations in lunar declination δ;

K_1 = diurnal tides, with frequencies close to one cycle per day;

K_2 = semi-diurnal tides, with frequencies close to two cycles per day.

Each set of tides (or tidal species) comprises a group of tides with slightly different periods. The equilibrium tide due to the Sun may be represented in an analogous form to Equation (4.11), but having somewhat different periods.

The characteristics of what are usually the main harmonics (or tidal constituents) are summarised in Table 4.1. The constituents M_2, S_2, K_1 and O_1 are usually predominant.

In practice the equilibrium tide assumptions do not apply fully. The tides are modified considerably from those predicted by equilibrium theory for several reasons, including:

1   the existence of continents
2   varying seabed topography

Table 4.1 The main tidal harmonics.

	Symbol	Period (h)	Description
Semidiurnal tides	M_2	12.42	Main lunar constituent
	S_2	12.00	Main solar constituent
	N_2	12.66	These two constituents between them allow for the changes in the Moon's distance due to its elliptic orbit round the Earth
	L_2	12.19	
	K_2	11.97	These two constituents together allow for the effect of the declination of the Sun and Moon and of changes in the Sun's distance
	T_2	12.02	
	μ_2	12.01	These four constituents together allow for perturbations of the Moon's orbit by the Sun
	$2N_2$	12.90	
	υ_2	12.63	
	λ_2	12.22	
	$2MS_2$	12.01	A semi-diurnal shallow-water constituent of the same speed as μ_2 produced by the interaction of M_2 and S_2.
Diurnal tides	K_1	23.93	Soli-lunar constituent
	O_1	25.82	Main lunar constituent
	P_1	24.06	With part of K_1 allows for the effect of the Sun's declination
	Q_1	26.87	These three constituents together allow for the effect of changes in the Moon's distance on K_1 and O_1.
	M_1	24.85	
	J_1	23.09	
Quarter-diurnal tides	M_4	6.21	The first shallow-water harmonic of M_2
	MS_4	6.10	The shallow-water constituent produced by the interaction of M_2 and S_2.

3 bed friction
4 inertia
5 wind stress
6 surface wave effects
7 density gradients.

Nevertheless, equilibrium tide theory is used widely as a basis for analysing records of water levels.

4.2.3 Spring-neap tidal variation

In a combined Earth–Moon–Sun system, the relative positions of the Moon and the Sun influence the height of the equilibrium tide in such a way as to cause an additional variation in tidal heights. This is known as the 'spring-neap cycle', which occurs approximately twice a month. When the Earth, Moon and Sun are collinear, the gravitational forces exerted by the Moon and Sun act along a single axis, distending the prolate spheroid in the axial direction. In this instance, the height of the equilibrium tide is increased, giving rise to what is known as a 'spring' tide. On the other hand, when the Earth, Moon and Sun are in quadrature, the gravitational forces act along lines 90° apart, distending the prolate spheroid to a lesser extent (Figure 4.5).

Under these circumstances the height of the equilibrium tide is reduced and is known as a 'neap' tide. Figure 4.6 illustrates a typical tidal trace containing a spring-neap cycle.

4.2.4 Tidal ratio

The relative importance of diurnal and semi-diurnal harmonics can be determined from the ratio, F, where

$$F = \frac{K_1 + O_1}{M_2 + S_2}$$

(4.12)

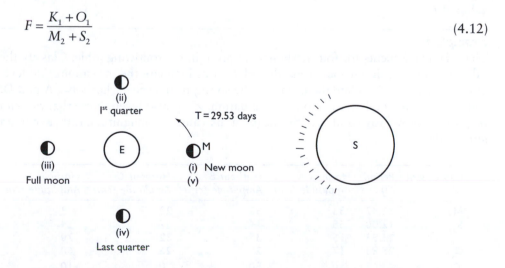

Figure 4.5 Positions of the Sun, Earth and Moon during spring-neap cycles.

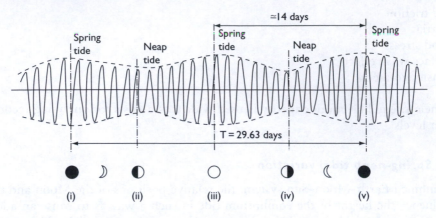

Figure 4.6 The spring-neap variation in tidal range with phases of the Moon.

and the tidal symbols denote the amplitudes of the respective tidal constituent. The diurnal inequality varies with the ratio F. The forms of tide may be classified as follows:

1   $F = 0.0–0.25$ (semidiurnal form). Two high and low waters of approximately the same height. Mean spring tide range is $2(M_2 + S_2)$.
2   $F = 0.25–1.50$ (mixed, predominantly semidiurnal). Two high and low waters daily. Mean spring tide range is $2(M_2 + S_2)$.
3   $F = 1.50–3.00$ (mixed, predominantly diurnal). One or two high waters per day. Mean spring tide range is $2(K_1 + O_1)$.
4   $F > 3.00$ (diurnal form). One high water per day. Mean spring tide range is $2(K_1 + O_1)$.

Figure 4.7 illustrates the change in the form of tidal behaviour with changes in the value of F.

Example 4.2
The tidal constituents for four harbours are given in the following table. Classify the tidal regime at each harbour using the tidal ratio. Estimate the maximum tide level at each harbour. Calculate the length of the spring-neap cycle at harbours A and D. The mean water level relative to the local datum, Z_0, is also given. Note that this can be positive or negative and, strictly speaking, its value is a magnitude rather than an amplitude.

Constituent	Period (h)	Harbour A Amplitude (cm)	Harbour B Amplitude (cm)	Harbour C Amplitude (cm)	Harbour D Amplitude (cm)
M_2	12.42	233	53	22	3
S_2	12.00	68	14	7	4
K_1	23.93	15	35	32	70
O_1	25.83	17	26	26	68
Z_0	–	0	50	0	−10

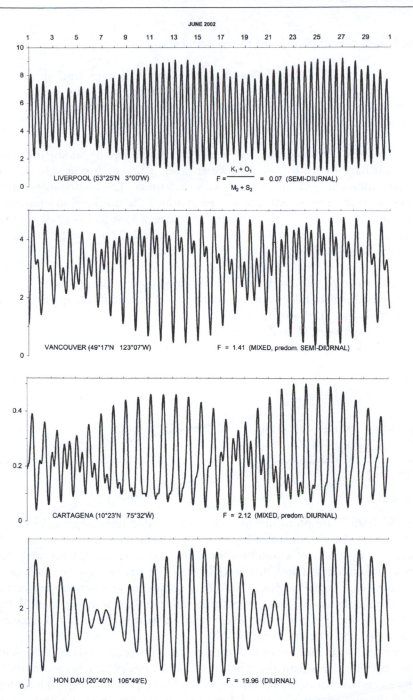

Figure 4.7 Tidal traces constructed from tidal harmonic amplitudes and phases quoted in the Admiralty Tide Tables 2002 for the month of June 2002 at Liverpool, Vancouver, Cartagena and Hon Dau. The tides at these ports are respectively semi-diurnal, mixed predominantly semi-diurnal, mixed predominantly diurnal and diurnal.

Solution
We calculate F from the amplitudes of the constituents:

> For Harbour A: $F = (15 + 17)/(233 + 68) = 0.11$, so tide is semi-diurnal.
> For Harbour B: $F = 0.91$, so tide is mixed, but predominantly semi-diurnal.
> For Harbour C: $F = 2.00$, so tide is mixed, but predominantly diurnal.
> For Harbour D: $F = 19.71$, so tide is diurnal.

Note that an estimate of the maximum tide level may be obtained by adding the amplitudes of the tidal harmonics to the mean water level. This corresponds to the case
when all the harmonics are exactly in phase with each other. The constituent denoted
by Z_0 is a fixed correction term that can be used to adjust the tide level to a particular
reference datum. The maximum tide levels are therefore: 233+68+15+17+0 = 333 cm
at Harbour A; 53 + 14 + 35 + 26 + 50 = 178 cm at Harbour B; 22 + 7 + 32 + 26 = 87 cm
at Harbour C and 3 + 4 + 70 + 68 –10 = 135 cm at Harbour D.

Note also that the spring-neap cycle is of slightly different period for semidiurnal
and diurnal tidal forms. The period can be determined by calculating the time taken
for the slower constituent to fall exactly one whole cycle behind its companion. Thus,
for the semidiurnal case, Harbour A, we have:

Period difference between M_2 and S_2 = 0.42 h
Number of S_2 cycles required for S_2 to lag one M_2 cycle = 12.4/0.42
$$= 29.52 \text{ cycles}$$
$$= 354.2 \text{ h } (12.00 \times 29.52)$$
$$= 14.8 \text{ days}$$

The corresponding spring-neap cycle period for diurnal tides, Harbour D, is 13.7
days.

4.3   Tide data

Lord Kelvin (William Thompson) invented the first reliable tide gauge in 1882. The
essential parts of his design remain today in the standard type of gauge. Figure 4.8
shows the main components of a common form of gauge used to measure variations
in water level. The underlying design was well known to engineers at the turn of the
twentieth century (Cunningham 1908). The gauge comprises an open-topped vertical
hollow tube with a hole near its base, sometimes referred to as a stilling well. A float
inside the tube is used to transmit water level variations within the tube to a recording
device such as a pen and rotating drum.

The small entrance near the base of the stilling well is designed to damp out oscillations that have periods of less than about a minute. Such devices have been deployed in
numerous docks and harbours around the world and have provided a good long-term
record of coastal water level variations. When connected with electronic data storage
systems a float gauge is perhaps one of the most reliable mechanisms for automatic
water level measuring stations. Floats or buoys are now also being used in conjunction with the satellite-based Global Positioning System (GPS) to measure water levels
in rivers (Moore *et al.* 2000). Submerged pressure transducers or pressure meters are

Rotating chart
on drum
Moving pen
Pulley
Float
Still-water
level
Stilling
well
Orifice

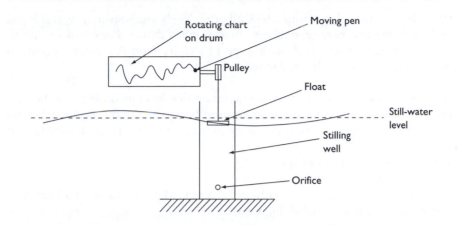

Figure 4.8 Float-stilling-well water level gauge.

also used to measure coastal water levels. They are usually mounted on or close to the seabed to avoid inaccuracies due to vertical accelerations of water associated with orbital wave motions.

As noted earlier, the tidal water level fluctuations measured at the coast vary from those expected from the equilibrium theory of tides. The distortion of the 'tidal long waves' by the continental shelf and nearshore bathymetry means that the equilibrium theory of tides can be of limited use in predicting water level fluctuations at coastal stations. Predictions are best made by analysing historical measurements obtained at specific locations. The mixture of historical analysis and theory has led to definitions of the state and level of the tide. These are shown in Figure 4.9, and are in widespread international use.

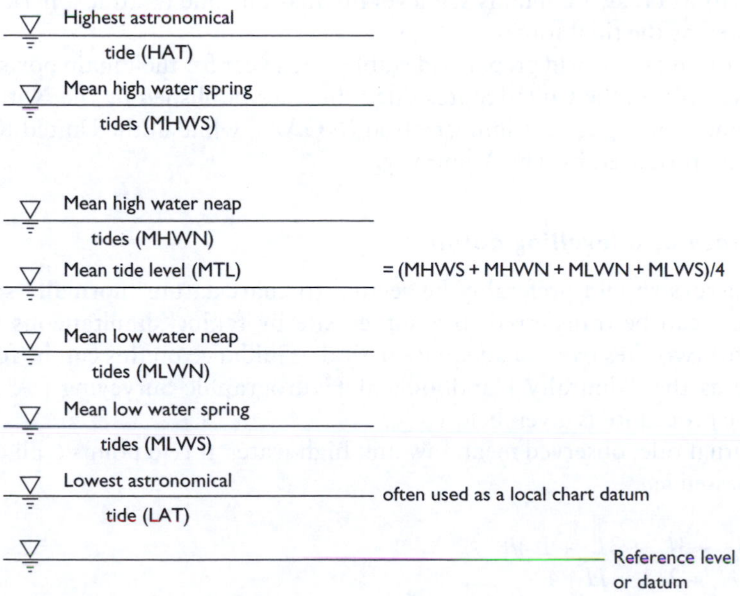

Highest astronomical
tide (HAT)
Mean high water spring
tides (MHWS)
Mean high water neap
tides (MHWN)
Mean tide level (MTL)
= (MHWS + MHWN + MLWN + MLWS)/4
Mean low water neap
tides (MLWN)
Mean low water spring
tides (MLWS)
Lowest astronomical
tide (LAT)
often used as a local chart datum
Reference level
or datum

Figure 4.9 Standard water level definitions for strongly diurnal tides.

The datum is a reference level from which all other levels are measured. Water levels and seabed levels are always quoted relative to a specified datum. For example, Admiralty Charts will usually quote levels with respect to 'chart datum', a datum specific to that particular chart. The tide levels defined in Figure 4.9 are:

- HAT – highest astronomical tide, the maximum tide level possible given the harmonic constituents for that particular location (theoretically this occurs once every 19.8 years, but may be exceeded once or twice a year due to meteorological conditions).
- MHWS – mean high water of spring tides
- MHWN – mean high water of neap tides
- MTL – mean tide level, the level midway between MLW(the mean of all low waters) and MHW (the mean of all high waters)
- MLWN – mean low water of neap tides
- MLWS – mean low water of spring tides
- LAT – lowest astronomical tide, the minimum tide level possible given the harmonic constituents for the location (frequency of occurrence as for HAT)

For tides with a strongly diurnal component:

- MHHW – average of higher (of two) daily high water levels
- MLHW – average of lower (of two) daily high water levels
- MHLW – average o higher (of two) low water levels
- MLLW – average of lower (of two) low water levels

Mean sea level (MSL) is calculated as the average level of the sea at a given site. MSL may be different to MTL as it contains sea level fluctuations due to atmospheric and wave effects as well as the tidal forces.

Governments around the world prepare tide tables each year for their main ports and harbours. For example, in the United States tide tables are published by the National Oceanographic and Atmospheric Administration (NOAA), while in the United Kingdom tide tables are produced by The Admiralty.

4.3.1   Sea surface as a levelling datum

Hydrographic surveys should preferably be related to chart datum (normally set as LAT). This datum can be transferred to a survey site by taking simultaneous tidal observations at the two sites over an adequate period. (Guidance on this can be found in manuals such as the Admiralty Handbook of Hydrographic Surveying.) A brief description of the procedure is given below.

For a semi-diurnal tide, observed mean low and high water at two points (call them 1 and 2), are assessed as

$$OMLW = (L_1 + 3L_2 + 3L_3 + L_4)/8$$
$$OMHW = (H_1 + 2H_2 + H_3)/4$$

where L_1, L_2, L_3 and L_4 are four consecutive levels of low water and H_1, H_2 and H_3

the intermediate high water levels. Then the observed mean range, OMR, is defined as OMHW – OMLW and the observed mean tide level, OMTL, is (OMHW + OMLW)/2. If complete sets of densely recorded measurements are available, observed mean sea level (OMSL) may be used instead of OMTL. The true mean tide level (TMTL) at gauge 1 can be computed directly from tide predictions, as the mean of MHWS and MLWS for example. The sounding datum at gauge 2 may be estimated by interpolation as:

$$d_2 = (OMTL)_2 - (OMTL - TMTL)_1 - [TMTL_1(OMR_2)/(OMR_1)]$$

Should $(TMTL)_1$ not be known the following alternative may be used:

$$d_2 = (OMTL)_2 - (OMTL)_1(OMR_2)/(OMR_1)$$

For a diurnal tide it is preferable to determine the major harmonic constituents from a harmonic analysis of one or more series of observations covering a period of at least 29 days. The chart sounding datum at the new location may then be obtained from:

$$d_2 = (OMTL)_2 - (OMTL - TMTL)_1 - TMTL_1(\Sigma A_2 / \Sigma A_1)$$

where ΣA_1 is the sum of the amplitudes of the tidal constituents at site 1 and ΣA_2 is the sum of the amplitudes of the tidal components at site 2. The sums are normally restricted to the four dominant constituents.

For construction work or surveys at sea, levels have to be related to the local surface of the sea and it is therefore necessary to calculate the surface level in order to determine heights in relation to a fixed datum. Where tidal variation is small and the site is close to shore it is often sufficient to use a tide curve from a tide recorder at a point on the coast in close proximity.

Example 4.3
Given the following information for observed ocean tide levels (semi-diurnal) at an established gauge and a new survey site, calculate the sounding datum for the new site.

	Established Gauge 1 (m)	New site Gauge 2 (m)
LW	0.5	1.2
HW	3.7	3.6
LW	0.8	1.4
HW	3.6	3.3
LW	0.4	1.0
HW	3.9	3.9
LW	0.7	1.3
MHWS	3.5	–
MLWS	0.5	–

Solution
Following the steps outlined in the section above:

$(OMLW)_1 = (0.5 + 3 \times 0.8 + 3 \times 0.4 + 0.7)/8 \quad = 0.51$
$(OMHW)_1 = (3.7 + 2 \times 3.6 + 3.9)/4 \qquad\qquad = 3.70$
$(OMR)_1 = 3.70 - 0.51 \qquad\qquad\qquad\qquad = 3.19$
$(OMTL)_1 = (3.70 + 0.51)/2 \qquad\qquad\qquad = 2.10$
$(TMTL)_1 = (3.5 + 0.5)/2 \qquad\qquad\qquad\quad = 2.00$
$(OMLW)_2 = (1.2 + 3 \times 1.4 + 3 \times 1.0 + 1.3)/8 \quad = 1.21$
$(OMHW)_2 = (3.6 + 2 \times 3.3 + 3.9)\ /4 \qquad\qquad = 3.52$
$(OMR)_2 \qquad\qquad\qquad\qquad\qquad\qquad\quad = 2.31$
$(OMTL)_2 \qquad\qquad\qquad\qquad\qquad\qquad\ = 2.36$

The correct sounding datum at the new site should be:

$$2.36 - (2.10 - 2.0) - (2.0 \times 2.31)/3.19 = 0.81\,\text{m}$$

Had $(TMTL)_1$ been unknown the next best estimate would have been

$$2.36 - (2.10 \times 2.31)/3.19 = 0.84\,\text{m}.$$

4.4 Harmonic analysis

Oceanic tides display an inherent regularity, due to the regularity of astronomical processes. As a result, certain species, or harmonics, can be identified easily from observations of tide levels. Harmonic analysis describes the variation in water level as the sum of a constant mean level, contributions from specific harmonics and a 'residual':

$$\eta = Z_0 + \sum_{i=1}^{n} a_i \cos(\Omega_i t - \phi_i) + R(t) \tag{4.13}$$

where
η = water level
Z_0 = mean water level above (or below) local datum
Ω_i = frequency of ith harmonic (obtained from astronomical theory)
a_i = amplitude of the ith harmonic (obtained from astronomical theory)
ϕ_i = phase of ith harmonic
n = number of harmonics used to generate the tide
t = time
$R(t)$ = residual water level variation.

Given a sequence of water level measurements Equation (4.13) may be used to determine a_i, ϕ_i and $R(t)$ for a selected group of i tidal harmonics. The numerical procedure involves fitting a sum of cosine curves to the measurements. Values of Ω_i are taken to be known from equilibrium theory and the a_i and ϕ_i are determined by choosing the values that give the best fit to the measurements. The error, or residual, is $R(t)$. This represents a combination of numerical errors arising from the fitting calculations, measurement errors, and water level fluctuations not attributable to the selected tidal harmonics. For example, wave set-up and storm surges are likely contributors to the residual. Figure 4.10 shows a typical set of measurements taken on the Norfolk (UK) coast, the reconstructed tidal trace determined from computed harmonics, and the

Table 4.2 Harmonic analysis for the main tidal harmonics.

Name	Frequency (Cycles/hour)	Amplitude (m)	Phase (degrees)
Z_0	0	2.78	—
M_2	0.081	1.56	160
N_2	0.079	0.30	136
S_2	0.083	0.52	206
K_2	0.084	0.14	204
O_1	0.039	0.16	116
K_1	0.042	0.15	285

residual. Table 4.2 provides a summary of the harmonic analysis for the harmonics listed. The full set is given in Appendix C.

Note that the residual, which is considered to be the contribution of all non-tidal effects on the total water level, can be both positive and negative. For example, a storm will be associated with low surface pressure and consequently a positive residual. Conversely, periods which are dominated by high surface pressure are likely to coincide with negative residual.

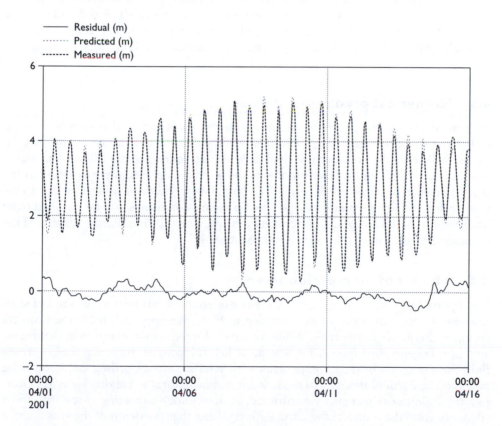

Figure 4.10 Water level time series, reconstructed tidal curves and residual.

The minimum length of record required to determine the main tidal harmonics varies with geographical location but a useful rule of thumb for a minimum length of record is one month – corresponding approximately to two spring-neap cycles. For strongly diurnal tides it is more satisfactory to have either a longer period or, if total duration of measurement is an issue, say two periods of two weeks separated by at least one month, in order to identify longer period components.

The accuracy of the calculations and the number of harmonics that can be identified reliably increases with the length of the record. Typically, to distinguish between two harmonics requires a record that contains a different whole number of cycles of each harmonic. Thus to distinguish two harmonics with very similar frequencies will require a much longer record than to distinguish between two harmonics with dissimilar frequencies. Analysis of records covering many tens of years has shown long period variations in tidal behaviour of the order of years and decades. Also, it is not necessary to have a complete continuous record of tidal elevation data in order to carry out an analysis. It is perfectly possible to obtain tidal constituents successfully from data that contains many gaps in the records. For strongly diurnal tides rather longer durations than one month of data are preferable. Once the amplitudes and phases of the main tidal harmonics at a site have been found they may be used to predict future tide levels by substituting the amplitude and phases into Equation (4.13), and setting $R(t) = 0$. If the amplitudes and phases have been determined from a short record and/or some years in the past, corrections may need to be made. The Admiralty Tide Tables, for example, provide an approximate tide prediction method that is updated on an annual basis.

4.5   Numerical prediction of tides

If detailed information on tidal elevations and flows is required over a large area, rather than at a few isolated points, then a numerical model can be set up. A tidal model will solve the equations of fluid flow to determine the tidal flows and elevation over a grid of points. Inputs to the model will be a detailed representation of the seabed surface over the model grid, and boundary conditions to specify the elevation or flows at open sea boundaries. The boundary conditions will be specified from observations at tide gauges, as described in the previous section. Further details of the numerical prediction of tides are given in Section 4.7.

4.6   Theory of long-period waves

Long-period waves, or 'long waves', were considered in Chapter 2 as the limit of shallow-water waves in water of uniform depth. Here, we approach the same problem from an alternative perspective. While we have adopted vector notation in the discussion of tide-generating forces, it is now more helpful to write the equations governing fluid flow in terms of their components with reference to an orthogonal coordinate system. The Earth is almost spherical, and it is thus natural to employ spherical coordinates θ (latitude) increasing northward, λ (longitude) increasing eastward, and r (distance from the centre of the Earth). Recognising that the depth of the seas is small in comparison with the radius of the Earth, s, we write

$$r = s + z \quad\quad with \quad\quad z \ll s \tag{4.14}$$

where z is the height above mean sea level. The three components of velocity are denoted by u (along lines of constant latitude), v (along lines of constant longitude) and w (vertical). We denote the rate of rotation of the Earth about its axis by Ω. Figure 4.11 shows the definition of these terms.

The equations of motion for an incompressible fluid in spherical coordinates are given below and a full derivation may be found in texts on fluid dynamics (e.g., Batchelor 1967):

$$\frac{\partial u}{\partial t} + \frac{u}{s\cos\theta}\frac{\partial u}{\partial \lambda} + \frac{v}{s}\frac{\partial u}{\partial \theta} + w\frac{\partial u}{\partial z} - \left(2\Omega + \frac{u}{s\cos\theta}\right)(v\sin\theta - w\cos\theta)$$
$$= -\frac{1}{\rho s\cos\theta}\frac{\partial p}{\partial \lambda} + V_\lambda \tag{4.15a}$$

$$\frac{\partial v}{\partial t} + \frac{u}{s\cos\theta}\frac{\partial v}{\partial \lambda} + \frac{v}{s}\frac{\partial v}{\partial \theta} + w\frac{\partial v}{\partial z} + \frac{wv}{s} + \left(2\Omega + \frac{u}{s\cos\theta}\right)u\sin\theta$$
$$= -\frac{1}{\rho s}\frac{\partial p}{\partial \theta} + V_\theta \tag{4.15b}$$

$$\frac{\partial w}{\partial t} + \frac{u}{s\cos\theta}\frac{\partial w}{\partial \lambda} + \frac{v}{s}\frac{\partial w}{\partial \theta} + w\frac{\partial w}{\partial z} - \frac{v^2}{s} - \left(2\Omega + \frac{u}{s\cos\theta}\right)u\cos\theta$$
$$= -\frac{1}{\rho}\frac{\partial p}{\partial z} - g + V_z \tag{4.15c}$$

The mass conservation equation is written as:

$$\frac{1}{\rho}\left(\frac{\partial \rho}{\partial t} + \frac{u}{s\cos\theta}\frac{\partial \rho}{\partial \lambda} + \frac{v}{s}\frac{\partial \rho}{\partial \theta} + w\frac{\partial \rho}{\partial z}\right) + \frac{1}{s}\frac{\partial u}{\partial \lambda} + \frac{1}{s\cos\theta}\frac{\partial}{\partial \theta}(v\cos\theta) + \frac{\partial w}{\partial z} = 0 \tag{4.16}$$

The terms on the right-hand sides of the Equation (4.15) containing Ω arise from the rotation of the Earth, and are sometimes called 'Coriolis force'. The terms V_λ, V_θ and

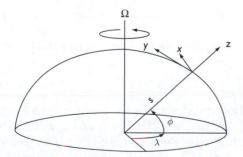

Figure 4.11 Definition of coordinate systems for describing fluid flows on a rotating sphere.

V_z denote the components of frictional force per unit mass. If the fluid motion does not range greatly with latitude then it is often helpful to adopt a local Cartesian set of coordinates (x,y,z) where $x = s\lambda\cos(\theta)$ is the eastward distance along the latitude circle and $y = s(\theta-\theta_0)$ is the distance poleward from some reference latitude θ_0. This approximation simplifies the above equations by removing many of the curvature terms without neglecting any of the primary physical processes. Using the assumption that the fluid is incompressible simplifies the mass conservation equation. This is equivalent to the condition that the divergence of the flow vanishes, or that the density is constant following the motion.

One other approximation that is often made concerns the equation governing motion in the vertical direction. For long wave motions the vertical pressure gradient term and the acceleration due to gravity dominate this equation. These two terms are almost equal and opposite; the hydrostatic balance expresses the approximation that they are in balance:

$$\frac{\partial p}{\partial z} = -\rho g \tag{4.17}$$

Typically, this balance breaks down for small-scale phenomena, such as flows near sharp seabed features. In what follows we will assume flows are in hydrostatic balance unless otherwise stated. With this approximation, we neglect terms involving w in the Coriolis force terms in the horizontal momentum equations Equation (4.15). The quantity $f = 2\Omega\sin(\theta)$ is twice the component of the Earth's angular velocity parallel to the local vertical, and is known as the 'Coriolis parameter'.

Three scales of motion can now be defined:

- *local scale* – where the effect of the Earth's rotation (viz. Coriolis terms) may be neglected and local Cartesian coordinates may be used;
- *regional scale* – where the effect of the Earth's rotation and curvature terms begin to become important. The scale of motion may require inclusion of the curvature terms; otherwise local Cartesian coordinates may be used with a suitable approximation of the Coriolis term;
- *global scale* – where the Earth's rotation and curvature terms are important and must be retained.

Example 4.4
Derive a suitable approximation for the Coriolis parameter for regional scale motions centred on latitude of θ_0.

Solution
We expand the Coriolis parameter in a Taylor series about the latitude θ_0 as $f = f_0 + \beta y + \ldots$ where $\beta = (df/dy)_{\theta 0}$, and $y = 0$ at θ_0. Thus, $\beta = 2\Omega\cos(\theta)/s$ and $f_0 = 2\Omega\sin(\theta_0)$. If the Taylor series is truncated after the first term then we have what is termed the 'f-plane' approximation which includes for the Coriolis effect but not its variation with latitude. Retaining the first two terms yields the 'β-plane approximation', which includes a simplified form of the variation of Coriolis parameter with latitude. Note

that in equatorial regions the second term becomes proportionately more important as $f_0 \to 0$ as $\theta_0 \to 0$.

One further approximation that can be adopted for small-amplitude waves is to neglect the non-linear terms, such as $w\partial u/\partial z$, in Equation (4.15).

As an example of the effect that approximations to the equations of motion can make on the physics, we apply the linear and hydrostatic approximations. First, we consider local-scale motions that are small, non-hydrostatic motions in a fluid with a free surface. The basic state is one of rest ($u = v = w = 0$) in which the undisturbed water depth is h and the hydrostatic relation holds for the undisturbed state.

Now, for the basic state we integrate Equation (4.17) with respect to z from $-h$ to 0, to give

$$\bar{p} = -\rho g h \tag{4.18}$$

The equations of motion governing the perturbation quantities become:

$$\frac{\partial u}{\partial t} = -g\frac{\partial \eta}{\partial x}$$

$$\frac{\partial v}{\partial t} = -g\frac{\partial \eta}{\partial y}$$

$$\delta\frac{\partial w}{\partial t} = -g\frac{\partial \eta}{\partial z} \tag{4.19}$$

$$\frac{\partial u}{\partial x} + \frac{\partial v}{\partial y} + \frac{\partial w}{\partial z} = 0$$

In the above, δ is set to 1 or 0 for non-hydrostatic or hydrostatic perturbations, respectively. For simplicity we now ignore variations in the y-direction. Next, we assume the perturbation quantities to be of harmonic form:

$$u = \phi(z)e^{ik(x-ct)}$$

$$w = \psi(z)e^{ik(x-ct)} \tag{4.20}$$

$$\eta = \varsigma(z)e^{ik(x-ct)}$$

Substituting Equation (4.20) into Equation (4.19) gives

$$-c\phi(z) + \varsigma(z) = 0$$

$$-ik\delta c\psi(z) + \frac{\partial \varsigma}{\partial z} = 0 \tag{4.21}$$

$$ik\phi(z) + \frac{\partial \psi}{\partial z} = 0$$

Eliminating $\varsigma(z)$ and $\phi(z)$ yields the following equation for ψ

$$\frac{\partial^2 \psi}{\partial z^2} - k^2 \delta \psi(z) = 0 \tag{4.22}$$

The solutions to Equation (4.22) are:

$$\begin{aligned} \psi(z) &= a_1 e^{kz} + a_2 e^{-kz} \\ \psi(z) &= a_3 z + a_4 \end{aligned} \tag{4.23}$$

where the a's are arbitrary constants to be determined from the boundary conditions. At the lower boundary we impose the condition of zero vertical velocity. Hence, for $\delta = 1$, $a_1 = -a_2$ and for $\delta = 0$, $a_4 = 0$ and a_3 is arbitrary, giving

$$\begin{aligned} \psi(z) &= a_1(e^{kz} - e^{-kz}) & \delta &= 1 \\ \psi(z) &= a_3 z & \delta &= 0 \end{aligned} \tag{4.24}$$

The remaining boundary condition is that the total pressure of a surface particle remains unchanged. This condition is applied at $z = h$ in linearised form, thus

$$\frac{\partial \eta}{\partial t} - \rho g w = 0 \qquad at \qquad z = h \tag{4.25}$$

Substituting Equation (4.24) into Equation (4.21), applying Equation (4.25) and simplifying yields the roots of the frequency equation as

$$\begin{aligned} c &= \sqrt{\frac{gL}{2\pi} \tanh\left(\frac{2\pi h}{L}\right)} & \delta &= 1 \\ c &= \sqrt{gh} & \delta &= 0 \end{aligned} \tag{4.26}$$

We thus retrieve the wave relations derived for Airy waves. In this context the approximation of hydrostatic perturbations can be seen as being equivalent to the 'long-wave' or 'shallow water' approximation of Chapter 2. Only by retaining the non-hydrostatic term do we ensure a fully accurate representation of wave propagation speed in all water depths.

The shallow-water equations are used widely for regional calculation of tides, storm surges and seiches in coastal regions. The equations of motion in these cases are discussed in Section 4.7 and 4.8. For instance, the Coriolis terms are important when considering flows in the North Sea. However, they can be neglected for narrow channels such as the Bristol Channel or the Bay of Fundy. In this case, the period of the oscillation and the geometry of the channel are of prime significance (see Section 4.8.3).

Long-period variations of the sea surface (typically with periods of between 30 s and 5 min), close to the shore were first observed by Munk (1949). Longuet-Higgins and Stewart (1964) proposed an explanation of surf beat based on non-linear effects in the incoming waves. They showed that incident wave groups could drive long-period waves that propagate with the wind wave group velocity. Unlike wind waves, these long waves are not dissipated significantly in the surf zone and can reflect from the beach propagating back into deep water. In turn this sets up a partial standing wave pattern, termed 'surf beat'. If waves approach the shore obliquely then 'edge waves' can also be generated. These waves propagate along the shoreline, and are 'trapped' to the coast. Their energy decays asymptotically to zero at large distances from the shoreline. Ursell (1952) showed that such waves would have a wave-number 'cut-off'. That is, for a given frequency, only waves with wave numbers less than a critical cut-off wave number can exist. Waves with wave numbers less than the cut-off are not trapped and radiate energy seaward. It is widely suggested that edge waves play an important part in generating quasi-periodic longshore features such as bars and cusps.

4.7   Tidal flow modelling

The equations for predicting flows due to astronomical tides were derived by Lagrange (1781). For tidal flow modelling for engineering applications, the effects of the Earth's rotation need to be included. The hydrostatic approximation may also be made, and we take the scale of motion to be such that a local Cartesian coordinate system is sufficient. Neglecting stratification effects, so that the density is assumed to be constant, the equations of motion are expressible in the form:

$$\frac{\partial u}{\partial t} + u\frac{\partial u}{\partial x} + v\frac{\partial u}{\partial y} - fv = -g\frac{\partial H}{\partial x} - \frac{\tau_{bx}}{\rho H} \tag{4.27a}$$

$$\frac{\partial v}{\partial t} + u\frac{\partial v}{\partial x} + v\frac{\partial v}{\partial y} + fu = -g\frac{\partial H}{\partial y} - \frac{\tau_{by}}{\rho H} \tag{4.27b}$$

$$\frac{\partial u}{\partial x} + \frac{\partial v}{\partial y} + \frac{\partial w}{\partial z} = 0 \tag{4.27c}$$

where $H = \eta + h$ is the total water depth, $\eta(x,y)$ is the surface elevation about the undisturbed water level ($z = 0$) and $h(x,y)$ is the seabed depth below the still water level. τ_{bx} and τ_{by} are the components of bottom stress along the directions of the x- and y- axes respectively. The assumption of constant density and the hydrostatic relation imply that the pressure force is independent of height. By assuming that the velocity field is initially independent of height, it will remain so; thus terms relating to vertical advection have been omitted from Equation (4.27). Integrating Equation (4.27c) over the depth of the fluid gives:

$$H\left(\frac{\partial u}{\partial x} + \frac{\partial v}{\partial y}\right) + w_\eta - w_b = 0 \tag{4.28}$$

The vertical velocity $w = dz/dt$ at the upper boundary represents the rate at which the free surface is rising, and so $w_\eta = d\eta/dt$. The vertical velocity at the lower boundary represents the rate at which fluid is flowing vertically in accordance with the requirement that there is no flow through the seabed surface. The seabed surface is taken to be fixed in time and so $w_h = dh/dt$. Thus the equation of continuity (Equation 4.28) may be written as:

$$\frac{\partial \eta}{\partial t} = -H\left(\frac{\partial u}{\partial x} + \frac{\partial v}{\partial y}\right) - \left(u\frac{\partial H}{\partial x} + v\frac{\partial H}{\partial x}\right) = 0 \tag{4.29}$$

Equations (4.27a), (4.27b) and (4.29) are the governing equations and form the basis of tidal flow prediction models. The bottom stress is related to the depth-mean current, $v = (u,v)$, again using a quadratic law:

$$\tau_b = C_B \rho |\mathbf{v}| \mathbf{v} \tag{4.30}$$

where ρ is the density of sea water ($=1025\,\text{kg/m}^3$), and C_B is the bottom friction coefficient, often taken to be equal to 0.0025 (Flather 1984). A slightly lower value of 0.002 has been suggested by the results of Mojfeld (1988) and Dewey and Crawford (1988).

The boundary conditions for solving the above equations take two forms. First, the condition that there is no normal flow at a land boundary and at an open sea boundary the outward component velocity normal to the boundary and the elevation can be set to the value predicted by harmonic theory. This does not allow waves to propagate out of the model domain and the following 'radiation condition', due to Flather (1984), is preferable:

$$v_n = v_n^T + \frac{c}{H}(\eta - \eta^T) \tag{4.31}$$

where v_n refers to the component of the depth-averaged current along the outward normal and $c = \sqrt{(gh)}$. From tidal theory v_n^T and η^T can be specified as in Equation (4.13).

Equations (4.27a), (4.27b) and (4.29), together with the boundary conditions Equations (4.30) and (4.31) are solved numerically. The origins of the numerical solution of these equations may be traced to the work of Leendertse (1964). Since that time much research has been undertaken into developing stable, accurate and efficient methods of solving the equations. Davies and Flather (1978) describe one scheme that includes the Earth's curvature. At a local scale, Falconer (1986) developed a formulation that accounted for non-uniformity of the vertical velocity profile. For local engineering applications it can be important to account for the wetting and drying process that occurs when tides ebb and flood over areas with shallow gradients, such as tidal flats. Falconer and Owens (1990) describe an evaluation of several methods of including this in a numerical scheme. An alternative method, used in conjunction with a curvilinear computational grid is described in Reeve and Hiley (1992). Shankar et al. (1997) describe the application of a boundary-fitted grid model to calculate tidal currents around the Straits of Singapore.

Tidal models can be used to investigate three important phenomena: short-term transport of pollutants and fine sediments; asymmetry in the tidal flows leading to a pattern of net long-term flows; and prediction of tide wave propagation in narrow seas and gulfs. In situations where there are sharp changes in seabed level, using the depth-averaged equations can result in inaccuracies. This is because in the vicinity of rapid changes in the height of the seabed significant vertical velocities can be generated. In this case Equations (4.15) and (4.16) (or their Cartesian equivalent) must be solved. An example of such a situation is given below (Shankar 2002, pers. comm.).

CASE STUDY – Numerical simulation of tidal flows around Singapore
The study area covers the main island of Singapore and its surrounding coastal waters defined by latitudes 0°59'N to 1°44'N and longitudes 103°18'E to 104°20'E as shown in Figure 4.12.

The orientation of the model is 7.5° counterclockwise relative to the geographical north. There are four open sea boundaries, namely W11, S11, S12 and E11, as shown in Figure 4.12. Some adjustments have been made near the open boundaries of the model in order to avoid instabilities. Part of the shallow beach north-west of Pulau Kukup at W11, the shallow beach north of Pulau Bintan at S12 and the narrow Straits east of Pulau Kepala Jernih at S11 have been ignored and simulated as land. On the other hand, some of the narrow channels amongst the many islands have been widened slightly or displaced laterally by one or two grid points to obtain a more regular grid representation.

As shown in Figure 4.14, the computational grid domain covers an area of 110 by 70 km. The area is represented by a two-dimensional rectangular grid (1 km by 1 km) in the horizontal plane. In the vertical direction it is divided into eight layers. By means of harmonic analysis, the water levels at the four open sea boundaries are prescribed and the program was run to simulate the tidal flows over a period of several weeks. Most models require an initial period to 'spin-up'. That is, to allow the effect of the imposed boundary conditions to be transmitted throughout the model grid. The flow fields from 14–18 August 1987 were obtained from a field survey campaign. The water surface elevations in the surface layer are plotted for comparison with the measured data. Note that the current speeds and directions were measured at a single

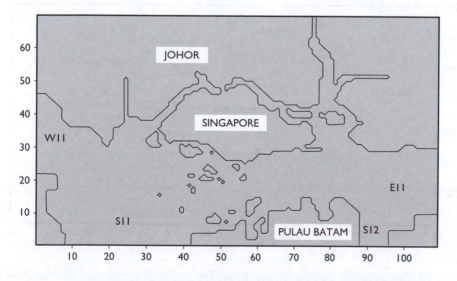

Figure 4.12 The open sea boundaries of the Singapore regional model.

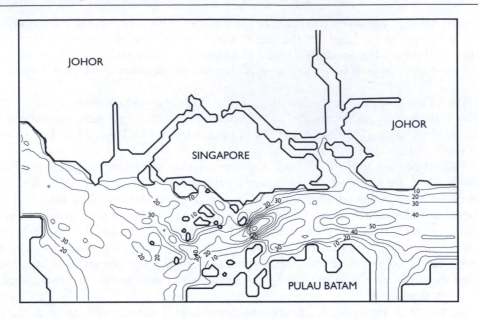

Figure 4.13 Bathymetry of Singapore coastal waters (1997).

point positioned approximately 10 m below the water surface. So for comparison with the numerical data model results in the third layer from the top are used.

Figure 4.14 Shows the computation domain of the model and the locations of the measuring stations. Figures 4.15(a), (b) and (c) and 4.16(a) and (b) show the comparisons between the time history records computed from the hydrodynamic model and

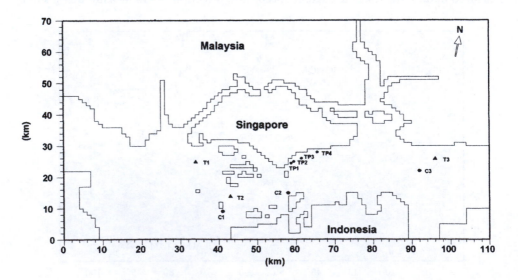

Figure 4.14 Computation domain of the Singapore regional model and the location of measuring stations.

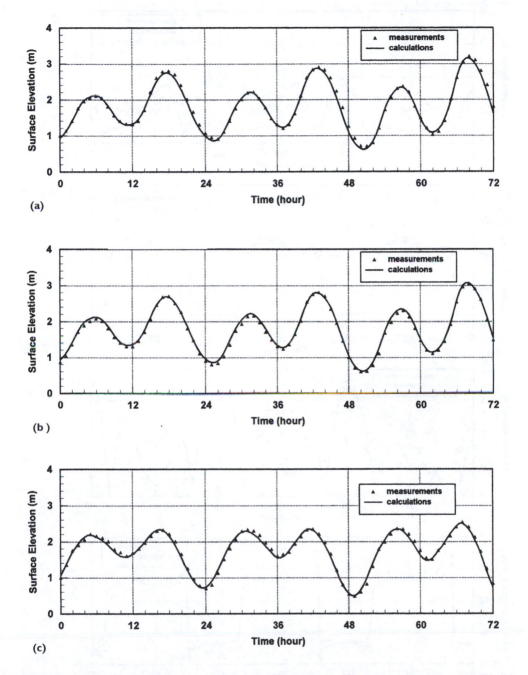

Figures 4.15 Comparison of measured and predicted elevations at stations T1 (a), T2 (b) and T3 (c).

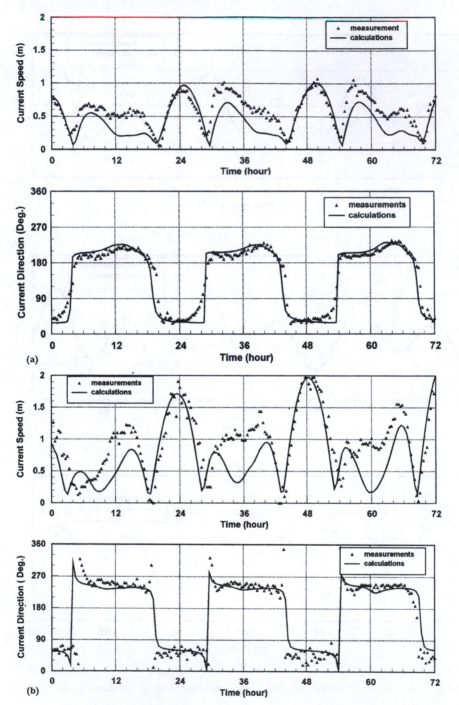

Figure 4.16 Comparison of measured and predicted tidal current speed and direction at stations C1(*a*) and C2(*b*).

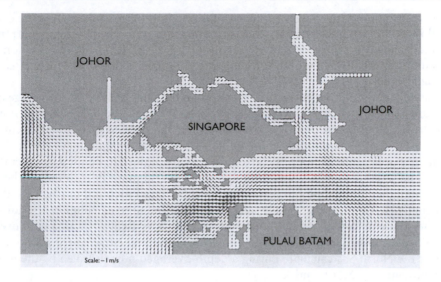

Figure 4.17 The computed surface flow during flood tide.

the measured data. It is observed that the water surface elevations from the numerical model are in very good agreement with the measured data. For the computed tidal currents, except for small differences in magnitude and phase between the measured and computed results, the tidal currents are generally well predicted and correlate well with the measured data.

Figures 4.17 and 4.18 show the computed surface flows during flood and ebb tide. The main flow on the flood tide is from east to west. The maximum current occurs at

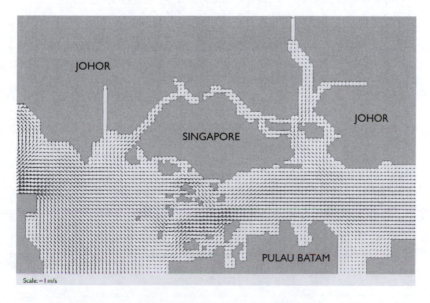

Figure 4.18 The computed surface flow during ebb tide.

the narrowest section of the Singapore Strait with a magnitude of 1.5 m/s. On the ebb tide the flow is stronger and the main flow is from west to east. The maximum current speed which also occurs at the narrowest section is 2.1 m/s. It can be noted that some local circulations are formed within the study domain. This is due to the local effects of the small islands.

Analysis of records of tidal currents has demonstrated that the long-term average of the measurements does not tend to zero, implying that the tidal currents have a time mean or 'residual' component. An understanding of the pattern of residual circulation in coastal regions is important for understanding the long-term movement of pollutants and sediments. Numerical simulation of tidal flows can be used to generate synthetic time series of currents and elevations at each grid point in the model. These time series can then be analysed to determine the tidal harmonics and to calculate the residual currents. This approach was used by Prandle (1978) to calculate the residual flow in the southern North Sea due to the M_2 tide, and by Reeve (1992a) who used additional tidal constituents. Figure 4.19 shows tidal residual currents for the southern part of the North Sea calculated using a depth-averaged model driven by the four tidal constituents M_2, S_2, O_1 and K_1.

The geometry of the land can have a significant effect on the propagation of the tide wave. Amplification of the tide wave can occur in much the same way it may occur for other long period waves, see Section 4.8.3. At the scale of a sea (such as the North Sea or Gulf of Mexico) the rotation of the Earth has an additional affect. The crest of the tide moves anti-clockwise (counterclockwise) around the sea. The usual way of

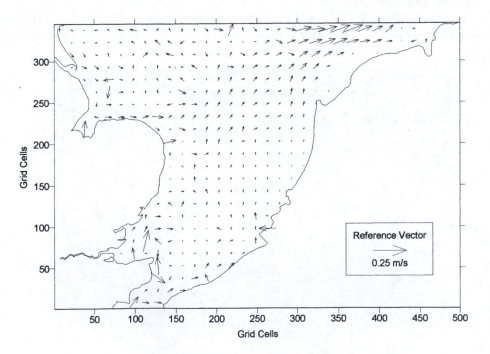

Figure 4.19 Computed tidal residual currents in the southern North Sea.

drawing tidal variations is in terms of the amplitude, A, and phase, ϕ, so that the tidal elevation is written as

$$\eta = A\sin(\omega t - \phi) \tag{4.32}$$

Contours of A are called 'corange lines', while contours of ϕ are termed 'cotidal lines' and the phase is usually given in degrees. Due to the Earth's rotation and the geometrical effects, tides in coastal regions can exhibit interference patterns. Points at which the tidal amplitude becomes close to zero are called 'amphidromic points'. They appear as bullseyes on tidal charts, at the centre of concentric corange lines. The cotidal lines will appear to meet at the amphidromic point, indicating that the tide wave propagates around the amphidromic point over a tidal cycle. Tidal charts are usually compiled on the basis of harmonic analyses of measurements or model output, and charts are drawn for each major tidal constituent. Further details may be found in Gill (1982) and DoE (1990). Figure 4.20 shows the cotidal chart for K_1 in the Gulf of Thailand derived from the results of a numerical model (Fang *et al.*, 1999).

Note the amphidromic point near the centre of the Gulf and the anti-clockwise progression of the tide wave indicated by the cotidal lines. Figures 4.21(a) and (b) show the amplitudes and phases of the M_2 tidal harmonic in the Persian Gulf (also some-

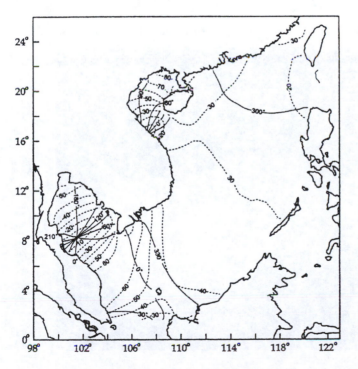

Figure 4.20 Cotidal chart for the tidal harmonic K_1 as determined from a numerical model simulation of flows in the Gulf of Thailand.

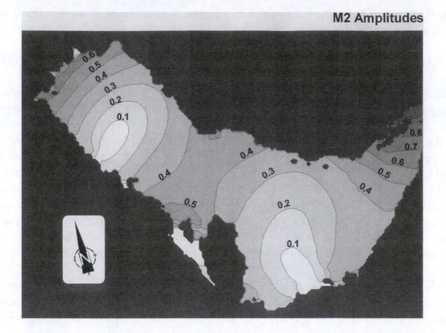

Figure 4.21(a) Amplitudes of the M_2 tidal harmonic determined from a numerical model simulation of flows in the Persian Gulf using depth-averaged equations.

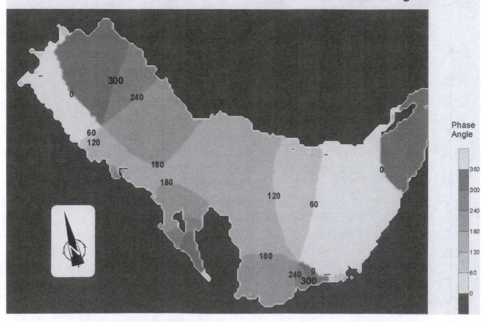

Figure 4.21(b)   Phases of the M_2 tidal harmonic determined from a numerical model simulation of flows in the Persian Gulf using depth-averaged equations.

times known as the Arabian Gulf), derived from model simulations (Osment 2002, pers. comm.). In this case there are two amphidromic points; the cotidal lines again indicate anti-clockwise propagation of the tide wave.

Figure 4.22 shows a cotidal chart for M_2 (DoE 1990), which is based on observations from a network of tide gauges. Due to the complex geometry of the seabed and the landmasses extremely complicated wave patterns are formed.

4.8   Storm surge

4.8.1   Basic storm surge equations

Storm conditions can increase the water levels at the coast, beyond the level predicted by tidal analysis. In mid-latitudes, storms are associated with low-surface pressure weather systems. These systems move bodily at speeds of ~20 kph. However, the

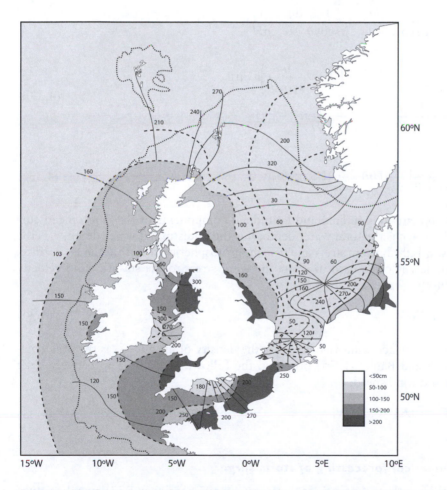

Figures 4.22 Cotidal chart for tidal harmonic M_2 around the British Isles (DoE 1990), determined from sea surface elevation measurements.

wind speeds within the system can be several times this value. Typically, water level variations arising from storms will be a combination of a local barometric effect (the low pressure causing a rise in local sea level), a kinematic effect due to the bodily movement of the weather system, and a dynamic effect of the wind-stress on the ocean surface. The resultant water level variation is often wave-like, and can often be treated successfully as long waves. The problem is complex because the forcing (or wave-generating mechanism) is moving and altering in strength, and the geometry of nearshore regions is often highly intricate.

The set of equations that has been used with some notable success for predicting storm surges is the non-linear hydrostatic equations, with terms accounting for energy generation and dissipation (surface and bottom stress respectively). These may be written as:

$$\frac{\partial u}{\partial t} + \frac{u}{s\cos\theta}\frac{\partial u}{\partial \lambda} + \frac{v}{s}\frac{\partial u}{\partial \theta} - \left(2\Omega + \frac{u}{s\cos\theta}\right)v\sin\theta$$
$$= -\frac{1}{\rho s\cos\theta}\frac{\partial p_a}{\partial \lambda} - \frac{1}{\rho s\cos\theta}\frac{\partial \eta}{\partial \lambda} + \frac{1}{\rho H}(\tau_{s\lambda} - \tau_{b\lambda})$$

(4.33a)

$$\frac{\partial v}{\partial t} + \frac{u}{s\cos\theta}\frac{\partial v}{\partial \lambda} + \frac{v}{s}\frac{\partial v}{\partial \theta} + \left(2\Omega + \frac{u}{s\cos\theta}\right)u\sin\theta$$
$$= -\frac{1}{\rho s}\frac{\partial p_a}{\partial \theta} - \frac{1}{\rho s}\frac{\partial \eta}{\partial \theta} + \frac{1}{\rho H}(\tau_{s\theta} - \tau_{b\theta})$$

(4.33b)

$$\frac{\partial \eta}{\partial t} + \frac{1}{s\cos\theta}\left(\frac{\partial}{\partial \lambda}(Hu) + \frac{\partial}{\partial \theta}(Hv\cos\theta)\right) = 0$$

(4.33c)

where now u,v are longitudinal and latitudinal components of the depth mean currents, v; $H = h + \eta$ is the total water depth, h is the undisturbed water depth, τ_s is the wind stress, τ_b is the bottom stress and p_a is the atmospheric pressure at the sea surface. A quadratic law is often used to relate the wind stress to the surface wind velocity (Flather 1984):

$$\tau_s = C_D \rho_a |V| V$$

(4.34)

where V is the surface wind velocity, ρ_a is the density of air and C_D is the drag coefficient. Smith and Banke (1975) proposed the following relationship between drag coefficient and wind speed:

$$1000C_D = 0.63 + 0.066|V|$$

(4.35)

4.8.2 Numerical forecasting of storm surge

In all but the simplest of situations the storm surge Equations (4.33) must be solved numerically in order to provide predictions. The details of the numerical solution of Equation (4.33) are beyond the scope of this book. However, the scale of calculations is usually such that the curvature of the Earth is important and computations are performed either in spherical coordinates, or the coordinates of a standard map projection, for example, Mercator coordinates. The usual means of solving the equations is by a finite difference scheme. Such schemes often divide the calculation into several distinct steps: an adjustment step that solves the continuity equation to obtain the elevation at one step into the future; an advection step that determines the depth-averaged velocities using the updated elevations; a 'physics' step that updates the variables to account for surface pressure gradients, wind stress and bottom stress. These three steps are then repeated to advance the prediction forward to the desired time.

The boundary conditions for storm surge models consist of several parts. First, there is the seabed levels and coastline in the area of interest. Second, the surface pressure and winds must be specified as a function of both space and time. As observations of future meteorological conditions are not available most surge forecasting models use the output of weather forecast models to specify surface pressures and winds. Third, the variations in elevation and velocity due to the astronomical tides must be specified. This is usually done by specifying the tidal input from knowledge of the main tidal harmonics. The tidal harmonics are determined from observations from coastal and offshore monitoring stations and from the output of larger-scale numerical models.

Two examples of numerical storm surge forecasting are now described. The UK Storm Tide Warning Service was established in the early 1980s with the responsibility for predicting and issuing warnings of situations likely to cause coastal flooding. It is run on an operational basis by the UK Meteorological Office to provide predictions of surge levels to regional and local authorities. The surge forecasting model comprises two nested models. The first, a coarse resolution model, covers the continental shelf around the UK. This model is run to provide the boundary conditions to smaller scale models that cover certain areas of the British Isles, such as the East Coast, The Thames Estuary and the Bristol Channel. Flather (1984) reports on the validation of this model against the major storm that affected the east coast of England and the Dutch coast on 31 January and 1 February 1953. Extensive flooding occurred and the resulting death toll in The Netherlands was over 1400 and was over 300 on the east coast of England. Figure 4.23 shows some of the comparisons of observations and numerical predictions obtained by Flather. On the open coast (Grimsby, Lowestoft) predictions and measurements agree very well. In the Humber Estuary (Hull , Immingham) predictions are not as good. Note that the difference between the predicted tide level and the actual measured level is termed the 'surge' or sometimes 'surge residual'.

Output from simulations of a set of major storms has been used to estimate the distributions of extreme depth-averaged currents associated with storm surges around the northwest European continental shelf (Flather 1987).

The Australian Bureau of Meteorology operates a similar surge forecasting service for the whole of Australia and continental shelf. The forecast is also based on a nested-model system. A coarse-grid model with resolution of ~30 km covers the whole of the continental shelf and a finer-grid model (resolution ~10 km) can be nested anywhere within the coarse-grid model. The coarse-grid model is based on a Lambert grid and

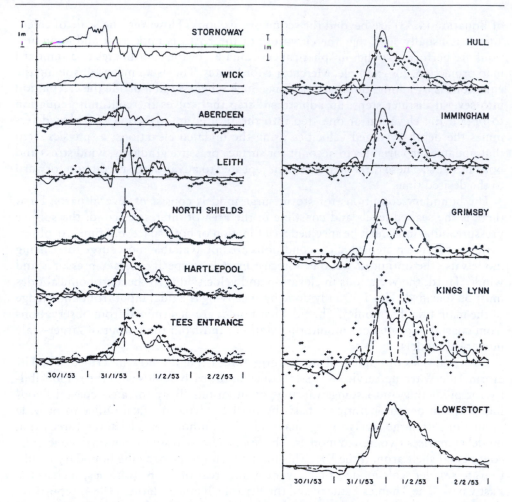

Figure 4.23 Comparisons between computed surge elevations from the coarse resolution model (full lines) and the finer resolution model (dashed lines) with surge residuals derived from observations (crosses) or taken from Rossiter (1954) (dots). The vertical line indicates the approximate time of maximum recorded water level at a port (after Flather 1984).

the finer-grid model uses a Mercator projection. Further details of this surge model are given in Hubbert *et al.* (1990).

In situations where it is important to be able to estimate the vertical distribution of wind-induced currents (for example offshore exploration), the three-dimensional equations of motion need to be solved. One computationally economic alternative to this has been proposed by Davies and Flather (1987). Under the assumption that variations about the depth-mean current quickly reach a steady state, a set of equations that describe the vertical profile of the currents can be derived.

4.8.3   Oscillations in simple basins

The nature of water level oscillations likely to be excited in simple bays or lakes can be interpreted in terms of linear wave theory and are strongly dependent on the geometry of the bay. In such situations the body of water will have natural or 'free' oscillations. If the wave-generating mechanism excites water level variations which are similar to a free mode of oscillation then resonance may occur. To begin, we consider a rectangular lake of depth h with sides of length A and B in the x- and y-directions respectively and assume that motions are hydrostatic. The governing equations are thus the equations of motion in the horizontal (Equations 4.19), the hydrostatic relationship and the equation of continuity. Eliminating u and v from these equations leads to a wave equation in two dimensions:

$$\frac{\partial^2 \eta}{\partial t^2} = c^2 \left(\frac{\partial^2 \eta}{\partial x^2} + \frac{\partial^2 \eta}{\partial y^2} \right), \qquad c = \sqrt{gh} \tag{4.36}$$

We seek solutions to Equation (4.36) that are periodic in time, with the surface elevation written as

$$\eta = \varsigma(x,y) e^{i\omega t} \tag{4.37}$$

Substituting Equation (4.37) into Equation (4.36) gives

$$\left(\frac{\partial^2 \varsigma}{\partial x^2} + \frac{\partial^2 \varsigma}{\partial y^2} \right) + k^2 \varsigma = 0 \tag{4.38}$$

where $k = \omega/c$. We now consider closed basins and impose the condition that the spatial derivative of the surface elevation vanishes along the edges:

$$\frac{\partial \varsigma}{\partial x} = 0 \quad at \quad x = 0 \quad and \quad x = A$$

$$\frac{\partial \varsigma}{\partial y} = 0 \quad at \quad y = 0 \quad and \quad y = B \tag{4.39}$$

The solution of Equation (4.38) subject to the boundary conditions given by Equation (4.39) may be written as a series of cosine functions:

$$\varsigma(x,y) = \sum_{m=0}^{\infty} \sum_{n=0}^{\infty} C_{mn} \cos\left(\frac{m\pi x}{A} \right) \cos\left(\frac{n\pi y}{B} \right) \tag{4.40}$$

Substituting this solution into Equation (4.38) yields

$$k^2 = \pi^2 \left(\frac{m^2}{A^2} + \frac{n^2}{B^2} \right) \tag{4.41}$$

From Equation (4.41) the periods of free oscillation in a rectangular lake are:

$$T = \frac{2}{\sqrt{gh}\sqrt{\dfrac{m^2}{A^2} + \dfrac{n^2}{B^2}}} \qquad (m = 0,1,...; n = 0,1,...)$$

(4.42)

where m and n are the along-basin and cross-basin mode numbers. The periods for the one-dimensional case can be retrieved by setting $n = 0$ in Equation (4.42). In the case of $m = 1$ and $n = 0$ the period is given by $T = 2A/\sqrt{(gh)}$. For open-ended bays and channels the periods of free oscillations in the along-channel direction are given by

$$T = \frac{4A}{\sqrt{gh}(2m+1)} \qquad (m = 0,1,...)$$

(4.43)

Example 4.5
Non-local forcing results in an oscillation of frequency ω_0 whose amplitude at the mouth of a channel is Q. If the channel is of constant depth H, and of length A, determine a general expression for the amplitude of the oscillation at the head of the channel. Calculate specific values for the case when $H = 20$ m, $\omega_0 = 0.001$ Hz, $Q = 0.5$ m and $A = 20$ km and for $A = 21$ km.

Solution
Governing equation is

$$\frac{\partial^2 \eta}{\partial t^2} = c^2 \frac{\partial^2 \eta}{\partial x^2}$$

where $c = \sqrt{(gH)}$. The solution that satisfies the condition of no flow across the closed end of the channel has the form of a standing wave (see also Section 2.3.10), or a superposition of them. Choosing the origin, $x = 0$, to be at the closed end the solution is

$$\eta = \eta_0 \cos(kx)\cos(\omega_0 t),$$

where $\omega_0 = kc$. Now, at $x = A$, $\eta = Q$. Substituting these values into the above solution gives the amplitude of the oscillation at the head of the channel as $\eta_0 = Q \sec(kA) = Q \sec(A\omega_0/\sqrt{(gH)})$.

Substituting the numerical values gives $\eta_0 = 3.5$ m for $A = 20$ km and $\eta_0 = 7.0$ m for $A = 21$ km. The increase in amplitude is very large (resonance occurs) when the frequency of the forcing is close to one of the natural frequencies of oscillation. The remarkable tidal ranges found in the Bristol Channel and Bay of Fundy can be attributed to the near resonant response. Resonant responses can also occur in ports and harbours, sometimes exacerbated by the neat geometrical outlines adopted in such cases.

Water level oscillations in basins, lakes, harbours and estuaries are sometimes referred to as 'seiches'. The crucial ingredient is that the body of water be partially constrained so that standing waves can form. Regular geometry is not a necessity as even harbours with highly irregular shapes can support seiches with very stable frequencies. The cause of seiches varies but can include the wind, earthquakes, long-period waves and low-pressure atmospheric storms.

4.9   Tsunamis

Tsunamis are perhaps one of the most deadly and devastating natural disasters that we face. This is because they contain a large amount of energy and this energy can be transferred huge distances, without much loss, at a very rapid rate and without being easy to observe. 'Tsunami' is a Japanese word meaning 'harbour wave', but is now used to describe ocean waves caused by underwater disturbances such as landslides, volcanic eruptions, earthquakes and nuclear explosions. There are several recent and well-documented examples of tsunamis caused by earthquakes. Earthquakes cause sudden vertical movements in the sea floor, which in turn cause a large volume of water to move. At that instant the water will no longer be in gravitational equilibrium with the surrounding ocean and the process of the water adjusting to restore equilibrium gives rise to gravity waves that move away from the source region. The details of the initial moments of tsunami generation are not well understood. However, the process very often leads to not one but a sequence of waves propagating outwards – much like the ripples spreading out on the surface of a pond after a stone has been thrown in, as illustrated in Figure 4.24. While the initiation and dissipation of tsunamis are

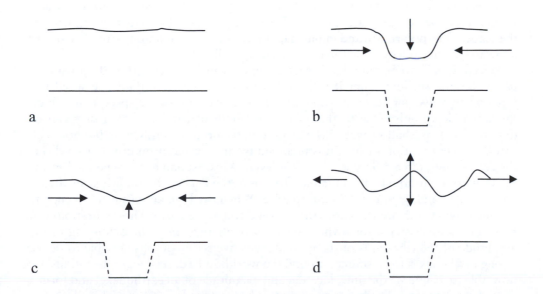

Figure 4.24 Illustrative diagram of a normal fault earthquake and formation of waves propagating outwards: (a) situation normal; (b) an earthquake results in a portion of the seabed dropping rapidly, the surrounding water starts to occupy the surrounding space; (c) water converges towards the sea surface above the epicentre; (d) the sea surface above the epicentre oscillates towards equilibrium and waves propagate outwards.

generally highly non-linear processes, the propagation of tsunamis from the typically deep ocean source to coastal waters can be described by the linear wave theory in Chapter 2 to a good degree of accuracy.

Tsunamis are often generated in the deep ocean where the water depth is many kilometres. Tsunamis are extremely long waves with wavelengths of the order of hundreds of kilometres and periods of between 5 min to an hour. Away from the generation area tsunamis are characterised by low wave height, O (10 cm), and are often difficult to recognise even by ships at sea. As a result of their small amplitude there is very little energy loss, so tsunamis represent an extremely efficient means of transferring a large amount of energy over great distances. It should be remembered that although the wave energy depends on the square of the wave height (Equation 2.7), this is energy per unit area and due to its large wavelength (of the order of a thousand times that of wind waves, a tsunami wave will contain a huge amount of energy). It is only in shallower coastal waters that shoaling causes amplification of the waves. Refraction and diffraction can cause localised focussing and defocussing of the wave front, leading to 'miraculous' escapes for some coastal sites and complete disaster for others. Depending on the steepness of the ocean floor and the distance from generation to landfall tsunamis may have broken before reaching the shoreline. A broken tsunami wave will likely have the appearance of a bore (bores such as the Severn Estuary bore are breaking tide waves). It will have dissipated some of its energy through the breaking process but, due to the highly turbulent nature of the flow, will be laden with sediments and other debris. An unbroken wave will appear not necessarily as a wave but as an accelerated, high frequency, tide-like change in water level. Importantly, the phrase 'tidal wave' should not be used to describe a tsunami. A tidal wave is the oscillation in water levels due to the tide-generating forces. A tsunami is caused by sudden movements of the seabed. The occurrence and initial magnitude of tsunamis cannot be predicted very accurately, while tides can be forecast extremely well.

Referring back to Section 2.2.4, surface waves can be described as deep water if $h/L > 0.5$ and shallow water if $h/L < 0.04$. In the former case the wave speed is a function of wave period only, while in the latter case the wave speed depends on the water depth only. Due to the great wavelength of tsunami waves they can be described well as 'shallow water' waves, even in the deep oceans. Often, but not in all cases, the approach of a tsunami is preceded by an unusual drop in water level (i.e. the sea recedes, often below the low tide level). Most tsunamis are generated in the active marine earthquake region along the rim of the Pacific Ocean, primarily affecting New Zealand, Japan and South America. When earthquakes occur they generate seismic waves which are vibrations that travel through the Earth. These vibrations can be recorded on seismographs which show an oscillating trace corresponding to the amplitude of the vibrations of the ground. Sensitive seismographs are able to detect strong earthquakes from sources around the world. Indeed, seismograph stations are now able to determine the time, location and magnitude of an earthquake, and form a crucial element of tsunami warning systems. The magnitude is often quoted as a number on a scale known as the Richter scale. This scale was proposed by Charles Richter in 1935 at the California Institute of Technology, and is a logarithmic measure of the amplitude of ground vibrations. The magnitude is usually quoted to one decimal place (e.g., 5.3). Due to the logarithmic scale, an increase in magnitude on the Richter scale of 1 corresponds to a 10-fold increase in measured ground vibration amplitude and

Table 4.3 Classification of earthquake magnitude.

Richter scale	Classification	Comments
< 2.0	Micro-earthquake	Not commonly felt by people; only detected locally
2.0–4.5	Minor earthquake	Felt by people. Minor structural damage
> 4.5	Major earthquake	Often several thousand such events in the world each year. Strong enough to be detected globally by seismograph stations
> 8.0	Great earthquake	Typically one such event occurs somewhere in the world each year

approximately 30-fold increase in energy. Table 4.3 summarises the classification of earthquakes by Richter scale.

The Richter scale has several drawbacks. The first is that it is a comparative rather than absolute scale. The second is that it was developed for conditions specific to southern California. An alternative measurement of earthquake magnitude, the Moment Magnitude scale (denoted by Mw), was designed to overcome the problems of the Richter scale. The Moment Magnitude is not comparative, has global applicability and gives magnitudes roughly equivalent to the Richter magnitudes. Both scales are still used, but the Moment Magnitude scale is gradually replacing the Richter scale and is preferred now by most seismologists. The numbers generated by the two scales are usually very similar (within 5 per cent).

A tsunami will typically comprise a group of waves with a spread of frequencies. There is a large body of work on tsunamis in the Japanese literature. In particular, researchers have sought to link the characteristics of earthquakes to those of related tsunamis. Iida (1959, 1969) determined a relationship between the strength of an earthquake as measured on the Richter scale and the height of the accompanying tsunami. Major tsunamis are due to large earthquakes that cause large vertical sea floor movements in relatively shallow water. Iida found that generally, earthquakes associated with the worst tsunami events had a magnitude greater than 6.5 on the Richter scale and a centre of less than 60 km below the seabed. Takahasi (1947) proposed a relationship between the earthquake strength and the dominant period of the tsunami. A detailed account of the 1964 Alaskan earthquake and tsunami has been provided by Wilson and Torum (1968), and more recent review of tsunami may be found in Camfield (1980). A summary of observations and regression results may be found in CIRIA/CUR (1991).

Example 4.6
An earthquake off the coast of Japan causes a tidal wave. Estimate how long it will take for this wave to reach the west coast of North America (The Pacific Ocean may be assumed to be 4000 km wide and 6 km deep on average).

Solution
On the basis of the difference in the horizontal and vertical scale we will use a shallow-water approximation. Thus the speed of the wave may be approximated as:

$$c = \sqrt{gh} = 242 ms^{-1}$$

The time of propagation can be calculated as:

$$time = \frac{distnce}{speed} = \frac{4000000}{242} = 16500s = 4.6 hours$$

Note the large phase speed of the wave.

Example 4.7
A wave in a tsunami has a period of 30 min and a height, H_0, of 0.5 m at a point where the ocean has a depth of 4 km deep. Calculate the phase speed, c_0, and wavelength, L_0, of this wave. Calculate its phase speed, c_i, wavelength, L_i, and height, H_i, in a coastal water depth of 15 m accounting for shoaling effects only.

Solution
We assume the wave is a shallow-water wave so that

$$c_0 = \sqrt{gh} = \sqrt{9.81 x 4000} = 198 ms^{-1}$$

$$L_0 = c_0 T = 198 x 30 x 60 = 356400 m$$

The depth to wavelength ratio is 4000/356 400 = 0.011, thus the wave is a shallow-water wave at a depth of 4000 m. The wave steepness is extremely small and is equal to $H_0/L_0 = 1.4 \times 10^{-6}$. To determine the inshore wave characteristics we assume there is negligible energy dissipation and equate the wave power in deep and nearshore sites. Thus, since we have

$$L_{i,0} = T \sqrt{gh_{i,0}}$$

equating wave powers gives

$$H_0^2 T \sqrt{gh_0} = H_i^2 T \sqrt{gh_i}$$

So at a depth of 15 m

$$H_i = 0.5 \left(\frac{4000}{15} \right)^{0.25} = 2.02 m$$

$$c_i = \sqrt{9.81 x 15} = 12.13 ms^{-1}$$

$$L_i = 12.13 x 30 x 60 = 21834 m$$

This simple example demonstrates two important points. First, the dramatic reduction in phase speed (by at least one order of magnitude). Second, the significant amplification of the height of the wave as it propagates into shallow water.

Tsunamis propagate as long gravity waves, and there are many records of such waves that have travelled across the Pacific or Atlantic Oceans. The amplification due to refraction and shoaling means that these waves can cause extensive damage and flooding in coastal areas. Some of the most-analysed records relate to the waves generated by the huge eruption of the volcano Krakatoa, in the Sundra Strait, on 26 and 27 August 1883. Analysis of such records allowed estimates of the mean depth of the ocean to be made. Under the assumption of shallow-water conditions and from knowing the travel times and distances over which waves had propagated, the wave speed could be determined. Thus, while Laplace had assumed the average depth of the oceans to be 18 000 m in his development of tidal theory, Bache computed the average depth to be approximately 4000 m in 1856 (Sverdrup 1945). This is remarkably close to recent estimates of the average depth of the ocean of 3800 m. A more recent example of this type of calculation is presented by Zhang *et al.* (2009) who used linear shallow-water theory to compute tsunami travel times and compared these with observed times for several different tsunamis, including the 2004 Indian Ocean event. There are several tsunami warning systems in place or under development. These include the US Pacific Tsunami Warning Center (PTWC) in Hawaii, the Japanese warning system, the US West Coast and Alaska Tsunami Warning System, the Australian Warning System and the UNESCO Indian Ocean Tsunami Warning System. In practice, these systems link tidal and seismographic stations to tsunami observation centres. Warnings are issued to directly to government agencies, and to the public via internet, radio and television broadcasts.

4.9.1   Indian Ocean Tsunami – 26 December 2004

One of the most deadly tsunamis in recent history, affecting millions of people across Southeast Asia, India and Africa, was the Indian Ocean tsunami. This event brought tsunamis into the consciousness of people worldwide due not only to its magnitude but also the speed at which information, photographs and video clips were disseminated through global telecommunications networks and the internet. The widespread and somewhat romantic notion of tsunamis as large curling waves was thoroughly dispelled. The number of fatalities is not known but estimates from various sources agree it was in excess of a quarter of a million people. Many of the features mentioned in the preceding section are clear from the extensive photographic and video records that can be found on the internet. For example: the recession of the sea prior to the arrival of the tsunami crests; the arrival at many locations of two, three or more waves in succession; the appearance of the tsunami waves as bores, are all evident in the records.

The epicentre of the main earthquake (i.e. the point on the Earth's surface directly above the focus of the earthquake) was at 3.316°N, 95.854°E, approximately 160 km away from the western coast of northern Sumatra, Indonesia. The focus of the earthquake was estimated to be ~30 km below mean sea level. The earthquake itself had a magnitude of between 9.1 and 9.3 on the Richter scale making it the second largest quake ever recorded on seismograph. Detailed analysis of observations by GPS stations suggests that the earthquake was a rupture that propagated northward from the epicentre along a segment of the Andaman–Sunda trench measuring over 1200 km. This caused subsidence of ~6 m on one side of the rupture and uplift of ~10 m on the other side, affecting a region 100–150 km wide across the area (Ioualalen *et al.* 2007).

Segur (2007) estimated that the net effect of the earthquake was to raise the ocean floor to the west of the epicentre and to lower it to the east. The region of seabed raised was about 100 km in the east–west direction and ~900 km in the north–south direction. The region lowered was of similar dimensions. Satellite observations indicated that the wavelength of the tsunami was between 500 and 800 km and the wave height in the open ocean was less than ~0.5 m.

Land closest to the epicentre, such as the city of Banda Aceh – the provincial capital of Aceh – had very little warning, but further afield an interesting distribution in travel times was observed. The tsunami reached Sumatra in a matter of minutes, while Sri Lanka and the east coast of India were hit after about 1.5–2 h. Thailand was also struck after about 2 h, despite being closer to the epicentre. The reason for this being that the tsunami travelled as a shallow-water wave and hence its speed depended upon depth only. It travelled more slowly in the shallow (c. 900 m) Andaman Sea to the west of Thailand than it did in crossing the deeper Pacific Ocean (c. 3900 m). The Pacific Tsunami Warning Centre (PTWC) obtained its first indications of a destructive tsunami at about 3.30 GMT from internet newswire reports of casualties in Sri Lanka.

Taking the wavelength of the tsunami to be ~600 km, the Andaman Sea and Pacific Ocean to be 900 m and 3900 m deep respectively, we can estimate some travel times. The first point to notice is that in both cases the tsunami may be considered a shallow-water wave as water depth/wavelength < 0.04. Thus, from Section 2.2.4, the wave travelling west had a speed $c = \sqrt{(gh)} = \sqrt{(9.81 \times 3900)} = 195.6$ m/s = 704 km/hr, while the wave travelling east towards Thailand moved more slowly, with $c = \sqrt{(gh)} = \sqrt{(9.81 \times 900)} = 94.0$ m/s = 338 km/hr. In passing it is interesting to note that the speed of sound in air is ~1000 km/hour and the tsunami wave crossing the Pacific was travelling at least as fast as a current day passenger jet. Arrival times of the tsunami at various locations have been reported by the PTWC. Selecting three sites, Colombo in Sri Lanka, Phuket in Thailand and Hurdiyo in Somalia, the time for the tsunami to travel to each of these sites can be estimated using shallow-water wave theory. Table 4.4 summarises the information and estimated results.

The level of agreement is perhaps somewhat fortuitous because we have taken no account of refraction and diffraction in our calculations. However, the simple linear theory provides a good first estimate and has the advantage of being very fast to calculate. Another process that is important in determining the amplitude of the tsunami wave at a point distant from the generation region is spreading. This is simply the fact that as the initial wave propagates further away from the source the length of each wave crest increases. As the wave receives no extra energy after the initial generation this same energy is spread along an ever-increasing length of wave crest. Thus the energy per unit crest length will decrease and thus also the wave height. The process of spreading was the main reason that the tsunami waves in Sri Lanka, Africa and Thailand were

Table 4.4 Indian Ocean tsunami travel times.

Location	Distance from epicentre (km)	Wave speed (km/hr)	Observed travel time (h)	Estimated travel time (h)
Hurdiyo	4998	704	7.1	7.1
Colombo	1783	704	2.5	2.5
Phuket	580	338	1.5	1.7

much smaller than those experienced in Indonesia. This tsunami was detected in tidal records as far away as Port Elizabeth in South Africa and the west coast of the UK, demonstrating the efficiency of tsunamis in transporting energy around the world.

4.9.2 South Pacific Islands Tsunami – 29 September 2009

The epicentre of the main earthquake was at 15.51°S, 187.97°E, approximately 175 km south of the Samoan Islands. The earthquake occurred on 29 September 2009 at 6:48am local time and its focus was estimated to be ~18 km below mean sea level. The earthquake itself had a magnitude of 8.3 Mw. Tsunami waves hit American Samoa, Western Samoa and the island of Niuatoputapu in Tonga a few minutes after the earthquake. As noted by EEFIT (2009), although the PTWC were able to issue their first warning only 16 minutes after the earthquake, the tsunami took less than 15 min to reach the islands. The tsunami waves were reported to have been up to 6 m in some places. The tsunami caused at least 143 fatalities in Samoa, 22 fatalities in American Samoa and 7 on Niuatoputapu, with numerous people injured. The estimated costs of repairing the damage to houses, roads and other infrastructure has been estimated at several hundreds of millions of pounds.

The United States Geological Survey (USGS) reported that a submarine earthquake occurred close to the northern end of the Pacific–Australia tectonic plate boundary. They also stated it was due to a normal fault rupture near the outer rise of the subducting Pacific plate, though data from tsunami detection buoys suggested the possibility of a reversed fault. The meaning of this is illustrated in Figure 4.25.

It seems that the September 2009 Samoa tsunami was generated by a specific form of earthquake that occurs near ocean trenches. Unlike the Indian Ocean earthquake that occurred due to a thrust fault that separates tectonic plates in a subduction zone, the Samoan earthquake was caused by what is termed a reverse fault in the outer-rise – the bulge in the down-going plate before it enters the subduction zone. There are few records of tsunamis generated in this way, but those that have occurred have been

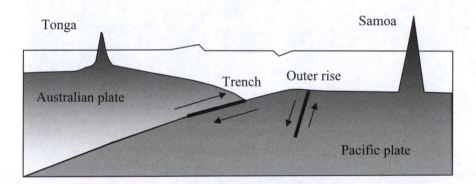

Figure 4.25 Illustrative diagram showing the main earthquake types in the Samoan tsunami. The movement of the plates at the trench is classified as a normal thrust fault where the subducting plate (the Pacific plate here) is pulled beneath a neighbouring plate. The other fault is a reverse fault in the outer rise, where the relative movement of the earth on each side of the fault is in the opposite sense to that in the thrust fault.

devastating. The 2009 Samoa outer-rise earthquake was the fourth largest outer-rise earthquake recorded since 1900. Subsequent analysis has suggested there were in fact two earthquakes, one in the outer-rise and one in the thrust zone. Arguments for which came first and whether one triggered the other are provided by Beavan *et al.* (2010) and Lay *et al.* (2010).

The EEFIT (2009) report provides an interesting assessment of the damage that occurred in Western Samoa, making observations about how local building methods and cultural influences affected the scale of fatalities and damage. One element of this was the layout of the main roads on the island which tended to follow the low-lying coastal areas. Their destruction by the tsunami was one reason recovery operations were made more difficult. Conversely, where roads had been built running normal to the shoreline the forest and ground cover had been cleared, allowing the tsunami wave to penetrate much further inland in these areas. As noted in other events a lot of damage arose not so much from the water itself but by impacts from debris picked up and carried along by the tsunami. Also, there was evidence of considerable scour (leading to whole or partial undermining of buildings) caused by the tsunami wave as it retreated and the flood waters flowed back to the sea. Figure 4.26 shows an example of the type of damage that occurred to buildings on Samoa. Evidence of the extent of erosion during the retreat of the tsunami waters is shown in Figure 4.27.

4.9.3 Japan Tsunami – 11 March 2011

More recently still, another earthquake (the Tōhoku earthquake) occurred off the eastern coast of Japan, triggering a huge tsunami. The earthquake was a 9.0 Mw

Figure 4.26 An example of the damage to buildings caused by the tsunami on Samoa. Note the undermining of the foundation due to scour as well as the damage to roof and walls. (Photograph courtesy of Dr Alison Raby.)

Figure 4.27 Post-tsunami evidence of scour near the top of the beach, Samoa. (Photograph courtesy of Dr Alison Raby.)

undersea megathrust earthquake that occurred at 14:46 JST (05:46 UTC) on Friday, 11 March 2011. The earthquake triggered huge tsunami waves (over 30 m height in some places according to local press) that struck Japan minutes after the quake, in some cases travelling up to 10 km inland. There was a huge loss of life and massive destruction of houses and infrastructure. Also, the nuclear power plant at Fukushima was damaged leading to the escape of radioactive compounds. At the time of writing the full extent of the damage was not yet confirmed, with a sequence of strong (~6.0 Mw) aftershocks hampering the rescue and restoration efforts. The event is notable because Japan is well known for being well protected against tsunamis. In this case the magnitude of the event exceeded the criteria used in the design of the defences, which may well be revised in the light of the consequences of the flooding caused by the tsunami.

Summarising:

- Tsunamis are caused by earthquakes but not all subsea earthquakes lead to destructive tsunamis.
- As tsunamis travel as small amplitude shallow-water waves they can transmit large amounts of energy vast distances in a short period of time.
- Given that many of the tsunami-generating earthquakes occur in the Pacific Ocean there are opportunities to provide warnings that can be used to evacuate people from danger, thereby reducing fatalities. The Pacific Tsunami Warning Centre is one such warning network. For an average velocity of a tsunami of 750 km/hr, a system such as the PTWC can provide a warning sufficient for adequate evacuation of coastal areas beyond ~750 km from the earthquake.
- Building tsunami defences in a manner akin to coastal flood defences against wind waves is possible. There are several examples in Japan, but these are extremely expensive.
- More immediately practical solutions would be to improve warning systems and have adaptation strategies to modify existing or new infrastructure to make it more resilient to damage by tsunami waves.

The authorities in some regions have prepared safety rules to advise the general populace. Examples of such rules are listed below:

- A strong earthquake felt in a low-lying coastal area is a natural warning of possible, immediate danger. Keep calm and quickly move to higher ground away from the coast.
- Not all large earthquakes cause tsunamis, but many do. If the quake is located near or directly under the ocean, the probability of a tsunami increases. If news states that an earthquake has occurred in the ocean or coastline regions, prepare for a tsunami emergency.
- A tsunami can occur at any time, day or night. They can travel up rivers and streams that lead to the ocean.
- A tsunami is not a single wave, but a series of waves. Stay out of danger until an ALL CLEAR is issued by a competent authority.
- Approaching tsunamis are sometimes heralded by a noticeable rise or fall of coastal waters. This is nature's tsunami warning and should be heeded.
- A small tsunami at one beach can be a giant a few miles away. Do not let the modest size of one make you lose respect for all.
- Never go down to the beach to watch for a tsunami! WHEN YOU CAN SEE THE WAVE YOU ARE TOO CLOSE TO ESCAPE. Tsunamis can move faster than a person can run!
- Homes and other buildings located in low lying coastal areas are not safe. Do NOT stay in such buildings if there is a tsunami warning.
- The upper floors of high, multi-story, reinforced concrete hotels can provide refuge if there is no time to move inland or to higher ground.

If you are on a boat or ship and there is time, move your vessel to deeper water (to at least 180 m).

4.10   Long-term water level changes

4.10.1   Climatic fluctuations

Evidence is growing that the world's climate is changing. Globally, 1998 was the hottest year ever recorded and all the years from 2000 to 2008 have been in the top 14 warmest years on record. Discussion continues between scientists involved in the global-warming debate, while new measurements are being gathered all the time and new global climate modelling studies are undertaken.

In the past, most coastal structures were designed to account for water level variations associated with tides or surges, but tacitly assume either in design or operation that the mean water level will be relatively stable over the life of the structure. With the benefit of our advances in understanding climate change, nowadays coastal structures are routinely designed to allow for some sea level rise.

It is sometimes easy to forget that climate change can mean that extreme water levels can go down as well as up. Lowering water levels can have a large impact too. Many installations such as ports, nuclear power stations, desalination plants and so on have been designed on the assumption of reasonably static mean water levels. The installations may well become unusable if water levels go up or down significantly. For example, if water levels fall, maintaining a navigable channel to a port will involve huge costs in dredging.

Most coastal structures have a design life of many decades (cf. Chapter 9), and it is important to consider changes in mean water level in the design. The reason for this is that structures are designed to withstand loads of a prescribed severity – for example, the 1 in 50-year water level which would be expected to be exceeded on average once in 50 years. This extreme water level would typically comprise both a tidal and surge component. However, if the mean water level is not constant but rising in the long term then what is a 1 in 50 year event at the time of construction may become say only a 1 in 5-year event by the time a few decades have elapsed. Figure 4.28(a) shows conceptually how this arises. Note that in this case only the mean water level is

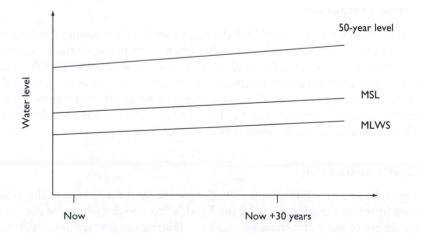

Figure 4.28(a) Underlying rise in mean sea level leads to what is an extreme water level becoming a much more common event in the future.

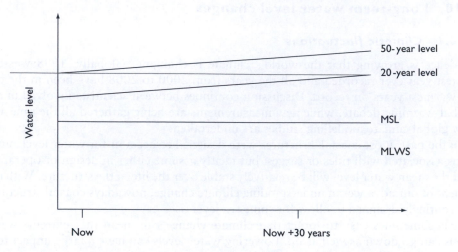

Figure 4.28(b) Increased storminess leads to larger surges so that what was an extreme water level now becomes a much more common event.

changing; we have assumed that the surge component remains unchanged. Figure 4.28(b) shows an alternative way in which climate change may reduce the effectiveness of our coastal defences. In this case the mean sea level remains unchanged but increased storminess causes an increase in the surge amplitude so that larger surges are experienced more often. In practice it is likely that we will experience some combination of these two processes. It might be noted that should average wind speeds increase by 10 per cent, (suggested as an allowance for in design), it follows that wave heights should be increased by 17 per cent and storm surge by 21 per cent using conventional formulations.

4.10.2 Eustatic component

Eustatic sea level change refers to a global change in sea levels resulting from thermal expansion of the water mass. Also included in this are phase changes such as melting or freezing of polar ice caps. The reader is referred to Carter (1988) for detailed descriptions. However, it is generally accepted that about 25 000 years ago sea levels were 150 m below their current level. The water levels rose at approximately 7 mm/year from then to about 3000 years ago. Since then, the eustatic rise has been relatively small. It has been estimated to be between 1 and 1.5 mm/year.

4.10.3 Isostatic component

Isostatic change refers to variations in the elevation of the land mass. One of the most common causes of this is the adjustment of the Earth's crust to the release of pressure exerted by ice sheets of great thickness (1 to 2 km). During the last ice age, thick ice sheets covered most land areas in the higher latitudes. These areas were depressed by the glaciers, while the regions just ahead of the glaciers rose slightly. With the end of the ice age the glaciers retreated and the land levels have been adjusting to the change in the distribution of pressure. By and large, those areas that were under glaciers in the ice age are now rising (or 'rebounding') while those areas ahead of the glaciers are sinking. This process continues today.

In mid-latitude regions isostatic rebound generally acts in the reverse sense to eustatic sea level rise, sometimes even reversing it in higher latitudes. For example, the relative sea level is dropping in Alaska and northern Norway, because the isostatic rise more than compensates for the eustatic rise.

4.10.4 Global climate change

Changes in the global climate will determine the eustatic sea level rise. Should global climate changes show a prolonged trend we may suffer ice ages or warm spells. In moving from one state to the other the sea levels are likely to change by many tens or even hundreds of metres. Evidence from geological records suggests that the transition from ice age to warm spell naturally occurs over the period of many human lifetimes, and certainly over a greater length of time than the typical design life of a coastal structure. However, recent research on the effects of human industrial activity indicates that the global climate may change sufficiently over the order of a century to have significant impacts on the coastal regions in which a significant proportion of human endeavour occurs.

The Intergovernmental Panel on Climate Change (IPCC) is a body established by the World Meteorological Organisation (WMO) and the United Nations Environment Programme (UNEP) in 1988. It consists of approximately 2500 scientists from all over the world, from a wide range of disciplines. The aim of the IPCC is to assess available scientific and socio-economic information on climate change and its impact, and to consider options for mitigating climate change. It is the world's leading body on the subject of climate change and is considered to be independent from national and political influence.

Since 1990, the IPCC has produced a series of reports, including four Assessment Reports in 1990, 1995, 2001 and 2007. The fourth Assessment Report consists of four volumes:

1   *Working Group I Report: The Physical Science Basis*
This volume presents the science of climate change. It looks at the factors that drive climate change, analyses the past climate and predicts future climate conditions and detects and attributes the influence of human activity on recent climate. The 6 years since the IPCC's Third Assessment Report saw large amounts of new data, the development of improved analysis techniques, advances in the representation of physical processes in climate models and greater investigation of uncertainty in model results. The scientific implications of these are discussed in this report.

2   *Working Group II Report: Impacts, Adaptation and Vulnerability*
This volume considers the environmental, social and economic consequences of climate change and potential responses. It considers the sensitivity, adaptive capacity and vulnerability of natural and human systems to climate change. It also addresses the potential impacts and adaptation options at regional and global scales.

3   *Working Group III Report: Mitigation of Climate Change*

This volume describes potential means of mitigating the effects of climate change. It considers the technological and biological options to mitigate climate change, their costs and ancillary benefits, and the barriers to their implementation. It also discusses policies, measures and instruments to overcome the barriers to implementation.

4   *The AR4 Synthesis Report*

This volume is based on the assessment carried out by the three Working Groups of the IPCC. It provides an integrated view of climate change and provides a synthesis intended for policymakers. It illustrates the impacts of global warming, discusses means by which society could adapt to mitigate the consequences of climate change, and presents an analysis of costs, policies and technologies to achieve this.

Some of the main environmental conclusions drawn from the fourth report are:

- The 100-year linear trend (1906–2005) of 0.74 [0.56 to 0.92] °C is larger than the corresponding trend of 0.6 [0.4 to 0.8] °C (1901–2000) given in the Third Assessment reports.
- Eleven of the last twelve years (1995–2006) rank among the twelve warmest years in the instrumental record of global surface temperature (since 1850).
- The linear warming trend over the 50 years from 1956 to 2005 of 0.13 [0.10 to 0.16] °C per decade is nearly twice that for the 100 years from 1906 to 2005.
- Global average sea level rose at an average rate of 1.8 [1.3 to 2.3]mm per year from 1961 to 2003 and at an average rate of about 3.1 [2.4 to 3.8]mm per year from 1993 to 2003.
- Average northern hemisphere temperatures during the second half of the twentieth century were *very likely* higher than during any other 50-year period in the last 500 years and *likely* the highest in at least the past 1300 years.
- Observations since 1961 show that the average temperature of the global ocean has increased to depths of at least 3000 m and that the ocean has been taking up over 80 per cent of the heat being added to the climate system.

As far as planning for sea level rise is concerned, the IPCC has identified three alternative responses (IPCC 1990, 1992):

- protection
- accommodation
- retreat

These have not been altered in the latest round, which discussed economic and policy development options in much greater detail.

In November 2009 the Copenhagen Diagnosis (www.copenhagendiagnosis.org/press.html), effectively an official consensus update on IPCC's 2007 report, concluded that updated estimates of the future mean sea level rise are significantly higher than IPCC projections from 2007 and indicated a rise of 0.4 to 1.2 m by 2100. The Executive Summary includes the following headlines:

- Global carbon dioxide emissions from fossil fuels were 40 per cent higher than those in 1990.
- Over the past 25 years temperatures have increased at a rate of 0.19 °C per decade, in very good agreement with predictions based on greenhouse gas increases.
- A wide array of satellite ice measurements demonstrate beyond doubt that both Greenland and Antarctic ice sheets are losing mass at an increasing rate.
- Summertime melting of Arctic sea ice has accelerated far beyond expectations of climate models and its area during 2007–2009 was about 40 per cent less than the average prediction from IPCC AR4 climate models.
- Satellites show recent global sea-level rise (3.4 mm/year over the past 15 years) to be about 80 per cent above past IPCC predictions (of glaciers and ice caps).
- By 2100 global sea level is likely to rise to at least twice as much as projected by Working Group 1 of IPCC AR4.

The IPPC has concluded that even if countries were able to maintain concentrations of the so-called 'greenhouse gases' at current levels, the world would still expect to see a further rise in temperatures of about 0.7 °C. The UK Government has published a set of indicators that are being used to monitor how the UK's climate is changing (DETR 1999).

Recent research undertaken for the UK Government suggests that sea levels could be between 26 and 86 cm above the current level in southeast England by the 2080s. At some sites this means that a sea level that now have a probability of 1/50 of occurring in a year could occur between 10 to 20 times more frequently by the 2080s.

Many cities and heavy industrial installations are located in coastal areas, and significant proportion of the population and wealth-generating infrastructure is vulnerable to rising sea levels. Governments around the world are putting in place the mechanisms for responding to sea level rise. In the UK, the Government has published its potential strategic adaptation priorities for the next 30 years (ERM 2000). These include:

- coastal and river flood defence programmes;
- enhancing resilience of buildings and infrastructure;
- coordinated approaches to planning;
- improved long-term and short-term risk prediction.

Practical steps towards addressing these issues have already taken. Notably:

- Allowances for sea level rise have been incorporated into guidance on project appraisal which has been adopted for all new and reconstructed coastal defences since 1989.
- Shoreline Management Plans have been prepared for England and Wales, which provide a planning framework for flood defence and coastal protection measures.

Recommended rates of sea level rise for England given by the UK Environment Agency are reproduced in Table 4.5. The corresponding recommended sensitivity ranges are given in Table 4.6.

The Environment Agency note that these allowances and sensitivity ranges were

Table 4.5 Recommended rates of sea level rise for England and Wales (UK Environment Agency, 2011).

Administrative or Devolved Region	Assumed vertical land movement (mm/yr)	Net sea level rise (mm/yr) 1990–2025	Net sea level rise (mm/yr) 2025–2055	Net sea level rise (mm/yr) 2055–2085	Net sea level rise (mm/yr) 2085–2115
East of England, East Midlands, London, SE England (south of Flamborough Head)	−0.8	4.0	8.5	12.0	15.0
South West and Wales	−0.5	3.5	8.0	11.5	14.5
NW England, NE England, Scotland (north of Flamborough Head)	+0.8	2.5	7.0	10.0	13.0

Table 4.6 Recommended sensitivity ranges for England and Wales (UK Environment Agency, 2011).

Parameter	1990–2025	2025–2055	2055–2085	2085–2115
Peak rainfall intensity (preferably for small catchments)	+5%	+10%	+20%	+30%
Peak river flow (preferably for larger catchments)	+10%	+20%	+20%	+20%
Offshore wind speed	+5%	+5%	+10%	+10%
Extreme wave height	+5%	+5%	+10%	+10%

developed before the UK's latest climate projections (UKCP09) were produced, but that they remain very reasonable estimates of change.

Finally, it is worth noting that there isn't unanimous agreement that humans cause climate change amongst climate scientists. Plimer (2009), for example, puts forward a detailed case for the contrarian viewpoint.

Further reading

Cartwright, D.E., 1999. *Tides: A Scientific History*, Cambridge University Press.

Copenhagen Analysis, 2009. *Updating the World on the Latest Climate Change Science*. Allison I. *et al.*, The University of New South Wales (CCRC), Sydney, Australia (http://www.copenhagendiagnosis.org/press.html)

Environment Agency, 2011. www.environment-agency.gov.uk/research/planning/116769.aspx

Graff, J., 1981. An investigation of the frequency distributions of annual sea level maxima at ports around Great Britain. *Estuarine, Coastal and Shelf Science*, 12, 389–449.

Institution of Structural Engineers, www.ipcc.ch/publications_and_data/publications_ipcc_fourth_assessment_report_synthesis_report.htm

www.istructe.org/knowledge/EEFIT/Documents/samoa_preliminary_report.pdf

Plimer, I., 2009. *Heaven and Earth: Global Warming – the Missing Science*, Quartet, London.

Pond, S. and Pickard, G., 2000. *Introductory Dynamical Oceanography*, Butterworth-Heinemann, Oxford.

Pugh, D. T.,1987. *Tides, Surges and Mean Sea Level*, John Wiley & Sons, Chichester.

UKCP09, http://ukclimateprojections.defra.gov.uk/

US Geological Survey, http://walrus.wr.usgs.gov/tsunami/samoa09/

Wells, N., 1997. *The Atmosphere and Ocean: A Physical Introduction*, John Wiley & Sons, Chichester.

West Coastal and Alaska Tsunami Warning Centre, http://wcatwc.arh.noaa.gov/safety.htm

Wikipedia, http://en.wikipedia.org/wiki/2004_Indian_Ocean_earthquake

Wikipedia, http://en.wikipedia.org/wiki/2009_Samoa_earthquake

Chapter 5

Coastal transport processes

5.1 Characteristics of coastal sediments

Sediment transport governs or influences many situations that are of importance to mankind. In rivers, estuaries and on coastlines, sediment movements can result in significant erosion or accretion over both local areas and on much wider geographic areas. This can take place on timescales of a few hours (resulting from storms or floods) to months and years (as a result of the seasonality in the waves and currents) and to decades and beyond (as a result of changing climate and natural and manmade influences). Important manmade facilities can have their operation impaired or destroyed by sediment deposition, for example by reducing the capacity of reservoirs, interfering with port and harbour operations and closing or modifying the path of watercourses. Erosion or scour may undermine structures on or in watercourses and coastlines. Thus the study of sediment transport is evidently of significant importance.

As previously mentioned in Chapter 1, most of our beaches today are composed of the remnants of sediments washed down the rivers in the last ice age, predominantly sands and gravels. Traditionally the sand and gravel sizes have been classified according to the Wentworth scale. This defines sand as being very fine (0.0625–0.037 mm), fine (0.037–0.25 mm), medium (0.25–0.5 mm), coarse (0.5–1 mm) and very coarse (1–2 mm). Material sizes larger than this are classified as gravel, subdivided into granular (2–4 mm), pebble (4–64 mm), cobble (64–256 mm) and boulder (>256 mm). Rounded gravel, typical of a significant number of UK beaches, is referred to as shingle.

There are several physical properties of sand and gravel beaches which are important in the study of coastal sediment transport. The first is the sediment density (ρ_s), typically 2650 kg/m^3 for quartz. The rest are required in recognition of the fact that a beach comprises a mixture of the beach material, interspersed with voids which may be filled with air or water. Thus the bulk density (ρ_b) is defined as the *in situ* mass of the mixture/volume of the mixture, the porosity (p_s) as the volume of air or water/volume of the mixture, typically about 0.4 for a sand beach, the voids ratio (e) as the volume of air or water/volume of the grains and finally the angle of repose (Φ), which is the limiting slope angle at which the grains begin to roll, typically 32° in air. In water this reduces to about 28°.

The material sizes on any particular beach will normally comprise a range of grain sizes, thus it is standard practice to measure the grain size distribution by a sieve analysis from which the percentage by weight of material passing through a range of sieve sizes is plotted against particle size. The median size is denoted by D_{50}, representing the diameter for which 50 per cent of the grains *by mass* are finer. The spread of sizes is often indicated by the values of D_{84} and D_{16} and their ratio is used to measure the degree of sorting. A well-sorted sample is one in which there is a small range of sizes ($D_{85}/D_{16}<2$), whereas a well mixed sample has a large range of sizes ($D_{85}/D_{16}>16$).

Beaches are further categorised according to the mixture of sands and gravels present, which has a significant influence on the beach slope at the shoreline. Four categories are shown in Figure 5.1, ranging from a pure sand beach through sand/shingle mixtures to pure shingle. The corresponding grain size distributions are also shown.

5.2   Sediment transport

In this section, the aim is to introduce the necessary concepts for a proper understanding of sediment transport processes, leaving consideration of the effects described above to succeeding chapters. Clearly, sediment transport occurs only if there is an interface between a moving fluid and an erodible boundary. The activity at this interface is extremely complex. Once sediment is being transported, the flow is no longer a simple fluid flow, since two materials are involved. Thus, the study of sediment transport involves many considerations and difficulties. The approach adopted in this chapter is to try to provide an understanding of the physics involved together with some useful knowledge of relevant equations used in practice to solve sediment transport problems. The treatment of the topic is directed towards sediment transport in the marine environment, rather than the riverine environment, although much of the underlying principles were originally derived for the latter. Only cohensionless particles are considered, comprising sands (grain size range 0.06–2 mm) and gravels, or shingle (grain sizes 2–256 mm). The movement of sediment can be effected by either waves or currents or a combination of the two. There are some important differences in the way that waves and currents move sediments and, accordingly, separate equations must be developed for these two cases and their combination. It is not always possible to provide rigorous proofs of equations. Even where this is possible, it may not be helpful to someone studying the subject for the first time, since some proofs are long and difficult. In general, proofs will be given only if they are reasonably simple. Where the development of an equation involves a complicated mathematical/empirical development, the equation will simply be stated with a brief outline of principles, the appropriate reference(s) and examples of its application. Only a limited selection of sediment transport equations can be given here. No special merit is claimed for this selection, but it is hoped that at least a path will have been cleared through the 'jungle' which will enable the reader to explore some of the more advanced texts.

5.2.1   Modes of transport

Sediment transport may be conceived of as occurring in one of two principal modes:

1   by rolling, sliding or hopping along the floor (bed) of the river or sea – sediment thus transported constitutes the bedload, with the hopping motion referred to as saltation;
2   by suspension in the moving fluid which is the suspended load.

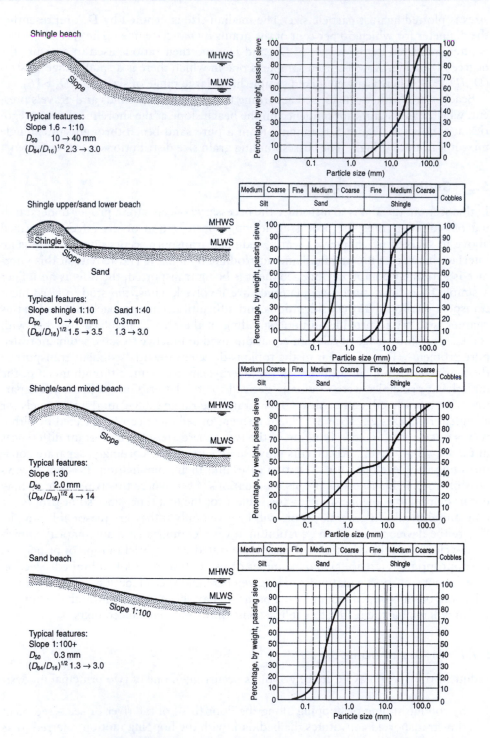

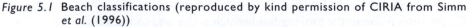

Figure 5.1 Beach classifications (reproduced by kind permission of CIRIA from Simm *et al.* (1996))

In addition to these two principle modes two further modes may be present:

3   The washload. This comprises very fine particles that are carried in suspension, but whose origin is not from the bed. Such particles typically enter the system from river tributaries. Their concentration cannot be predicted from the composition of the bed material.

4   Sheetflow. This comprises an extension to the bedload. At higher transport rates, more than one grain layer of particles is activated and thus the bedload comprises several layers of moving particles, all in contact with one another.

Bedload transport is the dominant mode for low velocity flows and/or large grain sizes. It is controlled by the bed shear stresses, as explained in the next section. Conversely, suspended load transport is the dominant mode for high velocity flows and/or small grain sizes. It is controlled by the level of fluid turbulence, as explained later. In the marine situation, gravel size fractions are typically transported as bedload, whereas sand-sized fractions are transported by both bedload and suspended load, with suspended transport occurring up to several metres above the bed. In this situation the suspended transport is often much larger than the bedload transport.

5.2.2   Description of the threshold of movement

If a perfectly round object (a cylinder or sphere) is placed on a smooth horizontal surface, it will readily roll on application of a small horizontal force. In the case of an erodible boundary, of course, the particles are not perfectly round, and they lie on a surface which is inherently rough and may not be flat or horizontal. Thus, the application of a force will only cause motion when it is sufficient to overcome the natural resistance to motion of the particle. The particles will probably be non-uniform in size. At the interface, a moving fluid will apply a shear force, τ_0 (Figure 5.2a), which implies that a proportionate force will be applied to the exposed surface of a particle. Observations by many experimenters have confirmed that if the shear force is gradually

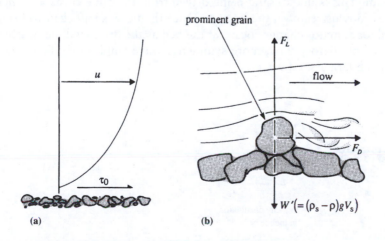

Figure 5.2 Fluid forces causing sediment movement.

increased from zero, a point is reached at which particle movements can be observed at a number of small areas over the bed. A further small increase in τ_0 (and therefore u) is usually sufficient to generate a widespread sediment motion (of the bedload type). This describes the 'threshold of motion' and the associated critical shear stress (τ_{CR}). After further increments in τ_0, another point is reached at which the finer particles begin to be swept up into the fluid. This defines the inception of a suspended load.

For most practical cases, current flows are turbulent. In the case of waves, a small turbulent boundary layer also exists. This means that the flow incorporates an irregular eddying motion, caused by the fluid turbulence. A close look in the region of the granular boundary at the bed would reveal the existence of a sub-layer comprising 'pools' of stationary or slowly moving fluid in the interstices. This sub-layer zone is not stable, since eddies (with high momentum) from the turbulent zone periodically penetrate the sub-layer and eject the (low momentum) fluid from the pools. The momentum difference between the fluid from the two zones generates a shearing action, which in turn generates more eddies, and so on. Grains are thus subjected by the fluid to a fluctuating impulsive force. Once the force is sufficient to dislodge the more prominent grains (Figure 5.2b) they will roll over the neighbouring grain(s), causing bedload transport. As sediment movement becomes more widespread, the pattern of forces becomes more complex as moving particles collide with each other and with stationary particles. As the bed shear stress increases further, granular movement penetrates more deeply into the bed. Bed movement may most simply be represented as a series of layers in relative sliding motion (Figure 5.3), with a linear velocity distribution (i.e. sheet flow).

5.2.3 Bedforms

Once the shear stress is sufficient to cause transport, the bed will begin to alter its form producing a variety of bedforms depending on the nature of the flow. For the case of a uniform current, 'ripples' will initially form in the bed. These ripples may grow into larger 'dunes'. In flows having quite moderate Froude Numbers, the dunes will migrate downstream. This is due to sand being driven from the dune crests and then being deposited just downstream on the lee side. Once the flow is sufficient to bring about a suspended load, major changes occur at the bed as the dunes will be 'washed out'. For the case of oscillatory flow, more symmetric wave ripples may be formed (Figure 5.4) and much larger sand waves.

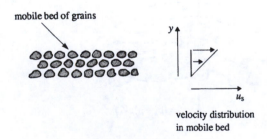

Figure 5.3 Idealised sheet flow layer.

Figure 5.4 Example wave ripples on a beach

Bedforms cause frictional resistance to the flow in addition to that caused by grain roughness and thus play a crucial role in the estimation of total bed shear stress. They also cause additional turbulence and thus also induce additional suspended sediment transport. A multitude of attempts have been made to develop relationships between the major parameters of the transport process (Froude Number, sediment properties, fluid properties, shear stress, bed roughness or dune size, and rate of sediment transport). Most of the equations in current use have been developed on the basis of a combination of dimensional analysis, experimentation and simplified theoretical models.

5.2.4   Estimation of bed shear stress

The total bed shear stress is composed of three contributions, namely:

- the skin friction or grain related friction (τ_{0s});
- the form drag (τ_{0f}) resulting from ripple/dune formation; and
- a sediment transport contribution (τ_{0t}) caused by momentum transfer to mobilise the grains.

Hence the total bed shear stress is given by:

$$\tau_0 = \tau_{0s} + \tau_{0f} + \tau_{0t} \tag{5.1}$$

Only the skin friction bed shear stress acts directly on the grains and thus this parameter must be used when calculating the threshold of motion, bedload transport and reference concentration, as will be detailed later in this section. However, the total shear stress is the parameter which determines the turbulent intensities which in turn governs suspended sediment transport. Furthermore, the determination of bed shear stresses depends on whether the flow is that for a steady current, a wave or a combination of waves and currents.

The general equation, relating bed shear stress to depth mean velocity ($\bar{U}$) is given by:

$$\tau_0 = \rho C_D \bar{U}^2 \tag{5.2}$$

This general equation can be used for all current flows and for total bed shear stress or skin friction shear stress. A very useful additional parameter, known as the friction or shear velocity (u_*), is related to τ_0 by:

$$u_* = \sqrt{\tau_0 / \rho} \tag{5.3}$$

Its application is described in succeeding sections.

Current skin friction bed shear stress

For river flow in the absence of bedforms, the skin friction bed shear stress can be simply related to the bed slope ($\tau_0 = \rho g h S_0$). By substitution into the Manning equation ($V = (1/n)h^{2/3}S_0^{1/2}$), the value for C_D is given by:

$$C_D = \frac{gn^2}{h^{1/3}} \tag{5.4}$$

However, in the presence of bedforms and for the case of tidal flows another approach is necessary. Skin friction bed shear stress is determined solely by the bed roughness, as quantified by either the Nikuradse roughness (k_s) or the roughness length (z_0), which is the height above the bed at which the velocity tends to zero. A widely used equation is that given by:

$$C_D = \left[\frac{0.4}{1 + \ln(z_0 / h)} \right]^2 \tag{5.5}$$

For hydraulically rough flow ($u_* k_s / v > 70$), commonly assumed for coarse sands and gravels,

$$z_0 = k_s / 30 \tag{5.6}$$

k_s is related to grain size and is usually given as:

$$k_s = 2.5 D_{50} \tag{5.7}$$

Current-generated ripples and dunes

Current-generated ripples will form on sandy beds for grain sizes up to about 0.8 mm. Their wavelength (λ_r) and wave height (Δ_r) can be estimated from $\lambda_r = 1000 D_{50}$ and $\Delta_r = \lambda_r/7$. Typical average measured values being $\lambda_r = 0.14$ m and $\Delta_r = 0.016$ m. Dunes and sandwaves are much larger, having dimensions (λ_s, Δ_s) which are dependent on both bed shear stress due to skin friction (τ_{0s}) and water depth (h). Wavelengths are typically tens of metres and wave heights a few metres. One set of equations, derived by Van Rijn (1984) are given by:

$$\lambda_s = 7.3h \tag{5.8a}$$

$$\Delta_s = 0 \quad \text{for } \tau_{0s} < \tau_{CR} \tag{5.8b}$$

$$\Delta_s = 0.11h \left(\frac{D_{50}}{h} \right)^{0.3} (1 - e^{-0.5T_s})(25 - T_s) \qquad \text{for } \tau_{CR} < \tau_{0s} < 26\tau_{CR} \tag{5.8c}$$

$$\Delta_s = 0 \quad \text{for } \tau_{0s} > 26\tau_{CR} \tag{5.8d}$$

where $\quad T_s = \left(\dfrac{\tau_{0s} - \tau_{CR}}{\tau_{CR}} \right)$

For unidirectional flow (e.g. rivers) these equations give reasonably accurate results. However, in tidal flow conditions where the current speed and direction are continuously altering, the bedforms may not be able to respond quickly enough to the changing conditions to establish the equilibrium forms given by the above equations. Where possible, therefore, measurements should be taken for tidal flow conditions.

Current total bed shear stress

Where bedforms are present, the ratio of total to skin friction shear stress is typically in the range 2–10. It is, therefore, very important to be able to calculate the bedform drag. This may be achieved by finding an appropriate value of z_0, including the bedform roughness, as well as the skin friction roughness. If the bedform wavelength and wave height are known or calculated (cf. Equations 5.8a to 5.8d) then an equation relating these to the bedform roughness height (z_{of}) is given by:

$$z_{of} = a_r \frac{\Delta_r^2}{\lambda_r} \tag{5.9}$$

where a_r is in the range 0.3–3 with a typical value of 1.

For sheet flow conditions, a further increase in z_0 arises due to turbulent momentum exchange between the particles (as noted above). Wilson (1989) gives an equation for this, given by:

$$z_{0t} = \frac{5\tau_{0s}}{30g(\rho_s - \rho)} \tag{5.10}$$

Hence the total roughness length (z_0) may be calculated as:

$$z_0 = z_{0s} + z_{0f} + z_{0t}$$

The corresponding total drag coefficient may be found using z_0 in Equation (5.5). This value of C_D may then be used in Equation (5.2) to estimate the total bed shear stress.

Wave skin friction shear stress

Under wave action the velocity at the bed varies rapidly in both magnitude and direction. Thus a very small oscillatory boundary layer develops (a few mm to a few cm thick). In consequence, the shear stress at the bed, τ_{ws}, is much larger than that developed under steady flow conditions with the equivalent free stream velocity. Equation (5.2) is adapted to read:

$$\tau_{ws} = \frac{1}{2}\rho f_w u_b^2 \tag{5.11}$$

where u_b is the bottom orbital velocity and f_w is the wave friction factor.

For rough turbulent flow an equation developed by Soulsby (1997) gives:

$$f_{wr} = 1.39\left(\frac{A}{z_0}\right)^{-0.52} \tag{5.12}$$

where $A = u_b T/(2\pi)$ = semi-orbital excursion

Wave-generated ripples

Waves generate ripples with wavelengths (λ_r) typically 1–2 times the wave orbital amplitude ($A = u_b T/(2\pi)$) and wave height (Δ_r) typically 0.1–0.2 times wavelength. These ripples are washed out in sheet flow conditions. One set of equations for regular waves, derived by Neilsen (1992) are given by:

$$\lambda_r = \Delta_r = 0 \qquad \text{for } \theta_{ws} < \theta_{CR} \tag{5.13a}$$

$$\Delta_r = (0.275 - 0.022\,\psi^{0.5})A \qquad \text{for } \psi < 156 \tag{5.13b}$$

$$\lambda_r = \Delta_r/(0.182 - 0.24\theta_{ws}^{1.5}) \qquad \text{for } \theta_{ws} < 0.831 \tag{5.13c}$$

$$\lambda_r = \Delta_r = 0 \qquad \text{for } \theta_{ws} > 0.831 \text{ or } \psi > 156 \tag{5.13d}$$

where $\quad \theta_{ws} = \dfrac{\tau_{ws}}{g(\rho_s - \rho)D} \qquad$ (see Section 5.2.5 for further details)

$$\psi = \frac{u_b^2}{g\left(\frac{\rho_s}{\rho} - 1\right)D}$$

Wave total shear stress

To determine the total shear stress at the bed under wave action, in the presence of bedforms, the same methodology as that described for currents may be applied. The only difference being that Equations (5.13a) to (5.13d) should be used to determine the ripple wavelengths and wave heights.

Bed shear stress under waves and currents

Where waves and currents co-exist, a non-linear interaction takes place between the wave and current boundary layers. The resultant bed shear stress cannot be simply found by the vector addition of the two bed shear stresses. Based on a comprehensive analysis of data and previous theoretical models, Soulsby (1995) derived algebraic expressions for the mean (τ_m) and maximum (τ_{max}) bed shear stresses as follows:

$$\tau_m = \tau_c\left[1 + 1.2\left(\frac{\tau_w}{\tau_c + \tau_w}\right)^{3.2}\right]$$

$$\tau_{max} = \left[\left(\tau_m + \tau_w\cos\phi\right)^2 + \left(\tau_w\sin\phi\right)^2\right]^{1/2}$$

where τ_c is the current bed shear stress, τ_w is the wave bed shear stress and ϕ is the angle between the wave and the current.

Soulsby (1997) gives a useful summary of the background and development of this work.

5.2.5   The entrainment function (Shields parameter)

A close inspection of an erodible granular boundary would reveal that some of the surface particles were more 'prominent' or 'exposed' (and therefore more prone to move) than others (Figure 5.2(b)). The external forces on this particle are due to the separated flow pattern (the lift and drag forces). Its resistance to motion is equal to W'$tan\phi$ (where W' is the submerged self-weight = $\pi D^3 g(\rho_s - \rho)/6$ for a spherical particle and ϕ is the angle of repose or internal friction). The number of prominent grains in a given surface area is related to the areal grain packing (= area of grains/total area A_p). As the area of a particle is proportional to the square of the typical particle size (D^2), the number of exposed grains is a function of A_p/D^2. The shear stress at the interface, τ_0, is equal to the sum of the horizontal forces acting on the individual particles, with the contribution due to prominent grains dominating, so the total force on each prominent grain in unit area may be expressed as

$$F_D \propto \tau_0 D^2 / A_p$$

At the threshold of movement $\tau_0 = \tau_{CR}$, so

$$\tau_{CR} \frac{D^2}{A_p} \propto (\rho_s - \rho)g \frac{\pi D^3}{6} \tan\phi$$

This can be rearranged to give a dimensionless relationship

$$\frac{\tau_{CR}}{(\rho_s - \rho)gD} \propto \frac{\pi A_p}{6} \tan\phi$$

The left-hand side of this equation is the ratio of a shear force to a gravity force, and is known as the entrainment function or Shields parameter (θ), that is

$$\theta = \frac{\tau}{(\rho_s - \rho)gD} \qquad (5.14a)$$

At the threshold of movement this becomes the critical Shields parameter:

$$\theta_{CR} = \frac{\tau_{CR}}{(\rho_s - \rho)gD} \qquad (5.14b)$$

The above analysis suggests that the critical entrainment function should be a constant.

In a classic investigation, Shields (1936), showed that the critical entrainment function was related to a form of Reynolds' Number, based on the friction velocity, that is $Re_* = \rho u_* D / \mu$. Shields plotted the results of his experiments in the form of θ_{CR} against Re_*, and proved that there was a well-defined band of results indicating the threshold of motion. The Shields threshold line has subsequently been expressed in a more convenient explicit form (Soulsby and Whitehouse (1997)), based on the use of a dimensionless particle size parameter, D_*, given by:

$$\theta_{CR} = \frac{0.3}{1 + 1.2D_*} + 0.055[1 - \exp(-0.02D_*)] \qquad (5.15)$$

where

$$D_* = \left[\frac{g(s-1)}{v^2} \right]^{1/3} D, \ s = \rho_s / \rho \text{ and } v = \text{kinematic viscosity of water} = \mu / \rho \qquad (5.16)$$

Equation (5.15) can, therefore, be used to determine the critical shear stress (τ_{CR}) for any particle size (D). Figure 5.5 is a plot of θ_{CR} against D_* showing data sets for waves, currents and combined waves and currents, together with both the original Shields curve and the Soulsby curve defined in Equation (5.15).

On a flat bed, if the bed skin friction shear stress (τ_{0s}) is known, then the value of

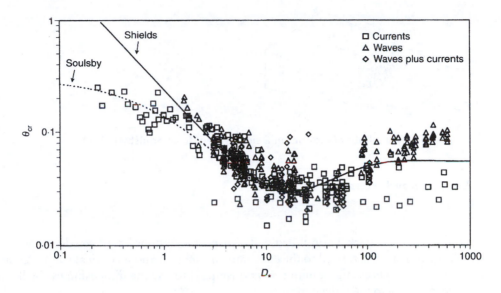

Figure 5.5 Threshold of motion of sediments beneath waves and/or currents from Soulsby (1997).

the Shields parameter ($\theta_s = \tau_{0s}/(g(\rho_s-\rho)D)$ can be calculated and used to determine the regime of sediment transport as follows:

- If $\theta_s < \theta_{CR}$ then no transport will occur.
- If $\theta_{CR} \leq \theta_s \leq 0.8$ then transport will occur with ripples or dunes.
- If $\theta_s > 0.8$ then transport will occur as sheet flow with a flat bed.
- If $u_{*_s} \leq w_S$ (the particle fall speed) then there will not be any suspended sediment transport.
- If $u_{*_s} > w_S$ then suspended sediment transport will occur.

The latter two conditions may be better understood by noting that the friction velocity can be related to the intensity of turbulence through the concept of the Reynolds' shear stress. This is the force resulting from the change of momentum associated with the fluctuating turbulent velocities (u', v'). Hence, it may be shown that $u_* = \sqrt{\overline{u'v'}}$

For the case of homogeneous turbulence u', v' have the same magnitude, hence $u_* = u' = v'$. It will be shown in the next section that for suspended sediment transport $u' = v' = w_S$ (and therefore $u_* = w_S$).

On a sloping bed, the critical bed shear stress ($\tau_{\beta CR}$) may be more or less than the critical bed shear stress on a flat bed τ_{CR}. If the bed is inclined at an angle β, and the flow is at an angle ψ to the upslope direction (refer to Figure 5.6), the two shear stresses are related by the following equation:

$$\frac{\tau_{\beta CR}}{\tau_{CR}} = \frac{\cos\psi \sin\beta + (\cos^2\beta \tan^2\phi - \sin^2\psi \sin^2\beta)^{1/2}}{\tan\phi} \tag{5.17}$$

where ϕ is the angle of repose of the sediment.

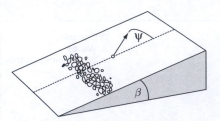

Figure 5.6 Threshold of motion on a sloping bed, from Soulsby (1977).

5.2.6   *Bedload transport equations*

Currents

Following Shields' work, various bedload transport equations have been developed, in which the transport is related to the entrainment function and its critical value. A convenient way to express the resulting relationships is to use the dimensionless bedload transport rate factor (Φ) given by:

$$\Phi = \frac{q_b}{\left[g(s-1)D^3\right]^{1/2}} \tag{5.18}$$

where q_b = volumetric bedload transport rate per unit width, with units of m³/m/s.
   An early formula, still commonly used, is that of Meyer–Peter Muller given by:

$$\Phi = 8\left(\theta_s - \theta_{CR}\right)^{3/2} \tag{5.19}$$

A more recent formula is that of Neilsen (1992), given by:

$$\Phi = 12\theta_s^{1/2}\left(\theta_s - \theta_{CR}\right) \tag{5.20}$$

This equation gives a good fit to a wide range of conditions. Soulsby (1997) presents these and other well-known formulae and provides references and further reading.

Waves

Here, the net bedload transport is zero, if the waves are symmetrical. Outside the surf zone, this is predominantly the case. However, for steep waves in shallow water the wave motion becomes asymmetrical, with high, short duration velocities under the crests and longer, lower velocities under the troughs. Under these conditions, net bedload transport will occur. Equation (5.20) may still be used to determine the net transport by integration over a wave cycle. Soulsby (1997) provides some resulting equations.

Waves and currents

In this case, the waves provide a stirring mechanism and the currents add to this *and* transport the sediment. Again, Equation (5.20) may be employed to integrate over the wave cycle. However, due to the non-linear interaction between the waves and currents, the instantaneous value of shear stress needs to be determined by the method given in Section 5.2.4 for combined waves and currents. Soulsby (1997) presents a set of equations resulting from such a calculation.

Example 5.1   Bedload sediment transport by a tidal current

Calculate the bedload sediment transport rate in a tidal current given the following data:

Depth mean current $\bar{u}$ = 2.0 m/s, grain size D_{50} = 0.4 mm, water depth h = 10 m, sea water density ρ = 1027 kg/m³ (@ 10°C and salt content 35 ppt), sediment density ρ_s = 2650 kg/m³, kinematic viscosity v = 1.36 × 10⁻⁶ m²/s

Solution

First calculate the roughness height and skin friction drag coefficient.
   From Equations (5.6) and (5.7)

$$z_0 = k_s/30 = 2.5D/30 = 2.5 \times 0.0004/30 = 3.33 \times 10^{-5} \text{ m}$$

Substituting into Equation (5.5),

$$C_D = \left[\frac{0.4}{1+\ln(z_0/h)}\right]^2 = \left[\frac{0.4}{1+\ln(3.33x10^{-5}/10)}\right]^2 = 1.187x10^{-3}$$

Now calculate skin friction shear stress from Equation (5.2) and shear velocity from Equation (5.3)

$$\tau_{0s} = \rho C_D \bar{U}^2 = 1027 \times 1.187 \times 10^{-3 \times} 2^2 = 4.875 \text{ N/m}^2$$

$$u_{*s} = \sqrt{\tau_{0s}/\rho} = \sqrt{4.875/1027} = 0.069 \text{ m/s}$$

It should be noted, at this point, that the equation for z_0 is only strictly applicable for hydraulically rough flow ($u_* k_s/v > 70$). In this case $u_* k_s/v$ = 50.7, thus Equation (5.6) is not strictly applicable. However, the resulting error in C_D is only about 1 per cent and hence Equation (5.6) is sufficiently accurate for practical purposes (see Soulsby (1997) for further details).

   The Shields parameter and critical Shields parameter can now be found from Equations (5.14a), (5.15) and (5.16).

$$\theta_s = \frac{\tau_{0s}}{g(\rho_s-\rho)D} = \frac{4.875}{9.81(2650-1027)0.0004} = 0.765$$

$$D_* = \left[\frac{g(s-1)}{v^2}\right]^{1/3} D = \left[\frac{9.81(\frac{2650}{1027}-1)}{(1.36x10^{-6})^2}\right]^{1/3} x0.0004 = 8.125$$

$$\theta_{CR} = \frac{0.3}{1+1.2D_*} + 0.055[1-\exp(-0.02D_*)] = \frac{0.3}{1+1.2x8.125} +$$
$$0.055[1-\exp(-0.02x8.125)] = 0.036$$

Now apply Equation (5.20)

$$\Phi = 12\theta_s^{1/2}(\theta_s - \theta_{CR}) = 12\times0.765^{1/2}(0.765-0.036) = 7.65$$

Substituting into Equation (5.18) gives

$$\Phi = \frac{q_b}{\left[g(s-1)D^3\right]^{1/2}}$$

Hence: $q_b = 7.65[9.81((2650/1027)-1) \times 0.0004^3]^{1/2} = 2.41 \times 10^{-4}$ m³/m/s

5.2.7 General description of the mechanics of suspended sediment transport

Mechanics of particle suspension

If sediment grains are drawn upward from the bed and into suspension, it must follow that some vertical (upward) force is being applied to the grains. The force must be sufficient to overcome the immersed self-weight of the particles. Consider a particle suspended in a vertical flask (Figure 5.7).

If the fluid is stationary, then the particle will fall due to its self-weight (assuming $\rho_s > \rho$), accelerating up to a limiting (or 'terminal') velocity w_s at which the self-weight will be equal in magnitude to the drag force, F_D, acting on the particle. If a discharge is now admitted at the base of the flask, the fluid is given a vertical upward

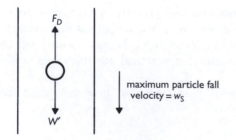

Figure 5.7 Forces acting on a falling particle.

velocity v. As $v \to w_S$, the particle will cease to fall and will appear to be stationary. If $v > w_S$ then the particle can be made to travel upward. From this argument, it must follow that the suspension of sediment in a current or wave flow implies the existence of an upward velocity component. As previously alluded to, this is provided by the fluctuating vertical (and horizontal) components of velocity, which are an integral part of a turbulent flow. Flow separation over the top of a particle provides an initial lift force (Figure 5.1(b)) which tends to draw it upwards. Providing that eddy activity is sufficiently intense, then the mixing action in the flow above the bed will sweep particles along and up into the body of the flow (Figure 5.8). Naturally, the finer particles will be most readily suspended (like dust on a windy day).

Clearly, we need to develop equations for the fall speed and a model for turbulence in order to predict suspended sediment transport. The first requirement may be approached by considering drag forces on a falling particle.

The general equation for a drag force is:

$$F_D = C_D \left(\frac{1}{2} \rho A U_\infty^2 \right)$$

where C_D is the drag coefficient and A is the cross-sectional area at right angles to the flow velocity U_∞

Hence: $(\rho_s - \rho) g V_s = C_D \left(\frac{1}{2} \rho A_s w_s^2 \right)$

where V_s and A_s are the volume and cross-sectional area of the particle. For spherical particles, the value of C_D is a function of Reynolds' number.

However, most sediments are not spherical and behave differently to spherical particles. For natural sands, Soulsby (1997) derived a simple, but accurate, formula which is of universal application. It is given by:

$$w_S = \frac{v}{D} \left[\left(10.36^2 + 1.049 D_*^3 \right)^{1/2} - 10.36 \right] \text{ for all } D_* \tag{5.21}$$

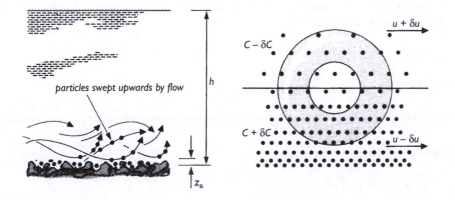

Figure 5.8 Mechanics of particle suspension.

At high sediment concentrations this fall velocity is reduced due to the interaction of the particles. According to Soulsby (1997), this is only of significance for concentrations greater than about 0.05, which usually only occur within a few millimetres of the bed.

Secondly, the Prandtl model of turbulence can be used as the basis for a suspended sediment concentration model. In this model a turbulent eddy can be conceived of as a rotating ring, superimposed on the mean flow. The tangential velocity of the ring is that of the fluctuating turbulent velocity (u'). The sediment concentration, C (= volume of sediment/(volume of sediment + fluid)) is assumed to vary as shown in Figure 5.8(b). This is further assumed to be an equilibrium condition in which the particles are held in suspension by the upward turbulent velocity balancing the fall velocity of the particles.

Under these conditions and by reference to Figure 5.8 the upward rate of transport is

$$(u' - w_s)\left((C - \delta z)\left(\frac{dC}{dz}\right)\right) \quad (as\ \delta C = -\delta z \frac{dC}{dz})$$

and the downward transport rate is

$$(u' + w_s)\left((C + \delta z)\left(\frac{dC}{dz}\right)\right)$$

For equilibrium the upward and downward transport rates must be equal. Hence, the net transport rate is

$$u' \delta z \frac{dC}{dz} + w_s C = 0 \tag{5.22}$$

To progress further, we must return to the Prandtl model of turbulence. At any depth of flow, a turbulent shear stress can be conceived of as that resulting from the exchange of momentum, via the turbulent eddy, from one level to the next. Consider the exchange of fluid through the annular ring shown in Figure 5.8. The discharge (δQ) through a section of area δA is $u'\delta A$ upwards and $u'\delta A$ downwards ie. $\delta Q = 2 u'\delta A$. The mass flow rate is therefore,

$$\dot{m} = \rho \delta Q = 2\rho u' \delta A$$

and the associated rate of change of momentum is

$$\delta \dot{M} = \dot{m} u' = 2\rho u'^2 \delta A.$$

Hence the turbulent shear stress is given by

$$\tau = \delta \dot{M}/(2\delta A) = \rho u'^2.$$

Also, as

$$u' = \delta u = \delta z \frac{du}{dz}$$

then

$$\frac{\tau}{\rho} = \delta z^2 \left(\frac{du}{dz}\right)^2.$$

Now, by analogy to laminar flow we can also say that for turbulent flow:

$$\tau = \varepsilon \frac{du}{dz} \tag{5.23}$$

where ε is the eddy viscosity,
  or

$$\frac{\tau}{\rho} = \frac{\varepsilon}{\rho} \frac{du}{dz} = \delta z^2 \left(\frac{du}{dz}\right)^2.$$

Hence

$$u' \delta z = \delta z^2 \frac{du}{dz} = \frac{\varepsilon}{\rho}.$$

This final equivalence may now be substituted into Equation (5.22) to yield the basic equation for suspended sediment given by:

$$\frac{\varepsilon}{\rho} \frac{dC}{dz} + w_s C = 0 \tag{5.24}$$

In this equation the first term represents the upward diffusion of the particles by turbulence and the second term the downward migration due to the fall velocity.

To solve this equation, we need to know if the eddy viscosity is a constant or whether it is a function of some other variable. Again returning to Prandtl's eddy model, he assumed that the eddy size (δz) varied as a constant x depth (i.e. $\delta z = \kappa z$). And hence, from the above development, we may write

$$\frac{\tau}{\rho} = \frac{\varepsilon}{\rho} \frac{du}{dz} = (\kappa z)^2 \left(\frac{du}{dz}\right)^2$$

or

$$\sqrt{\frac{\tau}{\rho}} = u_* = \kappa z \frac{du}{dz} \tag{5.25}$$

where κ is known as von Karman's constant (= 0.4 for water). We are now in a position to solve Equation (5.24), making use of Equation (5.23) and Equation (5.25), provided we can relate the shear stress τ (at any level z) to the shear stress at the bed, τ_0. This will depend on whether currents, waves or waves and currents are being considered.

5.2.8 Suspended sediment concentration under currents

In this case the shear stress may be assumed to vary as:

$$\frac{\tau(z)}{\tau_0} = 1 - \frac{z}{b} \tag{5.26}$$

The corresponding velocity profile is given by:

$$u(z) = \frac{u_*}{\kappa} \ln\left(\frac{z}{z_0}\right) \tag{5.27}$$

Substituting for du/dz from Equation (5.25) into Equation (5.23), then for τ from (5.26) gives

$$\frac{\varepsilon}{\rho} = \kappa z u_* \left(1 - \frac{z}{b}\right) \tag{5.28}$$

Substituting Equation (5.28) into Equation (5.24) and integrating yields

$$\frac{C(z)}{C_a} = \left[\frac{z_a(b-z)}{z(b-z_a)}\right]^{\left(\frac{w_s}{\kappa u_*}\right)} \tag{5.29}$$

where C_a is a reference concentration at height z_a. It should be noted that here both τ_0 and u_* are the total values.

Equation (5.29) is a simple mathematical model of suspended sediment transport. The model does appear to fit experimental results quite well, but this should be viewed with caution, since the value of the exponent ($w_s/\kappa u_*$) is difficult to estimate with confidence. The value for κ is often taken as 0.4, but this is for a clear fluid. There is no general agreement as to the effect of suspended sediment on the value of κ, though some experimental results are illustrated in Chang (1988) which indicate that κ is not a constant. More recently, Soulsby (1997) suggests that κ should be taken as 0.4 and that sediment induced effects on the velocity profile should be treated separately. Also the Prandtl turbulence model, used as the basis for the model, is only a rough approximation.

The exponent in Equation (5.29) is known as the Rouse number, or suspension parameter, and the resulting sediment concentration profile the Rouse profile. This can be applied to both rivers and the sea, although it is less accurate in the sea, as the eddy diffusivity in the Rouse profile reduces to zero at the surface. Observations indicate that this is incorrect. Van Rijn (1984) developed an alternative profile for application in the sea (see Soulsby (1997) for a summary). Figure 5.9 illustrates the predicted Rouse concentration profile for a range of Rouse numbers. As could be expected, for fine grains and high velocities the sediment is suspended throughout the water column.

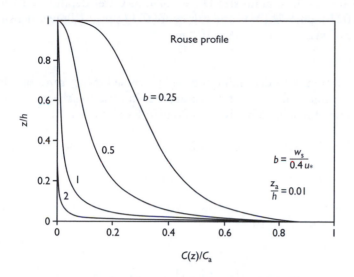

Figure 5.9 Suspended sediment concentration profiles from Soulsby (1997).

Conversely, for coarse grains and low velocities, the suspended sediment concentration rapidly reduces to zero above the bed.

To calculate the actual concentration $C(z)$, a value for C_a is also required, together with the corresponding reference height, z_a. Many expressions for C_a and z_a have been developed (see Chang (1988), Raudkivi (1990,) and van Rijn (1984)) for examples. More recently, a simple expression was derived by Zyserman and Fredsoe (1994) given by:

$$C_a = \frac{0.331(\theta_s - 0.045)^{1.75}}{1 + 0.72(\theta_s - 0.045)^{1.75}}$$ (5.30a)

$$z_a = 2D_{50}$$ (5.30b)

Finally, noting that the sediment transport rate per unit width, $q(z)$, at any height z is given by:

$$q(z) = u(z)C(z)$$

then the total suspended sediment rate (q_s) may be found by integration from:

$$q_s = \int_{z_a}^{h} u(z)C(z)dz$$ (5.31)

Example 5.2 Suspended sediment transport in a tidal current
    Calculate the total suspended sediment transport rate in a tidal current given the following data:

Depth mean current ū = 2.0 m/s, grain size D_{50} = 0.4 mm, water depth h = 10 m, sea water density ρ = 1027 kg/m³ (@ 10° C and salt content 35 ppt), sediment density ρ_s = 2650 kg/m³, kinematic viscosity ν = 1.36 × 10⁻⁶ m²/s

Solution
The first stage of the solution is to determine the regime of sediment transport, by calculating the Shields parameter and its critical value, the skin friction shear velocity and the fall velocity. Some of this has already been done in Example 5.1, i.e.

$$\theta_s = 0.765, \ \theta_{CR} = 0.036 \ D_* = 8.125 \ \text{and} \ u_{*s} = 0.069$$

Calculate the sediment fall velocity from Equation (5.21) as

$$w_S = \frac{\nu}{D}\left[\left(10.36^2 + 1.049D_*^3\right)^{1/2} - 10.36\right]$$

$$= \frac{1.36x10^{-6}}{0.0004}\left[\left(10.36^2 + 1.049x8.125^3\right)^{1/2} - 10.36\right] = 0.053m/s$$

The regime of sediment transport can now be found as follows:

As $\theta_{CR} \leq \theta_s \leq 0.8$ then transport will occur with ripples or dunes. As $u_{*s} > w_S$ then suspended sediment transport will occur. Thus we need to calculate the form drag contribution to the total bed shear stress before we can calculate the suspended sediment concentrations. For ripples use

$$\lambda_r = 1000D_{50} = 0.4 \ \text{m and} \ \Delta_r = \lambda_r/7 = 0.057 \ \text{m}$$

For dunes, first find the critical shear stress using Equation (5.14b):

$$\tau_{CR} = \theta_{CR} g(\rho_s - \rho)D = 0.23 \ \text{N/m}^2$$

Hence, as $\tau_{CR} < \tau_{0s} < 26\tau_{CR}$ then apply Equations (5.8a and 5.8c).
First find T_s

$$T_s = \left(\frac{\tau_{0s} - \tau_{CR}}{\tau_{CR}}\right) = \left(\frac{4.875 - 0.23}{0.23}\right) = 20.2$$

Hence

$$\Delta_s = 0.11h\left(\frac{D_{50}}{h}\right)^{0.3}(1 - e^{-0.5T_s})(25 - T_s) = 0.11x10\left(\frac{0.0004}{10}\right)^{0.3}$$

$$(1 - e^{-0.5x20.2})(25 - 20.2) = 0.253m$$

$\lambda_s = 7.3h = 7.3 \times 10 = 73$ m

Now find the additional roughness height due to bedforms, using Equation (5.9)

$$z_{of} = a_r \frac{\Delta_r^2}{\lambda_r}$$

Take $a_r = 1$.

For ripples $z_{0f} = 0.057^2/0.4 = 8.12 \times 10^{-3}$ m

For dunes $z_{0f} = 0.253^2/73 = 0.88 \times 10^{-3}$ m

It can be seen that the effect of the ripples on roughness height is much more signifi-cant than that of the dunes. However, add the two contributions to obtain the total $z_{0f} = 9 \times 10^{-3}$ m

Now find the total roughness height and calculate the total drag coefficient C_D:

$$z_0 = z_{0s} + z_{0f} = 3.33 \times 10^{-5} + 9 \times 10^{-3} = 9.033 \times 10^{-3} \text{ m}$$

$$C_D = \left[\frac{0.4}{1 + \ln(9.033 x 10^{-3} / 10)} \right]^2 = 4.44 x 10^{-3}$$

The total bed shear stress and total shear velocity can now be found:

$$\tau_0 = 1027 \times 4.44 \times 10^{-3 \times} 2^2 = 18.23 \text{ N/m}^2$$

$$u_{*0} = \sqrt{18.23 \big/ 1027} = 0.133 m/s$$

We can now (finally!) calculate the suspended sediment concentrations.

From Equations (5.30a, b) the reference concentration and reference height are:

$$C_a = \frac{0.331(\theta_s - 0.045)^{1.75}}{1 + 0.72(\theta_s - 0.045)^{1.75}} = \frac{0.331(0.765 - 0.045)^{1.75}}{1 + 0.72(0.765 - 0.045)^{1.75}} = 0.133$$

m³ sediment/m³ seawater

$$z_a = 2D_{50} = 2 \times 0.0004 = 0.0008 \text{ m}$$

From Equation (5.29) the concentration at any height (z) is given by:

$$\frac{C(z)}{C_a} = \left[\frac{z_a(h-z)}{z(h-z_a)} \right]^{\left(\frac{w_s}{\kappa u_*} \right)}$$

or $C(z) = 0.133 \left[\dfrac{0.0008(10-z)}{z(10-0.0008)} \right]^{0.053 \big/ 0.4 x 0.133}$

Finally, noting that the sediment transport rate per unit width, $q(z)$, at any height z is given by:

$$q(z) = u(z)C(z)$$

then the total suspended sediment rate (q_s) may be found by integration of Equation (5.31), i.e.

$$q_s = \int_{z_a}^{b} u(z)C(z)dz$$

in combination with Equation (5.24)

$$u(z) = \frac{u_*}{\kappa} \ln\left(\frac{z}{z_0}\right) = \frac{0.133}{0.4} \ln\left(\frac{z}{9.033 \times 10^{-3}}\right)$$

The integration must be performed numerically and is shown (in part) in Table 5.1. A small increment in z $(\Delta z = 0.1)$ is used and the value of $C(z)$, $u(z)$ calculated midway between each pair of depths. $q(z)$ is calculated at the midpoint and $q(z) . \Delta z$ summed through the water column. Using this method and depth increment the total suspended sediment transport rate is 0.00072 m³/s. The results are illustrated in Figure 5.10. It should be noted that the results are very sensitive to the method of numerical integration and the increment size (Δz), as $C(z)$ varies very rapidly with depth. If the calculations are carried out using a computer program, allowing choice of the increment size, then convergence of the solution may be found by sequential reduction of Δz. In this case, this does not occur until $\Delta z < 0.001$ m, for which $q_s = 0.0006$ m³/s. Also by reference to Example 5.1, it can be seen that in this case the suspended sediment transport is about 2.5 times that of the bedload transport.

Table 5.1 Numerical solution for suspended sediment transport.

z	z+Δz/2	u(z+Δz/2)	C(z+Δz/2)	q(z+Δz/2)	Σq(z+Δz/2)Δz
0.0008					
	0.0508	0.57	0.00216	0.0012	0.00012
0.1008					
	0.1508	0.94	0.00073	0.00068	0.00019
0.2008					
	0.2508	1.11	0.00044	0.00048	0.00024
0.3008					
	0.3508	1.22	0.00031	0.00038	0.00028
0.4008					
	0.4508	1.30	0.00024	0.00031	0.00031
0.5008					
.					
.		.	.	.	
	.	.	.	.	0.00072
10.0008					

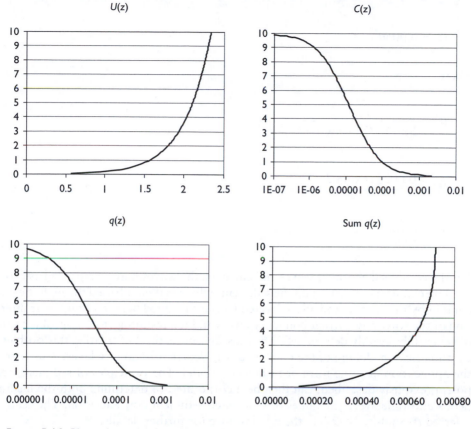

Figure 5.10 Plotted results from Table 5.1.

5.2.9 *Suspended sediment concentration under waves and waves with currents*

As previously noted, the wave boundary layer is very small. The suspended sediment is confined to the wave boundary layer. The eddy viscosity is often treated as being constant with height and the resulting concentration profile is thus given by:

$$C_z = C_0 e^{-z/l}$$

where C_0 is the reference concentration at the bed and l is the decay length scale. Neilsen (1992) has derived a set of equations for C_0 and l given by:

$$l = 0.075 \frac{u_b}{V_{FS}} \Delta_r \qquad \text{for} \quad \frac{u_b}{V_{FS}} < 18$$

$$l = 1.4\Delta_r \qquad\qquad \text{for } \frac{u_b}{V_{FS}} \geq 18$$

$$C_0 = 0.005\theta_r^3$$

where

$$\theta_r = \frac{f_{wr}u_b^2}{2(s-1)gD(1 - \pi\Delta_r / \lambda_r)^2}$$

$$f_{wr} = 0.00251\exp(5.21r^{-0.19})$$

$$r = \frac{u_b T}{5\pi D_{50}}$$

For the combined action of waves and currents, the waves may be regarded as providing a mechanism for stirring up the sediment, which is then diffused through the water column by the current and also advected by the current. There is a non-linear interaction of the wave and current boundary layers and the resulting velocity profile cannot, in general, be readily determined. The most recent method of treating this problem is by numerical modelling of the combined wave/current boundary layer. This requires the application of some form of turbulence model. The principles of sediment transport, previously described, can then be incorporated to determine both the instantaneous and time-averaged values of suspended sediment concentration. The reader is referred to Soulsby (1997) in the first instance for further details.

5.2.10   Total load transport formulae

General approach

In practice, virtually all sediment transport occurs either as bedload or as a combination of bedload and suspended load (suspended load rarely occurs in isolation, except for certain cases involving very fine silts). The combined load is known as a total load. It is possible to calculate the total load from the sum of the bedload and suspended load, as described in the preceding sections. However, this requires careful matching of the bed and suspended load transport equations at a well-defined height. Practically, it is very difficult to separate bed and suspended load. For this reason, some researchers have tackled directly the problem of total load. Some examples of total load formulae are now outlined.

Total load transport by currents

Two examples are given here. Both of these were originally derived for application to rivers, in which bed shear stress is related to the bed slope or water surface slope.

To apply them in the sea requires a reconsideration of the determination of the bed shear stresses, using the methods described in the preceding sections.

Ackers and White formula (White 1972) and in revised form in Ackers (1993)

Initially the underlying theoretical work was developed by considering the transport of coarse material (bedload) and fine material (suspended load) separately. Ackers and White then sought to establish 'transitional' relationships to account for the intermediate grain sizes. The functions which emerged are based upon three dimensionless quantities, G_{gr}, F_{gr} and D_*: G_{gr} is the sediment transport parameter, which is based on the stream power concept. For bedload, the effective stream power is related to the velocity of flow ($\bar{u}$) and to the net shear force acting on the grains. Suspended load is assumed to be a function of total stream power. The particle mobility number, F_{gr}, is a function of shear stress/immersed weight of grains. The critical value of F_{gr} (i.e. the magnitude representing inception of motion) is denoted by A_{gr}. Finally, the dimensionless particle size number, D_* (Equation 5.16), expresses the relationship between immersed weight of grains and viscous forces. The equations are then as follows:

$$G_{gr} = \frac{q_t}{uD}\left[\frac{u_*}{u}\right]^n = C\left[\frac{F_{gr}}{A_{gr}} - 1\right]^m \qquad (5.32a)$$

where

$$F_{gr} = \frac{u_*^n}{(g(s-1)D)^{1/2}}\left[\frac{\bar{u}}{\sqrt{32}\log(10h/D)}\right]^{1-n} \qquad (5.32b)$$

and q_t is the volumetric total transport rate per unit width (m³/ms)

The index n does have a physical significance, since its magnitude is related to D_*. For fine grains $n = 1$, for coarse grains $n = 0$, and for transitional sizes $n = f(\log D_*)$. The values for n, m, A_{gr} and C are as follows:

for $D_* > 60$ (coarse sediment with $D_{50} > 2$ mm):

$$n = 0, m = 1.78, A_{gr} = 0.17, C = 0.025$$

for $1 < D_* < 60$ (transitional and fine sediment, with D_{50} in the range 0.06–2 mm):

$$n = 1 - 0.56\log D_*$$

$$m = 1.67 + 6.83/D_*$$

$$A_{gr} = 0.14 + 0.23/D_*^{\frac{1}{2}}$$

$$\log C = 2.79\log D_* - 0.98(\log D_*)^2 - 3.46$$

The friction velocity ($u_* = C_D^{1/2}\bar{u}$, cf. Equations (5.4),(5.5)) should be determined by the White *et al.* (1980) alluvial friction method. However, for application to the sea, the methods outlined in Section 5.2.4 could also be used. The grain size (D) should be the D_{35} grain size, where a range of sediment sizes are present. The equations have been calibrated by reference to a wide range of data, and good results are claimed – 'good results' in this context meaning that for 50 per cent or more of the results,

$$\frac{1}{2} < \left(\frac{q_t estimated}{q_t measured}\right) < 2$$

Van Rijn (1984)

Van Rijn developed a comprehensive theory for sediment transport in rivers using fundamental physics, supplemented by empirical results. Van Rijn (1993) provides details of his full method. He also parameterised the results in a set of simpler equations, given by:

$$q_t = q_b + q_s \tag{5.33a}$$

$$q_b = 0.005\bar{u}h\left(\frac{\bar{u}-\bar{u}_{CR}}{\left[(s-1)gD_{50}\right]^{1/2}}\right)^{2.4}\left(\frac{D_{50}}{h}\right)^{1.2} \tag{5.33b}$$

$$q_s = 0.012\bar{u}h\left(\frac{\bar{u}-\bar{u}_{CR}}{\left[(s-1)gD_{50}\right]^{1/2}}\right)^{2.4}\left(\frac{D_{50}}{h}\right)(D_*)^{-0.6} \tag{5.33c}$$

where

$$\bar{u}_{CR} = 0.19(D_{50})^{0.1}\log\left(\frac{4h}{D_{90}}\right) \qquad \text{for } 0.1 \le D_{50} \le 0.5\text{mm} \tag{5.33d}$$

$$\bar{u}_{CR} = 8.5(D_{50})^{0.6}\log\left(\frac{4h}{D_{90}}\right) \qquad \text{for } 0.5 \le D_{50} \le 2.0\text{mm} \tag{5.33e}$$

with parameter ranges h = 1 to 20 m, $\bar{u}$ = 0.5 to 5.0 m/s, in fresh water @15° C.

Example 5.3   Total load transport by a tidal current
Calculate the total load sediment transport rate in a tidal current, using the Ackers and White method and the van Rijn method, given the following data:
    Depth mean current $\bar{u}$ = 2.0 m/s, grain size D = 0.4 mm, water depth h = 10 m, sea

water density $\rho = 1027$ kg/m³ (@ 10°C and salt content 35 ppt), sediment density $\rho_s = 2650$ kg/m³, kinematic viscosity $v = 1.36 \times 10^{-6}$ m²/s.

Compare these estimates with those previously calculated for bed and suspended load in Examples 5.1 and 5.2.

Solution
First use the Ackers and White method:
From Example 5.1 $D_* = 8.125$ hence calculate n, m, A_{gr}, C

$$n = 1 - 0.56 \log D_* = 0.49$$

$$m = 1.67 + 6.83/D_* = 2.51$$

$$A_{gr} = 0.14 + 0.23/D_*^{1/2} = 0.221$$

$$\log C = 2.79 \log D_* - 0.98(\log D_*)^2 - 3.46 = -1.733: C = 0.0185$$

Next calculate the particle mobility number F_{gr} from Equation (5.32b). From Example 5.2 the *total* shear velocity was found to be 0.133 m/s, hence:

$$F_{gr} = \frac{u_*^n}{(g(s-1)D)^{1/2}} \left[\frac{\bar{u}}{\sqrt{32} \log(10h/D)} \right]^{1-n} =$$

$$\frac{0.133^{0.49}}{\left(9.81 \left(\frac{2650}{1027} - 1 \right) 0.0004 \right)^{1/2}} \left[\frac{2}{\sqrt{32} \log(10 \times 10 / 0.0004)} \right]^{1-.49} = 1.177$$

Now calculate the sediment transport parameter G_{gr} from Equation (5.32a):

$$G_{gr} = C \left[\frac{F_{gr}}{A_{gr}} - 1 \right]^m = 0.0185 \left[\frac{1.177}{0.221} - 1 \right]^{2.51} = 0.735$$

Finally calculate the sediment transport rate q_t from Equation (5.32a):

$$G_{gr} = \frac{q_t}{\bar{u}D} \left[\frac{u_*}{\bar{u}} \right]^n$$

$$q_t = 0.735 \times 2 \times 0.0004 \times \left(\frac{2}{0.133} \right)^{0.49} = 2.22 \times 10^{-3} m^3 / s / m$$

Secondly use the van Rijn method. From Equation (5.33d):

$$\bar{u}_{CR} = 0.19(D_{50})^{0.1} \log\left(\frac{4h}{D_{90}}\right) = 0.19(0.0004)^{0.1} \log\left(\frac{4x10}{0.0004}\right) = 0.434 m/s$$

$$q_b = 0.005\bar{u}h\left(\frac{\bar{u} - \bar{u}_{CR}}{[(s-1)gD_{50}]^{1/2}}\right)^{2.4}\left(\frac{D_{50}}{h}\right)^{1.2} =$$

$$0.005x2x10\left(\frac{2 - 0.434}{\left[\left(\frac{2650}{1027} - 1\right)9.81x0.0004\right]^{1/2}}\right)^{2.4}\left(\frac{0.0004}{10}\right)^{1.2}$$

$$q_b = 6.9 \times 10^{-4} \text{ m}^3\text{/s/m}$$

From Equation (5.33c):

$$q_s = 0.012\bar{u}h\left(\frac{\bar{u} - \bar{u}_{CR}}{[(s-1)gD_{50}]^{1/2}}\right)^{2.4}\left(\frac{D_{50}}{h}\right)(D_*)^{-0.6} =$$

$$0.012x2x10\left(\frac{2 - 0.434}{\left[\left(\frac{2650}{1027} - 1\right)9.81x0.0004\right]^{1/2}}\right)^{2.4}\left(\frac{0.0004}{10}\right)8.125^{-0.6}$$

$$q_s = 3.57 \times 10^{-3} \text{ m}^3\text{/s/m}$$

From Equation (5.33a)

$$q_t = q_b + q_s = 4.26 \times 10^{-3} \text{ m}^3\text{/s/m}$$

Finally, from Examples 5.1 and 5.2 we have alternately

$$q_b = 2.41 \times 10^{-4}, q_s = 6 \times 10^{-4}$$

giving

$$q_t = 8.41 \times 10^{-4} \text{ m}^3\text{/s/m}$$

In summary, the three estimates of total load transport are:
Ackers and White method: $q_t = 2.22 \times 10^{-3}$ m³/s/m
van Rijn method : $q_t = 4.26 \times 10^{-3}$ m³/s/m
Bed + suspended load : $q_t = 0.84 \times 10^{-3}$ m³/s/m

Although on first sight these three estimates appear to be significantly different, all three estimates are nearly within a factor of 2 of their mean value, demonstrating their consistency within the known error bands.

Total load transport by waves and waves plus currents

Waves can cause a net sediment transport, provided that the waves are asymmetrical and/or generate currents (e.g. in the surf zone). Combinations of waves and currents will also produce net sediment transport, as previously noted. Probably the most widely used method, under these circumstances, is that of Bailard (1981). He derived these equations from the energetics approach, first put forward by Bagnold (1963, 1966). The general concept behind these methods is that the amount of energy (or work done) in transporting the sediment is some fixed proportion (ε) of the total energy dissipated by the flow. Separate efficiency factors are found for bed and suspended sediment transport. The method was originally derived for cross and longshore transport in the surf zone. It provides a point estimate of the total transport rate (q_t), which must be integrated through space to determine the total cross and longshore transport rates. The equations are given by:

$$q_t = q_{bo} - q_{bs} + q_{so} - q_{ss} \qquad (5.34a)$$

= bedload on a horizontal bed − slope effect + suspended load on a horizontal bed − slope effect

with

$$q_{bo} = \frac{c_f \varepsilon_b}{g(s-1)\tan\phi} \left\langle |u|^2 u \right\rangle \qquad (5.34b)$$

$$q_{bs} = \frac{c_f \varepsilon_b \tan\beta}{g(s-1)\tan^2\phi} \left\langle |u|^3 \right\rangle i \qquad (5.34c)$$

$$q_{so} = \frac{c_f \varepsilon_s}{g(s-1)w_s} \left\langle |u|^3 u \right\rangle \qquad (5.34d)$$

$$q_{ss} = \frac{c_f \varepsilon_s^2 \tan\beta}{g(s-1)w_s^2} \left\langle |u|^5 \right\rangle i \qquad (5.34e)$$

where

c_f = friction coefficient such that $\tau = \rho c_f |u|u$
τ = bed shear stress vector

u = total near-bed velocity due to combined waves and currents
$\tan\beta$ = bed slope
i = unit vector directed upslope
ε_B = efficiency of bedload transport (= 0.1)
ε_s = efficiency of suspended load transport (= 0.02)
$\langle\rangle$ is a time average over many waves

The height at which the velocity (u) is specified was not given in the original equations; Soulby (1997) recommends 0.05 m. Also the friction coefficient must be specified. Again Soulsby (1997) suggests 0.5 f_w or use of the methods given in Section 5.2.4 for combined waves and currents. The Bailard formulae have proved very popular with both numerical modellers and field scientists studying coastal sediment transport phenomena. However, the accuracy of this method is subject to very wide confidence limits (see Soulsby (1997) for a more detailed critique).

Several other formulae and more advanced numerical models (combining turbulence modelling of wave + current boundary layers with bed and suspended transport equations) have been recently developed. The reader is directed to Soulsby (1997) in the first instance for a discussion and summary of these techniques.

5.2.11   Cross-shore transport on beaches

Under constant wave conditions, any beach will tend to form an equilibrium beach slope on which the net sediment movement is zero. The equilibrium beach slope will increase with increasing grain size. Conversely, for a given grain size, the equilibrium beach slope will reduce with increasing wave steepness.

There are several known mechanisms of on- and offshore movement which may be explained as follows. Under swell conditions, the wave heights are small and their period long. When the waves break, material is thrown into suspension and carried up the beach (as bed and suspended load) in the direction of movement of the broken wave (the uprush). The uprush water percolates into the beach, so the volume and velocity of backwash water is reduced. Sediment is deposited by the backwash when the gravity forces predominate. The net result is an accumulation of material on the beach. In addition, the beach material is naturally sorted, with the largest particles being left highest on the beach and a gradation of smaller particles seaward. Under storm conditions, the waves are high and steep-fronted, and have shorter periods. Consequently, the volumes of uprush are much larger, and the beach is quickly saturated. Under these conditions the backwash is much more severe, causing rapid removal of beach material. Also, a hydraulic jump often forms when the backwash meets the next incoming wave. This puts more material into suspension, which is then dropped seaward of the jump. The net result is depletion of the beach.

Cross-shore transport is also affected by the wave shape and by undertow. In shallow water, waves become progressively more asymmetrical in form. Under the wave crests, the velocity is directed onshore and has a higher value than that under the troughs, which is directed offshore. However, the crest velocities persist for a shorter time than the trough velocities. Thus finer sediment migrates offshore and coarser sediment onshore. A strong undertow can also be generated in the surf zone. This is an offshore-directed flow near the bed, which results from the near surface onshore directed flow caused by the breaking waves. These flows carry suspended sediment shoreward and bedload seaward.

On sand beaches, the material moved offshore is often deposited seaward of the breaker line as a sand bar. During storm conditions, the formation of such a bar has the effect of causing waves to break at a greater distance from the beach, thus protecting the beach head from further attack. The subsequent swell waves then progressively transport the bar material back on to the beach in readiness for the next storm attack. Finally, the presence of long waves in the surf zone, briefly introduced in Section 2.6.6, can have a strong influence on surf zone sediment movements, producing complex three-dimensional features, such as beach cusps and bar systems.

All of the above description serves to highlight some of the difficulties in predicting cross-shore transport and the resulting beach evolution. Many attempts have been made to model these processes, some of which are described in Chapter 6. Some of the more recent developments have used a Boussinesq model for surf zone hydrodynamics coupled to various sediment transport equations (see Rakha *et al.* (1997), Rakha (1998) and Lawrence *et al.* (2001). For details of some of the most recent research, the reader is also directed to van Rijn *et al.* (2001) and Coast3D (2001).

5.2.12 Longshore transport ('littoral drift')

General description

Surf zone processes were introduced in Section 2.6, in which it was established that a longshore current is generated by oblique breaking waves. This current can then generate longshore transport. To some extent, the mechanisms associated with the longshore transport of sand may be differentiated from that of shingle. Thus for a sand seabed the oscillatory force due to the passage of a (breaking) wave will tend to stir the sediment into motion. The bed shear due to the longshore current can then transport the sand. Shingle beaches are much steeper than sand beaches. Thus, plunging breakers form more often. Under these conditions the surf zone is very small, with most transport taking place in the swash zone. The particles may undertake short trajectories or move as bedload. As the flow in the uprush is perpendicular to the wave crest and in the backwash is perpendicular to the beach contours, the shingle describes a sawtooth or zig-zag path along the beach.

Estimating longshore transport

Unfortunately, quantitative estimation of sediment transport rates is extremely difficult. Changes in beach volumes may be calculated from data derived from ground or aerial surveys. If surveys are carried out over several years a trend for accretion or depletion may be discernible. This is not necessarily a direct measure of the longshore transport rate along the coast. Rather it is an indication of any imbalances in the supply of sediment from one point to another. However, where marine structures are constructed which cut off the supply from further up the coast, comparisons of beach volumes before and after construction can give some indication of the longshore transport rates.

Direct measurement of longshore transport has been attempted using a variety of techniques, such as deposition of a tracer material (radioactive, dyed or artificial sediment) or installation of traps. A comprehensive review of field data for longshore sediment transport may be found in Schoonees and Theron (1993).

Longshore sediment transport equations

The equations presented here have been divided into four groups. The first category is the energy flux approach and the second is the stream power approach. The third category comprises equations derived by dimensional analysis and the fourth is that of force-balance methods.

The energy flux approach is based on the principal that the longshore immersed weight sediment transport rate, I_{ls}, is proportional to longshore wave power per unit length of beach, P_{ls}. The most widely used formula in this category is commonly known as the 'CERC equation' (US Army Corps of Engineers, 1984). The equation was derived from sand beaches and has been developed over a number of years. The formula is intended to include both bedload and suspended load and is usually given in the form of:

$$I_{ls} = KP_{ls} \qquad (5.35)$$

where P_{ls} is the longshore component of wave power per unit length of beach, given by:

$$P_{ls} = (EC_g)_b \sin\theta_b \cos\theta_b \qquad (5.36)$$

and where K is a dimensionless empirically derived coefficient. The volumetric transport rate, Q_{ls}, is related to I_{ls} by:

$$Q_{ls} = \frac{I_{ls}}{\Gamma} \qquad (5.37)$$

where

$$\Gamma = \frac{(\rho_s - \rho)g}{1+e} \qquad (5.38)$$

and e is the voids ratio.

It should be noted that for random waves, the choice of wave height used in the CERC equation (H_s or H_{rms}) must be correlated with the K value. Much confusion can arise, as some authors have used H_s and others H_{rms} without explicitly stating which one. For Rayleigh distributed waves, the K value using H_{rms} is twice that using H_s. A suggested value for K using H_{rms} is 0.77 for sand-sized sediments (US Army Corps of Engineers, 1984).

More recently, Schoonees and Theron (1993, 1994) fitted an energy flux expression (using H_s) to the 46 data points which best satisfied their selection criteria. The best-fit relationship for $D_{50} < 1$ mm was:

$$I_{ls} = 0.41P_{ls} \qquad (5.39)$$

This is equivalent to a $K = 0.82$, if H_{rms} is used.

The second category, that of stream power, was developed by Bagnold and extended later by Bailard (1981), as previously discussed in Section 5.2.10. In a later paper, Bailard (1984) integrated the local time-averaged longshore transport rate and introduced the following equation which produces a K value (to be used in conjunction with H_{rms}) which can be used in the CERC formula:

$$K = 0.05 + 2.6 \sin^2 2\theta_b + 0.007 u_b / w_S \tag{5.40}$$

Bailard concluded that this modification of the K coefficient extended the range of application of the CERC equation, which can also be applied to a range of sediment sizes (with grain size represented through its fall velocity, w_s).

Dimensional analysis methods were developed primarily from laboratory experiments and relate measured environmental parameters to volumetric transport rates. The resulting expressions bear a close resemblance to the energetic-based equations, but they were derived from mathematical relationships between groups of dimensionless variables, rather than from physical principles. Kamphuis *et al.*'s (1986) formula for longshore transport was developed for use on sand beaches. It was derived from an extensive series of laboratory tests and a broad set of field data. The formula is given by:

$$Q_K = 1.28 \frac{\tan \beta \, H_{sb}^{7/2}}{D} \sin 2\theta_b \tag{5.41}$$

where Q_K is the *immersed mass* transport (units kg/s per unit width).

The expression was refined later using a further series of hydraulic model tests (Kamphuis, 1991):

$$Q_K = 2.27 H_{sb}^2 \, T_p^{1.5} (\tan \beta)^{0.75} D_{50}^{-0.25} (\sin 2\theta_b)^{0.6} \tag{5.42}$$

For a typical sand, Kamphuis also expressed Equation (5.42) as an annual transport rate (Q_{LS}) with units of m³/annum, given by:

$$Q_{LS} = 6.4 \times 10^4 H_{sb}^2 \, T_p^{1.5} (\tan \beta)^{0.75} D_{50}^{-0.25} (\sin 2\theta_b)^{0.6} \tag{5.43}$$

Equation (5.42) was found to be valid for both laboratory and field sand transport rates. Kamphuis also investigated whether Equation (5.42) was applicable to coarse-grained beaches. He found that it over-predicted these results by a factor of 2 to 5, concluding that this was to be expected, since gravel beaches will absorb substantial wave energy by percolation and the motion of the larger grains is much closer to the critical Shields parameter. Neither of these factors was included in his dimensional analysis leading to Equation (5.42). Schoonees and Theron (1996) recalibrated Equation (5.42) using 123 data points from their field data sets to give:

$$Q_{ls} = 63433 \, x_{Kamphuis} [m^3 / annum] \tag{5.44}$$

$$\text{where} \quad x_{Kamphuis} = \frac{1}{(1-p)\rho_s} \frac{\rho}{T_p} L_0^{1.25} H_{sb}^2 (\tan \beta)^{0.75} \left(\frac{1}{D_{50}}\right)^{0.25} (\sin 2\theta_b)^{0.6} \quad (5.45)$$

It is useful to note here that Equation (5.45) as given by Schoonees and Theron (1996) is in fact incorrect. To convert from *immersed mass* transport to *volumetric* transport requires division by $(\rho_s - \rho)$ not ρ_s as given in Equation (5.45). In consequence, Schoonees and Theron's Equation (5.44) is almost identical to Kamphuis's Equation (5.43) after the latter is converted to m³/annum.

The fourth group of predictive longshore transport equations is usually known as force-balance formulae, where the sediment transport is related to the bed shear stresses associated with the longshore current. This method requires an appropriate hydrodynamic model to determine the wave-induced currents from the radiation stresses and is therefore more complex than the energetics or dimensional analysis-based methods. One of the earliest of these is that of Bijker (1971) who used a hydrodynamic model of the surf zone in combination with a shear stress based sediment transport equation, in which the local transport rates were integrated numerically across the surf zone. However, an analytical total longshore transport formula was not derived.

Damgaard and Soulsby (1996) used the force-balance method specifically for predicting total longshore bedload transport of shingle. The derivation of the formula is based on a bedload transport formula for combined waves and currents developed by Soulsby (1994). The second key element of the formula is that the shear stress vector is split up into a mean and an oscillatory part resulting from the incoming waves. Cross-shore integration of the volumetric sediment transport rate produces the total longshore transport rate, Q_{ls}. In order to perform this integration and produce an analytical expression, Damgaard and Soulsby made a number of simplifying assumptions. These included uniform beach conditions, shallow-water waves, constant breaking index, no further refraction in the surf zone and radiation stress gradient balanced by bottom shear stress. The resulting analytical expression for Q_{ls} is a combination of current-dominated transport, Q_{x1} and wave-dominated transport, Q_{x2}:

$$Q_{ls} = sign\{\theta_b\}\max\{|Q_{x1}|, |Q_{x2}|\} \quad (5.46)$$

The threshold condition is:

$$Q_{ls} = 0 \ for \ \theta_{max} \le \theta_{cr}$$

where:

$$\theta_{max} = \sqrt{(\theta_m + \theta_w \cos\phi)^2 + (\theta_w \sin\phi)^2} \quad (5.47)$$

The current- and wave-dominated parts of the transport are expressed as:

for $\sin 2\theta_b > \frac{5}{3}\theta_{cr}^*$

$$Q_{x1} = 0.21 \frac{\sqrt{g\gamma_b \tan \beta} H_b^{5/2}}{s-1} \left(\sin 2\theta_b - \frac{5}{3} \theta_{cr}^* \right) \sqrt{|\sin 2\theta_b|} \qquad (5.48a)$$

for $\sin 2\theta_b \leq \frac{5}{3} \theta_{cr}^*$

$$Q_{XI} = 0 \qquad (5.48b)$$

and for $f_{w,r} / f_{w,sf} > 1$

$$Q_{x2} = (0.25 + 0.051 \cos 2\phi) \frac{g^{3/8} D^{1/4} \gamma_b^{3/8} H_b^{19/8}}{T^{1/4} (s-1)} \sin 2\theta_b \qquad (5.49a)$$

for $f_{w,r} / f_{w,sf} \leq 1$

$$Q_{x2} = (0.050 + 0.010 \cos 2\phi) \frac{g^{2/5} \gamma_b^{3/5} H_b^{13/5}}{(\pi T)^{1/5} (s-1)^{6/5}} \sin 2\theta_b \qquad (5.49b)$$

where

$$\theta_{cr}^* = \theta_{cr} \frac{8(s-1)D}{\gamma_b H \tan \beta}$$

and

$$\phi = \frac{\pi}{2} - \theta_b$$

The friction factor for rough turbulent flows, $f_{w,r}$, is based on the analysis of a large data set and is approximated by:

$$f_{w,r} = (g\gamma_b H)^{-1/4} \sqrt{\frac{2D}{T}} \qquad (5.50)$$

For mobile beds, where sheet flow conditions may occur, the friction coefficient derived by Wilson (1989) was used:

$$f_{w,sf} = 0.0655 \left(\frac{\gamma_b H}{g} \right)^{1/5} (\pi (s-1) T)^{-2/5} \qquad (5.51)$$

Predictions from Damgaard and Soulsby's formula were compared with the transport rates calculated from beach profile data. They were found to over-predict transport by a factor of 12. Subsequent comparison against other field and the laboratory data suggested that the results of Equation (5.46), when divided by 12, produced reliable predictions of bedload transport on coarse-grained beaches. It should be noted that those results were produced using H_s in their equations.

   Chadwick (1991a, b) developed another (numerical) model, specifically for shingle beaches. The hydrodynamic module uses the non-linear shallow-water wave equations, which predict the instantaneous water levels and velocities throughout the surf and swash zones. These are combined with a sediment transport module based on Bagnold's stream power concept, as extended by McDowell (1989). Instantaneous transport rates across the surf and swash zones are subsequently summed in space and time to determine the total longshore transport rate. Thus, this model specifically includes a sediment threshold term and transport in the swash zone, both of which are of importance on shingle beaches. The model required calibration of only the friction coefficient, which was determined from field data. Subsequently, an algebraic formula (the Chadwick–Van Wellen formula) was derived from the numerical model results given by:

$$Q = 1.34 \frac{(1+e)}{(\rho_s - \rho)} H_{sb}^{2.49} T_z^{1.29} \tan \beta^{0.88} D_{50}^{-0.62} \sin 2\theta_b^{1.81} \qquad (5.52)$$

This equation was specifically designed for application to shingle beaches and has only been calibrated to the data from one field site. For further details of this and other longshore transport equations, applicable to shingle beaches, see van Wellen *et al.* (2000).

Example 5.4 Estimation of net annual longshore transport
Using the annualised wave climate data given in Table 5.2, estimate the net longshore transport rate for a natural beach site with a beach slope of 1 in 100 and a D_{50} grain size of 0.4 mm.

Table 5.2 Annualised wave climate.

H_{sb} (m)	T_p (s)	θ_b (deg.)	Frequency (%)
0.8	4.5	25	5
1.2	5.5	15	10
1.5	6	5	15
1.3	6	−5	12
1.1	5.5	−15	8
0.5	4	−25	5

Note: +ve and −ve wave angles refer to opposite sides of the beach normal.

Solution
Apply Equation (5.43) to each wave component, multiplying by the frequency of occurrence and then sum to find the net annual longshore transport rate. The results are tabulated in Table 5.3 and illustrated in Figure 5.11

   It can be seen from Table 5.3 that waves only occur for 55 per cent of the year,

Table 5.3 Longshore transport results.

H_{sb} (m)	I_p (s)	θ_b (deg.)	Frequency (%)	Q_{ls} (m³/annum)	ΣQ_{ls}
0.8	4.5	25	5	3725.5	3725.5
1.2	5.5	15	10	17536.9	21262.4
1.5	6	5	15	24829.5	46091.9
1.3	6	−5	12	−14919.8	31172.2
1.1	5.5	−15	8	−11788.7	19383.4
0.5	4	−25	5	−1219.6	18163.8
			$\Sigma = 55$	$\Sigma = 18163.8$	

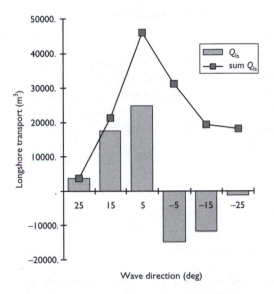

Figure 5.11 Plotted results from Table 5.3

representing the percentage of time for onshore winds and from Figure 5.11 that the net longshore transport rate is much less than the gross rates up and down coast.

5.2.13 *Concluding notes on sediment transport*

The treatment of sediment transport here is necessarily concise, and several important issues have not been discussed. For example it has been assumed that estimates of transport rates may be based on a single 'typical' particle size (say D_{50}). This may not give a realistic picture of what is actually occurring, since the bed will normally consist of a range of particle sizes. To try to meet this point it is possible to use a transport formula to estimate the transport rate for each of a series of size fractions. However, this introduces another complication. Firstly, the exposed particles of a given size will constitute only a fraction of the total bed area. Secondly, although each grain size has a theoretical threshold condition, some grains will be wholly or partly sheltered by surrounding grains and will therefore move less readily than others. To address this problem the concept of a hiding/exposure function has been introduced. The application of the size-fraction approach, coupled with a hiding/exposure function introduces other uncertainties in determining the correct parameters to use in calculating the threshold of motion. This subject is the focus of recent research (see Kleinhams and van Rijn (2002) for further details).

All of the equations for sediment transport presuppose that the number and size of the particles eroded from a given area are in equilibrium with the incoming particle deposits supplied from upstream. This is not always the case. For example, where the finer fractions are eroded and not replaced, the nature of the bed composition changes. The remaining, coarser, particles are less readily eroded, so the bed becomes more stable and the sediment load in the water is reduced. This process is known as 'armouring'. This problem has been studied in some depth for rivers (Pender and Li 1996), but not for coastal seas.

Research into the mechanisms governing sediment transport, and methods for predicting transport rates continues on an international scale. With the availability of computers, the use of numerical models is becoming more commonplace. These allow more sophisticated descriptions of the turbulent flow field to be used, though establishing the boundary conditions to the appropriate degree of accuracy can be a problem. Readers who wish to study the subject in more detail are referred in the first instance to van Rijn *et al.* (2001).

Further reading

Bagnold, R. A., 1980. An empirical correlation of bedload transport rates in flumes and natural rivers. *Proc. Roy. Soc.,* A372, pp. 453–73.

Bagnold, R. A., 1986. Transport of solids by natural water flow: evidence for a worldwide correlation. *Proc. Roy. Soc.,* A405, pp. 369–74.

Chadwick, A J., 1989. Field measurements and numerical model verification of coastal shingle transport. *Advances in Water Modelling and Measurement,* BHRA 1989, pp. 381–402.

Dyer, K. R., 1986. *Coastal and Estuarine Sediment Dynamics,* John Wiley & Sons, Chichester.

Fredsoe, J., Deigaard, R, 1992. *Mechanics of Coastal Sediment Transport.* World Scientific Publishing. Singapore, Advanced Series on Ocean Engineering, Vol. 3.

Einstein, H. A., 1942. Formulas for the transportation of bedload. *Trans. Am. Soc. Civ. Engrs,* 107, pp. 561–77.

Graf, W. H., 1971. *Hydraulics of Sediment Transport,* McGraw-Hill, New York.

Nnadi, F. N. and Wilson, K. K., 1996. Bed load motion at high shear stress: dune washout and plane-bed flow. *J. Hydraulic Engng,* 121(3), pp. 267–73.

Rouse, H. (ed.), *1950. Engineering Hydraulics,* John Wiley & Sons, New York.

Simons, D. B., Richardson, E. V., 1961. Forms of Bed Roughness in Alluvial Channels. *Proc. Am. Soc. Civil Engrs,* 87, no. HY3 p. 87.

White, W. R., Milli, H. and Crabbe, A. D., 1973. *Sediment Transport: An Appraisal of Existing Methods,* Rep. mt. 119. Hydraulics Research, Wallingford.

Yalin, M. S., 1977. *Mechanics of Sediment Transport,* 2nd edn, Pergamon, Oxford.

Yang, C. T., 1996. *Sediment Transport: Theory and Practice,* McGraw-Hill, New York.

Chapter 6

Coastal morphology
Analysis, modelling and prediction

6.1 Introduction

The previous chapter described how waves and currents transport sediment. The shape and orientation of a beach (its morphology) alters as a result of sediment transport. In this chapter, methods for analysing, modelling and predicting the change in coastal morphology are described. While the process of sediment transport may be of interest to the coastal scientist, it is the morphology of the shoreline that is of greater interest to those who live and work in the coastal zone. For example, when on holiday it is the slope, width and extent of the beach that is our primary concern (together with the aesthetic qualities such as the colour, grain size and cleanliness of the sand), rather than the quantity of sand moved by an individual wave.

Indeed, the morphology of the shoreline is a result of many individual sediment transport 'events' caused by a succession of waves. In this sense, the shape of the beach and nearshore region may be thought of as representing a form of averaging over time. The stability of a length of shoreline will depend on the difference between the volumes of sediment entering and leaving this section due to the net cross-shore and longshore sediment transport due to waves, currents and wind. The shoreline will be eroding, accreting or remaining in equilibrium. If equilibrium exists it is most likely to be a 'dynamic equilibrium'. This term is used to describe the situation in which the shoreline is evolving continuously in response to the varying winds, waves and currents. Nevertheless, the typical shoreline shape is relatively constant over a period of months or years, although the position of the shoreline at any particular time will vary about this average.

From the point of view of modelling the shoreline, in order to make predictions about its evolution, we are presented with something of a dichotomy. On one hand, if we could measure the position of the beach everywhere at a certain instant in time, and could predict the wind, waves and currents and resulting sediment transport, we could in principle predict exactly how the shoreline would evolve. On the other hand, it might be possible to make predictions useful for engineering purposes about changes in the overall shape and orientation of the shoreline by measurements to identify trends and cycles in time.

In the first case, it is not feasible to measure beach positions everywhere simultaneously, and sediment transport formulae are highly uncertain. Nevertheless, using partial information together with appropriate simplified dynamics it is possible to make predictions useful for engineering purposes in some situations. In the second case, if a strong trend or cyclic signature is present then predictions may be made on the basis of extrapolating this behaviour into the future. But an understanding of physical processes is not included, so this approach cannot predict changes in the assumed underlying variation.

When undertaking a study or scheme design for an area covering an individual beach there is rarely as much information available for making predictions of beach behaviour as one would like. There will also be constraints (of time and cost) on the scope for gathering additional data. The first step in any study must be to specify the requirements of the data analysis and modelling. This includes the type of beach change that is of interest, the location the period over which predictions are required and any specific wave and tide conditions of concern, and the form of answer that is sought. In some cases, for example, the stated requirements may be incompatible with the information available. This would be the case if seasonal variations in the beach were sought but only annual average wave conditions were available.

After specifying the problem (and the form of answer required), the next step is to review the existing information to provide a 'baseline' which may be used to guide the choice of approach. This is likely to include: the collection, checking and interpretation of past survey information; wave, current and tide measurements; previous work undertaken for the area of interest. The reason for doing this is to assist the selection of an appropriate method of analysis. For example, sophisticated numerical models often require much more information than simpler models, while a statistical method may be more appropriate if bounds on likely beach movement are sought.

The form of the answer required is another crucial influence on the choice of method. Broadly speaking, the choice usually turns on whether a general appraisal of the long-term, large-scale changes in coastal morphology is required or whether a short-term beach response to particular conditions is of greatest concern. The former is typical of strategic coastal management plans that may cover ~100 km of coastline and require predictions 50–100 years into the future, while the latter is more often associated with individual scheme design.

The key part of the process, which should have greatest influence on the method chosen for the study, is gathering and reviewing the baseline data. This is now covered in further detail.

6.1.1   Baseline review

The first step is to gather information that may be used for analysis or calibration and running a predictive model. This information is likely to take the form of:

- identification of the morphological features and their characteristics;
- measurements and qualitative descriptions of recent morphological changes in the area;
- details of past, present and proposed schemes;
- measurements of waves, winds, tides, currents and sediments that drive the morphological changes.

A practical difficulty in obtaining as much information as possible is that of cost, as not all data are held by public institutions. The accuracy of any predictions will be severely reduced if there is no historical information to carry out calibration or validation of a model. On a well-managed area a series of topographic and bathymetric surveys will be available, as well as measurements of beach levels along fixed transects (or profiles), and surface sediments. Measurements of the nearshore underwater beach profile and seabed levels are also valuable in establishing an accurate link between bathymetric (sea bed level) surveys and topographic (land level) surveys. The main features of a general beach profile are shown in Figure 6.1, which is similar to Figure 1.1 but provides some more detail.

It is also important to seek out any other information on past changes in morphology as may be gleaned from:

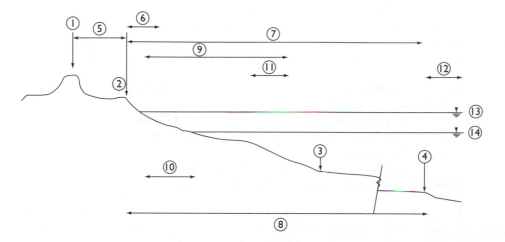

Figure 6.1 Beach profile taxonomy, showing the general features of a beach profile with geomorphological definitions: (1) Beach head – the cliff, dune or seawall forming the landward limit of the active beach; (2) Beach crest – the position of the normal limit of high tide wave run-up; (3) Storm limit – the limit of storm erosion, often identifiable on surveys by a change in the slope of the seabed; (4) Profile limit – the limit of wave-induced sediment motion, sometimes identifiable by a change in seabed slope and used to estimate the depth of closure; (5) Backshore – the section of beach between the beach head and beach crest, affected by waves occurring at high water during severe storms; (6) Swash zone – the region of wave action on the beach, which moves as water levels vary, extending from the limit of wave run-down to the limit of wave run-up; (7) Nearshore – the region that extends from the swash zone to the point marking the beginning of the offshore zone; (8) Beach face – the region between the beach crest and the profile limit; (9) Surf zone – the zone of wave action extending from the instantaneous water line to the seaward limit of the breaker zone; (10) Foreshore – the segment of the beach face between the highest and lowest tide levels; (11) Breaker zone – the zone within which waves approaching the shore commence breaking, typically in water depths of between 5 and 10 m; (12) Offshore – the region beyond the nearshore. It represents the zone where the influence of the seabed on surface waves has become small in comparison with the effect of wind; (13) High water – the normal highest water level experienced at the beach, typically well approximated by the Mean High Water Spring tide level; (14) Low water – the normal lowest water level experienced at the beach, typically well approximated by the Mean Low Water Spring tide level.

- geomorphological studies and reports;
- comparison of current and historical maps and charts;
- aerial photographs;
- repeated beach, shoreline or seabed surveys.

For both statistical analysis and long-term predictions it is particularly important to obtain as many past surveys as possible to identify long-term trends and reduce statistical errors. The value of a site visit and a 'walk-over' survey prior to starting any analysis should not be underestimated, and may be immensely important in identifying additional sources and sinks of beach material that will have to be estimated and included in any prediction. (Sediment transported and deposited by rivers, dredging and evidence of wind-blown sand are examples.)

A good representation of the processes driving beach evolution ('forcing') is required for predictions and can involve a significant study prior to any morphological calculations. Usually, wave conditions will be required and these may be provided as:

- measured or predicted time-series (values of wave parameters at a fixed location at regular intervals in time);
- seasonal or annual probabilistic distributions; often expressed as frequency of occurrence of selected wave height versus wave period or wave height versus wave direction combinations;
- long-term average 'climate' probabilistic distributions;
- specified conditions, for example, estimated extreme conditions.

Nearly all morphological models require wave heights (typically H_s), periods (T_p or T_z) and directions, but few require information on the full wave energy spectrum. Wave conditions can vary considerably and it may take a record as long as 20 years to establish an average annual longshore transport rate with reasonable accuracy. Unbroken records of measured wave conditions as long as 10 years are extremely rare and output from numerical wave hindcasting models are often used as a substitute.

In addition to wave conditions, water level variations and currents should be considered. Tides are more predictable than waves. However, the range in tidal amplitude between spring and neap tides may be substantial and this variation must be accounted for in any predictions for several tidal ranges, or modelling a full spring-neap tidal cycle to estimate net sediment transport rates.

Finally, information on past construction of sea defences, beach nourishment, dredging operations etc., that is, human intervention, should be gathered. It may be necessary for a full understanding of the historical changes to beach profile changes. Evidence of how a beach reacted to intervention in the past may hold useful clues as to how it may react to intervention in the future.

As is clear from the above, the prediction of beach morphology (even when simplified to consideration of the profile only) requires a significant amount of information about the physical processes driving shoreline change. Progress in this difficult field has been made in several generic areas:

- statistical analysis of past records;
- development of mathematical models that predict the evolution of beach morphology in response to changes in the wave and tidal 'climate';

- development of numerical models that describe sediment transport, and resulting beach change, on a 'wave-by-wave' basis.

An introduction to methods from each of these areas is given in the remainder of this chapter and is ordered as follows:

- key concepts for analysing beach profile measurements;
- introduction to the empirical orthogonal function (or EOF) technique;
- description of other, newer techniques, that have been used for analysing or characterising beaches; introduction to the equilibrium beach profile concept;
- numerical techniques for predicting beach profile evolution;
- stable beach plan shape;
- analytical and numerical methods for predicting beach plan shape evolution;
- prediction of three-dimensional nearshore morphology;
- issues arising when predicting morphological evolution over many years.

A brief summary of some key statistical concepts and terminology are provided in Appendix A for those readers not recently conversant with them.

6.2 Beach profiles

6.2.1 Analysis of beach profile measurements

A standard method of monitoring is to survey a beach along a fixed cross-section. Examining the shape and area of the cross-section or profile provides a quick check on the condition of the beach. However, storms or unusual weather conditions can produce significant changes in the profile. If a survey is performed shortly after such an event it can mask underlying, long-term trends in beach behaviour. In order to separate gradual changes from short-term fluctuations it is necessary to repeat surveys, preferably several times a year at regular intervals, over a period of many years.

Provided the surveys have been carried out consistently, beach profile surveys can provide a good source of information for studying the behaviour of beach morphology. Beach levels are normally reduced to a fixed datum and horizontal distance along the profile (termed 'chainage'), measured to a fixed point near the beach crest.

Given a set of beach profiles that have been recorded in a consistent manner and reduced to common datums a number of calculations are possible. The most obvious is to determine the mean profile shape, obtained by averaging the levels over time at each chainage. A measure of the variability along the profile is given by the variance at each chainage. It is also possible to calculate the area beneath each profile down to a fixed horizontal datum to estimate the total amount of sediment. For example, the amount of material on the beach may be relatively constant over time but substantial variations in the profile may occur. There are a number of software packages available now that allow beach survey information to be stored and analysed.

6.2.2 Empirical orthogonal function technique

The Empirical Orthogonal Function (EOF) method was developed by meteorologists to analyse variations in weather conditions at individual observation stations

(Lorenz 1956). They were particularly interested in identifying monthly, seasonal or annual cycles in behaviour that might lead to an improvement in weather forecasting. The observations were typically evenly spaced in time but irregularly spaced in geographical location. The EOF technique is powerful and robust, and is widely used in meteorology. It was introduced to the field of coastal engineering by Winant *et al.* (1975) who used EOFs to analyse a series of beach profile measurements for signs of seasonal behaviour. In coastal engineering applications the observations are often irregularly spaced in time but regularly spaced in distance, and the EOF method may be employed.

One advantage that the EOF method has over methods like Fourier analysis (see for example Bracewell 1986) is that it does not require data to be regularly sampled in both time and distance. Like Fourier analysis, EOF provide an expansion of the data, e.g. beach profile levels, in a series of functions that separates the spatial and temporal variation. The shape of these functions is determined by the correlations within the data set, in contrast to a Fourier series in which the shape of the functions is specified at the outset of the analysis. While the shape of the EOFs may suggest certain processes or timescales of change it should be remembered that as a purely statistical analysis it does not provide any means of ascribing physical processes to particular changes in morphology.

The theory behind EOFs is given in the following together with some example applications and interpretation.

Denoting the discrete beach level measurements by $g(\xi_l, t_k)$, where $1 \leq l \leq L$ and $1 \leq k \leq K$, we seek an expansion of the data in terms of a series of functions that depend on time or space individually. t_k denotes the times when surveys or observations are available at a set of fixed points ξ_l. In mathematical terms, the idea of EOF analysis is to express the data as

$$g(\xi_l, t_k) = \sum_{p=1}^{L} c_p(t_k) . e_p(\xi_l) \tag{6.1}$$

Where c_p are functions of time only and e_p are functions of space only. In fact, e_p can be determined as the eigenfunctions of the square $L \times L$ correlation matrix of the data and c_p are the coefficients describing the temporal variation of the p'th eigenfunction. In practice, many fewer than L eigenfunctions may be required to capture a large proportion of the variation in the data, and Equation (6.1) can provide an efficient means of identifying standing wave behaviour in the data. The correlation matrix, A, is calculated directly from the measurements and has elements

$$a_{mn} = \frac{1}{L.K} . \sum_{k=1}^{K} g(\xi_m, t_k) . g(\xi_n, t_k) \tag{6.2}$$

in the mth row and n^{th} column. The matrix A is real and symmetric and (from a result in linear algebra) has L real eigenvalues, λ_p, with $1 \leq p \leq L$. For each eigenvalue, λ_p, the corresponding eigenfunction, $e_p(\xi_l)$, satisfies the matrix equation

$$Ae_p = \lambda_p e_p \tag{6.3}$$

From another result in linear algebra, the eigenfunctions of a real $L \times L$ symmetric matrix are mutually orthogonal. It is common practice to normalise the eigenfunctions so they have unit length and thus

$$\sum_{l=1}^{L} e_p(\xi_l).e_q(\xi_l) = \delta_{pq} \tag{6.4}$$

where δ_{pq} is the Kronecker delta. This property of the eigenfunctions provides a means of calculating the coefficients c_p directly. The coefficients c_p may be calculated as follows. First multiply Equation (6.1) through by $e_m(t)$, for some m. Summing over l, using the orthonormality relation Equation (6.4), gives

$$c_p(t_k) = \sum_{l=1}^{L} g(\xi_l, t_k).e_p(\xi_l) \tag{6.5}$$

A useful check is to note that the sum of all the eigenvalues is equal to the mean of the sum of the squares of all the data.

Example – Illustrating the calculation of eigenvalues and eigenfunctions
Suppose beach levels are measured at three points across a profile on three occasions. This will provide three sets of measurements each containing three levels, say, $g(\xi_l, t_k)$ for $l = 1, 2, 3$ and $k = 1, 2, 3$. The correlation matrix can be computed as in Equation (6.2). For sake of argument, let us assume that the correlation matrix, A, is given by

$$A = \begin{pmatrix} 25 & 9 & 12 \\ 9 & 30 & 15 \\ 12 & 15 & 48 \end{pmatrix}$$

Calculate the eigenvalues, λ_p and corresponding eigenfunctions, e_p.
  We can find the eigenvalues by solving the characteristic equation. This is formed by setting the determinant of the matrix $(A-\lambda I)$ to zero (where I is the 3×3 identity matrix).
  In this case we have

$$\begin{vmatrix} 25-\lambda & 9 & 12 \\ 9 & 30-\lambda & 15 \\ 12 & 15 & 48-\lambda \end{vmatrix} = 0,$$

or

$$(25-\lambda)[(30-\lambda)(48-\lambda)-225]-9[9(48-\lambda)-180]+$$
$$12[135-12(30-\lambda)] = 0$$

This is a cubic equation for λ and will have three roots. These can be obtained from the formal solution of a general cubic equation or numerically using an iterative method (see e.g. Press *et al.* 1986). The solutions for λ, listed in decreasing order of magnitude, are 62.421, 22.442 and 18.137. The eigenfunction corresponding to a particular eigenvalue is determined by substituting the eigenvalue into Equation (6.3) and solving the resulting simultaneous equations for the values of the eigenfunction, e_p. The calculation for the first eigenvalue is done below. Let $e_1 = (x, y, z)$. Substituting $\lambda = 62.421$ in Equation (6.3) gives,

$$\begin{pmatrix} 25 & 9 & 12 \\ 9 & 30 & 15 \\ 12 & 15 & 48 \end{pmatrix} \begin{pmatrix} x \\ y \\ z \end{pmatrix} = 62.421. \begin{pmatrix} x \\ y \\ z \end{pmatrix}$$

Performing the matrix multiplication yields three simultaneous equations for the three unknowns x, y and z. In most practical applications the correlation matrix is much larger than 3x3 and the calculation of the eigenvalues and eigenfunctions is performed numerically using different techniques to the above for computational efficiency (see e.g. Press *et al.* 1986). These numerical routines usually provide the eigenfunctions in normalised form so that their length is one. Thus in the case above, if we find the solution $x = a$, $y = b$ and $z = c$ then the normalised eigenfunction is $(a/r, b/r, c/r)$ where $r = \sqrt{(a^2+b^2+c^2)}$. In the case above the normalised eigenfunctions corresponding to $\lambda = 62.421$, 22.442 and 18.137 are (0.37, 0.473, 0.8), (−0.43, −0.676, 0.599) and (0.823, −0.565, −0.047) respectively.

If the eigenvalues are arranged in decreasing order of magnitude, the first few terms may account for a substantial part of the total variance, so that the residual is small and can be neglected as representing contributions within observational and sampling error bounds. The EOF method is often a very efficient way of compressing a large part of the variance in the data onto a small number of functions. The technique has been used extensively in coastal engineering to investigate patterns of behaviour in beaches, particularly using beach profile measurements. Aranuvachapun and Johnson (1979) analysed beach profile measurements with EOF to investigate the differences in beach profile behaviour either side of a groyne. In Winant's analysis of beach profiles at Torrey Pines, California, the first three eigenfunctions accounted for over 90 per cent of the variance. The first spatial eigenfunction closely approximates the time mean beach profile. The corresponding first temporal eigenfunction will be almost constant over time. The second and subsequent sets of eigenfunctions represent manners of variation about the mean profile. Evidence of temporal oscillations will appear in these higher eigenfunctions. For example, Winant *et al.* (1975) found evidence to support a seasonal variation in beach profile shape. More recently, Wijnberg and Terwindt (1995) used the EOF method to analyse changes in the nearshore bathymetry of Holland over periods of several decades.

Figure 6.2 shows the eigenfunctions calculated by applying the method to a series of annual beach surveys undertaken on the Lincolnshire coastline in the UK covering the period 1977 to 1996. Beach levels were taken at fixed chainages across the beach and the series was interrupted between 1980 and 1984. The eigenfunction method can still be applied in this case. The three most important functions (corresponding to the three largest eigenvalues) are shown in Figure 6.2a. The first eigenfunction (shown by the full line) corresponds to the mean beach profile. This exhibits a small berm at the upper beach, a fairly constant slope to a chainage of 70 m beyond which the slope is

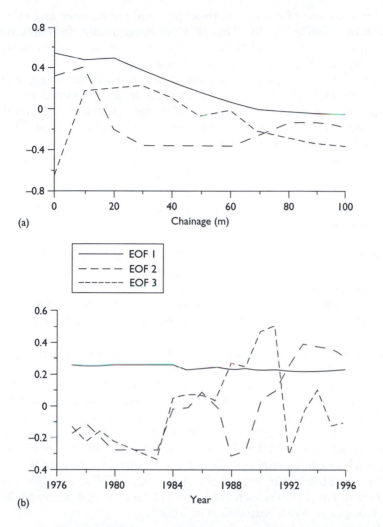

(a)

(b)

Figure 6.2 Eigenfunctions computed from annual beach profiles on the Lincolnshire coast (UK) from 1977 to 1996. (a) The first three normalised spatial eigenfunctions; (b) the corresponding non-normalised temporal eigenfunctions.

extremely small. The second and third eigenfunctions show the spatial characteristics of the variability of the beach profile about the mean. The corresponding temporal eigenfunctions are plotted in Figure 6.2b. As to be expected the first one is almost constant. The plots for the second and third temporal eigenfunctions show variations in sign and magnitude. Results such as these can be used to investigate the existence of oscillations in beach behaviour over time.

6.2.3   Other methods

Coastal morphology is driven by processes that are complex and often non-linear, and is characterised by a wide range of spatial and temporal scales. Reliable analysis requires high quality data sets of coastal morphology, and the number and extent of such datasets is now growing rapidly. This provides opportunities for extracting valuable information on morphological behaviour by means of more sophisticated techniques. A summary of such techniques is given below.

From the perspective of a statistical analysis, time series of observations (e.g. wave heights at a point, or beach levels along a profile at many intervals in time) may be considered to comprise a combination of a 'signal' and 'noise'. That is, the signal is the long-term trend or cycle, with the noise being the variability about the signal. Many statistical techniques provide different ways of writing the original time series in terms of a series expansion which, it is anticipated, can describe any signal in the first few terms and the noise in the remainder. The EOF method, described in the previous section, is a good example of this. The key to all such methods is how the signal is specified. This may be done using purely subjectively defined patterns or by optimising a particular statistical measure. For example:

- The random sine function analysis adopted by Pruszak and Rozynski (1998) involved fitting sine functions to describe beach changes in both time and space.
- EOFs are optimal in representing the variance of the observations.
- In canonical correlation analysis the expansion functions are determined by maximising the correlation between two simultaneous time series (Barnett and Preisendorfer 1987).
- Principal oscillation patterns and principal interaction patterns satisfy certain dynamical constraints (Hasselmann 1988).
- Singular spectrum analysis is a variation of the EOF method in which the data matrix contains values measured at a location lagged in time (Vautard et al. 1992, Rozynski et al. 2001).
- In complex EOF, or CEOF, the measurements are first used to construct a complex time series which is then analysed in an analogous manner to EOFs (Horel 1984). CEOF analysis can identify travelling wave features, whereas EOFs only pick up standing wave behaviour. In coastal engineering, CEOF analysis has been used to identify travelling features such as bars and channels (e.g. Bosma and Dalrymple 1996, Liang et al. 1992, Ruessink et al. 2000).

Other methods are related to what is known as 'chaos theory'. This theory originated from a numerical study of weather patterns by Lorenz (1963). Lorenz noticed that the numerical solutions to a simplified system of equations describing the atmospheric circulation settled down after a while but then, for some choices of parameter values, appeared to alternate between various states almost randomly hence the term 'chaos'. However, the equations being solved were deterministic! A key requirement for a system to exhibit chaos is for the governing equations to be nonlinear. The deterministic equations governing sediment transport are nonlinear and hence may, but not necessarily, exhibit chaotic behaviour. If beaches exhibit chaotic behaviour this clearly has an influence on how they are treated within the framework of a shoreline management plan. Several methods based on chaos theory have been adapted and applied to beach profiles by Möller and Southgate (2000) and Reeve et al. (1999). At present, the length of records has restricted this line of research to preliminary studies, but greater monitoring of coastal processes should lead to more detailed studies and findings in the near future.

6.2.4 Equilibrium profiles and the depth of closure

If the profile is considered over a longer time period of the order of years, rather than a timescale of the order of storm events or seasons, then it has been found that many ocean-facing coastlines exhibit a concave curve which becomes more gently sloped with distance offshore. Bruun (1954) and later Dean (1991) showed that this profile could be described by the equation

$$h = Ax^{2/3} \tag{6.6}$$

where h is the profile depth at a distance x from the shoreline and A is a constant which has been related to grain size by Dean ($A = 0.21D^{0.48}$ with D in mm). These equations predict that equilibrium beach slopes increase in steepness with increasing grain size. Dean also demonstrated how the profile equation could be related to physical principles, as follows.

The starting point is the assumption that an equilibrium profile will be such that uniform energy dissipation per unit volume (D_e) in the surf zone will exist. Hence we may write

$$\frac{1}{h}\frac{dP}{dx} = D_e \tag{6.7}$$

In shallow water the wave energy flux, P, is given by

$$P = \frac{1}{8}\rho g H^2 \sqrt{gh} \tag{6.8}$$

and assuming spilling breakers then, from Equation (2.35),

$$H = \gamma h \tag{6.9}$$

If Equations (6.8) and (6.9) are substituted into Equation (6.7) and then Equation (6.7) is integrated, with $h = 0$ for $x = 0$, then the result is Equation (6.6) in which

$$A = \left(\frac{24D_e}{5\rho g^{3/2}\gamma^2} \right)^{2/3} \tag{6.10}$$

This is a constant whose value can be related to grain size, as stated above. (A common source of confusion is the sign of the terms in Equation (6.7). The righthand side is positive, and so is the lefthand side. This is because the positive x-axis is taken to run towards the offshore direction. In the situation considered above, waves approaching the shoreline and gradually breaking correspond to a *positive gradient* in wave energy flux as the wave energy flux increases with increasing x.)

The depth of closure (d_c) is defined as the vertical distance between the still water level on the beach and the water depth at which waves can no longer produce any measurable change in the seabed profile. Where suitable records exist, this depth can be determined from profile data. In the absence of such records it has been shown to be of the order of $1.57H_{s12}$ (Birkemeier 1985), where H_{s12} is the significant wave height with a frequency of occurrence of 12 h/year. Of course, the depth of closure is not really constant, but will vary with the incident wave conditions. However, when considering timescales for morphological change it is a useful parameter.

Although Bruun's formula (Equation 6.6), and Dean's supporting physical argument are compelling in many ways, the mathematical form of the equilibrium profile has physical deficiencies:

1   The slope at the top of the beach (when $x = 0$) is infinite (this can be confirmed easily by differentiating Equation (6.6) with respect to x and setting $x = 0$).
2   As x tends to infinity, so the profile keeps getting deeper without limit, not tending towards an asymptote equivalent to a depth of closure.
3   The equation does not describe barred profiles.

These observations prompted Bodge (1992) and Komar and McDougal (1994) to propose the alternative form:

$$h(x) = d_c \left(1 - e^{-Kx}\right) \tag{6.11}$$

which asymptotically approaches a uniform depth d_c in the seaward direction, and has a slope of d_cK at the shoreline. Pruszak (1993) examined beach profiles at Lubiatowo on the Baltic Sea, which exhibit a multiple bar structure. The profiles on the Baltic Sea were measured over a period of 28 years. By fitting an equation of the form (6.6) he found the best fit was obtained by allowing the value of A to vary sinusoidally with time. Elsewhere, Inman *et al.* (1993) and Wang and Davis (1998) proposed models for strongly barred equilibrium profiles, which consisted of separate segments, each of which had parameters requiring data to specify them. A similar approach was adopted by Bernabeu *et al.* (2003) who developed an equilibrium profile description that extended the profile seaward from the surf zone into the 'shoaling zone'.

Example – Fitting an equilibrium profile to measured beach profiles
This example concerns a beach in Santa Marta in Colombia. A location map is shown below in Figure 6.3, with the profile line extending in an approximately north-westerly direction from the coast.

The beach has the following profile:

Chainage (m)	Depth (m)
0	+0.9
15	+0.09
15.41	0.0
20	−1.0
85	−2.0
155	−3.0
170	−4.0
345	−5.0
425	−6.0
690	−7.0

925	−8.0
1050	−9.0
1150	−8.79
1275	−10.0
1430	−11.0
1620	−12.0

Figure 6.4 shows a plot of the profile.
Sediment samples taken along the profile show the following values:

Chainage (m)	Depth (m)	D_{50} (mm)
0	Shoreline	0.25
155	−3.0	0.12
425	−6.0	0.20

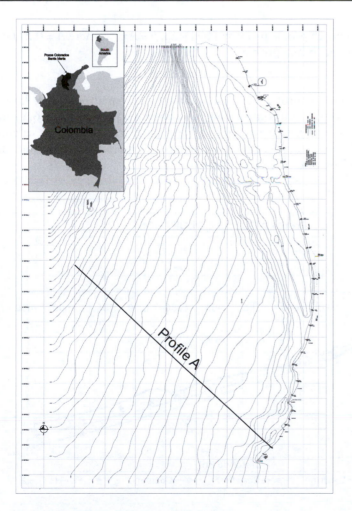

Figure 6.3 Location map for Santa Marta.

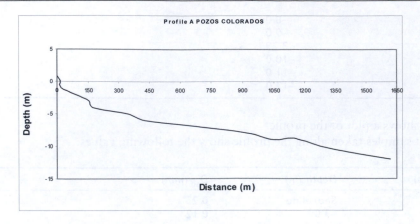

Figure 6.4 Beach profile at Santa Marta.

Using the equation of beach profile proposed by Dean, $h = Ax^{2/3}$, where $A = 0.21D^{0.48}$ with D in mm, calculate the beach profile.

Solution
We calculate the equilibrium profiles corresponding to the smallest (fine), largest (coarse) and mean value of D_{50} and compare with the observed profile. Table 6.1 summarises the sediment sizes and the corresponding Dean parameter value. The calculations can be performed in a spreadsheet and the corresponding equilibrium profiles are shown in Table 6.2 and are also plotted in Figure 6.5.

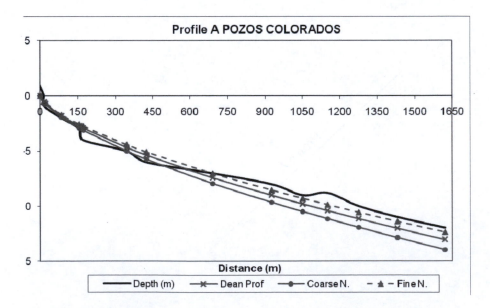

Figure 6.5 Measured and predicted beach profiles.

Table 6.1 Beach parameters.

Distance (m)	Depth (m)	D_{50} (mm)	D_{50} Coarse (mm)	D_{50} Fine (mm)	DEAN A parameter	DEAN A parameter	DEAN A parameter
0	shoreline	0.25	0.22	0.17	0.107951952	0.101527	0.089709
155	−3	0.12	0.22	0.17	0.075897294	0.101527	0.089709
425	−6	0.2	0.22	0.17	0.096987037	0.101527	0.089709
Mean		0.19	0.22	0.17	0.094628297	0.101527	0.089709

Table 6. 2 Measured and predicted profile.

Distance (m)	Depth (m)	Dean Prof	Coarse N.	Fine N.
0	0.90	0	0	0
15	0.09	−0.576	−0.61751	−0.546
15.41	0.00	−0.586	−0.62879	−0.556
20	−1.00	−0.697	−0.74806	−0.661
85	−2.00	−1.829	−1.96273	−1.734
155	−3.00	−2.73	−2.92957	−2.589
170	−4.00	−2.904	−3.11565	−2.753
345	−5.00	−4.655	−4.99415	−4.413
425	−6.00	−5.349	−5.73907	−5.071
690	−7.00	−7.389	−7.92772	−7.005
925	−8.00	−8.984	−9.63851	−8.517
1050	−9.00	−9.776	−10.4884	−9.267
1150	−8.79	−10.39	−11.1442	−9.847
1275	−10.00	−11.13	−11.9377	−10.55
1430	−11.00	−12.01	−12.8866	−11.39
1620	−12.00	−13.05	−14.0042	−12.37

The computed profiles agree pretty closely with the trend in the observations. Further offshore the observed profile is closest to the curve corresponding to the finest material – but this is found much closer to shore. Such differences can be expected because the equilibrium profile may be considered as representing the longterm average, whereas the observations represent a snapshot of the beach at a particular time at which the beach may not be close to its equilibrium form.

In the light of the above example it is reasonable to ask whether the concept of an equilibrium profile could be extended to a beach that has spatially varying sand size. That is, where A is a function of cross-shore position. The answer is a qualified yes. Given a set of observed values of grain size across a profile, the profile can be split into segments in which *it is assumed* that the grain size is constant. In each segment a separate equilibrium profile equation for depth will apply, with a separate origin. The whole profile is found by linking up the solutions for each segment so that the profile is continuous (see e.g. Dean and Dalrymple 2004).

The concepts of an equilibrium profile and depth of closure have proved extremely useful in the design of beach nourishment schemes and in modelling shoreline evolution. For shingle beaches research has focussed more on predicting the profile response to storm conditions, and is discussed in the next Section 6.2.5.

Bruun rule for beach erosion resulting from sea level rise

This is a simple geometric relationship between shoreline recession (Δx) which results from a rise in sea level (ΔS) first proposed by Bruun (1962 and 1983). The principle is that an initial equilibrium profile of length l for a given depth of closure d_c will re-establish itself further landward and higher by a depth ΔS after the sea level rise (as the depth of closure remains constant). This implies that the material eroded on the upper part of the profile is deposited on the lower part of the profile. Hence

$$\Delta x d_c = l \, \Delta S \qquad (6.12)$$

or

$$\Delta x = \frac{\ell}{d_c} \Delta S \qquad (6.13)$$

As l is, in general, much larger than d_c, the shoreline recession will also be much larger than the rise in sea level.

6.2.5   Numerical prediction of beach profile response

As discussed earlier in this chapter the methods available for predicting the evolution of beach profiles in response to winds, waves and water level variations may be divided into three groups:

1   Deterministic process models that simulate the redistribution of sediment due to a succession of specified conditions.
2   Morphological models that include a representation of physical processes but whose aim is to predict changes in the overall shape of the beach.
3   Statistical models that are based on extrapolating past observations.

Examples from each of these groups are discussed in this section.

Statistical methods

As mentioned in previous sections, predictions using statistical methods rely on extrapolating trends and cycles, observed in historical measurements, into the future. Some link with physical processes may be introduced by postulating a correlation between beach movement and the environmental forcing.

An example of this approach is the study, by Masselink and Pattiaratchi (2001) of seasonal changes in beach morphology along the coastline of Perth in Western Australia. Figure 6.6a shows the coastline at the study site. It is characterised by large seasonal variations in wave height and the local beaches exhibit a corresponding change in morphology. However, the morphological changes are better explained by changes in the littoral drift direction (see Section 6.3) than by the changes in incident wave energy. In summer (December, January, February), the net littoral drift is northward. As a result beaches to the south of groynes and headlands accrete, becoming wider. In contrast, beaches to the north of obstacles erode and become narrower (Figure 6.6b).

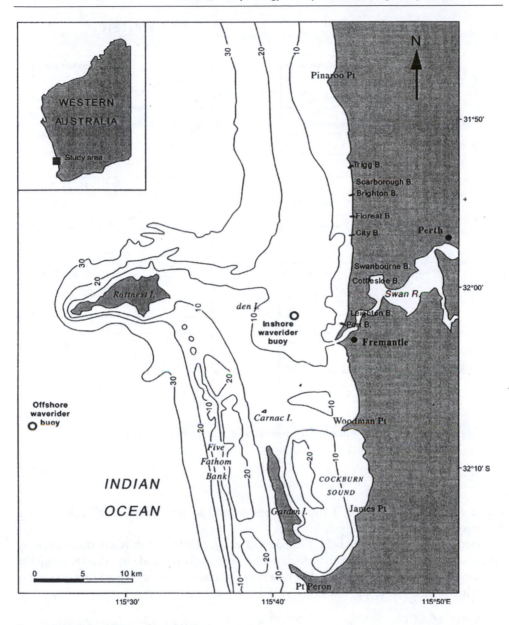

Figure 6.6(a) Location of study site.

In the winter months (June, July, August) the situation is reversed with the net littoral drift being southward.

Wave energy thresholds have been defined by several researchers to predict the occurrence of beach movement. These thresholds tend to be site-specific. Parameters that include additional information have more general applicability (e.g. Dean 1973; Sunamura 1987; Kraus *et al.* 1991; Dalrymple 1992). One such parameter is the dimensionless fall velocity

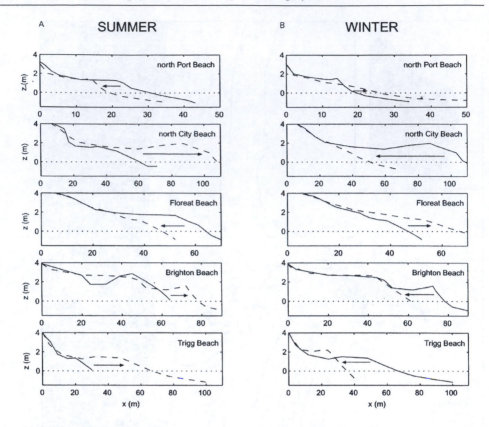

Figure 6.6(b) Location and typical summer and winter beach profiles.

$$W = \frac{H}{w_s T} \qquad\qquad (6.14)$$

where H is the wave height, T the wave period and w_s the sediment fall velocity (see Equation 5.21).

Based on the analysis of field data, Kraus *et al.* (1991) established the following criteria to determine whether the beach profile was likely to erode or accrete, using the significant wave height and period to determine the wave steepness and dimensionless fall velocity (the subscript 'o' denoting offshore values):

$S_o/W_o^3 > 0.0014$	*or*	$W_o < 2.4$	*accretion is highly probable*
$S_o/W_o^3 > 0.0027$	*or*	$W_o < 3.2$	*accretion is probable*
$S_o/W_o^3 > 0.0027$	*or*	$W_o \geq 3.2$	*erosion is probable*
$S_o/W_o^3 > 0.0054$	*or*	$W_o < 4.0$	*erosion is highly probable*

Other results from field experiments indicate that a value of less than 1.5 to 2 is characteristic of beaches with no bar while larger values of W are typical for beaches with a bar (Allen 1988; Wright *et al.* 1987).

Masselink and Pattiaratchi (2001) calculated values of W over several years and compared these with measured changes in beach profile. They discovered that W

fluctuated about the threshold value of 1.5~2 most of the time, showing coherent changes over periods of weeks and seasons. These variations indicate that the near-shore morphology is not in equilibrium with the prevailing hydrodynamic conditions (although they may be in dynamic equilibrium when considered over a period of several years). Consequently W, while a valid predictor of beach condition in principle, is of little use in predicting the form of the beach profile at that particular site.

Hashimoto and Uda (1980) suggested an alternative approach. Taking the results of an EOF analysis of beach profile measurements they used the Iribarren number to estimate future values of the significant temporal eigenfunctions. This has the advantage that it accounts for changes in wave height, wave period and beach slope. It does assume that the spatial eigenfunctions will continue to give a good description of the beach profile. The method has been refined several times and has performed reasonably well against field measurements.

Parametric methods

Parametric methods combine observational evidence with either statistical fitting methods, simplified analytical models, or both. As such, they tend to be limited in the range of their applicability. Examples of parametric models are described below for three types of feature: shingle beaches; barrier beaches and dunes.

SHINGLE BEACHES

The term 'shingle beach' is commonly used in the UK and can be considered synonymous with the more general term 'coarse-grained beach'. The mean grain size on a typical shingle beach can range from 10–40 mm, yet despite the large grain size, the sediment is highly mobile. Changes in the vertical profile of the order of 0.5 m are commonly found over just one tidal cycle for moderate to large waves. Coarse-grained beaches are widespread around the world and in the UK about one-third of the coastline is protected by shingle beaches.

The characteristics of a shingle beach are very different from those of a sand beach. Most notably, shingle can support a steep gradient typically of the order of 1 in 8 with a tendency to form a near vertical berm towards the high water mark. The steep beach gradient allows waves to propagate much closer inshore before breaking, often resulting in a single plunging breaker. An important consequence of this is that the swash zone can be of similar width to the surf zone and hence the sediment transport within the swash zone is of more significance than on sand beaches. Another distinguishing characteristic is the high permeability of shingle which, compared to a sand beach, increases the potential for infiltration during the swash uprush and exfiltration during the swash downrush. The existence of a berm often found at the maximum swash run-up is generally considered to be due to this process.

As can be appreciated from the foregoing, the prediction of the expected profile for a shingle beach is very different from the concepts used for equilibrium profiles on a sand beach. The problem was first comprehensively addressed by Powell (1990), who undertook an extensive series of scaled model tests using lightweight model sediment (anthracite). This work resulted in the development of a parametric profile model, which described the profile as a set of three curves: from the crest to the still-water level, from the still-water level to a transition point and from the transition point to the base of the profile. The details may be conveniently found in the *Beach Management Manual* (CIRIA 2010).

More recent work on coarse-grained beaches was performed in the EU project 'Large Scale Modelling of Coarse Grained Beaches' which was undertaken in 2002. The experiment was conducted at full scale, thus avoiding the sediment scaling issues inherent in small-scale models. The results of this work may be found in López de San Román-Blanco *et al.* (2006). Using these results a new numerical model for predicting the profile response of coarse-grained beaches was developed (see Pedrozo *et al.* 2006). This model gave very promising results, but further developments coupling the effects of infiltration/exfiltration with sediment movement are still required. The recent results reported by Jamal *et al.* (2010) indicate that including this process leads to significant improvements in prediction.

BARRIER BEACH PROCESSES

Barrier beaches are a common geomorphological feature across the world. Their essential features comprise a narrow, elongated ridge of sand or gravel existing slightly above the high tide level. The ridge generally extends parallel to the shore, but is separated from it by a wetland, lagoon or a tidal flat. Barrier beaches act as natural means of coast protection. In addition, wetlands and lagoons formed behind barrier beaches provide shelter for many coastal habitats and are therefore of considerable environmental significance. One explanation of their formation is that they have been formed by landward migration of submerged sand/shingle banks with rising sea levels since the last ice age.

Barrier beaches are constantly evolving in response to short- and long-term processes. Short-term changes in barrier beaches are related to local wave and current climate, tidal variations, frequency and magnitude of storm events, barrier geometry and type of beach sediment and permeability. Over longer terms, the primary factors for change and modification of barrier beaches are sea level rise, longshore sediment transport and changes in sediment sources and/or sinks according to Orford *et al.* (1995).

A barrier beach can respond to these factors by landward or seaward migration, reshaping and re-alignment and crest breakdown or build-up (Carter *et al.* 2003). The *episodic* processes of over-washing, over-topping and associated breaching are the primary phenomena behind long-term evolution.

Previous studies providing predictive equations for such episodic events are scarce. However, one such study, Bradbury (2000), does provide some very useful results. Bradbury carried out a series of 3-D mobile bed laboratory tests on barrier crest response to hydrodynamic conditions and initial barrier geometry. Based on the model investigations, several categories of barrier response to hydrodynamic conditions were identified and underlying characteristics were qualitatively defined. These included crest raised by overtopping, crest lowered due to undermining of crest but with no overtopping, crest raised by over-washing with roll-back, crest lowered by over-washing with roll-back and finally, no change to the crest elevation with profile contained to seaward of the barrier crest.

He developed an expression for an over-washing threshold of barrier crests, based on regression analysis. It is a function of wave steepness (H_s/L_m), barrier free board (R_c) and barrier cross-sectional area B_a (above the still water line), given by:

$$\frac{B_a R_c}{H_s^3} \le 0.0006 \left(\frac{H_s}{L_m}\right)^{-2.54} \quad \text{for over-washing.}$$

The expression was validated against field data gathered at Hurst Spit, UK and found to be consistent with the field data. In addition, a conceptual model for barrier over-washing was formulated. According to this model, the beach will initially attempt to reach a dynamic equilibrium; if the critical barrier inertia ($\frac{B_a R_c}{H_R^B}$) is exceeded then the crest will be lowered by over-washing.

A comprehensive assessment of the historical development and contemporary processes affecting a significant barrier beach system in the UK (Slapton Sands) may be found in Chadwick *et al.* (2005). This paper includes assessment of the effects of sea level rise on barrier migration and its susceptibility to over-washing and overtopping, *inter alia*.

DUNE EROSION

In a series of papers Vellinga (1982, 1984a, b) presented a parametric model describing the dune and upper beach shape in response to storm waves. He used results based on extensive 2-D and 3-D experiments in the Delta flume to define equations for the cross-sectional profile of the dune/beach. His proposed formula is:

$$y = 0.7 \left(\frac{H_o}{L_o}\right)^{0.17} v_s^{0.44} x^{0.78}$$

$$\textit{where} \qquad v_s = 4.36 D_{50}^{0.5} \qquad \textit{for} \quad D_{50} > 1 \text{ mm} \qquad (6.15)$$

$$v_s = 273 D_{50}^{1.1} \qquad \textit{for} \quad 0.1 \text{ mm} < D_{50} < 1 \text{ mm}$$

$$v_s = 1.1 \times 10^6 D_{50}^{0.5} \quad \textit{for} \quad D_{50} < 0.1 \text{ mm}$$

and is applicable to sandy duned beaches. The location of the reference point ($x = 0$) is determined by considerations of mass conservation, requiring an iterative process. The beach profile is similar to that given by Equation (6.6) but can become noticeably deeper in the outer surf zone.

Larson *et al.* (2004) presented an analytical model to predict dune erosion caused by wave impact, with this being based on impact theory. It describes the amount of dune erosion that will occur in a storm event and is given in Equation (6.15). To apply the analytical model, Larson *et al.* (2005) state that simplifications are required, in describing the governing processes, forcing, and initial and boundary conditions. An outline of the meaning of each coefficient is given in Table 6.3.

$$\Delta V_E = 8 \frac{C_s}{T} \left[\left(\frac{T_s}{2} - t_L\right)\left(\frac{1}{2} R_T^2 + Z_D^2\right) + R_T^2 \frac{T_s}{4\pi} \sin\left(2\frac{\pi t_L}{T_s}\right) - 2 R_T Z_D \frac{T_s}{\pi} \cos\left(\frac{\pi t_L}{T_s}\right) \right] \qquad (6.16)$$

Table 6.3 Coefficients in the dune erosion model.

C_s = Empirical transport coefficient	T_s = Duration of the storm (seconds)
T = Period at which waves hit the dune (seconds)	t_L = Time when waves start hitting the dune (seconds)
$R_T = R_a + z_a$ (m)	$z_D = z_i - R_i$ (m)
R_a = Amplitude of the sinusoidal run-up variation (m)	z_i = Vertical distance to the dune foot (m) when $t = 0$
z_a = Amplitude of the sinusoidal water level variation (m)	R_i = The run-up height (m) when $t = 0$

The model has a number of variables, which must be determined to enable its accurate application. Larson *et al.* (2004) validated the model against four data sets, which cover both monochromatic and random waves. The model has empirical coefficients that required calibration from the data set. The run-up height (R) was found using estimates from the data set, using a general least-squares fit procedure, to be:

$$R = 0.158\sqrt{H_o L_o} \qquad (6.17)$$

where H_o is the wave height and L_o is the wavelength. Using the predictions established for the run-up height, the optimal values for C_s were derived through a least-square fit procedure. The mean value for the transport coefficient was found to be:

$C_s = 1.4 \times 10^{-4}$ with a standard deviation of 0.74×10^{-3}.

Larson *et al.* (2005) state that a C_s value within the range of 10^{-4} to 2×10^{-4} is generally appropriate. The situation being modelled is illustrated in Figure 6.7. D_s represents the dune crest elevation, from the water level to the top of the dune (in metres). As the bore reaches the shore and before the up-rush starts, it is travelling at a speed u_s (m/s). As the bore reaches the dune face it is travelling at a speed u_o (m/s), with the bore reaching a height h_o (m). z_o is the elevation difference between the dune foot and the location where the up-rush starts (m). The slope of the foreshore is $\tan(\alpha)$.

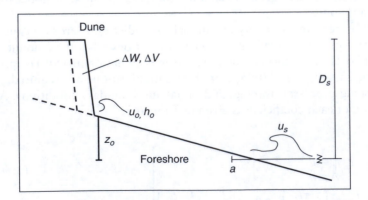

Figure 6.7 Illustrative diagram for the dune erosion model (after Larson *et al.*, 2004).

ΔV represents the eroded volume (m³/m), and ΔW represents the weight of the eroded volume (N/m).

This erosion model uses an analytical approach, applying simplifications that enable closed form solutions to be derived.

Morphological methods

Morphological profile models predict the evolution of the shape of a beach profile over time, specifically over periods of 1–100 years. In principle, the behaviour of coastal morphology at all scales must obey the sediment continuity equation. For a cross-shore profile this may be written as

$$\frac{\partial h}{\partial t} = \varepsilon \left(\frac{\partial c}{\partial t} + \frac{\partial q_x}{\partial x} \right) - \frac{\partial q_y}{\partial t} \tag{6.18}$$

where h is the level of the profile at each point, x is distance from the shore, ε accounts for sediment density and porosity. The cross-shore flux of material is denoted q_x, and the longshore flux by q_y (y is taken to be parallel to the shore), as shown in Figure 6.8. The total concentration of sediment suspended above the bed is denoted by c. The terms c, q_x and q_y are depth integrated quantities.

Equation (6.18) states that the beach profile can change through material being picked off the bed and thrown into suspension, by being moved along the profile or by the addition or removal of sediment being moved alongshore. Morphological models use Equation (6.18) coupled with simplifying assumptions about c, q_x and q_y. One such model is the 'behaviour-oriented model' proposed by Stive et al. (1991). This

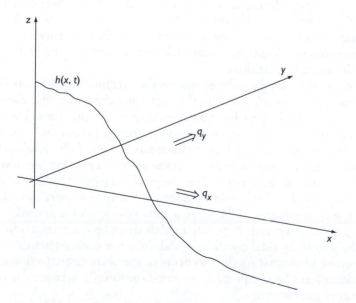

Figure 6.8 Morphological models for beach profiles. Definition of coordinates and sediment fluxes.

simplifies Equation (6.18) to a diffusion-type equation by considering the profile to be made up of a number of horizontal layers. The length of each layer can change with time due to sediment exchange with neighbouring layers or through the addition of material from outside the profile. This is very similar to the *n*-line beach plan model (see Section 6.3). Further physics can be included in the model by using appropriate transport formula to calculate the cross-shore transport due to waves and gravity (Stive and de Vriend 1995). Niedoroda *et al.* (1995) use Equation (6.14) as the basis for deriving an equation for the fluctuation of the profile about an assumed equilibrium form. In this case the profile equation takes the general form

$$\frac{\partial h'}{\partial t} = a(x)\frac{\partial h'}{\partial x} - \frac{\partial}{\partial x}\left(b(x)\frac{\partial h'}{\partial x}\right)$$

(6.19)

where $h' = h - h_o$, $h_o(x)$ is the equilibrium profile. The first term on the right-hand side represents advection of material while the second term describes diffusive processes. Niedoroda *et al.* (1995) discuss the time scales of these two terms and suggest that the inclusion of an advective term is necessary for realistic morphological predictions.

It should be remembered that the purpose of these morphological models is to predict the change in beach shape over the medium to long term. They are not designed to predict beach response over the short term, that is, changes in beach shape due to an individual or cluster of storms.

Deterministic process models

Deterministic process models seek to describe changes in profile morphology by calculating the cumulative effect of a series of water level and wave conditions. Such models usually take offshore wave conditions as input and include a description of wave transformation to shallow water, wave breaking and dissipation. The more sophisticated models will include calculations of wave set-up and wave-induced currents and their effect on wave propagation. Tidal elevation and currents can also be included though usually as depth-averaged quantities.

One of the key mechanisms causing profile erosion is the offshore current near the seabed, commonly termed 'undertow'. Outside the surf zone there is a shoreward steady flow near the bed, caused in part by the asymmetry in the (nonlinear) wave orbital velocities. At the transition zone, waves transform from a non-breaking state to one resembling a turbulent base in which turbulence has become fully developed. The transformation does not occur instantaneously at the point of breaking but develops over a distance beyond the breaking point. Two ways of modelling this transition zone are described by Nairn *et al.* (1990). Wave-induced sediment transport formulae typically involve one or more moments of the wave orbital velocity at the seabed.

The hydrodynamic part of deterministic profile models describe the interaction of the incoming wave, the prevailing tidal conditions and the wave-induced set-up and currents, with the objective of estimating the moments of the wave orbital velocities at the seabed. This calculation often requires some iteration to fully account for the interaction between waves and current.

At this point sediment transport in the longshore and cross-shore directions can be determined from a combination of the sediment continuity equation and the sediment transport rates determined from the hydraulic calculations. The resulting changes in sediment distribution allow the beach profile to be updated and the whole cycle is repeated (see e.g. Nairn and Southgate 1993; Southgate and Nairn 1993). The modelling process is summarised in Figure 6.9.

In practice the time steps for the hydraulic and sediment transport parts of the calculation may be different. Further details of this type of model may be found in the special

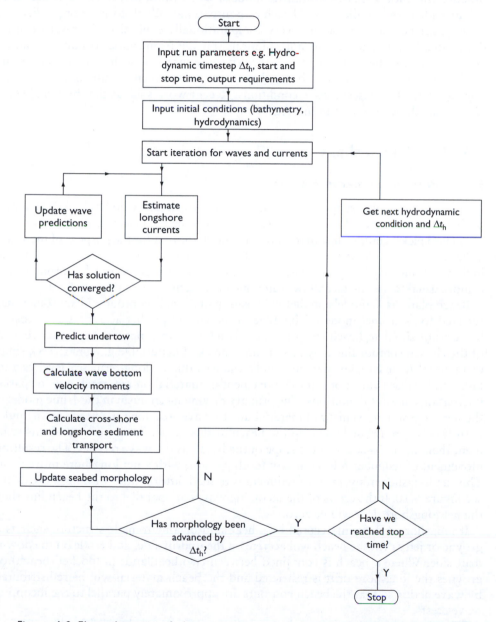

Figure 6.9 Flow diagram of deterministic morphological modelling (adapted from Southgate and Nairn 1993).

issue of *Coastal Engineering* published in 1993, entitled Coastal Morphodynamics: Processes and Modelling. Cross-shore models have not proved particularly successful at describing evolutionary beach processes (Schoones and Theron, 1995), principally because most models neglect or drastically simplify the swash hydrodynamics.

Although describing more of the detailed physical processes than the other types of model described in this section, assumptions are still made in this type of model. In practice this means that deterministic models, when calibrated for a particular site, can provide useful predictions of beach response to individual storms. For predictions over longer periods they become very computationally expensive, the results can be dominated by numerical errors and solutions starting from similar conditions can be very different. Perhaps the greatest difficulty is the inability of these models to accurately simulate the erosion of a beach during storms and then its subsequent recovery during relatively quiescent wave conditions. Recent work such as that by Jamal *et al.* (2010) has shown promise in this regard.

6.3   Beach plan shape

6.3.1   Plan shape measurements

The plan view, or plan shape, of a beach is familiar to anyone who has looked at a map of the coast or a chart of coastal waters. When defining the beach shape (coast line), it is usual to pick a contour line that represents the shape best for the purpose of the study in hand. For example, if sea defence structures are the main interest then selecting the high water line may be appropriate. For investigating changes in beach morphology it is more usual to use the mean low water line or sometimes the mean sea level.

Beach plan shape models predict the position of a single contour and are sometimes referred to as '1-line' models. This type of model can predict accretion and erosion but cannot simulate beach steepening or flattening. On a natural beach, movement of the chosen contour line may arise from sources of material (e.g. from rivers), sinks of material (embayments/marshes), bulk changes due to long-term sea level rise or fall, and localised variations in the movement of material on the beach due to spatial fluctuations in wave conditions. The primary driving mechanism in the 1-line model is the net longshore movement of material due to waves reaching the beach at an angle, with waves carrying sediment up the beach at this angle. Some portion of this sediment then moves back down the slope of the beach as the wave recedes. The resulting movement of sediment follows a saw-tooth pattern, with a net longshore movement. This net longshore movement of sediment is termed 'littoral drift' or 'longshore drift', see Figure 6.10. If the crests of the incoming waves are parallel to the beach line then the net longshore drift will be zero.

If longshore drift is intercepted by a headland or a man-made structure such as a groyne or jetty then the beach will accrete on the updrift side, and erode on the downdrift side. Where a beach is contained between two headlands (a 'pocket' beach) or groynes the longshore drift is restricted and the beach material will be redistributed by wave action so that the beach contours are approximately parallel to the incoming wave crests.

When considering beach plan shape prediction it is essential to define a 'baseline' or reference line from which to measure the distance to the chosen contour line. Unlike

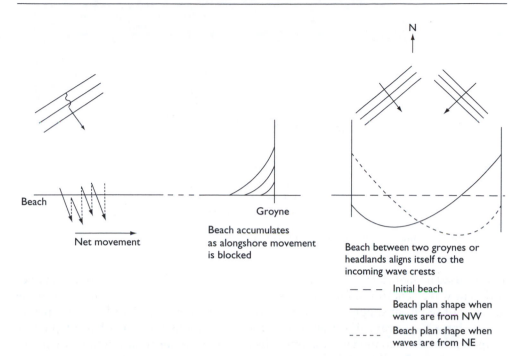

Figure 6.10 Zig-zag sediment paths, accretion at a groyne and movement of pocket beach line.

beach profile measurements, beach plan shape observations are rarely taken on a regular basis. The best sources of information are usually:

- specifically commissioned surveys;
- aerial photographs (which may be of limited use for obtaining the position of low water unless the state of the tide is known);
- historical maps and charts;
- remote-sensing techniques such as the video-based ARGUS system (Holland *et al.* 1997), from which it is possible to derive information on plan shape, profile shape and some wave characteristics.

In the remainder of this chapter some of the more widely used techniques for predicting beach plan shape are presented. These include:

- equilibrium beach forms;
- derivation of the beach plan shape equation;
- analytical and numerical solutions of the plan shape equation.

6.3.2 Equilibrium forms

Where an erodible coastline exists between hard, stable headlands a bay will form. The shape of the bay will depend on the wave climate and supply of sediments. Silvester

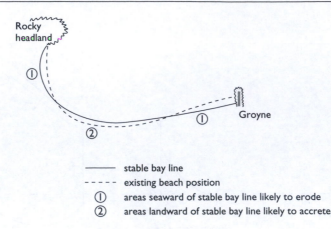

Figure 6.11 Illustrative use of stable bay concept.

(1974) used a laboratory wave tank to investigate the equilibrium shape of bays under different wave conditions. On the basis of these experiments he suggested that in the absence of sediment supply a stable bay would form adopting a half-heart or cardiod shape, for a fixed wave direction. Under these conditions the beach has adapted its shape so that the incoming wave crests, which are curved due to diffraction, are parallel to the shore. The littoral drift is therefore zero and the bay stable.

These results are significant for several reasons. Firstly, it provides a relatively simple way of predicting the stable bay shape and orientation given characteristic wave conditions. When designing new beaches between natural or artificial headlands this technique can provide a useful rule-of-thumb. Secondly, for a natural bay the method may be used to determine the equilibrium shape of the bay. If this shape does not coincide with the current shape the bay is not stable and can be expected to evolve. If the existing bay lies seaward of the stable bay line then either the bay is receding, or there is a source of sediment maintaining the bay. This is illustrated in Figure 6.11. Possible sources are sediment supply from a channel or river and sediment bypassing the headland. Bypassing may be intermittent, for example occurring only under specific storm conditions, but be of sufficient volume to maintain the bay. Conversely, if the bay is landward of the stable bay line, this suggests a sink of sediment (e.g. dredging) is present or that the coast has an accretionary tendency.

More recent work on this method may be found in Hsu *et al.* (1989), Silvester and Hsu (1997) and Gonzalez and Medina (2001). Further discussion of the equilibrium bay shape concept may be found in Chapter 9 (Section 9.2.2).

6.3.3 *Beach plan shape evolution equation*

If a natural beach has an adequate supply of sand or shingle then it may remain in stable equilibrium over an extended period (Figure 6.12a). However, if the sediment transport is intercepted (by a natural or artificial feature) then the beach will accrete on the updrift side of the feature (Figure 6.12b). In the case of a large structure, such as a breakwater, it is possible that all sediment will be trapped, and that the coastline on the downdrift side will be starved of sediment and will deplete. Now, transport

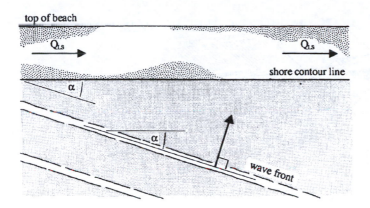

Figure 6.12(a) Equilibrium plane beach.

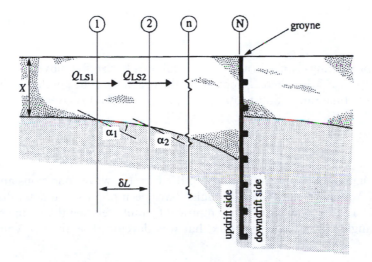

Figure 6.12(b) Accretion and erosion near a groyne.

rate is a function of angle α between the wave front and the beach contour. However beach accretion alters the line of the beach contour, the angle α is no longer constant, so sediment transport rate will vary with position along the shoreline.

A simple mathematical model of this situation can be developed, based on the concept of an equilibrium profile extending to the depth of closure. Consider the element of beach between boundaries 1 and 2 in Figure 6.12(b), shown in sectional elevation in Figure 6.12(c). Applying the continuity equation, in a time interval δt the change in the volume of sediment in the element is equal to the volume entering less the volume leaving. Hence

$$A\delta x - \left(A + \frac{\partial A}{\partial t}\delta t\right)\delta x = Q\delta t - \left(Q + \frac{\partial Q}{\partial t}\delta x\right)\delta t \tag{6.20}$$

where A is the cross-sectional area of the beach profile.

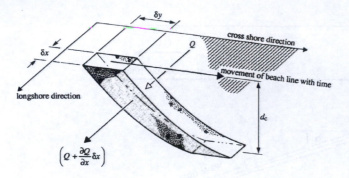

Figure 6.12(c) Definitions for 1-line model.

Simplifying

$$\frac{\partial Q}{\partial x} = \frac{\partial A}{\partial t} \tag{6.21}$$

For an equilibrium profile any change in area must result in a horizontal movement of the profile, δy, $\delta A = (d_c + d_b)\,\delta y$, where d_b represents the beach berm height above the still water line plus the tidal range. Substituting in Equation (6.21) this gives

$$\frac{\partial Q}{\partial x} = (d_c + d_b)\frac{\partial y}{\partial t} \tag{6.22}$$

To make further progress we require a relationship between the wave conditions and sediment transport. A number of different formulae have been proposed and are discussed in Section 5.2.12. For now we use the empirical formula relating the sediment transport to wave angle and wave energy flux that was developed by the US Army Corps (1984):

$$Q = Q_0 \sin (2\alpha_b) \tag{6.23}$$

where Q_0 is the amplitude of the longshore transport rate (m³/s), given by

$$Q_0 = \frac{\kappa P}{2g(\rho_s - \rho)(1 - p_s)} \tag{6.24}$$

where P is the wave energy flux, p_s is the porosity, ρ_s is the density of the sediment (kg/m³), ρ is the density of sea water and κ is a dimensionless coefficient which is a function of particle size. Evaluating the wave energy flux using linear wave theory yields the following expression for Q_0:

$$Q_0 = \frac{\rho}{16} H_b^2 c_{gb} \frac{\kappa}{(\rho_s - \rho)(1 - p_s)} \tag{6.25}$$

where the subscript b denotes values at the point of breaking, and c_g is the wave group velocity. The quantity α_b is the angle between the wave front and the shoreline, and may be written as

$$\alpha_b = \alpha_0 - \tan^{-1}\left(\frac{\partial y}{\partial x}\right) \tag{6.26}$$

where α_0 is the angle between the wave front and the x-axis. Substituting Equation (6.26) into (6.23), and assuming both α_0 and $\partial y/\partial x$ are small, yields

$$Q \approx Q_0\left(2\alpha_0 - 2\frac{\partial y}{\partial x}\right) \tag{6.27}$$

Substituting this into Equation (6.22) gives

$$\frac{\partial y}{\partial t} = K\frac{\partial^2 y}{\partial x^2} \tag{6.28}$$

where $K = 2Q_0/(d_c + d_b)$. Equation (6.28) has the form of a linear diffusion equation, where K is a parameter that depends on the wave climate and beach material and has the role of a diffusion coefficient. In practice, K will be a function of both time and position, in which case a more complicated governing equation results (Larson *et al.* 1997). This method only gives the evolution of a single shoreline contour and is known as a 1-line model.   ·

6.3.4   *Analytical solutions*

The 1-line equation, Equation (6.28), while apparently straightforward may be employed to investigate the beach response to a wide variety of situations. In fact, one mathematical tool that is routinely used for prediction of long-term shoreline evolution is the 1-line model. In comparison with other models of shoreline evolution, i.e. complex three-dimensional process-based models, 1-line models are computationally cheap, have reasonable data requirements for calibration and validation, are physics based and thus have a level of generality, and have performed well in numerous projects of long-term shoreline evolution (e.g. Hanson 1987; Reinen-Hamill 1997; Dabees and Kamphuis 1998).

Solutions to the 1-line model can be analytical or computational and are based on the concept of the continuity of sediment. Analytical solutions involve small-angle approximations which reduce the continuity of sediment equation to a diffusion type equation as described in Section 6.3.3. Analytical solutions have been derived in a number of studies for different cases of shoreline change using simple wave-driven sediment transport models (e.g. Pelnard-Considère 1956; Grijm 1961; Le Méhauté and Soldate 1977; Wind 1990; Larson *et al.* 1997). Apart from the assumptions of a smooth shoreline and a small angle of wave approach, analytical solutions are further limited by the common assumption that waves are constant in time and in space (i.e. the diffusion coefficient is constant).

In practice, the constraints of a natural beach can rarely be simplified to the extent to allow an analytical solution of Equation (6.28) and a numerical solution is necessary. Numerical 1-line models solve the continuity, sediment transport and wave angle equations simultaneously, stepping forward in time and may include, amongst others, elements of wave prediction, wave refraction and diffraction, and beach slope variation (see e.g. Gravens *et al.* 1991); they are well suited for engineering practice and are discussed further in Section 6.3.5.

Nevertheless, analytical solutions to Equation (6.28) can give a useful guide for the engineer as well as providing tests for validating numerical models. In this section, different beach situations are considered and the solution to the idealised problem is presented.

Analytical solutions

Analytical solutions to Equation (6.28) require initial conditions and boundary conditions to be specified. These conditions determine the nature of the problem being solved. Solutions are obtained by using integral transform techniques and the interested reader is referred to the cited papers for additional details. Some selected examples are shown in this section to provide an overview of the scope of the problems that can be solved with analytical methods.

1   Straight impermeable groyne
  Boundary conditions: $Q = 0$ at $x = 0 \rightarrow \tan(\alpha_0) = \dfrac{dy}{dx}$ at $x = 0$

$$y(x,0) = 0, \, y(x,t) \rightarrow 0 \text{ at } x \rightarrow \infty$$

Solution:

$$y(x,t) = \tan \alpha_0 \sqrt{\frac{4Kt}{\pi}} \left\{ e^{-\frac{x^2}{4Kt}} - \frac{x\sqrt{\pi}}{2\sqrt{Kt}} \, erfc\left(\frac{x}{2\sqrt{Kt}}\right) \right\} \tag{6.29}$$

Example solutions are shown in Figure 6.13.

2   *Straight permeable groyne, length L*

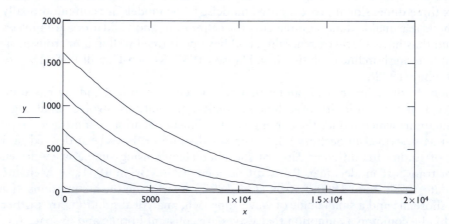

Figure 6.13 Analytical solution for an infinite breakwater positioned at $x = 0$ at selected times (t = 0.1, 10, 20, 50 and 100 years), for K = 500 000 m²/year and a wave angle of 0.2 radians.

Boundary conditions $Q = 2 Q_0 \alpha_0 \left(\dfrac{y}{L}\right)$ at $x = 0$

$$y(x,0) = 0, \ y(x,t) \to 0 \text{ as } x \to \infty$$

Solution:

$$y(x,t) = L\, erfc\left(\frac{x}{2\sqrt{Kt}}\right) - L e^{\alpha_0\left(\frac{y}{L}\right) + \alpha_0^2\left(\frac{Kt}{L^2}\right)} erfc\left(\alpha_0 \frac{\sqrt{Kt}}{L} + \frac{x}{2\sqrt{Kt}}\right) \tag{6.30}$$

Example solutions are shown in Figure 6.14.

3   *Sediment by passing a breakwater of length*

Boundary conditions: $y(x,0) \begin{cases} L & x \langle 0 \\ 0 & x \rangle 0 \end{cases}$

Solution: $y(x,t) = L\, erfc\left(\dfrac{x}{2\sqrt{Kt}}\right)$ (6.31)

4   *Point source of sediment at origin*
Boundary conditions: $y(x,0) = 0$
Source: $Q\delta(x)$

Solution: $y = \dfrac{Q}{\sqrt{\pi 4Kt}}\, e^{-\left(\frac{x}{2\sqrt{Kt}}\right)^2}$ (6.32)

5   *Rectangular nourishment*

Boundary conditions: $y(x,0) = \begin{cases} V & |x| < a \\ 0 & |x| > a \end{cases}$

Solution: $y = \dfrac{V}{2}\left[erf\left(\dfrac{a-x}{2\sqrt{Kt}}\right) + erf\left(\dfrac{a+x}{2\sqrt{Kt}}\right)\right]$ (6.33)

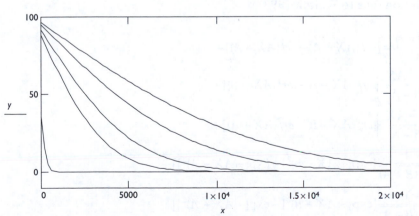

Figure 6.14 Analytical solution for a permeable breakwater of length 100 m under the same conditions as in Figure 6.13.

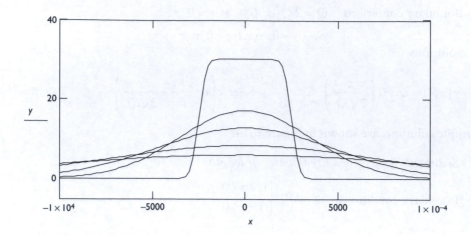

Figure 6.15 Analytical solutions for a beach nourishment (rectangular at $t = 0$ with width 5000 m and depth 30 m. Solutions are shown for $t = 0.1, 10, 20, 50$ and 100 years, and $K = 500\,000$ m²/year.

Example solutions are shown in Figure 6.15.

6   *Tapered rectangular nourishment*

$$
y(x,0) = \begin{cases}
V & |x| & < a \\
\dfrac{v}{b-a} \ (b-x) & a < x & < b \\
\dfrac{v}{b-a} \ (b+x) & -a > x & > -b \\
0 & & Otherwise
\end{cases}
$$

Solution (due to Walton 1993):

$$
\frac{2y(x,t)}{V} = [erf(AX + A) - erf(AX - A)] +
$$

$$
\left(\frac{B-AX}{B-A}\right)[erf(AX - A) - erf(AX - B)] +
$$

$$
\left(\frac{B+AX}{B-A}\right)[erf(AX + B) - erf(AX + A)] + \tag{6.34}
$$

$$
\frac{1}{\sqrt{\pi(B-A)}}[\exp\{-(AX - B)^2\} - \exp\{-(AX - A)^2\}] +
$$

$$
\frac{1}{\sqrt{\pi(B-A)}}[\exp\{-(AX + B)^2\} - \exp\{-(AX + A)^2\}]
$$

where $A = a/(2\sqrt{(Kt)})$, $B = b//(2\sqrt{(Kt)})$ and $X = x/a$.

7   *Free evolution of arbitrary initial beach shape with beach fixed at x = 0.*
    Boundary conditions. $y(x,0) = f(x)$, $y(0,t) = 0$

$$y(x,t) = \frac{1}{2\sqrt{\pi Kt}} \int_{-\infty}^{\infty} f(x')e^{-\frac{(x-x')^2}{4Kt}}\, dx' \tag{6.35}$$

Example solutions are shown in Figure 6.16.

8   *Evolution near a detached breakwater*
    Solutions assume $Q = 0$ at midpoint behind breakwater (see Figure 6.17).

More recent developments of analytical solutions to the 1-line equation have included treatment of time-varying wave conditions. Time variation was introduced by Larson *et al.* (1997) who allowed for a sinusoidally time-varying breaking wave angle at a single groyne and at a groyne compartment. The solution method involved Laplace transforms. Time variation was specified as a known function and was constrained at the location of the boundaries (groynes). Dean and Dalrymple (2002) describe a method for investigating beach nourishment longevity, allowing wave conditions to vary arbitrarily in time through a time-varying diffusion coefficient. Using Fourier decomposition of the shoreline position an analytical solution for an individual Fourier component was found, assuming an initially straight, undisturbed shoreline. Reeve (2006) presented a more general analytical solution for the case of a single groyne exposed to arbitrary varying wave forcing, arbitrary initial shoreline shape and specified source terms. The Fourier cosine transform technique was used to derive the solution in terms of closed-form integrals, which require numerical evaluation for an arbitrary sequence of wave conditions. The solution has been further developed to describe open coasts and groyne compartments by Zacharioudaki and Reeve (2010).

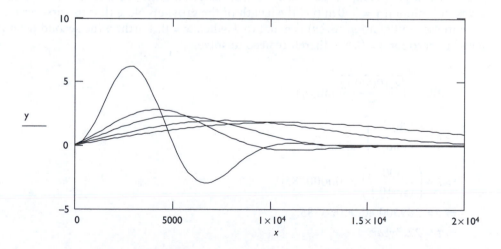

Figure 6.16 Analytical solutions at selected times (t = 0.1, 10, 20, 50 and 100 years), and K = 500 000 m²/year. The initial beach shape is specified as $f(x)$ = 10exp[$-(x/5000)^2$]sin($2\pi x/1000$).

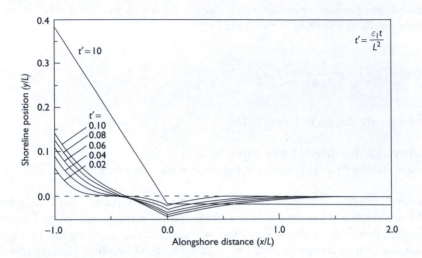

Figure 6.17 Solutions for evolution behind a detached breakwater under normal wave attack (from Larson et al. 1997).

Example – Use of analytical solutions to estimate groyne life

An impermeable groyne of length 300 m is placed on an initially straight beach. Given that $K = 100\,000$ m^2/year and the prevailing wave direction is $\alpha = 10°$, determine how long it takes until the groyne is full.

Solution

This situation is described by the solution given in Equation (6.29). Without loss of generality we may take the groyne to be positioned at $x = 0$. (This simplifies the algebra while not affecting the answer.) We use Equation (6.29) to find the time, t, that gives the solution $y = 300$ m (i.e. the length of the groyne). Note that the first term in parentheses in Equation (6.29) is equal to 1 when $x = 0$. Further, the second term is equal to zero for $x = 0$. We therefore need to solve:

$$\therefore 300 = \sqrt{\frac{4.100000.t}{\pi}}\tan(\alpha_b)$$

$$\Rightarrow t = \left(\frac{300}{\tan(\alpha_b)}\right)^2 \frac{\pi}{400000}$$

$$\Rightarrow t = \left(\frac{300}{\tan(10°)}\right)^2 (0.00000785)$$

$$\Rightarrow t = 2895000(0.00000785)$$

$$\Rightarrow t = 22.7\,years$$

So the groyne will be full (i.e. the beach on the updrift side will reach the tip of the groyne) after 22.7 years in this case.

6.3.5 Numerical solutions

Before discussing numerical procedures for solving the 1-line equation a brief introduction to numerical techniques for solving partial differential equations is given.
This is intended to provide the basics for the reader and to raise some of the crucial
ideas. For more detailed description of numerical methods the reader should consult a
specialist textbook (e.g. Ames 1977).

Discretisation

Consider the 1-line equation for the evolution of the position of a beach contour.
Boundary values of y at A and B are prescribed as well as the initial beach contour
position. To solve this equation with a computational method we need to evaluate the
function at a discrete set of points. The first step is to divide the domain into discrete
elements by placing a number of nodes (or grid points) between A and B. These are
normally chosen to give an even distribution of points so that the distance between
adjacent points is the same; equal to δx say (see Figure 6.18).

The points may be considered as defining the ends or the centres of line segments of
length equal to δx. In Figure 6.18 the nodes have been numbered from 1 to $N + 1$ as
we move from A to B. The values of the ordinate at the grid points are written as x_1,
$x_2, \ldots, x_{N+1}$, and the corresponding values of beach position are $y(x_i)$ for $i = 1, 2, \ldots$
$N + 1$. Note that $x_i = (i - 1)\delta x$. By employing a Taylor expansion we can develop an
approximation to the derivative of y at any position along the x-axis. So, for example,
we write

$$y(x_i + \delta x) = y(x_i) + \delta x \frac{\partial y}{\partial x}\bigg|_{x=x_i} + \quad higher \quad order \quad terms \tag{6.36}$$

Ignoring the higher order terms and rearranging the remaining terms yields an approximation for the derivative of y at the point x_i:

$$\frac{\partial y(x_i)}{\partial x} \approx \frac{y(x_i + \delta x) - y(x_i)}{\delta x} = \frac{y(x_{i+1}) - y(x_i)}{\delta x} \tag{6.37}$$

The approximation in Equation (6.37) is termed 'first-order' accurate as terms up to
and including those involving δx raised to the first power are retained. It is also termed

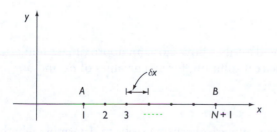

Figure 6.18 Discretisation of a line segment.

a 'forward difference' because the derivative at grid point i is determined from values of y at grid points with index i or greater. Higher order accuracy can be obtained through approximations involving additional terms. As the value of y is specified at grid points, derivatives are often specified at the same points. However, it is possible to define the derivatives at any intervening location by interpolating between the values of y at adjacent grid points. This can have some desirable computational properties and 'staggered grid' methods, as they are sometimes termed, involve calculations at grid points $x_1, x_2, \ldots, x_{N+1}$ and 'half-points' $x_{3/2}, x_{5/2}, \ldots, x_{N+1/2}$, which are simply the midpoints between the original grid points. For example, by expanding the values of shoreline position at $x_i + \delta x/2$ and $x_i - \delta x/2$ as in Equation (6.37), and subtracting the second from the first yields the approximation

$$\frac{\partial y(x_i)}{\partial x} \approx \frac{y(x_{i+1/2}) - y(x_{i-1/2})}{\delta x} \tag{6.38}$$

which is accurate to second order because the terms in $(\delta x)^2$ cancel. This approximation is known as a 'central difference' as it is symmetric about the point x_i.

We still have the matter of how we estimate the value of y at half-points. One option is to use linear interpolation so that

$$y\left(x_{i+\frac{1}{2}}\right) \approx \frac{y(x_i) + y(x_{i+1})}{2}$$

Substituting this expression into Equation (6.38) gives the central difference in terms of values at grid points:

$$\frac{\partial y(x_i)}{\partial x} \approx \frac{y(x_{i+1}) - y(x_{i-1})}{2\delta x}$$

as illustrated in Figure 6.19. Second derivatives may be estimated in a similar manner.

The main concepts that are used to analyse numerical methods are convergence, accuracy and stability.

Convergence

A numerical scheme is convergent if the numerical solution approaches the real solution as and tend to zero.

Accuracy

The accuracy of a numerical scheme describes how close an approximate numerical solution can be expected to be to the real solution, for given values of δx and δt.

Stability

Analysis of the stability of a particular numerical scheme seeks to determine whether the method will find a solution.

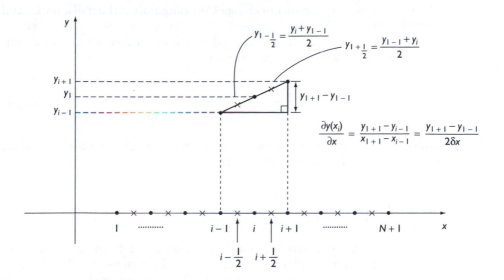

$$y_{1-\frac{1}{2}} = \frac{y_i + y_{1-1}}{2}$$

$$y_{1+\frac{1}{2}} = \frac{y_{1-1} + y_i}{2}$$

$$y_{1+1} - y_{1-1}$$

$$\frac{\partial y(x_i)}{\partial x} = \frac{y_{1+1} - y_{i-1}}{x_{1+1} - x_{i-1}} = \frac{y_{1+1} - y_{1-1}}{2\delta x}$$

Figure 6.19 Staggered grid and finite difference approximation.

Methods for determining the accuracy, stability and convergence properties of a wide range of numerical schemes may be found in Ames (1977). Approximating derivatives with respect to time also requires some care. For example, the 1-line equation used for analytical solutions may be approximated by

$$y(x_i, t_{j+1}) \equiv y_{i,j+1} \approx y_{i,j} + \frac{\delta t K}{(\delta x)^2}(y_{i+1,j} - 2y_{i,j} - y_{i-1,j})$$

where we have used the forward difference approximation

$$\frac{\partial y_{i,j}}{\partial t} \approx \frac{y_{i+1,j} - y_{i,j}}{\delta t}$$

and the subscripts i and j refer to location and time respectively.

This type of scheme is termed 'explicit' because the values of the dependent variables at the new time ($t = (j + 1)\,\delta t$), may be found solely in terms of values at the current time ($t = j\delta$.). Explicit schemes are generally straightforward to program but require small time steps and grid spacing for stability. In contrast, 'implicit schemes' do not impose such severe limitations on time step. In this type of scheme the new values of y at time $t = (j+1)\,\delta t$ along the shoreline are calculated simultaneously. The values of the dependent variables at $t= (j+1)\,\delta t$ are found in terms of their values at time $t = j\,\delta t$. and $t = (j+1)\,\delta t$. Implicit schemes are more difficult to program than explicit schemes but allow larger time steps to be taken. However, the time step cannot be increased indefinitely without compromising the accuracy of the computed results. If solutions are sought over periods of many years, a trade-off between stability, accuracy and computational effort may be necessary. This will almost certainly require 'trial and error' computations to assess the robustness of the results to changes in computational details, initial and boundary conditions.

Equations (6.21, 6.22, 6.25) can be rearranged into finite difference forms in a number of ways. A simple explicit numerical scheme is:

$$y_{i+1,j+1} \equiv y(x_{i+1}, t_{j+1}) \approx \frac{1}{d_c}\left(\frac{Q_{i+1/2,j} - Q_{i-1/2,j}}{\delta x}\right)\delta t + y_{i+1,j} \tag{6.39}$$

The subscripts i and j refer to location and to time respectively. As the values of y alter, there will be a corresponding change in α; so that

$$\alpha_{i,j+1} = \alpha_0 - \tan^{-1}\left(\frac{(y_{i+1,j+1} - y_{i,j+1})}{\delta x}\right) \tag{6.40}$$

for instance. The new transport rate $Q_{i,j+1}$ can be calculated (e.g. from Equation 6.23), and its values at half-points obtained by interpolation. Equations (6.39) and (6.40) may be used as the basis for a computer program. This can be applied to a simple problem in which the waves approach the shore from one direction. The solution is started with initial values of y and Q at initial time t. The effect of an intercepting feature is to reduce $Q_{i,j+1}$, say to zero. The equations are solved sequentially for all grid points at $t + \delta t$, $t + 2\delta t$ and so on.

Where the direction of the incident waves varies (as will be the case in most real situations) the above approach requires modifications. Some care is also needed in selecting the magnitudes of the distance and time differences (δx, δt) to avoid problems of numerical stability in explicit schemes. A description of an implicit numerical scheme for solving the one-line equation may be found in Kamphuis (2001).

The 1-line model has proved to be remarkably robust, despite its simplicity, and has been used widely for research and design in numerous applications (see e.g. Bakker *et al.* 1970; Ozasa and Brampton 1980; Kriebel and Dean 1985; Hanson *et al.* 1988; Chadwick 1989b; Fleming 1990a,b; Kamphuis 1991; Hanson *et al.* 1997).

However, there are many situations in which understanding the beach plan shape is just one piece of the information required for successful scheme design. The contribution of cross-shore sediment transport cannot be neglected in many cases, especially in schemes that have a shore-normal element such as a groyne. One way of including cross-shore transport is to extend the 1-line model to predict the position of two or more contour lines simultaneously. Multi-line models have been developed (see e.g. Fleming 1994; Hanson *et al.* 1999), but they require detailed information about the distribution of the sediment transport rates, and this is not always available. An alternative approach is to include the effects of cross-shore transport as an additional source term in the 1-line model, rather than modelling cross-shore transport explicitly. Techniques following this method have been proposed and tested against historic shoreline position measurements by Hanson *et al.* (1997) and Reeve and Fleming (1997).

Situations that support strong variations of wave height along the shoreline segment of interest require a modification of the volumetric rate of sediment transport. Such cases occur in the vicinity of detached breakwater schemes in which the wave heights (and directions) are deliberately manipulated in order to control the nearby shore-line. The modifications are discussed in detail by Hanson (1987), Hanson and Kraus (1989) and Fleming (1990b), and replace Equation (6.23) with

$$Q = \left(H_s^2 c_g\right)_b \left(a_1 \sin(2\alpha_b) - a_2 \cos(\alpha_b) \frac{\partial H_s}{\partial x} \right)_b$$

where a_1 and a_2 are empirical parameters dependent upon sediment size, sediment density, beach slope and so on. Further details can be found in the online Coastal Engineering Manual. In modelling practice a_1 and a_2 are very often treated as calibra-tion parameters. The modified equation recognises that there are two components comprising the total sediment transport:

1   transport due to wave breaking obliquely onto the shoreline;
2   transport by currents caused by wave height gradients.

In the case of an offshore breakwater the wave height gradient creates currents into the lee of the structure. These, combined with reduced wave heights, result in deposition of material.

This formulation was extended and adapted by Hanson *et al.* (2006) to include the effects of tidal and wind driven currents in shallow water. The formula uses an averaged concentration and longshore current velocity for the surf zone, where the current may originate from breaking waves, wind, tides and alongshore gradient in wave height, and reads

$$Q = \frac{\varepsilon}{(\rho_s - \rho)(1-p) gw} P_b \cos\theta_b$$
$$\left[K_1 \left(\frac{5}{32} \frac{\pi\gamma}{c_f} \sqrt{g} A^{3/2} \sin\theta_b + \overline{V}_t + \overline{V}_w \right) - K_2 \left(\frac{\pi}{c_f \gamma^2} \sqrt{gh_b} \frac{\partial H_s}{\partial x} \right) \right]$$

where K_1 and K_2 are calibration coefficients adjusting to local conditions, $\overline{V}_t$ and $\overline{V}_w$ are the surf zone average tidal current and wind-driven current respec-tively, h_b is water depth at breaking point, C_f is bottom friction coefficient in the range of 0.005 to 0.01, w is the fall velocity, γ is the breaker index, $A = 9/4(w^2/g)^{1/3}$ is the sediment shape parameter and $P_b = \frac{1}{8}\rho g \left(H_s^2 C_g\right)_b$. is the shoreward wave energy flux and the coefficient ε is the portion of total amount of work stirred by wave breaking. This formulation was used by Wang and Reeve (2010) to simulate the beach response to the construction of a scheme at Happisburgh, UK, consisting of nine detached breakwaters.

6.4   Nearshore morphology

6.4.1   Introduction

In the previous Sections 6.2 and 6.3 we have seen how changes in the plan shape and cross-sectional profile of a beach may be predicted. In reality, beaches evolve in a three-dimensional manner with the plan shape and profile at any point along the shore altering simultaneously.

The simplified profile and plan shape models may be extended to simulate three-dimensional beach evolution.

In three-dimensional models the hydrodynamic and continuity equations for fluid and sediment are written in three dimensions. Three-dimensional models aim to predict the interaction between sediment transport and the hydrodynamics in three dimensions (longshore, cross-shore and depth directions) plus time, and involve much computer time and memory.

Three-dimensional models may be simplified into two-dimensional models describing evolution of the profile with depth and cross-shore position. Further details of this type of model may be found in de Vriend *et al.* (1993a, 1993b), Roelvink (1991), Sato and Mitsunobu (1991), Broker *et al.* (1991), Roelvink and Broker (1993), Nicholson *et al.* (1997). Figure 6.20 shows schematically how we move from one type of model to another.

If, in the depth-averaged model, we replace the cross-shore variation with a fixed, representative profile shape that does not vary with time or longshore position, the 1-line model is obtained.

The robustness, simplicity and numerical ease of the 1-line model has made it attractive to practitioners and researchers alike. This has led to the reverse process to that described above – namely, 'complication' as opposed to simplification. The 1-line idea

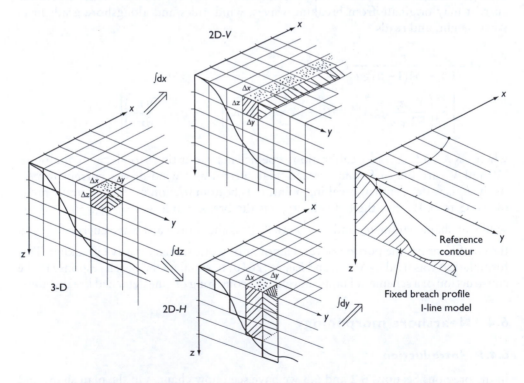

Figure 6.20 Relationship between 3-D, 2-D and 1-line morphological models (after Kamphuis 2001).

has been extended to simulate changes in beach profile. Describing the beach profile by two or more contour levels, treating the movement of each contour in the manner of a 1-line model and accounting for the cross-shore movement of sediment is the essence of N-line models. In an N-line model each of the N contours describing the beach profile is related to the others by a cross-shore transport calculation. The resulting computation is on a two-dimensional grid, which adapts in time since the location of the grid points move in the cross-shore direction. (e.g. Perlin and Dean 1983; Johnson and Kamphuis 1988; Fleming 1994; Dabees and Kamphuis 2000). Early examples of this approach were the analytical 2-line models of Bakker (1968) and Le Méhauté and Soldate (1978). These rely on a simplified formula governing cross-shore transport and an assumption that the profile is continually drawn (or relaxed) towards an equilibrium shape.

In an analogous way two-dimensional beach profile models can be linked together to provide a description of three-dimensional evolution of nearshore morphology. In this case, the grid flexes in the alongshore direction as the profiles alter orientation to remain perpendicular to the local shoreline. Dales and Al-Mashouk (1991) describe the application of such a model to the Norfolk coast in the UK.

Similarly, the statistical techniques for analysing beach profiles can be extended to cover observations of the nearshore seabed and beach levels. This is covered in Section 6.4.2.

Fully three-dimensional morphological models are not currently in common use by practitioners. There are several reasons for this. Firstly, such models require large amounts of measurements to calibrate and validate for a given site. Secondly, they are very computationally demanding. Thirdly, purely process-based models must account for sediment movements over the order of a wave period (i.e. seconds) whereas solutions may be required over periods of many years. The cost of such computations is prohibitive. Fourthly, the governing equations are non-linear and small changes in initial conditions can lead to very different solutions, particularly when many time steps are taken. Finally, the equations used for sediment transport are themselves approximate and there are considerable uncertainties associated with any choice of sediment transport formula.

The net result is that research has also moved towards developing 'regional scale' (up to ~10 km), morphological models that involve some means of smoothing or averaging the sediment movements over small time steps. This allows larger model time steps to be taken, at the cost of losing some resolution in space and time.

6.4.2   EOF methods for beaches and the nearshore bathymetry

The EOF method used for analysing beach level changes along a line may be extended to cover variations in an additional dimension. However, there is not a unique way to do this. Three methods for which results have been published are:

1   Repeated expansion
Here, the seabed levels $h(x,y,t)$ are expanded as the product of two sets of functions $A(x,t)$ and $B(y,t)$ such that

$$h(x,y,t) = \sum_n A_n(x,t)B_n(y,t) = \sum_n \left\{ \left(\sum_m d_{nm}(x)e_{nm}(t) \right) \cdot \left(\sum_l f_{n\rho}(y) \cdot g_{n\rho}(t) \right) \right\} \quad (6.41)$$

Each $A_n(x,t)$ is a function of x and t and may be expressed as an expansion in terms of functions depending on x and t, as for beach profiles. The $B_n(y,t)$ may, similarly, be expressed in terms of functions of y and t.

The functions A_n and B_n at any time $t = t'$ are determined by applying the methods of Section 6.2.2 to expand $h(x,y,t')$ as functions of x and y. This procedure is repeated for each time interval to generate the set of functions $A_n(x,t)$ and $B_n(y,t)$. These are then analysed to separate the time and space dependence following the method in Section 6.2.2. One drawback of this method is that a large number of eigenfunctions are produced and there are two sets of functions describing the time dependence of the measurements.

2   Multiple expansion

In this case $h(x,y,t)$ is expressed as the product of three sets of functions:

$$h(x,y,t) = \sum_n a_n(x)\, b_n(y)\, c_n(t) \tag{6.42}$$

a_n are determined as the eigenfunctions of the correlation matrix obtained by averaging over the longshore (y) and time, t. The b_n are calculated similarly by averaging over the cross-shore (x) and time, t. The $c_n(t)$ are found by substitution of a_n and b_n in Equation (6.42) and using the orthonormality of the eigenfunctions (see Section 6.2.2). Hsu *et al.* (1994) describe the application of the repeated and multiple expansion methods to a set of beach measurements along the Redhill Coast, Taiwan. Both methods produced useful results. However, they found the EOFs computed with the second method easier to interpret and to relate to the observed changes in the beach and wave conditions.

3   Extended one-dimensional expansion

In this method $h(x,y,t)$ is expressed as the product of two sets of functions; one containing the spatial variation and one the temporal variation.

$$h(x,y,t) = \sum_n u_n(x,y)\, v_n(t) = \sum_n w_n(\xi)\, v_n(t) \tag{6.43}$$

Values of h are known at a discrete set of (x,y) coordinates, say (x_p,y_q) with $1 \leq P$ and $1 \leq Q$. These points can be covered using a single index, r, such that $r = p + (q-1)p$ for example, as shown in Figure 6.21. Using such an ordering of points allows the 2-D spatial dependence to be represented as a 1-D set of values. These may then be analysed in the same manner as a beach profile. The correct spatial pattern of each spatial EOF is obtained by plotting the 1-D set of values in their correct (x,y) position according to the ordering formula adopted at the outset.

Reeve *et al.* (2001) used this method to analyse the changes in morphology of offshore sandbanks on the east coast of the UK over periods of many decades. Good convergence properties were shown on this dataset.

The first method has the advantage that it uses the 1-D EOF technique but may produce an expansion that does not converge quickly, making it more difficult to identify important patterns in behaviour. Method 2 is more complicated than the 1-D EOF technique but overcomes the disadvantages of method 1. Method 3 also has the

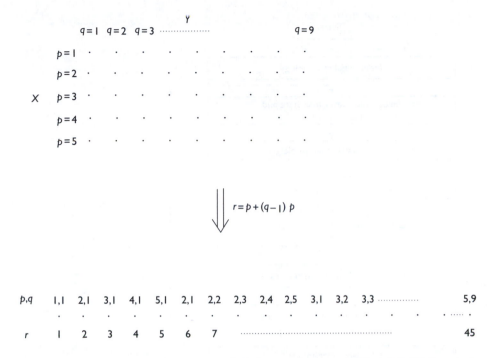

Figure 6.21 Enumeration of points in a 2-D dataset for the modified 1-D EOF analysis.

advantage of using the 1-D EOF technique but requires some careful bookkeeping so that results are ordered correctly.

6.4.3 Combined wave, tide and sediment transport models

A deterministic, process-based approach to predicting morphological evolution demands an explicit description of sediment transport due to waves and tidal currents and maybe also wind. We consider the first two processes only here. One of the difficulties with this approach is that there is a large range in the timescales of sediment transport due to waves and tides. In principle, it would be possible to set up a hydrodynamic model that described tidal and wave motion, and link this via sediment transport formulae, to an equation that described the changes in seabed elevation. In practice this is neither a feasible nor efficient approach due to the reasons outlined in Section 6.4.1. Further discussion of these issues is given by de Vriend (1997).

Progress has been made by averaging the effect of some of the processes over time. Nevertheless, the time-varying hydraulic and morphological conditions are calculated separately and dealt with in a 'sequential' manner. This is shown schematically in Figure 6.22. In some models, as suggested in Figure 6.5, an extra check is made before updating the morphology. If the changes are less than a pre-set amount then the morphological updating is omitted. This test checks to determine whether the bathymetry has changed sufficiently to warrant recalculating things like wave transformation,

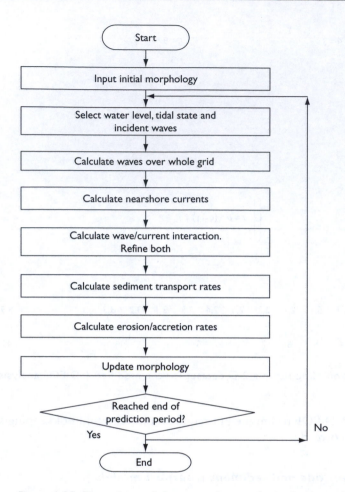

Figure 6.22 Flow chart of the general procedure of a coastal morphological model.

so there can be several iterations for which the changes are accumulated before a complete remodelling iteration is performed, thus saving processing time.

Given the initial morphology and environmental conditions the sediment transport over a time step is calculated. The corresponding erosion and accretion is calculated and the morphology updated accordingly. This new morphology is then taken as the initial condition for the next step, for which new wave and water level conditions are specified. The sediment transport and morphological updating over this new step is calculated as before to complete the second step. The prediction continues in this manner until the equations have been advanced over the desired period. The forcing conditions (wave and tidal state), and the morphology are considered constant over the duration of a time step. The length of the time step must be determined carefully; for it must be small enough to resolve important changes in the forcing conditions and to ensure numerical stability and accuracy, but not so small that the computational effort is excessive.

As this type of model treats the forcing and morphology as steady over a time step, the time step must also be sufficiently small to resolve significant changes in morphology. The link between changes in bathymetry and the effect it may have on the steady forcing conditions is critical. What constitutes a significant change in bathymetry will depend on the water depth at the location, the state of the tide and the wave conditions. This dependence is nonlinear, so halving the time step may not necessarily halve the sediment transport or morphological change. 'Resolution invariance' of the stability of the solution to changes in time step is a desirable property of any model and provides a useful check on the robustness of the predictions. (Resolution invariance does not guarantee you have obtained a correct answer, but lack of invariance suggests that the physical processes being modelled are not being adequately resolved.)

Such models make an implicit assumption that the forcing conditions are slowly varying in comparison to the time step. Hence considering a sequence of steady conditions, for example at different stages during a tidal cycle, provides an adequate representation of time variation. Examples of the application of this type of model can be found in Nicholson *et al.* (1997), who compared the performance of five commonly used area models in simulating a physical model experiment and an idealised benchmark test.

6.5   Long-term prediction

6.5.1   *Limits on predictability*

The previous sections have alluded to the presence of practical limitations on the scale and duration of using deterministic process models to predict over periods of months, seasons or years. These limitations arise through the potential for errors associated with the numerical approximation of derivatives to accumulate and eventually dominate the procedure rendering the solution useless. Also, integrating a model with a time step of a few seconds over the period of years would need a prohibitive amount of computer time. However, with the continuing advances in computer processing power this will become less of a constraint. Setting aside these two issues, serious doubts remain as to whether such models can accurately describe the long-term behaviour (or morphological 'climate') of the shoreline. The processes described in these models have been validated against short-term coastal response to major forcing, such as severe storms. To what extent these formulae can reproduce the underlying long-term variability is an open question.

In addition to these practical difficulties there are also theoretical arguments that suggest there may be an inherent uncertainty or limit of predictability in the equations used for process modelling. Coastal hydrodynamics and sediment transport are strongly nonlinear, that is, the forcing and response are not simply related. For example, the impact of a 4 m high wave cannot be predicted by simply scaling the response to a 2 m high wave. It is the nonlinearity of the equations that can (but not necessarily) lead to slightly different bathymetry configurations evolving to radically different states under the same wave conditions, irrespective of the time step or spatial resolution used in the model. Similar behaviour has been observed in laboratory and field conditions. Newe *et al.* (1999) used constant wave conditions at a fixed water level and found that the beach profile evolved until it reached one of two equilibrium states, depending on the imposed initial morphology. Lippman *et al.* (1993) observed nearshore bar systems switching between quasi-stable configurations. This type of behaviour has been termed 'deterministic chaos' or simply chaos. Baas (2002) has discussed the concepts of strange attractors, fractal dimensions, chaos theory, fractals and self-organisation, providing a useful review of their application in many areas of geomorphological research. Southgate *et al.* (2003) discuss a number of methods developed in the science of nonlinear process and explain how these can be applied in the context of coastal engineering. While such methods may not be mainstream techniques in coastal engineering, they are potentially of more than just academic interest, and the practitioner should at least be aware of their existence. It is in this context that a little space is devoted here to describe the problem that led to the 'discovery' of chaos and the development of chaos theory. Readers for whom the mathematical details are of little interest should nevertheless take the main point which is that even with three simple-looking equations, incredibly complicated and unpredictable behaviour can result. The equations are much simpler than those frequently used to describe hydrodynamic flows and sediment transport, but are not so dissimilar to simpler models used to describe morphodynamic evolution. If the system of equations used for morphological prediction exhibits chaos then this has major implications for the period over which useful deterministic predictions can be made. Even with perfect information about the nearshore bed levels and environmental forcing the predicted evolution will have an apparently random character. In practice, uncertainties in initial conditions, measurement errors and numerical errors are likely to be exacerbated and lead to a reduction in the period over which useful predictions can be obtained.

Deterministic chaos was discovered in the field of numerical weather prediction by Lorenz (1963). His work has been the subject of many studies by applied mathematicians and physicists. A fascinating historical account of the discovery and development of 'Chaos theory' may be found in Lorenz (1993). A brief description of his model is given here as an illustration of the nature of deterministic chaos. Lorenz was interested in developing a simplified model of the Earth's weather climate and derived a set of three nonlinear equations that described the change in global temperature and wind field distributions over time.

The governing equations for the three variables x, y and z are

$$\frac{dx}{dt} = -\sigma(x + y) \tag{6.44}$$

$$\frac{dy}{dt} = -xz + rx - y \tag{6.45}$$

$$\frac{dz}{dt} = xy - bz \tag{6.46}$$

The three constants b, σ and r determine the behaviour of the system. When the constants have the values $b = 8/3$, $\sigma = 10$ and $r = 28$ the system exhibits what has been termed 'chaotic' behaviour. This is perhaps an unfortunate terminology, with its implications of randomness. In fact, the solutions to the equations exhibit highly complex structure. Equations (6.44), (6.45) and (6.46) represent a deterministic dynamical

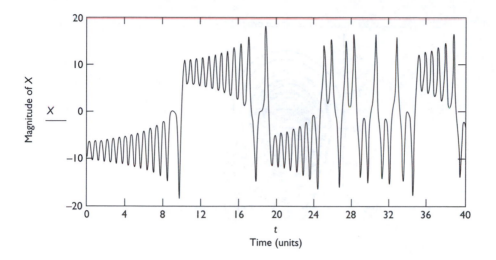

Figure 6.23 Chaotic solution for the variable x in the Lorenz equations.

system. That is, for given initial values of x, y and z, the subsequent behaviour of the system is, in principle, determined exactly. However, no general analytical solution to Equations (6.44) to (6.46) is available and so the equations are solved by numerical means. Here, the equations have been solved with the values of the constants quoted above using a fourth-order Runge–Kutta scheme, with a time step equal to 0.018. Figure 6.23 shows a time series up to t = 40 (consisting of 5000 points) for the variable x as it evolves from the initial conditions $x = -10$, $y = -10$, $z = 30$. After an initial stage in which x adjusts towards its 'equilibrium' value there follows episodic periods of fluctuation in an apparently random manner.

In order to classify the behaviour of dynamical systems the concept of 'phase space' has been introduced. For a dynamical system with n variables the solution at time t defines a point in n-dimensional space (the coordinate axes measure the value of each of the n-variables at any time). As t increases continuously from an initial value, so the point representing the solution traces out a path in this space. This path is referred to as the trajectory of the initial point, and the n-dimensional solution space is termed the 'phase space' of the system. For example, in the Lorenz system there are three variables and the trajectory of any initial point will be a curve in three-dimensional space. The projection of this trajectory onto each of the coordinate planes can be obtained by plotting the coordinate pairs $x(t)$, $y(t)$; $y(t)$, $z(t)$; and $x(t)$, $z(t)$.

In many dynamical systems the trajectory of any initial point that starts in some region B of n-dimensional phase space eventually limits on a fixed subset A of phase space, known as the attractor. The behaviour of the system is reflected in the structure of the attractor. In many cases the attractor will occupy only a small portion of the phase space. Methods for analysing the behaviour of dynamical systems and time series of observations have therefore focussed on the structure of trajectories and attractors.

An example of a solution trajectory in the three-dimensional phase space of the Lorenz system is shown in Figure 6.24. This has been constructed from the numerical

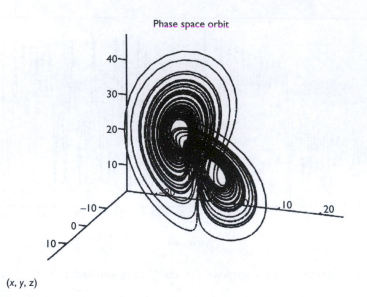

Phase space orbit

(x, y, z)

Figure 6.24 3-D trajectory of chaotic solution to the Lorenz system with parameters as described in the text.

solution described above and consists of the 5000 (x,y,z) coordinates joined in chronological order. Projections of this trajectory onto the x-y and x-z planes are shown in Figure 6.25, exhibiting the familiar 'figure of eight' and 'butterfly' patterns respectively.

Imperfectly known initial conditions automatically introduce errors into the solution. For certain choices of the constants, such as those used to produce Figure 6.23, these errors amplify so that arbitrarily close initial conditions will eventually lead to widely different solutions. An example of two diverging solution trajectories is shown in Figure 6.26. Equations (6.44), (6.45) and (6.46) have been integrated from the initial conditions $x = -10$, $y = -10$, $z = 30$ and $x = -10.05$, $y = -10$, $z = 30$ using the same method and step size as described above. The two sets of (x,z) values have been plotted as two curves for the first 600 steps. For the first 500 steps or so the solution curves are extremely close. After this stage the two solution curves diverge dramatically. In Figure 6.26 this occurs when the two curves approach the origin. One turns southwest back into the left-hand quadrant while the other turns to the northeast into the top right-hand quadrant.

The separation between the trajectories at a given time can be measured in several ways. Here, we define the separation by the Euclidean distance between the points on each trajectory. Thus if $x_i(t)$, $y_i(t)$, $z_i(t)$ denotes the point on the i'th trajectory at time t then the separation between trajectories i and j, s_{ij}, is given by

$$s_{ij}(t) = ((x_i(t) - x_j(t))^2 + (y_i(t) - y_j(t))^2 + (z_i(t) - z_j(t))^2)^{0.5}$$

(6.47)

The separation between the two trajectories initially increases with time although for any particular pair of trajectories there are occasions when they become closer

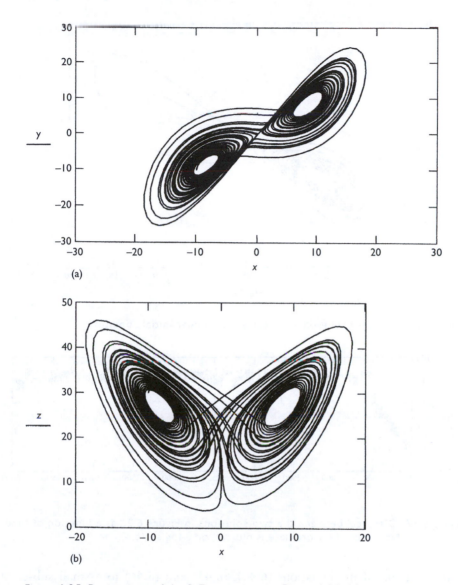

Figure 6.25 Projections of the 3-D trajectory in Figure 6.24: (a) projection onto the x–y plane; (b) projection onto the x–z plane.

again. The separation between the two solutions shown in Figure 6.26 is plotted in Figure 6.27 as a function of time. Note that for the initial few hundred steps the solutions are fairly close.

The distance between the solutions increases approximately exponentially for about the first 500 steps. Subsequently, the distance between the solutions continues to vary but does not continue to grow exponentially. Rather, the distance appears to vary about a value of approximately 20 units. The limit on the distance between the solutions of similar initial conditions is explained by the existence of an attractor that

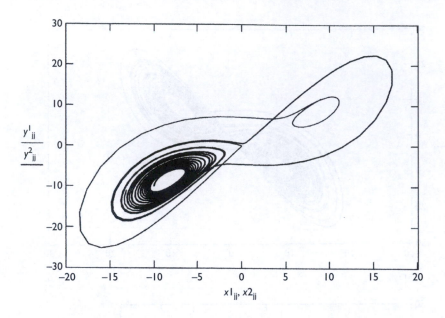

Figure 6.26 Diverging solution trajectories for similar initial conditions.

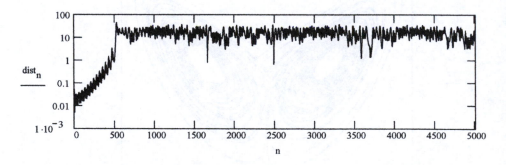

Figure 6.27 'Distance' between the two solutions in Figure 6.26 as a function of time. Note that the y-ordinate is plotted on a log scale.

limits the solution of the Equations (6.44), (6.45) and (6.46) to a small subset of three-dimensional phase space. That is, the solution along either trajectory will be constrained to the surface shown in Figure 6.24. The maximum separation will occur when points on the two trajectories are at opposite ends of the surface.

This form of behaviour is certainly analogous to that of the atmosphere and also ostensibly to that of coastal morphology. The divergence of solutions from almost identical initial conditions is reminiscent of the 'forecasting problem'. The forecasting problem arises when solving the governing deterministic non-linear equations from an imperfectly defined initial state to make forecasts of a future state of a system. (It is well known that the atmosphere is not predictable in this sense (see e.g. Palmer 1999) – a numerical weather prediction generally loses any skill after about a week in most circumstances.) That the same is true for coastal morphology has yet to be established (e.g. de Vriend 1997). However, the form of the governing equations is similar and so the potential for analogous behaviour can be anticipated.

Phase space reconstruction

When the equations governing a dynamical system are known the system's behaviour can be determined through numerical integration. In practice, a time series of measurements of a particular quantity is often the only information available. Neither the trajectory nor the governing equations are likely to be accessible. Further, the measured quantity is usually dependent on a number of other variables. For example, the sea surface elevation at a point will depend on several factors including astronomical tide, 'surge' due to atmospheric pressure variations and surface wave activity.

In this situation it may be possible to reconstruct the essential features of the trajectories and attractors from the observations alone. A number of methods to do this have been devised, including the 'Ruelle-Takens method' and Singular Value Decomposition (SVD). For further details the reader is referred to the papers of Takens (1981) and Broomhead and King (1986). Reeve *et al.* (1999) have applied these ideas to beach profile measurements as a means of classifying the temporal behaviour of beach morphology. Baas (2002) discusses how these types of ideas can be used to understand the formation of dunes. Southgate *et al.* (2003) give some examples of applying these type of techniques to beach profiles and Gunawardena *et al.* (2008) describe the application of fractal methods to analyse changes in beach morphology.

Determining whether a set of equations can support chaotic behaviour is one thing, but establishing whether a system is exhibiting deterministic chaotic behaviour on the basis of analysing measurements is quite another. To date, although a number of techniques have been developed to investigate the presence of chaotic behaviour in noisy data (e.g. Broomhead and King 1986) no methods for establishing the equations or even their number governing observed behaviour have been developed.

6.5.2 Probabilistic methods

Accepting that there will be uncertainties in any prediction, but that predictions are required to manage development and conservation in the coastal zone, prompts a probabilistic approach, that is, the environmental forcing and the morphological response are treated as stochastic processes (see Section 6.2.1). Probabilistic models for coastal engineering are at an early stage of development, and this is an area of active research. Some methods are outlined in this section, but it is not meant to be an exhaustive review of such methods.

From a probabilistic perspective, the output of a deterministic model is treated as one possible realisation of the beach evolution process. To obtain useful and meaningful results in this way it is necessary to:

1   run the model many times to generate a set of realisations;
2   calculate sample statistics from the realisations to infer characteristics of the whole population of possible outcomes;

3   choose the conditions for creating the realisations so that the set of realisations
can give a significant and unbiased estimate of the population statistics.

This procedure is termed Monte Carlo simulation and is shown in Figure 6.28.

The output of this approach is not a single, well-defined solution for the coastal
bathymetry at a given time. Rather, it gives the statistics of the solution; for example,
the average and variance. This can be extremely useful information for shoreline management, but must be interpreted with care because the mean might not be a member
of the population. As a simple example, when throwing a die the result will be 1, 2,
3, 4, 5 or 6, but never 3.5 which is the average value. Conversely, each realisation is
a valid outcome and examination of individual realisations can provide some insight
into beach response.

The input data requirements for this type of model are also different. Firstly, they
require sequences of environmental forcing events that are representative of the period
to be simulated. Knowing the probability distribution of wave heights, for example, it

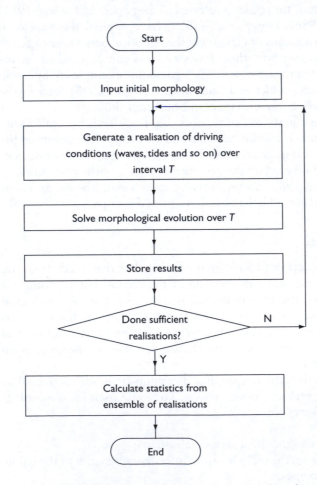

Figure 6.28   Flow chart of the general procedure for a Monte Carlo simulation.

is possible to generate realisations of time sequences of wave heights with the appropriate statistics. For more complicated models the distributions and correlations between the forcing variables are required – for instance, the correlations between wave height and wave period, or wave height and direction. This form of information is not routinely available and so, to date, the Monte Carlo simulation method has been used with simplified morphological models. An example is shown in Figure 6.29 from Vrijling and Meijer (1992) who used a Monte Carlo simulation with a 1-line model to estimate the likely bounds on beach position near a port development.

To introduce further realism to such simulations it is necessary to consider the characteristics of the changes in the driving forces with time. The frequency of storms and calm period and their relative positioning in a time series of events will be important. A healthy beach will be much more able to resist the effects of a storm than one that has yet to recover from the effects of an earlier storm. The sensitivity of predictions to the temporal correlation or 'chronology' of the wave climate has been investigated by several researchers. Southgate (1995) used a deterministic profile model to simulate the evolution of an initially uniform 1:80 slope. The driving conditions were a uniform tidal variation and a measured four-month sequence of significant wave heights (H_s) and zero-crossing period T_z at three-hourly intervals. The wave data was split into five segments and the model run with different orderings of the five segments (in all 120 sequences). The predictions were then treated as realisations of a Monte Carlo simulation to calculate profile statistics including range, mean and standard deviation. The spread in results is shown in Figure 6.30. The range is of the same order as the erosion of the mean from the initial state and Southgate and Brampton (2001) suggest this indicates that chronology can be important.

The studies by Dong and Chen (1999) investigate this further by including random temporal variability in a Monte Carlo model based on beach plan shape models

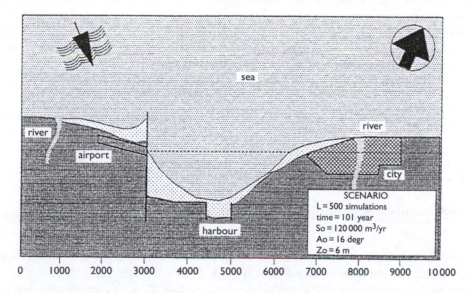

Figure 6.29 Example of a Monte Carlo simulation with a 1-line model, after Vrijling and Meijer (1992).

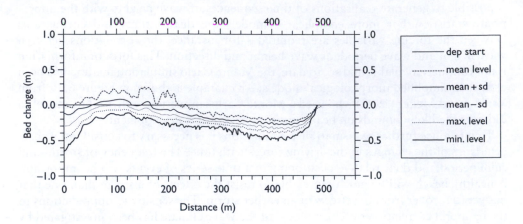

Figure 6.30 Probability of exceedance of the maximum seaward extent of the shoreline during a five-year period, after Lopez de San Roman and Southgate (1998).

modified to account for some cross-shore sediment exchange. On the basis of their simulations, they conclude that chronology is important but suggest it becomes less so as the period of prediction increases. An alternative approach to Monte Carlo simulation has been proposed by Reeve and Spivack (2001). Using the 1-line beach plan equation they develop expressions for the first and second moments of beach position by performing the averaging over realisations analytically. The inputs to this model are the initial beach configuration, the probability distribution of the forcing and its temporal correlation function.

In situations where beach measurements are available but wave and water level information is not it may still be possible to adopt a quasi-probabilistic approach. For example, consider the 1-line beach plan shape equation. Considering this to describe the instantaneous evolution of the beach we can write the beach position, y, as the sum of a time average, $\bar{y}$, and the deviation from the average, y'.

Thus

$$\frac{\partial \bar{y}}{\partial t} + \frac{\partial y'}{\partial t} = \left(\bar{K} + K'\right)\left(\frac{\partial^2 \bar{y}}{\partial x^2} + \frac{\partial^2 y'}{\partial x^2}\right) + \bar{S} + s' \tag{6.48}$$

where the external source/sink term $S = \bar{S} + s'$, and the diffusion coefficient $K = \bar{K} + K'$. Taking time averages of Equation (6.48) leads to

$$\frac{\partial \bar{y}}{\partial t} = \bar{K}\frac{\partial^2 \bar{y}}{\partial x^2} + \bar{S} + \overline{\frac{K'\partial^2 y'}{\partial x^2}} \tag{6.49}$$

where the last term represents the time-mean contribution of short-term fluctuations (turbulence) to the time-averaged morphology. (Time averages are considered to be taken over a finite period that is sufficiently large so that $\bar{y}'$, $\bar{K}'$, $\bar{S}'$, are all zero, but not so large that $\frac{\partial \bar{y}}{\partial t}$ is negligible.)

If $\bar{y}(x,t)$ is known or can be estimated at two distinct times t_1 and t_2, say, the inverse methods described by Spivack and Reeve (2000), may be used to estimate $\bar{K}$ and $\bar{S} + K'\frac{\partial^2 y'}{\partial x^2}$ averaged over the interval $t_2 - t_1$. The latter is the distribution of sediment sources/sinks required for the observed beach position to obey the 1-line equation.

If $\bar{y}(x,t)$ is known at a number of times then this procedure may be used repeatedly to calculate a series of integrated sediment source terms. These source functions may, under suitable assumptions, be used to characterise a source term in the forward integration of Equation (6.47) to produce predictions of future position, and estimates of the mean and bands of variation about this mean (see Reeve and Fleming 1997).

There are similarities between this approach and that of the chronology method of Southgate. Both rely on using past measurements to generate the driving mechanism for generating predictions. However wave chronology involves changing the sequence of wave events to generate new realisations of wave series, while the inverse method uses past measurements to generate a set of different sediment source functions. As a consequence, neither method can predict major changes in the morphological regime represented in the historical measurements; although the inverse method will automatically extrapolate any long-term trends that are present in the measurements.

6.5.3 Systems approach

Another alternative for modelling long-term morphological evolution is the systems approach, so-called because it treats a coastal region as being a system of linked elements. Each element is a morphological feature such as a channel, sandbank, ebb delta and so on. In general terms the model consists of establishing a sediment budget such that total sediment is conserved. Subsidiary equations describe sediment exchange mechanisms between elements. An example is shown in Figure 6.31 which illustrates a five-element system. It represents a tidal inlet, comprising a barrier, tidal flats, a channel, an ebb delta seaward of the barrier and an element that denotes the region external to the inlet.

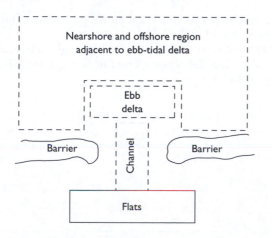

Figure 6.31 Systems approach to an ebb-tidal delta/inlet.

This type of model can be helpful in studying the dynamics of sediment exchanges between different elements in response to external forcing. Van Goor *et al.* (2001) have used a systems model to investigate the impact of sea level rise or coastal inlets, while Niedoroda *et al.* (2001) studied the impact of the discharge of sediment by a river at an open coast. These ideas have since been developed further, and in the UK a Coastal System Mapping (CSM) methodology was piloted to proof of concept stage for the Environment Agency (Whitehouse *et al.* 2009). CSM is a new approach to developing conceptual models of large-scale coastal geomorphic systems, by finding a formalisation of geomorphological and engineering knowledge into a map of the interactions between them.

Modelling morphological systems (i.e. with the focus being on the behaviour of the system rather than small elements within it) is at an early stage of development. Current techniques provide a way of organising information and thinking in a structured manner. Techniques tend to be qualitative or, if quantitative, then at a very coarse level of resolution. Nevertheless, it is an area of active research showing some promise for practical applications.

Further reading

Dearing, J.A., Richmond, N., Plater, A.J., Wolf, J., *et al.*, 2006. Modelling approaches for coastal simulation based on cellular automata: the need and potential. *Phil. Trans. Roy. Soc.*, A 364, pp. 1051–1071.

French, J.R., Burningham, H. and Whitehouse, R.J., 2010. Coastal system mapping: a new approach to formalising and conceptualising the connectivity of large-scale coastal systems. *AGU Fall Meeting* 2010, EP23A-0765.

Gunawardena, Y., Ilic, S., Southgate, H.N. and Pinkerton, H., 2008. Analysis of the spatio-temporal behaviour of beach morphology at Duck using fractal methods, *Marine Geology*, 252(1–2), p. 38–49.

Karunarathna, H. and Reeve, D.E., 2008. A Boolean Approach to Prediction of Long-term Evolution of Estuary Morphology, *Journal of Coastal Research*, 24(2B), pp. 51–61.

Reeve, D.E. and Karunarathna, H., 2009. On the prediction of long-term morphodynamic response of estuarine systems to sea level rise and human interference, *Continental Shelf Research*, 29, pp. 938–950.

Southgate, H.N. and Moller, I., 2000. Fractal properties of coastal profile evolution at Duck, North Carolina, *J. Geophys. Res.*, 105(C5), pp. 11 489–11 507.

Walkden, M.J. and Rossington, S.K., 2009. *Characterisation and Prediction of Large Scale, Long-term Change of Coastal Geomorphological Behaviours: Proof of Concept Modelling*, UK Environment Agency, Report SC060074/PR1.

Chapter 7

Design, reliability and risk

7.1 Design conditions

7.1.1 Introduction

Coastal regions have always been a popular place for commerce, recreation and habitation. In many countries, land adjacent to the sea is much more valuable than inland. However, low-lying coastal plains are subject to flooding and cliffs are vulnerable to erosion. The prospect of accelerating sea level rise and storminess associated with climate change has heightened public awareness of the hazards faced by those living and working in coastal areas.

Economic and social pressures have led to the construction of defences to protect against flooding and erosion. There is a high degree of uncertainty in the conditions that may be experienced by a coastal structure, and strong economic pressure to restrict the cost of defences. As a result, defences are typically designed to withstand conditions of a specified severity (for example the storm conditions encountered once every 50 years on average), judged to provide an appropriate balance between cost on one hand and the level of protection on the other.

This automatically introduces a probabilistic approach from which concepts of uncertainty, reliability and risk arise naturally. In this chapter the necessary statistical techniques are introduced and used to describe some of the methods now used in the design of sea defences. Descriptions and applications of design formulae are covered in Chapter 9.

7.1.2 Extreme values and return period

In previous sections we saw how design conditions are expressed in terms of the structure's performance under unusual or 'extreme' conditions. It is often the case that a sea defence is to be designed to resist a condition so unusual that no similar condition may be found in available measurements or records. One way to proceed is to fit a probability distribution to the measurements and extrapolate this to find the conditions corresponding to the rarity of the required event.

The range of probability distributions that are used for design and the methods for fitting them to the measurements have been the subject of much study and the following sections provide an introduction and guide to some of the main ones currently used in accepted practice.

First, we need to make some definitions and introduce concepts from probability theory. The main variables we will be dealing with are water level, wave height, wave period, wave direction, beach level and beach slope. These control the conditions at the structure. There are derived quantities (sometimes termed structure functions), such as run-up and wave overtopping that also involve parameters describing the structure (e.g. wall slope, roughness and crest level).

All these variables are continuous functions of time. However, measurements are taken at fixed intervals resulting in a discrete set of values over time or time series. Typically, wave and water level records are available at hourly or three-hourly intervals, although this can vary according to the instrument and processing adopted. Suppose we have a time series of significant wave heights (Figure 7.1a). Each point of the time series can be considered to be an individual event with duration equal to the interval between successive points. Repeatedly throwing a die and noting the sequence

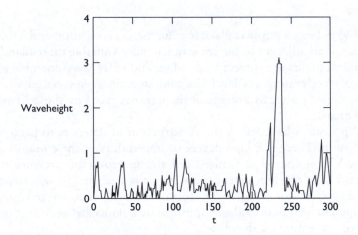

Figure 7.1(a) Time series.

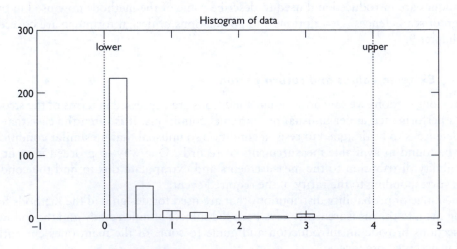

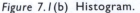

Figure 7.1(b) Histogram.

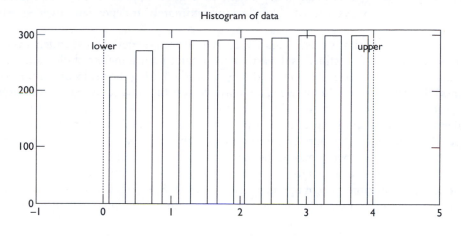

Figure 7.1(c) Cumulative histogram.

of results would have a similar output, with two important differences. First, the numbers on a die may take only certain fixed values (1, 2, 3, 4, 5 or 6) whereas wave heights may have any non-negative real value. Secondly, while throws of a die can be considered to be independent (that is the value obtained on one throw of the die is not related to previous values), consecutive measurements of wave height may not be. For many of the statistical manipulations that will be used later we require the events to be independent. It is therefore important to check that time series are independent. A convenient check is to calculate the autocorrelation function of the time series (see Appendix A). If this drops rapidly from its value of 1 at zero lag (full line in Figure 7.2), then the assumption that events are independent is reasonable.

If the autocorrelation drops more slowly (dashed line in Figure 7.2) this shows a similarity or correlation between consecutive values. Correlation of wave records has been investigated and for UK waters, measurements separated by more than 12

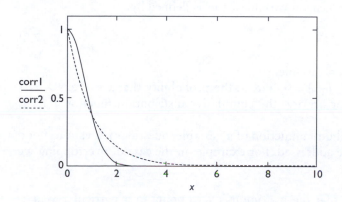

Figure 7.2 Autocorrelation functions: Gaussian (full line) and Exponential (dashed line).

hours may be treated as being independent to a good approximation (HR 2000). At a specific site, wave records may be approximately independent over shorter intervals.

Let us suppose the time series in Figure 7.1(a) comprises a series of independent events. Extreme events may be defined as those that are greater than some threshold value. The statistics of the values over the threshold may be studied using the peaks over threshold method described later. Alternatively, we may convert the time series into a histogram plot, such as in Figure 7.1(b), showing the number of occurrences of waves within a particular wave height band expressed as a proportion of the total number of events. The longer the time series, the more events and the narrower the bands can be made. In the limit of an infinite number of points, the bands can be made infinitesimally small and the frequency of occurrence tends to the probability, and the histogram tends to the probability density function (*pdf*) of the time series variable. Mathematically, the probability (*Pr*) that the significant wave height lies within some finite band may be written as:

$$Pr(h_1 < H_s < h_2) \approx n/N.$$

where N = total number of events and n = number of events for which $h_1 < H_s < h_2$.

$$\lim_{\Delta h \to 0} Pr\ (h_1 < H_s < h_1 + \Delta h) \quad = \quad Pr(H_s = h_1) = f_H\ (h_1) \tag{7.1}$$

where f is the probability density function.

The *pdf* is non-negative for all values and it is normalised so that its integral over all values is 1, that is:

$$\int_{-\infty}^{\infty} f(h)dh = 1 \qquad and \qquad f(h) \geq 0 \text{ for all } h \tag{7.2}$$

In practice, time series have a finite number of points and we may fit a *pdf* to the histogram heights using the mid-band point as the ordinate value. The *pdf* gives us the probability of an event of a given magnitude. For design, we are interested in the probability of a particular value being exceeded. This is most easily obtained from the cumulative distribution. For a general variable x this is defined by

$$F_x(X) = \int_{-\infty}^{X} f(t)dt = Pr(x \leq X) \tag{7.3}$$

where t is a dummy integration variable.

Note that $F_x(\infty) = 1$ and $F_x(-\infty) = 0$, and so the probability that x exceeds X is $1 - F_x(X)$ (see Figure 7.1c). For brevity, the cumulative distribution function $F_x(X)$ is sometimes written as $F(x)$.

Using the cumulative distribution function of a variable, questions relevant to flood or damage prediction may be addressed. For example, in the case of overtopping we might wish to know:

1   What is the probability that the maximum overtopping in a particular year is more than q m³/s/m?

2 In the next n years what is the probability that the highest annual maximum over-topping rate in the n years will be more than q m³/s/m?
3 What is the 1 in N-year annual maximum overtopping rate?

If $F(q)$ is the distribution function of the annual maxima of overtopping then the answers are (1) $1 - F(q)$; for (2) the distribution function of the largest n observations is given by $F_n = 1 - (1 - F(q))^n$. For (3), we need the concept of a return period.

A usual measure of the rarity of an event is the return period, R. The R-year event is the event that has a $1/R$ chance of being exceeded in any given year. It may be linked to the distribution function as follows. Suppose we have a time series containing N independent values of wave height $H_1, H_2, H_3, ..., H_i, ..., H_N$ at intervals of Δt, and we have a wave height threshold of H_t. Of the N values, assume m are greater than H_t. As before, we may approximate the probability of H_i exceeding H_t by $m/N \approx 1 - F(H_t)$. Equivalently, we may interpret this as meaning there will be m exceedances every N events (or over $N.\Delta t$ units of time), on average. That is, one exceedance every N/m events or $\dfrac{N.\Delta t}{m}$ units of time. $\dfrac{N.\Delta t}{m}$ is the return period and is usually expressed in units of years.

Example
A time series of wave heights is sampled at 3-hourly intervals and contains 292 200 independent values. Of the values only two exceed a given threshold wave height, H_t. What is the return period corresponding to the wave height H_t?

Solution
We have $\Delta t = 3$ hours, $N = 292\ 200$ (corresponding to 100 years), and $m = 2$. The return period is

$$\frac{292200}{2} \times 3x \frac{1}{24x365.25} = 50 \text{ years}$$

The last term on the left-hand side converts units of hours to years (24 hours per day and 365.25 days per year)

It follows that there is a finite chance that the design conditions will be exceeded during the life of a scheme. This probability of exceedance is usually referred to as the 'return period'. An event with a return period of T years is likely to be exceeded, *on average*, once in T years. The return period should not be confused with the design life of a scheme. For example, if the return period of an extreme event is the same as the design life, then there is a ~63 per cent chance that the extreme event will be exceeded during the period of the design life. Considering annual maxima the probability of exceedance and the return period are related by:

$$P = 1 - \left(1 - \frac{1}{T}\right)^L$$

where L is the design life. Figure 7.3 plots return period against duration for fixed values of probability of exceedance.

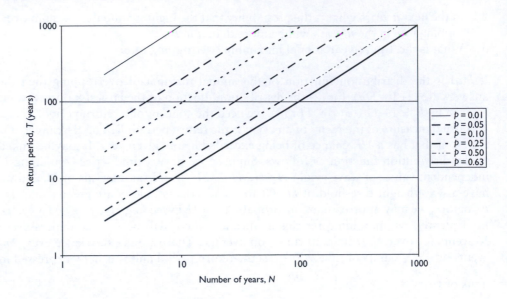

Figure 7.3 Exceedance probability as a function of event return period and duration.

A related concept that is sometimes used is the degree of security, defined as the probability of a given condition not being exceeded over a fixed period. (This is of course just 1 minus the probability of the condition being exceeded.) Graphs of the return period against the degree of security for fixed periods of 2 to 100 years are drawn in Figure 7.4. So, for example, a structure designed to withstand the 1 in 100-year conditions provides a degree of security of ~ 0.73 over a period of 30 years.

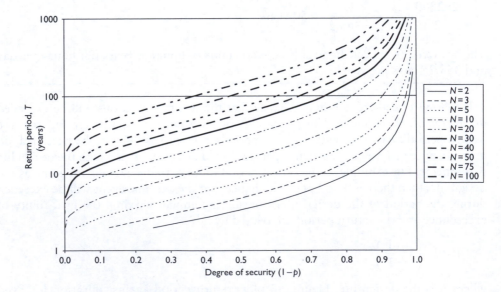

Figure 7.4 Duration as a function of event return period and degree of security.

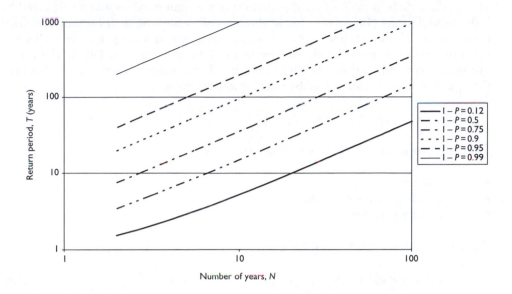

Figure 7.5 The degree of security as a function of return period and duration.

Figure 7.5 plots the return period against duration for fixed levels of the degree of security. This type of plot is useful when an engineer has to prepare a design on the basis of a specified probability of failure over the (known) design life. The required return period of extreme event against which to design can be read from the graph. For example, if a degree of security of 0.9 is required for a structure with design life of 10 years then the structure must be designed against 1 in 100 year conditions.

In some cases only the maximum value over a given period is recorded. For example often only the annual maximum water level is recorded. We have, therefore, one event per year and this event could have occurred at any time throughout the year.

To answer the last question (3), the N-year return value, q_N, is calculated from the equation

$$F(q_N) = \begin{cases} 1 - \frac{1}{N} & for \quad maxima \\ \\ \frac{1}{N} & for \quad minima \end{cases} \tag{7.4}$$

where $F(q)$ is the cumulative distribution function of the annual maximum overtopping values.

7.1.3   *Distribution of extreme values*

Consider a time series of values that we take as being random and independent, and which have the same distribution at each time point. The distribution of the maximum of a sequence of size N as $N \to \infty$, is the generalised extreme value (GEV) distribution.

The asymptotic behaviour of the distribution of maximum values was investigated by Fisher and Tippett (1978) who found three types of limiting distribution. The GEV encompasses all three types. The only requirement is that values that are well separated in time are approximately independent (see Leadbetter *et al.* 1983). These conditions are typically satisfied by all sea and beach condition variables. If X obeys the GEV(μ, σ, δ) distribution it has the distribution function

$$\Pr(X \le x) = \exp\left\{-\left(1 - \delta\left(x - \mu\right)/\sigma\right)^{1/\delta}\right\} \tag{7.5}$$

The three parameters of this distribution are

- μ – a location parameter
- σ – a scale parameter ($\sigma > 0$)
- δ – a shape parameter

The level, x_p, exceeded with probability P, i.e. $Pr(X > x_p) = P$, is given by

$$x_p = \mu + \sigma\left\{\left[\log(1 - P)^\delta + 1\right]/\delta\right\} \tag{7.6}$$

so x_p is the return level for return period *1/P* units of time (normally years). Two limiting distributions that are often used instead of the GEV distribution, but which are special cases of the GEV distribution are the Weibull and Gumbel distributions. These correspond to $\delta > 0$ and $\delta = 0$ respectively. If x follows the 2-parameter Weibull distribution it obeys

$$F(x) = 1 - e^{-\left(\frac{x}{\lambda}\right)^\kappa} \tag{7.7}$$

and if it follows the Gumbel distribution it obeys

$$F(x) = e^{-e^{-(x-u)/\xi}} \tag{7.8}$$

and if it follows the 3-parameter Weibull distribution it obeys

$$F(x) = 1 - e^{-\left(\frac{x-\theta}{\lambda}\right)^\kappa} \tag{7.9}$$

In the above, u and θ are location parameters; λ and ξ are scale parameters that are greater than zero; and κ is a shape parameter.

The theoretical justification for the GEV provides a basis for extrapolation beyond the data to long return period events; however, its biggest drawback is that it is wasteful of data when applied to annual maxima. The number of data points on which to fit the distribution corresponds to the number of years of data; often a few tens of points. Smith (1986) developed a means of using the largest values in a year to mitigate this problem to some extent. An alternative approach is to use all the large values in the sequence, not just the annual maximum observations. This method is known as the 'threshold method', the 'peaks over threshold' method, or simply the 'POT' method. The basis for this method is shown in Figure 7.6 below. A threshold value, *u* say, is selected and a distribution is fitted to the values above *u* while ignoring all the values

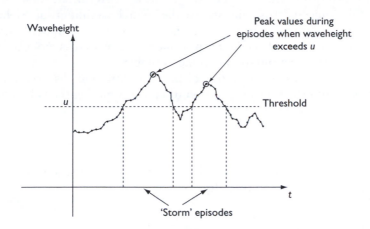

Figure 7.6 Illustration of the peaks over threshold method.

below u. The main aspects of this method are the threshold and the distribution used to fit the exceedances.

Again, the key assumption is that data values are independent. For this reason, often only the peak value during each episode for which values exceed the threshold is selected, although this is not a requirement of the method. Using essentially the same arguments that justified the GEV asymptotic form for maximum values, it may be shown that the natural family of distributions to describe the exceedances of a threshold is the generalised Pareto distribution (Pickands 1975). Thus, the conditional distribution of the random variable X, given that $X > u$, is

$$\Pr(X \le x \mid X > u) = \frac{F(x) - F(u)}{1 - F(u)} \qquad x > u \tag{7.10}$$

Where F is the distribution function of X. For u sufficiently large this can be well approximated by the generalised Pareto distribution, GPD(σ, ξ),

$$F(X \le x \mid X > u) = 1 - \left\{1 - \xi (x - u)/\sigma\right\}^{1/\xi}, x > u \tag{7.11}$$

Again, σ is a scale parameter ($\sigma > 0$ and ξ is a shape parameter. If $\xi = 0$, we retrieve the exponential distribution. It should not be surprising if the GEV and GPD distributions have some similarities. Indeed, Davison and Smith (1990), showed that if there are n observations in a year, each with the probability P of exceeding the threshold u, and the exceedances obey a GPD(σ, ξ) distribution then the distribution of the annual maxima of the observations is GEV($(u + \sigma(nP)^\xi - 1)/\xi$, $\sigma(nP)^\xi$, ξ).

Thus, knowing the GPD parameters we may estimate the corresponding GEV parameters, using all the large values rather than just the annual maximum values.

In practice, if the standard errors associated with fitting annual maxima to a GEV distribution are sufficiently small to allow specification of design conditions within acceptable bounds then this method will be used. As mentioned earlier, some recordings, particularly tide gauge records, only archive the annual maximum values. In this case using the GEV distribution is the practical choice. However, with the introduction of digital recordings, the POT method is now increasingly used as more continuous time series records of longer duration become available.

In the remainder of this section methods of plotting data and distributions are described. The process of fitting the distribution (that is, estimating the values of the parameters), is described in the next section.

Probability plots

We start by considering the Normal or Gaussian distribution to illustrate some ideas. Suppose the variable X obeys the Normal distribution

$$\Pr(X < x) \equiv P = \frac{1}{\sigma\sqrt{2\pi}} \int_{-\infty}^{X} e^{-\frac{1}{2}\left(\frac{X-\mu}{\sigma}\right)^2} dx \equiv \Phi\left(\frac{X-\mu}{\sigma}\right) = \Phi(z) \tag{7.12}$$

Where μ is the mean value of X, σ its standard deviation, z is the standard normal variate (that is it has zero mean and unit variance).

Φ is shorthand notation for the integral in Equation (7.12) and its values are tabulated for a range of z values in many books and tables (e.g. Abramowitz and Stegun 1964), and efficient routines to compute it directly have been developed (e.g. Press *et al.* 1986). If z is known then such tables yield $P = \Phi(z)$ as in Equation (7.12). The tables may also be used in reverse to find the value of z corresponding to a particular P. Symbolically,

$$z = \Phi^{-1}(P).$$

As z is a linear function of X, by plotting z against X should yield a straight line with slope $1/\sigma$ and intercept $-\mu/\sigma$. This procedure is illustrated in Figure 7.7, which shows a plot of Indian Monsoon annual rainfall using the 'reduced variate' as the x-axis and rainfall along the y-axis. The reduced variate is simply the values of the cumulative distribution function determined empirically from the data (see Appendix A) converted to standard normal form using Equation (7.12). The points do not fall on a straight line and we may therefore draw the qualitative conclusion that the Normal distribution does not describe our data as well as the GEV or Weibull distributions.

Now suppose the variable X obeys a Gumbel distribution (Equation 7.8), that is

$$\Pr(X < x) = P = e^{-e^{-(x-u)/\xi}}$$

Taking logarithms of both sides twice gives

$$-\ln\left(-\ln\left(P\right)\right) = \frac{x-u}{\xi} \tag{7.13}$$

If we plot x against $-\ln(-\ln(P))$ we should retrieve a straight line with slope ξ and intercept u.

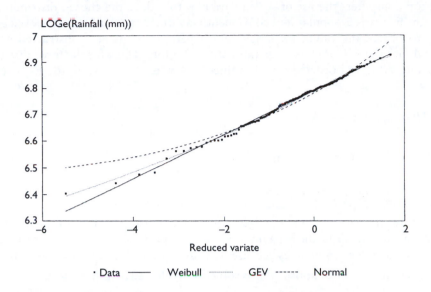

Figure 7.7 Weibull Q-Q plot showing monsoon rainfall data and best fit Weibull, Normal and GEV curves.

Similarly, if X obeys a three-parameter Weibull distribution

$$\text{Prob } (X < x) = P = 1 - e^{-\left(\frac{x-\theta}{\lambda}\right)^{\kappa}}$$

we can obtain:

$$\left(\ln\frac{1}{1-P}\right)^{\frac{1}{\kappa}} = \frac{X-\theta}{\lambda} \qquad (7.14)$$

Again by plotting x against $\left(\ln\dfrac{1}{1-P}\right)^{\frac{1}{\kappa}}$ we should obtain a straight line with slope λ and intercept θ. In this case however the value of κ has to be guessed before plotting. If no guess is available the plots may be constructed for a series of values of κ and the one that gives the closest plot to a straight line can be used.

For the two-parameter Weibull distribution (set $\theta = 0$ in above), we can take logarithms twice giving

$$ln(-ln(1-p)) = \kappa ln(x) - \kappa ln(\lambda)$$

Plotting $ln(x)$ against $ln(-ln(1-p))$ gives a straight line with slope $\dfrac{1}{\kappa}$ and intercept $ln(\lambda)$.

Plotting the fitted distribution against the data point provides a quick visual check on how the distribution fits the data. Prior to the widespread availability of desktop computers the use of scaled variables to reduce the chosen distribution to a straight line was a popular and convenient way of extrapolating the curve (a straight line) to obtain the values for larger return periods. Plots are constructed either as probability plots, P–P plots or quantile (Q – Q) plots. Thus, for the fitted distribution function, $H(x)$ and ordered sample values $x_1 \le x_2 \le \ldots \ldots \le x_N$ in the P – P plot we graph

$$H(x_i) \text{ against } \frac{i}{N+1} \text{ for } i = 1,\ldots..N \tag{7.15}$$

And for the Q – Q plot we graph

$$x_i \text{ against } H^{-1}\left(\frac{i}{N+1}\right) \text{ for } i = 1,\ldots..N \tag{7.16}$$

Departures from a straight line suggest deficiencies in the fit or the ability of the chosen distribution to describe the behaviour of the data. Q–Q plots are more informative for assessing fits to extreme values because they highlight discrepancies in the upper tail of the distribution. The quantity $\frac{i}{N+1}$ is sometimes referred to as 'the plotting position' in the literature and a number of alternative expressions have been proposed (Chambers *et al.* 1983). One popular alternative is $\frac{i-0.5}{N}$ which is known as Hazen's formula. An example of a Q–Q plot is shown in Figure 7.7. The plot is taken from Reeve (1996) and shows the best-fit Normal, 2-parameter Weibull and GEV distributions to the All Indian Monsoon Rainfall from 1871 to 1991.

Figure 7.8 shows Q–Q plots for water levels at Workington (UK), and the extreme distribution determined using (a) the peaks over threshold method and (b) the GEV distribution fit to the annual maxima. There are more points for fitting the curve in the POT analysis. It would not be recommended practice to use such a short record (~6.5 years) to determine extreme water levels for design purposes.

7.1.4 Calculation of marginal extreme values

Design criteria very often depend on more than one variable. Where we are concerned with only one of these variables we consider the extremes of that variable only. Such extreme values are termed 'marginal extremes' to acknowledge that no account has explicitly been taken of any dependence on the other variables.

Various distribution functions have been described in the previous section. In this section methods of fitting a distribution to observed data are described.

The basic methods include:

1   the method of moments
2   the method of maximum likelihood.

All the methods start from the premise that the general form of the distribution is known, or postulated, and that its parameters are to be estimated. The different methods correspond to different estimators for the parameters. Estimations are generally

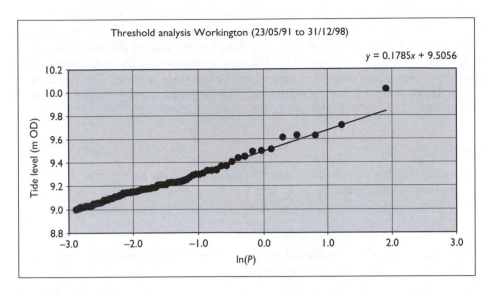

(a)

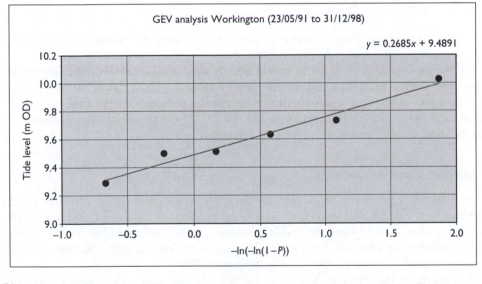

(b)

Figure 7.8 Q-Q plots for water levels at Workington using (a) the POT method (b) the GEV distribution fit to annual maxima only.

sought that provide unbiasedness, consistency and efficiency (Feller 1957). No estimator has all these properties, and in practice the choice of method (therefore estimator) is governed by the nature of the problem.

Graphical procedures

For simple distributions it is possible to plot the cumulative distribution function of a variable X for different values of X as a straight line by scaling the ordinates appropriately. The slope and intercept of the line give estimators of the parameters of the distribution. This is the method described in the previous section. The line may be drawn by hand or determined by regression. This provides a quick way of estimating parameters for distributions with one or two parameters.

The method of moments

Let the variable X have probability density function f_x, with parameters $\alpha_1, \alpha_2.....\alpha_n$. The ith moment of X is given by

$$m_i = E\left(X^i\right) = \int_{-\infty}^{\infty} x^i f_x(x)dx \tag{7.17}$$

The ith moment is clearly a function of the parameters of the distribution. Using Equation (7.17) to generate the first n moments, m_n, provides n equations in the n unknown parameters. However, the sample moments of a random sample of the variable X of size n ($x_1, x_2, ..., x_n$ say) are given by

$$M_j = \frac{1}{n}\sum_{i>1}^{n} x_i^j \text{ for } j = 1, 2, ...n \tag{7.18}$$

Estimates of the parameter values are obtained by equating the moments of X (the m_i), and the sample moments, M_j.

So, for example, in the case of fitting a GEV distribution we would compute the first three sample moments and equate them to their expected values under the GEV distribution. However, the sample third moment is a poor estimator of the population third moment in moderate sample size and for $\delta > 1/3$ the third population moment is not defined. If used, this method must be employed with care.

Method of maximum likelihood

This method is generally more difficult to apply than graphical procedures or the method of moments, but maximum likelihood estimators of distribution parameters can be shown to have some desirable properties (e.g. Cox and Hinkley 1974). For those not familiar with the maximum likelihood technique, an introduction, through a series of worked examples, is given in Appendix B.

As before, let the variable X have probability density function f_x, with parameters $\alpha_1, \alpha_2.....\alpha_p$ that are to be determined. In addition, assume that a random sample of the variable X has been obtained ($x_1, x_2, ..., x_n$, say). The likelihood function of this sample is defined as

$$L(\alpha_1, \alpha_2, ...\alpha_p) = \prod_{i=1}^{n} f_x\left(X_i \,|\, \alpha_1, \alpha_2....\alpha_p\right) \tag{7.19}$$

L expresses the relative likelihood of having observed the sample as a function of the parameters $\alpha_1, \alpha_2.....\alpha_p$. The maximum likelihood estimators $\hat{\alpha}_1, \hat{\alpha}_2....\hat{\alpha}_p$ are defined as those values of $\alpha_1, \alpha_2, \ldots \alpha_p$ which maximise the likelihood or equivalently and more conveniently, the logarithm of L. The evaluation of $\hat{\alpha}_1, \hat{\alpha}_2 \ldots \hat{\alpha}_p$ therefore requires the solution of the set of p equations

$$\sum_{i=1}^{p} \frac{\partial}{\partial \alpha_j} \log\left(f_x\left(x_i, \alpha_1, \alpha_2, \ldots \ldots \alpha_p\right)\right) = 0 \; for \; j = 1, 2, \ldots \ldots p \tag{7.20}$$

taking account of any constraints on the parameter values. In practice, this is usually done by applying Newton numerical methods to maximise the log likelihood.

Extensions of this approach also provide a means of estimating the standard errors associated with the data and fitting procedure. Confidence intervals can be calculated using these standard errors and assuming a normal distribution for the estimated parameter values (Efron and Hinkley 1978).

Alternatively, resampling techniques such as the Bootstrap method (Efron 1982) may be used in conjunction with maximum likelihood estimation to derive confidence intervals (e.g. Reeve 1996; Li et al. 2008).

Confidence limits can be a useful, but not infallible, check on the reliability and utility of extrapolated return values. For example, if the standard error is of the same order as the difference between the estimated 100-year and 200-year return values, there is cause for concern. Extrapolations are also dependent on the assumption that the chosen distribution function remains valid when we extrapolate, so the total uncertainty will be larger than that indicated by confidence intervals alone.

The return period for wave heights can also be obtained through the closed form solution based on the rule of equivalent triangular storms (Boccotti 2000). This gives

$$R = \frac{b(h)}{hf(h) + 1 - F(h)} \tag{7.21}$$

Where $f(h)$ is the *pdf* of the wave heights, $F(h)$ is the *cdf* of the waveheights and $b(h)$ is the regression duration-heights of equivalent triangular storms. Figure 7.9 shows the equivalent triangular storm concept.

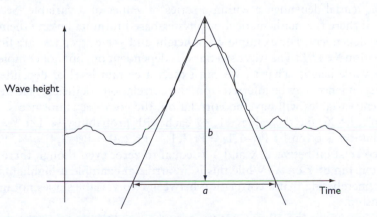

Figure 7.9 Triangular storm concept and definitions.

A real individual storm is considered to be equivalent to a storm with a triangular profile with duration a and height b. The aim is to obtain the regression $b(h)$ relating durations to amplitudes or equivalent triangular storms. This regression was obtained by Boccotti (2000) for a number of locations including the central Mediterranean Sea, the Northwest Atlantic and the North Eastern Pacific. He proposed the form

$$b(h) = K_1 \, \overline{b} \, e^{K_2 \, \frac{a}{\overline{a}}} \tag{7.22}$$

where $\overline{a}$ is the average a of the set of N strongest storms in the time span under examination, and $\overline{b}$ is the average b of this set. $\overline{a}$ and $\overline{b}$ are site specific, while K_1 and K_2 depend on the geographic region. For the Mediterranean Sea, Boccotti gives $K_1 = 1.12$ and $K_2 = -0.115$. The appropriate value of b for a specific site has to be obtained from regression analysis of wave measurements as mentioned above.

7.1.5 Dependence and joint probability

Design conditions may not always be specified in terms of an exceedance level of a single variable. For example, one of the performance criteria of a sea defence might be to limit the amount of overtopping under conditions with a 50-year return period. Several issues arise. First, overtopping is a function of many quantities apart from the wave height, and will also be dependent on the shape and material of the sea defence. Secondly, the wave conditions at the structure used to estimate overtopping are also a function of several variables. For example, both the beach slope and the water depth may affect the wave height through their influence on wave shoaling and breaking. The water depth will depend on the water level (which may have tidal and surge components) and the beach level. Neither of these is fixed and variations such as these must be accounted for in the design of structures. As another example, wave height is likely to depend on direction, particularly in situations that are fetch-limited. Seasonal variations (e.g. monsoon circulations) can also lead to a strongly directional wave climate.

How can you tell whether two variables are dependent? (Note this is different to looking at sequential dependence within a series of values of a variable (Section 7.1.2).) Clearly, if there is a mathematical or physics-based formula linking them the variables will be dependent. For example wave height and wave steepness are linked through the equation $S = H/L$. The wave steepness is dependent on, but not completely specified by the wave height. That is, we can expect a certain level of dependence. The degree of dependence may be inferred from the correlation coefficient. However, while independent variables will have no correlation, the converse is not necessarily true. For example, let X take the values $\pm 1, \pm 2$ each with probability ¼. Let $Y = X^2$. The joint distribution is given by $\Pr(-1,1) = \Pr(1,1) = \Pr(2,4) = \Pr(-2,4) = ¼$. From symmetry, the correlation between X and Y is equal to zero, even though there is a functional relationship of Y on X. While this is an artificial example it highlights the importance of remembering that a correlation between two variables does not imply a causal relationship.

Indeed, what constitutes the 50-year return period event in relation to the design performance criterion is not a straightforward question. Present design and analysis methods for coastal structures are essentially deterministic, based on individual values and response functions. The structure is set to resist conditions greater than the design loading by a margin of safety that is selected to take into account uncertainty and variability in the design parameters. The choice of safety factor is based largely on experience rather than on quantification of the uncertainties. As a result, it is very difficult to determine the current performance of a structure in relation to its original specification. More recently, research has been undertaken in probabilistic design and risk assessment for coastal structures to try to quantify the uncertainties in design (see e.g. CUR/TAW 1990; Meadowcroft *et al.* 1995a, b; Reeve 1998; Environment Agency 2000; Oumeraci *et al.* 2001).

Some good progress has been made in describing the joint occurrence (or probability) of extreme wave heights and water levels and we cover this in the remainder of this section.

Suppose we have a simultaneous time series of water levels and wave heights at regular intervals over a period of several years. We can then use the same procedure as for a single variable (Section 7.1.2) to construct a two-dimensional histogram or frequency table. If the two variables are independent (like two dice) then the probability of the joint event that the significant wave height $H_s = h$ and water level $wl = w$ is equal to the product of the probability that $H_s = h$ and the probability that $wl = w$. In symbols,

$$Pr_{H.wl}(H_s = h; wl = w) = Pr(H_s = h) \, Pr(wl = w) \qquad (7.23)$$
$$\text{or} \quad f_{H.wl}(H_s, w) = f_H(h) f_{wl}(w)$$

The individual (or marginal) probabilities for water level and wave height may be found as in the previous sections for single variables. It is then a simple check to determine whether Equation (7.23) is a good approximation or not. Figure 7.10 shows a typical result of a two variable frequency plot from HR Wallingford (2000). In this instance the tidal variation has been removed from the water level (leaving the water level variations due to non-tidal effects or the 'surge residual'). Two cases are shown, illustrating positive and negative correlations. The joint probability of extreme events can be found as follows. First convert the number of occurrences of each combination of wave height and water level to a frequency of occurrence by dividing by the total number of events. Next convert the frequencies to a probability and thence a return period as in Section 7.1.2. Contours of return period may then be drawn on the table, extrapolating as necessary. Note that this gives not a single extreme event but a multiplicity of wave height – water level combinations with the same return period. Which of these represents the harshest test of the proposed sea defence design will depend on what criterion is being used.

If the two variables X_1 and X_2 are completely dependent, i.e. knowledge of one of them allows you to specify the value of the other, then the probability of the joint event that $X_1 = x_1$ and $X_2 = x_2$ is equal to the probability that $X_1 = x_1$ and this will be equal to the probability that $X_2 = x_2$. A simple example of this type of relationship would be the water level at a gauge measured to different datum levels, such that $X_1 = X_2 + \Delta$ where Δ was the constant difference in the two datum levels.

There are a number of techniques in current practice that are used to account for dependence between variables used for design. The method employed will depend on

SURGE RESIDUAL LEVEL (M)

HS (METRES)	-1.25 – -1.00	-1.00 – -0.75	-0.75 – -0.50	-0.50 – -0.25	-0.25 – 0.00	0.00 – 0.25	0.25 – 0.50	0.50 – 0.75	0.75 – 1.00	1.00 – 1.25	NO. IN ROW
4.00–4.50	0	0	0	0	0	0	0	0	0	0	0
3.50–4.00	0	0	2	0	0	0	1	0	0	0	3
3.00–3.50	1	1	0	6	4	0	0	0	0	0	12
2.50–3.00	1	4	8	22	26	14	2	0	0	0	77
2.00–2.50	0	3	19	59	106	54	6	1	0	0	248
1.50–2.00	0	3	16	91	350	195	33	0	0	0	688
1.00–1.50	0	1	15	133	921	631	78	14	2	0	1795
0.50–1.00	0	3	10	133	1898	1798	192	35	7	1	4077
0.00–0.50	0	1	0	65	1949	2628	369	74	13	1	5100
NO DATA/CALMS	0	1	0	7	212	217	11	1	0	0	
NO. IN COLUMN	2	16	70	509	5254	5320	681	124	22	2	

Contours of no. of events

1
10
100
1000

RECORDS ANALYSED FROM DAY   1 OF MONTH   2 OF 1971 TO DAY 31 OF MONTH   3 OF 1990
NUMBER OF RECORDS ANALYSED   12449

(a)

SURGE RESIDUAL LEVEL (M)

HS (METRES)	-1.00 – -0.70	-0.70 – -0.40	-0.40 – -0.10	-0.10 – 0.20	0.20 – 0.50	0.50 – 0.80	0.80 – 1.10	1.10 – 1.40	CALMS/ NO DATA	NO. IN ROW
6.50–7.00	0	0	0	0	1	1	0	0	0	2
6.00–6.50	0	0	0	0	1	1	0	0	0	2
5.50–6.00	0	0	0	0	1	1	0	0	0	2
5.00–5.50	0	0	0	0	3	4	0	0	0	7
4.50–5.00	0	0	0	7	11	5	3	0	0	26
4.00–4.50	0	0	2	20	22	15	6	2	0	67
3.50–4.00	0	0	6	32	54	19	2	0	0	113
3.00–3.50	0	0	8	64	56	30	2	1	0	161
2.50–3.00	0	0	30	185	154	33	4	0	0	406
2.00–2.50	0	0	44	329	227	27	1	0	0	628
1.50–2.00	0	0	78	491	246	17	1	0	0	833
1.00–1.50	0	0	189	1023	277	13	2	0	0	1504
0.50–1.00	0	4	530	2066	353	16	0	0	0	2969
0.00–0.50	0	6	973	3005	263	10	0	0	0	4258
NO DATA/CALMS	0	2	167	556	83	4	0	0		
NO. IN COLUMN	0	10	1860	7223	1669	192	21	3		

Contours of no. of events

1
10
100
1000

RECORDS ANALYSED FROM DAY   1 OF MONTH   1 OF 1978 TO DAY 31 OF MONTH   3 OF 1990
NUMBER OF RECORDS ANALYSED   11790

(b)

Figure 7.10 Joint (two-way) frequency analysis of wave heights against surge residual showing (a) negative correlation (at Hythe, UK) and (b) positive correlation (at Christchurch Bay, UK).

time, budget, the purpose of the calculations and the amount of field data available. Outlines of the methods are given below:

1   *Range of dependence*   By making the assumption of complete independence and complete dependence of the design variables will give bounds on the behaviour of your variables for a particular failure mode. That is, if as is likely there is some partial dependence between variables then the answer will lie between the answers obtained using the assumptions of complete dependence and independence. These bounds may not provide a *useful* constraint on the variables to allow you to proceed with design; in which case an alternative method should be employed.

2   *Intuitive assessment*   In the absence of substantial measurements, intuitive assessment, based on general experience and information gleaned from a site visit, is possible. For example, if only a modest dependence is adduced, then the N-year joint return period is likely to be more towards the independent rather than completely dependent bound. Conversely, a strong dependence suggests that the return period is closer to the complete dependence bound. This is the basis of the method described in CIRIA (1996).

3   *Empirical frequency analysis*   If budget and data permit, an empirical frequency analysis may be performed. For two variables this requires constructing a frequency table (as in Figure 7.10). Knowing the sampling rate of the data and the total number of data allows the frequencies to be converted first into probabilities, and then into return period (as in the single variable case). Contours of return period may then be drawn on the table. Extrapolation is normally done along each column and each row individually and then constructing contours by joining combinations with equal joint exceedance return period. This procedure is usually performed on wave heights and water levels. Any dependence on wave period is normally accounted for by assuming it is completely dependent on wave height; equivalent to assuming waves have a constant steepness.

4   *Direct extrapolation of the design variable*   An alternative is to use the time series data to generate a time series of the design variable (sometimes termed the structure function). For example, time series of wave run-up or overtopping can be constructed from time series of waves and water levels, together with information on the beach level, beach slope, geometry and material of the defence. The time series of the design variable may then be treated in the same way as calculating marginal extremes. The disadvantages of this method are that it is site and structure specific, so a new time series would have to be constructed for each sea defence option for example. Also, the method relies on the extrapolated variable being of the same form as within the body of the distribution. It would not be applicable where wave overtopping gives way to structural failure or weir overtopping for instance. The advantages are that all dependences between the variables (including those characterising the structure) are automatically taken into account. In addition, the extrapolation method relies on well-tested and reliable techniques.

5   *Extrapolation of joint probability density*   As an alternative to extrapolating the design variable we may extrapolate the joint distribution of two or more variables. Where two variables are involved combinations with a given return period will lie on a line or contour. For three variables, such combinations will lie on a surface. A method of performing the extrapolation of the joint distribution of two variables is given in

HR Wallingford (2000). The technique has been tailored specifically to analyse waves and water levels.

The key elements in the procedure are:

1   preparation of input data that are independent records of wave height, wave period and water level;
2   fitting distributions to wave heights, water levels and wave steepness;
3   fitting the dependence between wave heights and water levels and between wave heights and steepnesses;
4   creating a large sample of wave height, wave period and water level values with the fitted distributions;
5   evaluating the structure function for the combinations derived in step (4) to estimate extreme values.

This method has the advantage of being based on sound statistical theory. In principle, it may be applied to offshore conditions representative of a region. However, because wave transformation can have a significant effect it may not be assumed that the severest conditions at the shore correspond to the severest conditions offshore. As such, it is best suited for site-specific studies, in which step (5) must be repeated for each defence option, as when extrapolating the design variable directly. Further details may be found in Coles and Tawn (1994) and Owen *et al.* (1997)

6   *Simple bounds for systems*   When considering a system that contains a number of elements it can be helpful to use the concepts developed further in Section 7.2.1. We consider the different walls, embankments etc. that protect an area as a system comprising a number of elements. Certain elements may fail without causing the system as a whole to fail, whereas if other elements fail the whole system fails. In the former case we may consider the elements to be connected in parallel in analogy to connections in an electrical circuit. In the latter case the elements are connected in series.

Now, each element will have a certain probability of failing. Let the probability of failure for element i be P_i. Upper and lower bounds on the probability of system failure can be obtained from the assumption that (a) the elements are all perfectly correlated and (b) there is no correlation between any pair of elements.

For series systems the lower bound corresponds to the assumption of perfect dependence, while the upper bound corresponds to the assumption of no pairwise correlation between elements. The converse is true for parallel systems. These results are summarised in Table 7.1. In practice these bounds may be so wide as to provide no useful constraint. Other bounds have been suggested in the literature, such as those proposed by Ditlevsen (1979) for series systems.

Table 7.1 Simple bounds for systems.

	Lower bound	Upper bound
Series system	$\max\limits_{i=1,n} P_i$	$1 - \prod\limits_{i=1}^{n}(1 - P_i)$
Parallel system	$\prod\limits_{i=1}^{n} P_i$	$\min\limits_{i=1,n} P_i$

7.2   Reliability and risk

7.2.1   Risk assessment

When performing a risk assessment it is important to relate any mathematical analyses to the practical aspects of the physical system being assessed. It should always be borne in mind that it is impossible to design a structure that will never fail and is therefore completely safe. Through professional design methods and controlled construction the likelihood of failure may be reduced to an acceptably small value. In the case of a flood defence system, this cannot be expected to prevent flooding with perfect certainty, because the height of the defence will have been designed to resist conditions of a specified return period. If a storm occurs that generates conditions in excess of the specified return period then the flood defence could not reasonably be expected to prevent flooding.

In the recent past, and the last couple of decades in particular, two developments have had a major influence on the way engineers approach design. Firstly, the concept that certainty was attainable in engineering was questioned. There has been recognition that predictions provided by engineering models, however mathematically sophisticated, are not perfectly accurate; so that the results predicted from a numerical or empirical model are unlikely to be realised exactly. For example, if a geotechnical engineer predicts that a sea defence embankment will settle by 0.3 m there is some uncertainty associated with this calculation. The prediction might be better presented as a band, for example 0.2 to 0.4 m, to reflect the uncertainty.

Secondly, methods to deal with uncertainty in engineering calculations have been developed. Scientists had already used statistics to describe many natural phenomena such as water levels, rainfall, wave heights and so on. However, Benjamin and Cornell (1970) demonstrated how statistical techniques could be introduced to engineering calculations so that uncertainties were represented by probability density functions. CIRIA (1977) presented the methods in a widely accessible form. Subsequent reports have specialised the approach for coastal structures, beaches and tidal defences (CIRIA/CUR 1991; CIRIA 1996; EA 2000; Oumeraci *et al.* 2001).

The application of probabilistic calculations to engineering design, or 'probabilistic design', has become an integral part of design guidance for sea defences in many countries (e.g. MAFF 2000). This is particularly so in The Netherlands (e.g. Vrijling 1982), where much of the inhabited land is below mean sea level.

To illustrate some of the key concepts required for probabilistic design and risk assessment consider the schematic sea defence system shown in Figure 7.11. An inhabited area is protected by dunes against the sea and also by a flood embankment along the tidal reach that separates this area from an industrialised harbour precinct. This is protected by a hard sea wall on the seaward side and from flooding by a railway embankment. There is also a slipway onto the river to provide boating access, together with a mechanical floodgate that must be closed to maintain the same level at the slipway as the flood embankment.

Looking at the system; dunes, embankment or gate, it may be seen that if any part of the system fails then there will be disastrous flooding and inundation of the inhabited area. There will be some *probability that failure occurs*, p_f. The *consequences of failure* occurring will depend on the geographical extent of the inundation, its duration

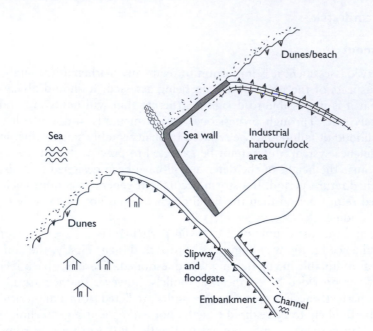

Figure 7.11 A schematic sea defence system.

and the nature of whom and what are adversely affected. The *risk* of inundation is the combination of the probability of failure and an evaluation of the consequences:

$$\text{risk} = (\text{probability of failure}) \times (\text{consequences}) \tag{7.24}$$

The *probability of failure* is sometimes termed the *hazard*. Consequences may be measured in many forms but are often converted to a monetary value so risk has units of rate of expenditure (e.g. $/year, £/month, €/quarter). A *failure* occurs when an item stops performing its desired function. For sea defences, that function will be defence against flooding (as in this example) or erosion. Note that a flood defence can fail even though it itself is not damaged. For example, this can happen if water levels rise above the crest of the defence but do not erode or breach the defence.

Associated with the concept of risk is the idea of *reliability*. Reliability is a characteristic of an item, usually expressed as the probability that the item will perform its required function under given conditions for a specified time interval. That is, the reliability is the probability of no failures over a stated time interval. A single item is characterised by the distribution function $F(t) = Pr\,(\tau \leq t)$ of its failure-free operating time τ. Its reliability function, $R(t)$, is given by

$$R(t) = Pr\,(\text{no failure in } (0,t)) = 1 - F(t) \tag{7.25}$$

The mean time to failure, MTTF, can be computed from

$$\text{MTTF} = \int_0^\infty R(t)dt \tag{7.26}$$

For systems composed of many items, such as a flood defence, determining the reliability function is more complicated. A possible approach is based on the load-strength method or Level 1 Method (see Section 7.2.4). If S is the load and R is the strength then a failure occurs at the time t for which $S > R$ for the first time. If S and R are considered as deterministic values the ratio R/S is termed the *safety factor*. Figure 7.12 shows an illustrative sequence of loads and strength of a structure over its lifetime. The strength of a structure will decrease gradually over time but may be reduced significantly by individual storms and increased by maintenance and repairs.

Risk assessment is the process through which the reliability of different components of a system are analysed and combined with an evaluation of the consequences of failure. Risk problems are typically interdisciplinary and may involve collaboration between engineers, scientists, politicians and psychologists. An appropriate weighting between probability of occurrence and consequences must be established. (The multiplicative rule, Equation 7.24, is the most common is coastal applications.) It is also important to consider different causes and effects of a failure. Risk assessment has been defined as 'the integrated analysis of risks inherent in a product, system or plant and their significance in an appropriate context' (Royal Society 1992).

Underlying any assessment should be recognition that risks cannot be eliminated. If hazards are naturally occurring (for example high winds or large waves) it may not be feasible to limit these. However, it may be possible to reduce the consequences of a failure by, for example, setting up a flood warning system. This leads to the concepts of *risk management* and *risk acceptance.*

Risk management involves procedures to limit risks as far as is practicable. Using the example in Figure 7.11, various risk management strategies to limit the flood risk to properties might include: limiting any further development; installation of a flood warning system; construction of additional embankments to create 'cells' within the area containing a small number of properties; relocation of occupants from the area. Risk acceptance has more to do with the psychological nature of human behaviour.

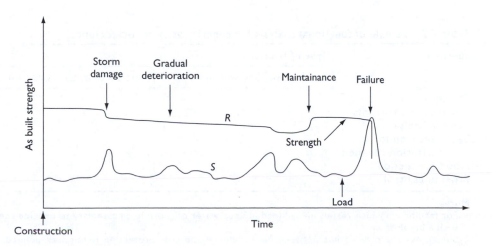

Figure 7.12 Illustrative time evolution of strength and load during the lifetime of a structure.

For example, the risks taken on by an individual (e.g. the risk of injury due to driving a car, smoking) are generally easier for the individual to accept than those that occur externally (e.g. flooding), even though the probability of occurrence may be similar.

When designing sea defences as part of a larger coastal management strategy it is important to be cognisant of the risk acceptance of the community. The level of risk acceptance will be reflected in the choice of design conditions. There are a number of techniques that have been developed for undertaking risk assessment and probabilistic design (see Birolini 1999). We review here a number of these that have been used for sea defence structures.

Functional analysis

A key step in the design process is to identify the functions that the structure has to fulfil. The outcome of the functional analysis is identification of any unstated elements of the requirements and a set of requirements for the planned structure. Table 7.2 provides an example of functional analysis for generic types of rock defence.

Further details of this type of analysis are given in CIRIA (1996).

Reliability block diagram (RBD)

The reliability block diagram is a visual way of answering the question 'which elements of the system under consideration are necessary for the required function and which can fail without affecting it?' To construct an RBD it is necessary first to partition the system into elements that have clearly defined functions. Elements that are necessary to fulfil the function are connected in series while elements that can fail with no affect on the function are connected in parallel. The ordering of the series elements is arbitrary. Each required function has a corresponding RBD.

For example in Figure 7.11 the system consists of the elements: I dunes; II flood embankment; III sea wall; IV railway embankment and, say V an access road on top

Table 7.2 Example of functional analysis for generic types of rock defence.

Function	Type of structure			
	Breakwater	Sea wall groyne	Offshore breakwater	Gravel beach
Shelter from waves and currents for vessels	*		(*)[1]	
Sediment trap for navigation channels	*	*		
Flood protection	*		(*)[2]	
Erosion control	*	*	*	*

Notes
1 For facilities in which vessels unload/load in deep water offshore it is customary to provide shelter with a breakwater.
2 Sometimes the dividing line between flood defence and cost protection is not well defined. For example, the detached breakwater scheme at Happisburgh, UK is classified as a flood defence (by its primary function), although it achieved this by altering the littoral drift to build beach levels (Gardner et al. 1997; Fleming and Hamer 2001).

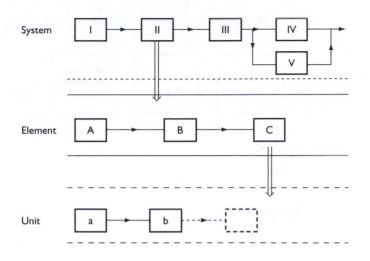

Figure 7.13 Illustrative reliability block diagram, showing the hierarchy of system, element and unit.

of the flood embankment. The procedure for setting up the RBD for the inhabited area is as shown in Figure 7.13, where *A*, *B* and *C* could be underlayer, core and armouring of the embankment and *a*, *b* could be the placement and integrity of the armour units.

Fault tree

For a series system a fault tree may be used to analyse the reliability of each element and unit of the system in a logical manner. In its strictest definition, a fault tree is a description of the logical interconnection between various component failures and events within a system. A fault tree is constructed from events and gates. Gates are logical operators used to combined events to give an event at a higher level. Gates are built from the logical operators (sometimes known as Boolean operators) AND, OR and NOT. The highest level of event is known as the TOP event, which would normally be chosen to correspond to failure of the system as a whole.

The main purpose of constructing a fault tree is to establish the logical connection between different sets of component failures, assign values to component reliability and thus calculate the probability of the TOP event occurring.

Fault tree analysis is strictly only applicable to systems in which components can be in one of two states: working or failed. Where the processes studied do not obey this behaviour then fault trees on their own are less appropriate. Sea defences are a good example of such systems. Nevertheless, fault tree analysis can be a useful step in a qualitative assessment of risk. Figure 7.14 shows the fault tree concept applied to the example of Figure 7.11. We begin with the event of system failure and work backwards, identifying how this event could come about through failure of different parts of the system. The tree can be extended further downward from system level to element level to unit level and so on.

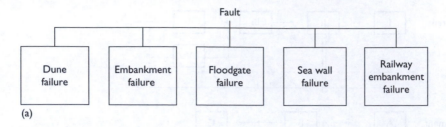

(a)

Figure 7.14(a) System level fault tree.

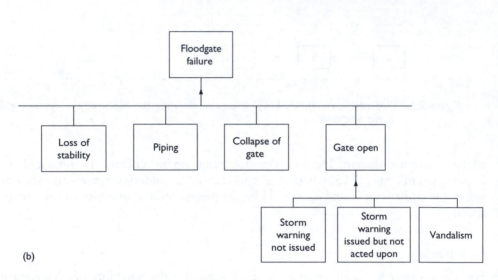

(b)

Figure 7.14(b) Fault tree for flood gate.

The power of the fault tree is evident in Figure 7.14 where the floodgate is analysed. A number of technical failure mechanisms may be identified for which design equations are available. This, together with knowledge of the environmental and design conditions enables the probability of these events to be estimated using the techniques in Sections 7.2.4 to 7.2.6 for example. In the case of a flood gate a non-negligible contribution to the probability of failure arises from human error or intervention. The reliability of this element of the system is influenced by the possibility of management failure that cannot be expressed by classical engineering calculations. The fault tree at least identifies this issue. Fault trees for excessive waves behind a rubble breakwater and inundation due to failure of a flood defence have been developed by CIAD (1985) and CUR/TAW (1990), respectively.

Event tree

Event trees are similar in concept to fault trees except the starting point is an event. We then identify possible ramifications arising from the event that could contribute to changes in the strength of the system. As an example, Figure 7.15 shows an event

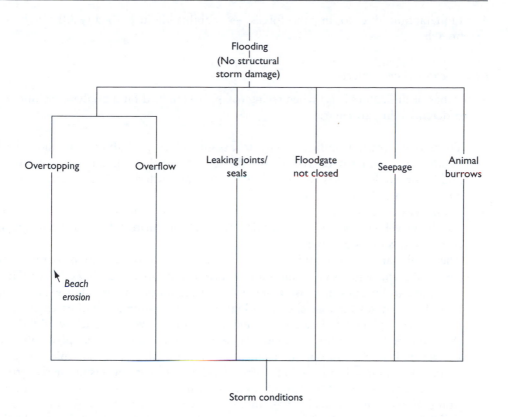

Figure 7.15 **Event tree for storm trigger: no storm damage.**

tree for the case of storm conditions that lead to flooding without causing structural damage to the defence.

Again, the event tree approach assumes a binary (working or failed) behaviour of all components. A less formal approach to event tree analysis for sea defences was proposed in EA (2000). In this case chains of events were identified which could be analysed independently. Each event chain is a sequence of events that comprise a *failure mode*. As an example, the following event chains lead to an initial breach in a flood defence. The initiating event will normally be a storm defined as a combination of waves and water levels that results in extreme loading on the structure. The event chains, or failure modes, are sequences of *ordered* events. Some event chains for breach initiation are listed below:

1 Overtopping → erosion of crest → lowering of crest level → breach
2 Erosion to seaward toe → slip failure of seaward face → damage to seaward slope → erosion of core → breakthrough → breach
3 Overtopping → erosion of landward face by overtopping flow → erosion of core → loss of stability leading to breakthrough → breach
4 Damage to seaward face → erosion of core → breakthrough → breach
5 Seepage through internal layer → internal erosion → piping → breach

6   Liquefaction due to impact forces → stability failure → breakthrough → breach

Cause–consequence charts

A number of difficulties arise when trying to apply standard fault analysis techniques to sea defences. In particular:

* There is generally insufficient data to assign failure probabilities to individual components with any degree of confidence because the number of sea defences and failure of defences is very small. (Contrast this with the manufacture of printed circuit boards, for example.)
* Fault trees are essentially binary in character, that is, components either work or fail. However, components of a sea defence undergo various degrees of damage in response to storms of different severity.
* The combination of probabilities of different events to obtain the probability of the TOP event requires an assumption that the events are mutually exclusive. That is, failure will occur due to only one of the branches on the tree. For example, it would be easy for the case shown in Figure 7.8 to assume that failure of each of the elements of the flood defence system are exclusive events, which would allow the combined probability to be calculated by adding the individual probabilities. However, it is possible for the elements to fail together. For example, the embankment and flood gate could fail simultaneously if water levels in the river rose above their crest level. More problematically, during the course of a storm, failure of individual components is often influenced by the occurrence of the failure of a different component. For example, scour at the front toe of the defence is likely to affect the likelihood of failure of the front armour layer. The fault tree approach has difficulty in representing this behaviour. This has prompted the use of cause-consequence diagrams (e.g. Townend 1994). These show possible changes in the system and link these to consequences, thereby providing a more complete description of the system. They allow representation of some degree of recursion and dependence between chains of events. Figure 7.16 shows an example cause-consequence diagram for coastal flooding.

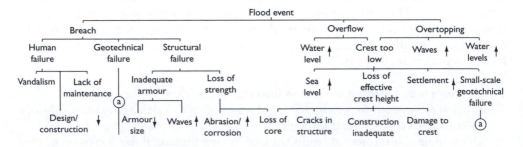

Figure 7.16 Example of a cause-consequence tree. Up and down arrows indicate conditions that are respectively larger or lower than anticipated. The 'a's denote a link that has not been drawn with a line. (Adapted from Townend 1994.)

Tiered approach

When assessing a large number of existing defences applying detailed assessment methods to each structure would be prohibitively time consuming. In the case where the assessment is done primarily to guide expenditure of capital or maintenance works a tiered approach to risk assessment can prove effective. For example, Meadowcroft *et al.* (1995a) described a scheme with three levels of detail. First a screening test is applied to the defences to identify those that are 'low risk' and remove them from further consideration. Secondly, an indicative risk calculation is performed on each of the remaining structures to determine whether detailed analysis is required, and if so, on which failure modes the analysis should concentrate. The calculations typically involved Level I and Level II type methods (see Sections 7.2.4, 7.2.5). Finally, for those structures that demand it, a detailed risk assessment is performed. This accounts for all hazards and potential failure modes, including a detailed description of wave and water level conditions at the defence.

7.2.2   Structures, damage mechanisms and modes of failure

In Section 7.1.1 it was explained how the requirements of a structure are described by reference to design conditions. That is, a set of conditions is defined that describe the severest loads that the structure must withstand to perform its function. For coastal structures not all these loads are expressed in terms of forces, but rather, in terms that describe some aspect of the fluid movement of seawater. For example, an estuarine flood defence may be required to withstand a certain still-water level, while an open coast flood defence might be required to withstand a specified still-water level and to limit wave overtopping rates to a prescribed level under given conditions.

Generation of design ideas is based both on the function requirements and the experience and creativity of the designer. An important factor in considering alternative structure types is their respective failure risk and ways in which they might fail.

In most situations there is a choice between the type of structure as well as the form and materials. Table 7.3 summarises various types of structure that are often used for sea defence. A sea defence structure may provide protection against flooding of low-lying land by the sea (a flood defence function), or it may also provide protection to the coastline against erosion (a coast protection function).

The components making up a sea defence will vary with the type of defence, as indicated in Table 7.3. When considering the risk associated with each type of defence it is often helpful to enumerate the ways in which the components of the structure may be damaged and lead to failure of the defence to perform its function. Common damage mechanisms for different types of sea defence are also listed in Table 7.3.

Two frequently used types of structure are rock-armoured revetments and sea defence revetments with rock armour and a wave wall. Illustrative cross-sections of these two types of defence are shown in Figure 7.17(a) and (b).

Sea defences may be classified by factors other than their function. A classification system for sea walls in the UK was given in CIRIA (1986) and is shown in Figure 7.18.

A wall is first classified according to the slope of its seaward face. The next level of classification distinguishes between porous and non-porous walls, and the final levels are concerned with detailed aspects of the materials and form of construction.

Table 7.3 Damage mechanisms for sea defences.

Type	Function	Components	Damage mechanisms
Seawall/ Rock armoured revetment	flood protection, sea defence protection to erodible coastline. Protection to reclamation bunds. Rehabilitation/reinforcement of existing wall	rock armour layer filler layer beach or subsoil	slip failure of subsoil liquifaction of subsoil erosion at toe unstable armour erosion of subsoil cracking of armour outflanking
Seawall with armoured revetment and wave return wall	flood protection, sea defence protection to erodible the coastline. Protection to reclamation bunds. Rehabilitation/reinforcement of existing wall	rock armour layer filler layer beach or subsoil core for revetment wave return wall	as for sea wall plus cracking/ shattering of wave wall erosion of core material damage to joints between rock armour and wave wall
Seawall with blockwork revetment	flood protection, sea defence protection to erodible the coastline. Protection to reclamation bunds. Rehabilitation/reinforcement of existing wall	embankment core filler layer armour layer	as for rock armoured revetment plus settlement of crest or toe beams structural failure of crest or toe beams
Unprotected embankments	sea defence	outer layer core subsoil	slip failure of subsoil or core seepage fissuring erosion of front face erosion of crest erosion of rear face overtopping
Natural embankments	sea defence	as for unprotected embankments e.g. sand dunes or shingle ridges	as for unprotected embankments
Protected embankments	sea defence	as for unprotected embankment plus armour layer filter layer	as for unprotected embankments plus erosion of material from beneath armour vandalism
Control structure and gates	flood defence water management	barriers flaps valves boom hinges locks bolts	unstable armour cracking of armour mechanical failure operational error vandalism

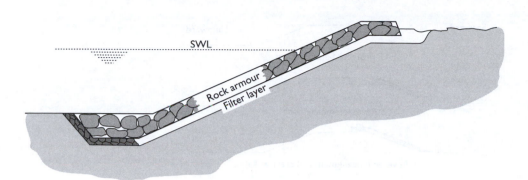

Figure 7.17(a) Typical cross-section of a rock armoured revetment.

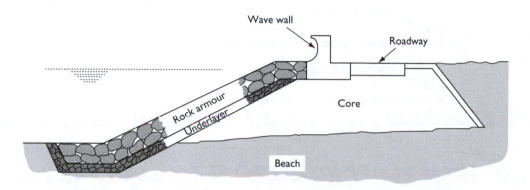

Figure 7.17(b) Typical cross-section of a sea defence revetment with rock armour and wave wall.

Other classification systems have been used – for example Department of Environment (1980), Environment Agency (2000) and Thomas and Hall (1992). A classification system developed specifically for sea and tidal defences was developed by Halcrow for the Association of British Insurers in 1993, and is described in Meadowcroft *et al.* (1995b). The system is designed to assist a tiered approach to risk assessment described in the previous section. Thus the first level is based upon generic type, the second level defines the general form of construction and the third level identifies individual components. Failure can be attributed at all levels. Figure 7.19 illustrates this classification scheme. For any individual structure there will be components that may include some or all of the following: foreshore, toe, beam, front face, crest, crest wall and back face. These fall into the third level of classification.

Reported damage and failures

In order to quantify the probability of failure of a seawall or one of its components it is necessary to identify the processes by which failure occurs and the frequency at which similar components fail under comparable conditions. EA (2000) provides a

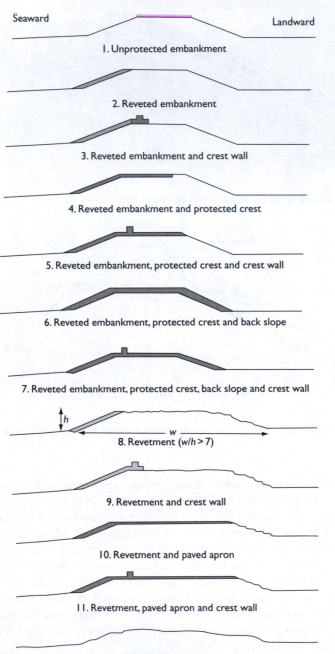

Seaward

Landward

1. Unprotected embankment

2. Reveted embankment

3. Reveted embankment and crest wall

4. Reveted embankment and protected crest

5. Reveted embankment, protected crest and crest wall

6. Reveted embankment, protected crest and back slope

7. Reveted embankment, protected crest, back slope and crest wall

8. Revetment ($w/h > 7$)

9. Revetment and crest wall

10. Revetment and paved apron

11. Revetment, paved apron and crest wall

12. Dune/shingle ridge

Figure 7.18 Classification of structure types based on susceptibility to geotechnical and hydraulic instability.

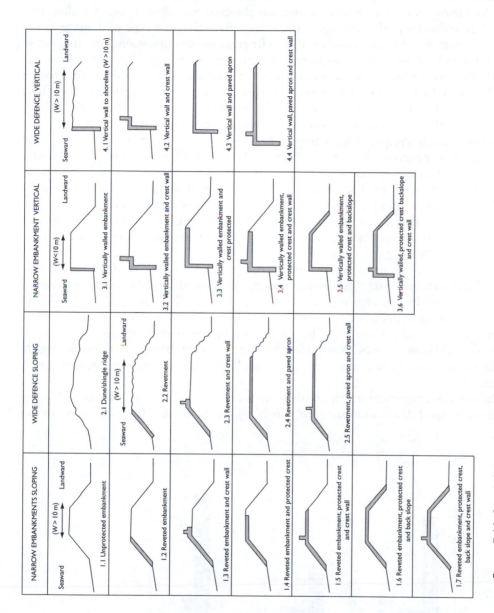

Figure 7.19 Sea defence classification scheme.

brief synopsis of a survey of reported damage and failures in the UK. By and large the reports are qualitative and rarely describe the failure in detail. The number of recorded failures is small and insufficient to ascribe a probability of failure based on a frequency count. A survey of the performance of seawalls was published by CIRIA (1986) and covered seawalls for which wave action was the dominant design consideration. Figure 7.20 summarises the findings of the survey that relate the type of damage to the type of seawall. Information was gathered by response to a questionnaire. Responses included 188 incidences of damage, representing approximately 37 per cent of the seawalls for which returns were received. The most common type of damage was erosion at the toe. Other types of damage included partial crest failure, removal of armour, wash-out of fill, concrete disintegration, collapse or breach, structural member failure, landslip, spalling of concrete, concrete cracking and uplift of armour units.

Some caution is required when interpreting Figure 7.20. The reported incidences of damage do not necessarily represent failures of the seawall. For example, damage can occur to a structure whose primary function is to limit overtopping without excessive overtopping having occurred. Also, the types of damage reported may not be independent. For example, wash-out of the fill may be the result of erosion at the toe. While information on damage to sea defences is helpful it is of limited use for a formal risk assessment or probabilistic design without further information on failure and the mechanism(s) leading to failure.

The process of constructing diagrams such as the event chains described in Section 7.2.1 can be a useful technique in thinking through possible ways in which a sea defence might fail. Figure 7.21 shows the principal failure mechanisms (or failure modes) for rock structures as identified by CIRIA/CUR (1991). Unfortunately, only a very few of the failure modes can be precisely quantified without current knowledge.

Similar diagrams can be drawn for other type of sea defence. For example, Figure 7.22 shows some failure modes for a shingle bank that lead to the formation of a breach.

In this case, empirical beach profile models can be used in conjunction with wave and water level data to estimate the narrowing of the bank. However, there are no

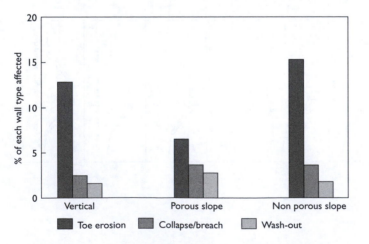

Figure 7.20 Summary of reported seawall damage.

Overtopping   Settlement

Wave overtopping   Slip circle outer slope

Slip circle inner slope   Liquefaction

Micro instability   Drifting ice

Piping   Ship collision

Sliding   Erosion outer slope

Tilting   Erosion fore shore

Figure 7.21 Failure modes for rock structures.

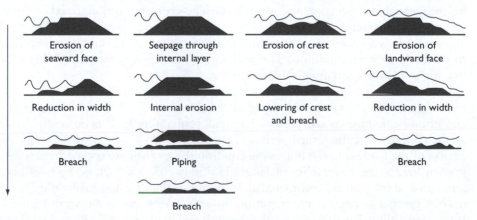

Water level below crest   Water level above crest

Erosion of seaward face   Seepage through internal layer   Erosion of crest   Erosion of landward face

Reduction in width   Internal erosion   Lowering of crest and breach   Reduction in width

Breach   Piping   Breach

Breach

Figure 7.22 Possible failure modes for a shingle bank (after MAFF 2000).

standard methods to predict the probability of breaching the narrowed bank under various wave and water level conditions, although Meadowcroft *et al.* (1995a, b) describe approaches to overcome this difficulty.

The main ways in which sea defences have failed, 'failure modes', are summarised by EA (2000) as:

- excessive overtopping without structural failure;
- failure of surface protection leading to crest level reduction which in turn leads to increased overtopping, washout and breaching;
- geotechnical failure of structure or foundation leading to reduction of crest level and breaching;
- seepage or piping and internal erosion leading to breaching.

This is a somewhat simplistic description of failure modes. It reflects the lack of detailed information on and understanding of sea defence behaviour. The failure modes listed above are themselves the result of failure of one or more components of the sea defence. It is likely that structural failure is the result of combinations of these failure modes, and formal risk analysis methods will be difficult to apply. Our understanding of the failure of sea defences is incomplete and requires further research to provide reliable predictive models for quantitative design and assessment.

Treating the design or assessment of sea defences at different levels or tiers allows progress to be made. The sophistication of the description of the loads and response of the sea defence system can be similarly tiered according to the amount and quality of data available. This is an area of active research and tiered methods for linking flooding to damage have been developed that allow authorities to treat flood plain and flood defences from the perspective of asset management.

7.2.3 Assessing the reliability of structures

Strength (R) and loading (S) were introduced in Section 7.2.1 as a means of assessing the performance of a structure under design conditions. Strength and loading are usually both functions of many variables. The load variables normally include wave height, period and direction and water level. The geometry and material of the structure and the characteristics of the beach are typical strength variables. When no damage or excess is allowed then the condition $R = S$ is applied. This is sometimes referred to as the limit state condition. The probability of failure is the probability that the loading exceeds the strength, that is, that $S > R$.

In the traditional assessment or design approach a limit state condition is set in accordance with the accepted loading of the structure. Exceedance of the limit state condition (ie 'failure') is accepted with a small probability P_f. P_f is normally expressed as the reciprocal of the return period of exceedance ($P_f = 1/T_R$ where T_R is the return period of the loading in the limit state condition). For the case where there is a single known load s, the probability of failure is simply $P(R - s \le 0)$, or $F_R(s)$. Where the strength and the load are considered as random variables then he probability of failure may be given a geometric interpretation (in one dimension) as shown in Figure 7.23. If the probability distributions for the strength and loading are $F_R(R)$ and $F_s(S)$ respectively then the probability of failure is given by:

$$P_f = \int_{-\infty}^{\infty} F_R(x)f_S(x)dx \tag{7.27}$$

under the condition that R and S are independent. This equation is best understood by plotting the density functions of R and S, as shown in Figure 7.23. Equation (7.27) gives the probability of failure as the product of the probabilities of two independent events summed over all possible occurrences. $F_R(x)$ is the probability that R is less than x, and $f_S(x)dx$ is the probability that S lies close to x, within in an interval of length dx.

In the traditional approach, characteristic values of strength $\bar{R}$, and load, $\bar{S}$ are used to ensure that $\bar{R}$ is sufficiently greater than $\bar{S}$ to meet the design requirements. In the probabilistic design approach the probability of failure is estimated directly through evaluation of the area of the overlap of the distributions.

In practice, the problem will involve many variables and the evaluation of the probability of failure will involve integrating over a volume in many dimensions. An additional complication can arise if there is dependence between strength and load variables, such as through the effect that beach levels can have on wave conditions at a structure. In general, a reliability function G is defined as

$$\begin{aligned} G &= R - S \\ &= R(x_1, x_2, x_m) - S(x_{m+1}, \ldots, x_n) \end{aligned} \tag{7.28}$$

and the probability of failure (i.e. the probability that $G < 0$) is evaluated from

$$\int_{G<0} f_G(g)dg \quad or \quad 1 - \int_{G>0} f_G(g)dg \tag{7.29}$$

where the integral is over a volume defined in n-dimensions.

A common assumption in much reliability analysis is to take the valuables $x_1, x_2, \ldots, x_m$ to be independent so that the integral in Equation (7.29) reduces to a multiple integral

$$\int\int\int\int_{G>0} f_{x_1}(x_1) \cdot f_{x_2}(x_2) \cdot \ldots\ldots\ldots \cdot f_{x_n}(x_n)dx_1dx_2\ldots\ldots dx_n \tag{7.30}$$

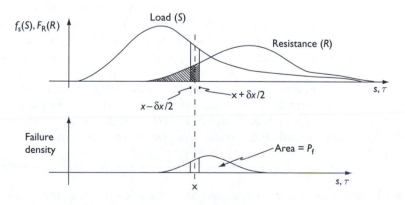

Figure 7.23 Probability of failure definitions.

where $f_{x_1}(x_1), f_{x_2}(x_2), \ldots \ldots, f_{x_n}(x_n)$ are the marginal probability density functions of the loading and strength variables. Even with the assumption of independence the integral can be very difficult to evaluate. This has prompted the development of various approximate methods that are often classified in the manner below:

Level 0: Traditional methods that use characteristic values of strength and loading.

Level 1: Quasi- probabilistic methods that assign safety factors to each of the variables to account for uncertainty in their value. Some Level 2 methods are also known as first order risk methods (FORM).

Level 2: Probabilistic methods that approximate the distribution functions of the strength and load variables to estimate Equation (7.29).

Level 3: The most complex probabilistic methods that estimate Equation (7.29) either directly or through numerical simulation techniques.

7.2.4 Level I methods

Level I methods are design methods in which appropriate measures of structural reliability are provided by the use of partial safety factors that are related to predefined characteristic values of the major loading and structural variables.

Probabilistic design techniques are based on the Limit State Equation (Equation 7.31) in which G is the failure function, R is the resistance (or strength) of the structure and S is the design load. For a structure to resist a specified load S we require $R \geq S$, or $R = \Gamma S$ where Γ is a number greater than or equal to 1. Γ is the factor of safety against failure, included to account for uncertainty. More generally we may write

$$G = \frac{R}{\Gamma_r} - \Gamma_s S = 0 \tag{7.31}$$

where, Γ_r is a safety coefficient relating to the resistance (sometimes called the performance factor), and Γ_s is a safety coefficient relating to the load. The product $\Gamma_r \Gamma_s$ is the (global) factor of safety, Γ.

In Level I methods R and S are assigned characteristic or mean values. The safety factors are normally specified for a discrete set of values of R and S, being based on laboratory or prototype tests. For many types of failure function the resistance and loading will depend on several variables, say N. Typically partial safety factors will be tabulated for each variable and so the global safety factor will be the product of N partial safety factors. In standard structural design, partial safety factors are provided in building codes and the like, and are based on a large body of designs and tests. A similar volume of accurate measurements is not normally available for coastal structures and hence the level of confidence in partial safety factors has not been as great. Safety factors for Level I design may be found in PIANC (1992) and Burcharth and Sorensen (1998) for rubble mounds and vertical breakwaters respectively.

Example

A rock-armoured breakwater has been designed deterministically, using the Van Der Meer equations (see Section 9.4.3) with the following design parameters:

$H_s = 3.0m$

$T_m = 6.0$

$P = 0.1$

$\cot \alpha = 2$

$\Delta = \dfrac{\rho_r}{\rho} - 1 = 1.59$

$N = 2000$

$S_d = 2$

The resulting necessary rock size, using the above is:

$D_n = 1.3m$

Use the partial Safety Factors for Stability Failure of Rock Armour, Plunging Waves, Van Der Meer Formula, Design Without Model Tests, to determine the failure probability for this rock armoured breakwater, at ultimate limit state collapse for which

$S_d = 8$

Solution
Design equation taken from the CEM (cf. Table VI-6-5):

$$G = \frac{1}{\Gamma_r} 6.2 S_d^{0.2} P^{0.18} \Delta D_{n50} \varepsilon_m^{-0.5} N^{-0.1} - \Gamma_s H_s$$

The partial safety factors given in Table VI-6-5 are reproduced below. It may be seen that they are a function of both the probability of failure and the coefficient of variation of the design wave height ($\sigma'_H = \sigma/\mu$). The values of D_{n50} also given in this table are found using the above design equation.

Pf	$\sigma'_H = .05$			$\sigma'_H = 0.2$		
	Γ_s	Γ_r	D_{n50}	Γ_s	Γ_r	D_{n50}
0.01	1.6	1.04	1.59	1.9	1	1.82
0.05	1.4	1.02	1.36	1.5	1.06	1.52
0.1	1.3	1	1.24	1.3	1.1	1.37
0.2	1.2	1	1.15	1.2	1.06	1.22
0.4	1	1.08	1.03	1	1.1	1.05

Hence it can be seen that the originally selected D_{n50} results in an ultimate limit state failure probability of about 7.5 per cent for a coefficient of variation of 5 per cent or about 15 per cent for a coefficient of variation of 20 per cent.

7.2.5 Level II methods

Level II Methods introduce the concept of probability distributions to the calculations. The main features of Level II analyses are:

- an assumption that the basic variables can be adequately described by a Gaussian distribution;
- the failure function is a linear function of the basic variables;
- the choice of expansion point in the case where the failure function is non-linear and is approximated by its truncated Taylor expansion.

With suitable precautions the first assumption can be relaxed, Level II methods give an estimate of the probability of failure and also an influence factor for each variable, indicating the variable's importance to the final result.

For the case of a linear failure function G and Gaussian basic variables Cornell (1969) defined the reliability index β as

$$\beta = \frac{\mu_G}{\sigma_G} \tag{7.32}$$

where μ_G and σ_G are the mean and standard derivation of G, and the probability of failure, P_F, is given by $P_F = \Phi(-\beta)$

If G can be written as

$$G = R - S$$

Where R and S are each functions of a single variable, and are uncorrelated then the Gaussian distributions for R and S, say P_r and P_s, can be combined into a single Gaussian distribution for G (Ang and Tang 1984) with

$$\mu_G = \mu_R - \mu_S$$
$$\sigma_G^2 = \sigma_R^2 + \sigma_S^2 \tag{7.33}$$

At failure, $G = 0$ and the probability of failure, P_F, is equal to the area of the shaded region in Figure 7.24.

Thus,

$$P_F = \Phi\left\{\frac{0 - \mu_G}{\sigma_G}\right\} = \Phi\left(\frac{\mu_G}{\sigma_G}\right) = 1 - \Phi\left(\frac{\mu_G}{\sigma_G}\right) = \Phi(-\beta) \tag{7.34}$$

The reliability index β may be given a simple geometric interpretation in this case. Consider the standardised variables

$$R' = \frac{R - \mu_R}{\sigma_R} \text{ and } S' = \frac{S - \mu_S}{\sigma_S}$$

In terms of R' and S' the failure function becomes

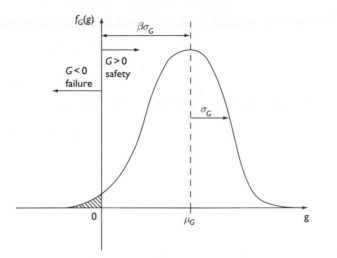

Figure 7.24 Illustration of the reliability index β.

$$G = R - S = \sigma_R R' - \sigma_S S' + (\mu_R - \mu_S) \tag{7.35}$$

For $G = 0$, this equation describes a line in the plane, as shown in Figure 7.25.

The shortest distance from the origin to the failure 'surface' is equal to the reliability index.

Example
The crest level of embankment over a reach is described by a Gaussian distribution with mean 5 and standard deviation 0.5. This is often written as $N(5, 0.5)$. Monthly maximum water levels along the reach obey $N(3, 1)$. What is the probability of flooding?

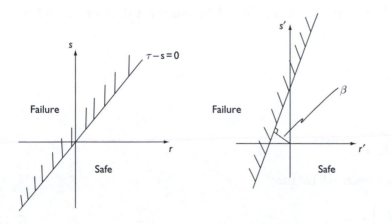

Figure 7.25 Geometric interpretation of the reliability index.

Solution
Flooding occurs when water level > crest level. So the failure function can be written as

$$G = CL - WL$$

The variables are Gaussian and independent so (from Equation 7.33):

$$\mu_G = 5 - 3 = 2, \sigma_G^2 = 0.5^2 + 1^2 = 1.25 \Rightarrow \beta = \frac{2}{\sqrt{1.25}} \qquad \text{(by Equation 7.32)}$$

From Equation (7.34),

$$P_F = \Phi\left(\frac{0-2}{\sqrt{1.25}}\right) = \Phi(-1.79) = 0.037 \approx 4\%$$

Thus the probability of failure is approximately 4 per cent per month.

If the failure function is non-linear then approximate values for μ_G and σ_G can be obtained by linearising the failure function. Let

$$\begin{aligned} G &= R(x_1, x_2, x_3, \dots, x_m) - S(x_{m+1}, x_{m+2}, \dots, x_n) \\ &= g(x_1, x_2, \dots, x_n) \end{aligned} \qquad (7.36)$$

We expand this function in a Taylor series about the point

$$(x_1, x_2, \dots, x_n) = (X_1, X_2, \dots, X_n) \qquad (7.37)$$

Retaining only linear terms gives

$$G = f(x_1, x_2, \dots, x_n) \approx f(X_1, X_2, \dots, X_n) + \sum_{i=1}^{n} \frac{\partial f}{\partial x_i}(x_i - X_i) \qquad (7.38)$$

where $\dfrac{\partial f}{\partial x_1}$ is evaluated at $(X_1, X_2, \dots, X_n)$. Approximate values of μ_G and σ_G are obtained from

$$\mu_G \approx f(X_1, X_2, \dots X_n) \qquad (7.39)$$

and

$$\sigma_G^2 \approx \sum_{i=1}^{n} \sum_{j=1}^{n} \frac{\partial f}{\partial x_i} \frac{\partial f}{\partial x_j} \cdot \text{cov}(x_i, x_j) \qquad (7.40)$$

If the variables are uncorrelated then

$$\sigma_G^2 = \sum_{i=1}^{n} \left(\frac{\partial f}{\partial x_i}\right)^2 \sigma_{x_i}^2 \qquad (7.41)$$

The quantities $\left(\dfrac{\partial f}{\partial x}\sigma_{x_i}\right)^2$ are termed the 'influence factors' and are denoted by α_i.

Three variants of Level II methods that are in current use are now described.

Mean value approach (MVA)

In this case we take $(X_1, X_2, ..., X_n) = (\mu_{x_1}, \mu_{x_2}, ... \mu_{x_n})$ That is, we expand the failure function about the mean values of the basic variables. The mean and standard deviation of the failure function can then be evaluated directly from the equations above.

Example (adapted from CIRIA/CUR 1991)

Derive an expression for the probability of failure of a rock armour revetment. Use this to calculate the probability of failure for the specific conditions given below with the MVA method.

Solution

We take as the response function Van der Meer's (1988a) formula for armour stability under deep-water plunging waves. For a given damage level, S_d, the formula provides an estimate of the required nominal median stone size D_{n50}. The failure function may be written as

$$G = R - S = aP^b S_d^{0.2} \Delta D_{n50} \sqrt{\cot(\theta)} \left(\frac{g}{2\pi}\right)^{-\frac{1}{4}} - H^{0.75} T_m^{0.5} N^{0.1}$$

where H_s is the significant wave height; T_m is the mean wave period; Δ is the relative mass density $(\rho_a - \rho_s)/\rho_s$, ρ_s is the density of seawater; ρ_a is the density of rock; θ is the angle of the face of the rubble mound; P is a permeability parameter; N the number of waves, $a = 6.2$ and $b = 0.18$. We take a, b, Δ, H_s, T_m and D_{n50} as random variables. The first step is to calculate the partial derivatives. Performing this analytically gives:

$$\frac{\partial G}{\partial a} = R/a$$

$$\frac{\partial G}{\partial H_s} = -0.75 S/H_s$$

$$\frac{\partial G}{\partial T_m} = -0.5 S_d/T_m$$

$$\frac{\partial G}{\partial D_{n50}} = R/D_{n50}$$

$$\frac{\partial G}{\partial b} = bR/P$$

$$\frac{\partial G}{\partial \Delta} = R / \Delta$$

With $a = N(6.2, 0.62)$, $b = N(0.18, 0.02)$, $\Delta = N(1.59, 0.13)$, $H_s = N(3, 0.3)$, $D_{n50} = N(1.30, 0.03)$, $T = N(6, 2)$ $S_d = 8$, $\cot(\theta) = 2.0$, $P = 0.1$ and the number of waves equal to 2000, the probability of failure may be estimated. Table 7.4 summarises the results of the MVA calculations.

Table 7.4 Application of MVA.

Variable	Mean	Standard deviation	Partial derivative	α_i^2
a	6.2	0.62	2.63	2.64
b	0.18	0.02	29.2	0.34
Δ	1.59	0.13	10.2	1.76
D_{n50}	1.30	0.03	12.5	0.14
T_m	6.0	2.0	−0.67	1.77
H_s	3.0	0.30	−2.0	0.36

$$\mu_G = \mu_R - \mu_s = 4.29$$

$$\sigma_G^2 = \sum_{i=1}^{6} \alpha_i^2 = 7.01$$

so $\sigma_G = 2.65$

Therefore the reliability index $\beta = \dfrac{\mu_G}{\sigma_G} = 1.62$ and the probability of failure is $\Phi(-\beta) = 0.053$.

   In this relatively simple case we have calculated the probability of failure taking into account uncertainty in the parameter values of an empirical equation, construction materials and the random nature of waves. In passing, it is interesting to note that the result is sensitive to the rock size and density, not just the wave conditions. It may also be noted that the rock size of 1.3 m was originally determined using the standard deterministic method using the no damage value of $S_d = 2$. In estimating the failure probability at $S_d = 8$, corresponding to complete failure of the primary amour layer, it can be seen that is quite low (e.g. about 5 per cent).

   The MVA method is relatively easy to use but can be inaccurate if the failure function is strongly non-linear. The method also relies on accurate estimates of the mean and variance of the key variables. In the example above these have been assumed to be known exactly, but in practice there is likely to be considerable uncertainty in estimating both the mean and the variance. Where parameters have been measured in a series of experiments then the sample mean and variance could be used in the absence of other information. Experience shows that this approach should not be used in isolation. If in doubt, results should be checked against other methods.

Design point approach (FDA)

A serious objection to the MVA method is that the point about which the failure function is linearised is not necessarily on the failure surface. Hasofer and Lind (1974) introduced a modified form of reliability index based on expanding about a point in the failure surface. As before, we start with a failure function which is a function of a Gaussian independent random variables, $x_1, x_2, ..., x_n$. The first step is to map these into standard form by

$$z_i = \frac{x_i - \mu_{xi}}{\sigma_{x_i}} \quad i = 1, 2, ..., n \tag{7.42}$$

so that $\mu_{z_i} = 0$ *and* $\sigma_{z_i} = 1$

Hasofer and Lind's reliability index is defined as the shortest distance from the origin to the failure surface in the standardised z-coordinate system. This is shown for two dimensions in Figure 7.26.

The point A is known as the design point. In general, the reliability index can be found from

$$\beta = \min\left(\sum_{i=1}^{n} z_i^2\right)^{\frac{1}{2}} \tag{7.43}$$

for z_i in the failure surface. The calculation of β may be performed in several ways but generally involves iteration, which is suited to numerical schemes; see e.g. Thoft-Christensen and Baker (1982), Melchers (1999) and Reeve (2010).

Approximate full distribution approach (AFDA)

In this case we use the FDA approach but allow the variables to be non-Gaussian. Equating Gaussian and non-Gaussian distribution functions at the design point, allows the requirement of Gaussian variables for the FDA method to be relaxed. Specifically,

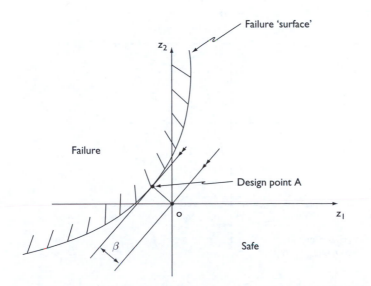

Figure 7.26 Geometric interpretation of Hasofer and Lind's reliability index.

if the failure function depends on a non-Gaussian variable Y this can be rewritten in terms of Gaussian variables through the transformation.

$$Z = \Phi^{-1}\left(F_Y(y)\right) \tag{7.44}$$

Where $F_Y(y)$ is the distribution function of Y and Φ^{-1} is the inverse normal distribution function. This transformation is shown pictorially in Figure 7.27.

Point moment techniques

Point estimate methods, as developed by Rosenblueth (1975) and Li (1992), can be applied to the failure function G to obtain estimates of the moments of the distribution function of G. The method requires the function G to be evaluated at a specific set of values of the basic variables, the statistical moments of the basic variables and their correlations. If analytical forms for the moments of the distribution exist, they may be solved simultaneously to estimate the parameters of the distribution. The nth moment of G is the expected value of $G^n(R,S)$ where R and S are the strength and load variables, dependent on the basic variables $x_1, x_2, \ldots\ldots, x_n$. Estimates of the nth moments of G may be obtained from a truncated Taylor expansion of G about chosen values of R and S. If R and S are chosen as their mean values the method is similar to MVA. Formulae for functions of up to three correlated variables have been given by Rosenblueth (1975). As an example, suppose we suspect the failure function obeys a Weibull distribution

$$F_G(g) = 1 - e^{\left(-(g/\lambda)^\kappa\right)} \ for \ g > 0 \tag{7.45}$$

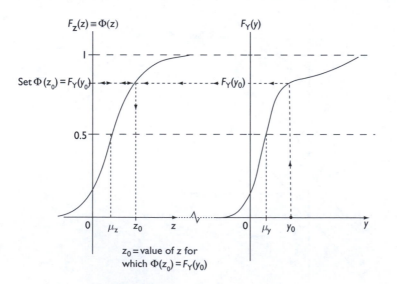

Figure 7.27 The transformation of non-Gaussian variables to equivalent Gaussian variable at the design point.

This has mean μ_G and variance σ_G^2 given by

$$\mu_G = \lambda \Gamma(1 + \tfrac{1}{\kappa})$$

$$\sigma_G^2 = \lambda^2 \left[\Gamma\{(1 + \tfrac{2}{\kappa})\} - \{\Gamma(1 + \tfrac{1}{\kappa})\}^2 \right] \tag{7.46}$$

where $\Gamma(x)$ = gamma function (see e.g. Abramowitz and Stegun 1964).

The point estimate method gives us estimates of μ_G and σ_G^2. Substituting these estimates into the left-hand sides of Equations (7.45) and (7.46) gives two simultaneous equations which may be solved to obtain the two unknown parameters λ and κ.

Figure 7.28 illustrates a comparison of various Level II methods and a Level III simulation for the case of wave overtopping of a simple sea wall. The plots show the distribution function of overtopping discharge as determined using different assumptions. As expected the assumption of complete dependence between waves and water levels provides an upper bound. The Level III result, obtained by generating a time series of overtopping rates from the time series of waves and water levels and then performing a univariate extremes analysis on the series, provides the least conservative result. Distributions derived using PEM (two dimensions using wave height and water level and three dimensions using wave height, wave period and water level) lie between them. Further details may be found in Reeve (1998, 2003) and references therein.

7.2.6 Level III methods

Level III methods are the most general of the reliability techniques. The approach in Level III methods is to obtain an estimate of the integral in Equation (7.29) through

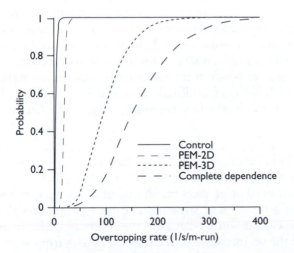

Figure 7.28 Cumulative probability curves of overtopping computed from a 30-year synthetic time series data using: (i) Weibull fit to data; (ii) PEM applied to extreme wave heights and water levels; (iii) PEM applied to extreme wave heights, periods and water levels; (iv) the assumption of complete dependence between wave heights and water levels.

numerical means. The complexity of the integral (in general) means that numerical rather than analytical methods are used. There are two widely used techniques:

1   Monte Carlo integration;
2   Monte Carlo simulation.

The first method may be used if you have a closed analytical form for the probability distribution of the reliability function and a failure region that is well-defined in terms of the basic variables. Monte Carlo integration evaluates the function at a random sample of points, and estimates its integral based on that random sample (Hammersley and Handscomb 1964). This method becomes less straightforward as the number of integration variables increases and the complexity of the failure region becomes greater.

In the second method a set of values of the basic variables are generated with the appropriate probability distribution and values of the reliability function determined. By repeating this process many times and storing the results the integral may be estimated as the proportion of the results for which the reliability function is negative. In symbols, if X_n is the nth simulation then the Monte Carlo estimate of the integral is

$$\frac{\textit{number of points } X_n \, (n = 1, \ldots\ldots, N) \textit{ in the failure region}}{\textit{total number of simulated points} (= N)} \tag{7.47}$$

Clearly, increasing N improves the precision of the answer and in practice N should correspond to at least 10 times the length of the return period of interest. Evidently large sample sizes are required for the most extreme events, which can be computationally demanding. There are methods available for improving precision without increasing N. These use a disproportionate number of extreme conditions at the simulation stage, but the manner in which this is done is not straightforward.

In either method any dependence between variables must be taken into account in the specification of the probability distribution functions. As a result, the assumption that the basic variables are independent is sometimes made, where this can be justified. Otherwise a means of specifying a non-Gaussian joint distribution function with appropriate cross-correlation properties between the variables is required, as well as a means of generating samples with the correct distribution. Further details of this approach for dependence between water level and wave height are given by Coles and Tawn (1994) and HR (2000).

Example   Level III (Monte Carlo simulation) prediction of damage
to a rock armour structure (from Meadowcroft et al. 1995b)
The response function is the equation that predicts the degree of damage, S, to rock armour under plunging waves as a function of structure and load parameters (Van de Meer 1988a). To simplify the calculation we consider the response under design storm conditions. Uncertainty in the performance of the structure arises from sources such as variability in rock armour size, errors in estimating design wave height, the approximate empirical nature of the design equation. For this example we take the distribution functions of the basic variables to be known and to be Gaussian, and are given in Table 7.5. In practice the choice of distributions and their parameters should be estimated against observations.

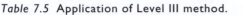

Table 7.5 Application of Level III method.

Basic variable	Distribution	Mean	Standard deviation (% of mean)
Significant wave height(m)	Normal	3.0	10
Slope angle (°)	None	0.5	—
Rock density hg/m³	Normal	2650	5
Nominal rock diameter(m)	Normal	1.3	5
Permeability parameter	None	0.1	—
Wave steepness	Normal	0.05	10
van der Meer parameter a	Normal	6.2	10
van der Meer parameter b	Normal	0.18	10

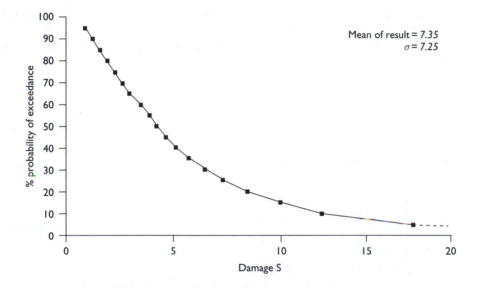

Figure 7.29 Probability of exceedance of the predicted damage for a structure designed for minor damage (S = 2).

The resulting probability distribution shows the predicted damage for a structure designed for minor damage (S = 2), as a probability of exceedance. For example, the probability of the damage exceeding 6 is about 10 per cent.

7.2.7   Accounting for dependence

If you suspect dependence to be important, then Level III techniques provide a means of accounting for this. Dependence between basic variables can also be accounted for in the less computationally demanding Level II methods, and is described in this Section.

Consider a set of n correlated variables $X_1, X_2, ..., X_n \equiv \underline{X}$. In some cases the basic variables can be chosen so that they are statistically independent. If this is the case, the individual variables will be uncorrelated, and can be individually mapped into unit standard normal variables, z, through the transformation

$$z_i = \Phi^{-1} \qquad i = 1, 2,...n \tag{7.48}$$

where $F_{X_i}(x_i)$ is the cumulative distribution function for variable X_i and $\Phi^{-1}()$ is the inverse normal distribution function. However, in cases where the individual variables are not statistically independent they can be represented only through their joint distribution function $F_{\underline{x}}(\underline{x})$,

$$F_{\underline{x}}(\underline{x}) = \text{Prob} \left[(X, \leq x,) \text{ and } (X_2 \leq x_2) \text{ and } \text{ and } (X_n \leq x_n) \right] \tag{7.49}$$

Where sufficient data are available the joint probability methods described in Section 7.1.5 may be used. However, in many situations sufficient data and other information are not available to determine the form of $F_X(\underline{x})$ with any certainty. Often, the most that can be expected is that the marginal distributions $F_{X_i}(x_i)$ can be determined, together with their correlation matrix.

For a pair of jointly distribution random variables X_1, X_2 the marginal distribution function is defined as

$$F_{X_1}(x_1) = \int_{-\infty}^{x_1} f_{X_1}(t)dt = \int_{-\infty}^{x_1} \int_{-\infty}^{\infty} f_{X_1,X_2}(t,x_2)dx_2 dt \tag{7.50}$$

and analogously for n jointly distributed variables where $f_{X_1}(t)$ is the density function of X_1 and so on.

The correlation matrix $\underline{R}$ is given by

$$\underline{R} = \begin{pmatrix} \rho_{11} & \rho_{12} & \cdot & \cdot & \rho_{1N} \\ \rho_{21} & \rho_{22} & \cdot & \cdot & \cdot \\ \rho_{31} & \cdot & \rho_{33} & \cdot & \cdot \\ \cdot & \cdot & \cdot & \cdot & \cdot \\ \rho_{N1} & \cdot & \cdot & \cdot & \rho_{NN} \end{pmatrix} \tag{7.51}$$

where ρ_{ij} is the correlation coefficient between variables X_i and X_j with $i, j = 1, 2, ..., N$. However, if $F_{\underline{x}}()$ or $f_{\underline{x}}()$ are not known, the marginal distributions cannot be obtained from Equation (7.50) and they plus the correlation matrix must be obtained directly by fitting from data. In practice, this corresponds to the situation where data on individual variables may have been collected over a period of time, but no attempt has been made to obtain meaningful joint samples.

For the purposes of undertaking a reliability analysis it is necessary to transform the set of correlated basic variables into a set of uncorrelated standard normal variables.

Correlated normal variables

For the simplest case where the basic variables are normal and correlated the method given by Thoft-Christensen and Baker (1982) may be used or, the set of independent standard normal variables, $\underline{z}$, may be found from

$$\underline{Z} = \underline{L}^{-1} \underline{D}^{-1} (\underline{X} - \mu_{\underline{x}}) \tag{7.52}$$

where $\underline{D}$ is the diagonal matrix of standard deviations σ_i of the basic variables X_i, $\underline{L}$ is a lower triangular matrix obtained from the correlation matrix $\underline{R}$ such that $\underline{R} = \underline{L}\underline{L}^T$, where $\underline{L}^T$ is the transpose of $\underline{L}$ and $\mu_{\underline{x}}$ is the vector of mean values of $\underline{X}$.

The matrix $\underline{L}$ can be found using standard matrix algebra techniques, such as Cholesky decomposition (see e.g. Press *et al.* 1986). In this case the set of basic variables $\underline{X}$ occurring in a realiability function may be replaced by

$$\underline{X} = \underline{D}\,\underline{L}\,(\underline{Z}) + \mu_{\underline{x}} \tag{7.53}$$

Non-normal marginals with known correlation matrix

In the event that the full joint distribution function $F_{\underline{x}}(\underline{x})$ is unavailable, but the individual marginal distributions of the basic variables $\underline{X}$ and their correlation matrix $\underline{R}$ can be estimated, the method proposed by Der Kiureghian and Liu (1986) may be used. An outline of the method is given here.

A joint density function is assumed which is consistent with the known marginal distributions and correlation matrix:

$$f_{\underline{x}}(\underline{x}) = \phi_n(\underline{y}, \underline{R}')\ \frac{f_{x_1}(x_1)f_{x_2}(x_2)........f_{x_n}(x_n)}{\phi(y_1)\phi(y_2)............\phi(y_n)} \tag{7.54}$$

where $\phi_n(\underline{y}, \underline{R}')$ is the n-dimensional normal probability density function with zero means, unit standard deviations and with correlation matrix $\underline{R}'$.

$\phi(y_i)$ is the standard univariate normal density function

$f_{x_i}(x_i)$ is the marginal density function for basic variable X_i.

The elements ρ_{ij} of the correlation matrix R' are related to the known marginal densities $f_{x_i}(x_i)$ and $f_{x_j}(x_j)$ and the correlation coefficient ρ_{ij} between the basic variables X_i and X_j through the relationship

$$\rho_{ij} = \int_{-\infty}^{\infty}\int_{-\infty}^{\infty}\left(\frac{x_i - \mu_i}{\sigma_i}\right)\left(\frac{x_j - \mu_j}{\sigma_j}\right)\phi_2\left(y_i, y_j; \rho'_{ij}\right)dy_i\,dy_j \tag{7.55}$$

where $\phi_2(.,.;.)$ is the bivariate normal density with correlation coefficient ρ_{ij} and x_i is given by

$$x_i = F_{x_i}^{-1}(\Phi(y_i)) \tag{7.56}$$

Solution of Equation (7.54) usually has to be performed iteratively to determine ρ'_{ij} and needs to be solved for each pair of values of i and j.

Once the new correlation matrix $\underline{R}'$ has been obtained the steps are as follows:

1) Obtain a set of correlation normal variables with zero mean and unit standard deviation y_i from

$$\underline{Y} = \Phi^{-1}\left[F_{\underline{x}}(\underline{x})\right] \qquad \underline{Y} = (Y_1, Y_2, ...Y_n) \tag{7.57}$$

2) Compute $\underline{L}$ from

$$R' = \underline{L}\,\underline{L}^T \tag{7.58}$$

3) Obtain a set of uncorrelated unit standard normal variables by

$$\underline{Z} = \underline{L}^{-1}\underline{Y} \tag{7.59}$$

4) The set of correlated, non-normal basic variables $\underline{X}$ occurring in a reliability function may be replaced by

$$X_i = F_{Y_i}^{\,-1}[\Phi(Y_i)] \qquad i = 1, 2 \ldots\ldots n \tag{7.60}$$

with Y_i being obtained from Equation (7.58).

Example
Given that: X_1 is normal with mean $\mu_{x_1} = 100$ and standard deviation $\sigma_{x_1} = 15$; X_2 is Gumbel with mean $\mu_{x_2} = 5$ and standard deviation $\sigma_{x_2} = 1$; their correlation $\rho_{x_1} = 06$; and they are involved in a reliability function of the form $G = X_1 - 1.5X_2$, determine the reliability function in terms of uncorrelated normal variables. This form of the reliability function arises from, for example, consideration of wave run-up on a rock breakwater where the surf similarity parameter is large (see Section 9.4.1) and X_1 would be an acceptable run-up level and X_2 the significant wave height.

We have $\underline{R} = \begin{pmatrix} 1 & 0.6 \\ 0.6 & 1 \end{pmatrix}$

and $F_{X_2}(x_2) = \exp(-\exp(-\alpha(x_2 - u)))$

where $\alpha = \pi / (\sigma_{x_2}\sqrt{6}) = 1.283$

and $u = \mu_{x_2} - \dfrac{\gamma}{\alpha} = 4.55$

from the properties of the Gumbel distribution.

This gives $F_{X_2}(x_2) = \exp(-\exp(-1.283(x_2 - 4.55)))$

Or $x_2 = 4.55 - \{\ln(\ln(F_{X_2}(x_2)))/1.283\}$

Therefore

$$\rho_{12} = 0.6 = \int_{-\infty}^{\infty}\int_{-\infty}^{\infty} \left(\frac{x_1 - \mu_{x_1}}{\sigma_{x_1}}\right)\left(\frac{x_2 - \mu_{x_2}}{\sigma_{x_2}}\right)\Phi_2(y_1, y_2; \rho'_{ij})\,dy_1\,dy_2$$

$$\int_{-\infty}^{\infty}\int_{-\infty}^{\infty}\left\{\frac{4.55 - \{\ln\,(-\ln(\Phi(y_2)))/\alpha\} - 5.0}{1}\right\}\Phi_2(y_1, y_2; \rho'_{ij})\,dy_1 dy_2$$

$$\int_{-\infty}^{\infty}\int_{-\infty}^{\infty} y_1 \cdot \{-0.45 - \ln(-\ln(\Phi(y_2)))/1.283\}\Phi_2(y_1, y_2; \rho'_{ij}) dy_1 dy_2$$

Iterative solution of the above integral gives $\rho'_{12} \approx 0.62$ by numerical integration.

Hence, $\underline{R'} = \begin{bmatrix} 1 & 0.62 \\ 0.62 & 1 \end{bmatrix}$

Cholesky decomposition (determined numerically) gives

$$\underline{L} = \begin{bmatrix} 1 & 0 \\ 0.62 & 0.79 \end{bmatrix} \text{ and } \underline{Y} = \underline{L}\,\underline{Z} = \begin{bmatrix} 1 & 0 \\ 0.62 & 0.79 \end{bmatrix}\begin{bmatrix} Z_1 \\ Z_2 \end{bmatrix}$$

Therefore $X_1 = F_{x_1}^{-1}[\Phi(Z_1)] = \mu_{x_1} + Z_1\sigma_1 = 100 + 15Z_1$

and $X_2 = F_{x_2}^{-1}[\Phi(0.62Z_1 + 0.79Z_2)]$

$$= \{u - \ln(-\ln(\Phi(0.62Z_1 + 0.79Z_2)))\}/\alpha$$

$$= \{4.55 - \ln(-\ln(\Phi(0.62Z_1 + 0.79Z_2)))\}/1.283$$

Finally, $G = X_1 - 1.5X_2$ becomes
$G = \{100 + 15Z_1 - 4.55 + 1.5\ln(-\ln(\Phi(0.62Z_1 + 0.79Z_2)))\}/1.283$
i.e. $G = \{95.45 + 15Z_1 + 1.5\ln(-\ln(\Phi(0.62Z_1 + 0.79Z_2)))\}/1.283$

Example
A new sea wall is to be designed to limit wave overtopping under severe conditions to below a critical value Q_c. Write down a failure function and make appropriate simplifying assumptions to derive an approximate failure function that is dependent on water level and wave height only.

Solution
We use the formula due to Owen (1980) for overtopping (See Section 9.4). Substituting the dimensional expressions for freeboard and overlapping into Owen's equation gives

$$G = Q_c - Q = Q_c - gT_m H_s A e^{-B\left(\frac{CL - WL}{rT_m\sqrt{gH_s}}\right)} \tag{A}$$

where A and B are constants depending on sea wall geometry, CL is the crest level, WL is the still water level and r is wall roughness.

One current approach is to take T_z as being directly related to H_s through the assumption that storm waves have a similar (i.e. constant) wave steepness. That is,

$$T_z = \left(\frac{2\pi H_s}{gS}\right)^{\frac{1}{2}} \tag{B}$$

where S denotes wave steepness. Assuming a JONSWAP wave spectrum we have $T_m = 1.073T_z$ (see Section 3.4.3).

$$\text{and so } T_m = 1.073\left(\frac{2\pi H_s}{gS}\right)^{\frac{1}{2}} \equiv aH_s^{1/2} \tag{C}$$

The distribution of T_m is thus completely determined by the distribution of H_s, and we may substitute Equations A, B and C to eliminate one variable in the reliability function.

There is likely to be dependence between H_s and water level because of wave breaking due to depth limitation. However, there is unlikely to be much, if any, physical cause for T_m and WL to have strong dependence. Due to wave generation and propagation processes we might expect some dependence of both H_s and T_m on wave direction. A common way of accounting for wave direction is to undertake a series of 'conditional' calculations, one for each direction sector of interest. The results for each sector can be considered in turn and the worst case(s) used for design purposes. As construction of a new sea defence is being considered we will take CL, r, A and B as being known values, although they could also be taken as random variables with known probability distributions. The failure function thus becomes

$$G = Q_c - gaH_s^{3/2} A e^{-B\left(\frac{CL-WL}{r\sqrt{gaH_s}}\right)}$$

which is a function of two dependent variables H_s and WL.

7.2.8 Accounting for uncertainty

This section summarises types of uncertainty relevant to the assessment and design of sea defences. As seen in the example in Section 7.2.6, we may account for uncertainty by allowing what might have been treated as a known parameter as a stochastic variable, with a specified probability distribution. It is important to acknowledge uncertainty, wherever it lies in the design or assessment process. When applied correctly, a probabilistic approach to design allows uncertainties to be quantified, even if not removed.

Sources of uncertainty include:

- *Incompleteness* – if not all possible failure mechanisms have been identified then the risk assessment will be incomplete. For coastal structures detailed observations of failures are scarce, due to the relatively small number of failures and the difficulty of taking measurements during the physical conditions under which failures occur.
- *Empiricism* – the behaviour of most coastal structures is predicted by design equations that are generally empirical, based on experiments performed at laboratory scale. Such experiments are rarely exactly repeatable, giving scatter in the results and errors in fitting an equation to the data.
- *Extrapolation* – in determining design conditions, observations are used to specify the parameters of a probability distribution. There are statistical errors associated

with this procedure but in addition there is uncertainty when design values are estimated from extrapolating the distribution curve.

- *Measurement error* – the observations used for design will themselves have uncertainties due to the accuracy of measurement equipment. The accuracy of the measurements will affect the estimates of the design loads (e.g. water levels and wave heights) and the strength of the structure (e.g. measurements of soil parameters and geotechnical properties of an earth embankment).
- *Compound failure mechanisms* – coastal structures in particular can be difficult to assess in terms of separate failure mechanisms. That is, failure may occur through a particular sequence of partial failures. For example, seepage through permeable foundation layers could lead to piping at the landward toe of a structure. In turn, this could lead to erosion of the landward toe, slipping of the landward face, a consequent reduction in dimensions (and therefore strength) and through erosion lead to a breach of the defence. Analysis of this type of chain of events is made difficult because design equations are formulated to represent a single mechanism. While the first 'link' in the chain may be identifiable, subsequent events can be difficult to identify qualitatively and almost impossible to define quantitatively. In practice, designs are governed by a small number of mechanisms that are treated independently;
- *Stationarity* – the design loads and corresponding structure derived from these have, in the past, been taken as being applicable for the duration of the structure's design life. That is, there is an assumption that the statistics of say, wave heights, remains constant over time. However, the effects of sea level rise and long-term climate change have caused a reappraisal of this assumption. If there is a long-term underlying trend (such as a gradual rise in the mean level of the sea), or if the variance of a variable changes over time (such as changes in typical storm intensity, duration or frequency), then these can have a significant effect on the design life of the structure. Design guidelines in the UK now specify an allowance for sea level rise that must be included in the design of new defences so that they provide protection against the required level of design conditions (e.g. 50-year return period) at the end of their design life as they do at the beginning. This takes into account the fact that over the design life the conditions corresponding to a given return period are expected to change due to the underlying long-term changes in sea level.

Further reading

Benjamin, J.R. and Cornell, C.A., 1970. *Probability, Statistics, and Decisions for Civil Engineers*, McGraw-Hill Book Company, New York, NY.

Burcharth, H. F., Liu, A., d'Angremond, K. and van der Meer, J.W., 2000. Empirical formula for breakage of Dolosse and Tetrapods. *Coastal Engineering*, 40(3), pp. 183–206.

Melchers, R. E., 1999. *Structural Reliability Analysis and Prediction*, 2nd Edition, John Wiley & Sons, Chichester, 437 pp.

Reeve, D.E., 2010. *Risk and Reliability: Coastal and Hydraulic Engineering*, SPON Press, London, 304 pp.

Smith, R. L., 1990. Extreme Value Theory, in *Handbook of Applicable Mathematics* (ed. W. Lederman), Vol. 7, John Wiley & Sons, Chichester.

Thoft-Christensen, P. and Baker, M.J., 1982. *Structural Reliability Theory and its Applications*, Springer, Berlin.

Chapter 8

Field measurements and physical models

8.1 The need for field measurements and physical models

Previous chapters have introduced many aspects of the known physics of the coastal zone and how to incorporate the physics into numerical models. Such models are now in common use in investigating coastal processes and the design of coastal engineering schemes. To the casual observer, such techniques may appear sufficient to cover all eventualities. However, on closer inspection, this can easily be shown to be untrue. The current state of the art involves a (sometimes) subtle interplay between the use of numerical models, physical models and field measurements. This interplay is important for both research studies and for coastal engineering design. To illustrate the individual usefulness and the interdependencies of these three approaches, their own particular benefits and drawbacks are now briefly summarised.

Numerical models may be used to predict both the spatial and temporal variation of the wave, current and sediment transport fields. This can be achieved quickly and (relatively) cheaply in many cases. However, the accuracy of their predictions is limited primarily by the known physics and secondly by the assumed boundary and initial conditions. Currently, there are many aspects of the true physics which are either unknown or have not yet been included in numerical models. Physical models, on the other hand, can be conceived of as being an analogue model of the true physics, without us necessarily knowing what the true physics is. Thus, in principle, they should provide more accurate predictions. Experiments using physical models can also be undertaken using controlled conditions, thus allowing investigation of each controlling parameter independently. Physical models, of necessity, are normally smaller scale versions of the real situation. This requires a theoretical framework to relate model measurements to the real (prototype) situation. Unfortunately, the outcome of this theoretical framework is that scaled physical models are unable to simultaneously replicate all of the physical processes present in the prototype in correct proportion. Thus we return to nature, by way of field measurements. Such measurements obviously do contain all the real physics, if only we knew what to measure and had instruments to do so. Such measurements, as are possible, have to be taken in an often hostile environment, at considerable relative cost and under uncontrolled conditions. Thus, it can be appreciated that the three approaches all suffer from drawbacks, which preclude their exclusive use. On the other hand, it can also be seen that each approach can benefit from results gleaned from the others. In terms of the development of our understanding and the incorporation of that understanding in the design process, field studies and physical model studies are required to improve both our knowledge of the physics and to calibrate and verify our numerical models. In addition, current design methodology often makes joint use of all three approaches.

8.2 Field investigations

Field investigations are often carried out for major specific coastal defence projects. Typically, measurements are made of waves, currents, water levels and beach profiles. Standard commercial measurement systems are available to carry out these measurements. Such measurements are often used to derive the local wave climate, current circulation patterns, extreme still-water levels and beach evolution through the use of numerical models which are calibrated and take their boundary conditions from the measurements. However, such measurements are generally not sufficient to validate the numerical models, or to discern the fundamental coastal processes and their interactions. Thus a second category of field investigations is required to address these issues, namely research-based field campaigns. Over the last twenty years a number of such major investigations have been carried out in the USA, Canada, Europe and elsewhere. In these studies, very detailed measurements of waves, currents, sediment transport, shoreline evolution and beach morphology have been undertaken. Well-known campaigns include the Nearshore Sediment Transport Study, DUCK, SUPERDUCK, DELILAH, DUCK 94, SANDYDUCK, all in the USA, and the Canadian Coastal Sediment Transport Study. These studies have brought to light many aspects of coastal processes not previously well understood or indeed even recognised (see Dean and Dalrymple (2002) for further details and references). In Europe, the most recent major study is COAST3D (2001), which involved intensive field measurements at two European sites to study coastal morphology and the performance and validation of numerical morphological models. Part of this study involved development of guidelines on the selection of coastal zone management tools. One set of such tools comprises measurement equipment. The set of measurement tools described in Appendix 2 of the report (see Mulder *et al.* 2001) is both comprehensive and informative, comprising descriptions of equipment to measure bathymetry/topography, seabed characteristics/bedforms, water levels/waves, velocities, suspended sediment concentrations, morphodynamics/sediment transport and instrument carriers/frames/ platforms. The appendix also contains guidelines on the use of such equipment and examples of results at the COAST3D field sites. This appendix is reproduced here for ready reference, together with three photographs illustrating a device called the inshore wave climate monitor, deployed at one of the COAST3D field sites at Teignmouth, UK.

This appendix is reproduced with permission of Rijkswaterstaat/RIKZ and the Environment Agency from the report 'Guidelines on the selection of CZM tools', J.P.M. Mulder, M. van Koningsfeld, M.W.Owen and J. Rawson, Report RIKZ/2001.020, Rijkswaterstaat, April 2001. It was compiled by members of the EC MAST COAST3D project, funded by the European Commission (MAS3-CT97-0086), MAFF (UK), Environment Agency R&D Programme (UK) and Rijkswaterstaat Research Programme Kust × 2000 (NL).

Appendix 2 (from Mulder et al. 2001) Tools used in COAST3D: measurement equipment

Name	Brief description and/or Teignmouth	Guidelines on how to use	Examples of results for Egmond
Bathymetry/Topography			
Total Station Levelling	Method of surveying the beach and inter-tidal area, using a laser levelling system	Commercially available instrumentation, used extensively for ground surveys. For surveying the intertidal and supra-tidal beach, levels are usually measured along regularly spaced cross-shore transects, from a landward reference point down to about 1 m below the low water mark (depending on wave conditions). Surveys are best carried out at about low water of high spring tides to give best coverage of beach. For beach monitoring purposes, permanent markers should be established to enable accurate relocation of transects, and the beach levels should be related to a standard datum.	Beach levels at Egmond have been regularly monitored for decades. At Teignmouth, beach levels were monitored each month at three transects for three years prior to the Coast3D measurement campaign
Differential Global Positioning System (DGPS)	Method of fixing absolute position (three coordinates), based on calculated distance from at least four geostationary satellites	Commercially available instrumentation. The method uses 2 DGPS receivers, one in fixed location (base station), the second being moved between each measurement. The base station needs to be very stable, and must be referenced in the local geodesic system with great accuracy. The second receiver can be stand-alone, or mounted on a mobile platform, e.g. cross-country vehicle. The accuracy of the horizontal co-ordinate is about ±20 mm. Overall the vertical accuracy is about 50 mm on relatively flat and smooth areas, rising to about 100 mm on steep sloping faces of bars, and on areas with sediments that are less well consolidated.	At Egmond, used on the WESP (by RWS) to measure cross-shore profiles. At both Egmond and Teignmouth, the second receiver was mounted on a quad-bike (by UCa), allowing a very large area to be surveyed in a very short time. Surveys were carried out at about low water, to give repeat measurements of beach topography. At Egmond, 21 surveys were carried out, each covering an area of about 300 by 100 m. Three surveys were carried out at Teignmouth during the main campaign, covering an area of about 900 m (alongshore) by 50–150 m (cross-shore), depending on the tidal range and wave activity (wave set-up).

Echo Sounder Surveys	Method of surveying the seabed, using a standard maritime echo sounder	Data logged at 10 Hz gives a reading every 0.2 m at a ship speed of 2 m/s (speed depends on sea-state). A 208 Hz echo sounder has a quoted accuracy of 0.5% of indicated depth (minimum error 25 mm). Final result depends on sea-state, technology for heave-compensation, skill of surveyor and sophistication of post-processing. Frequent (twice-daily) bar checks of the echo sounder are required	Bathymetric charts were produced for Teignmouth at a scale of 1:2500 based on parallel track lines 25 m apart, plus cross-check lines perpendicular to these. A zig-zag line in the estuary channel gave the estuary channel form. The area covered was approximately 1.5 km × 1 km, plus part of the estuary.

Seabed characteristics/Bedform

Van Veen Grab	A method of obtaining samples of subtidal seabed material either for visual analysis or for quantitative particle size distribution analysis.	Deployed by hand over side of survey vessel. The primed grab is triggered when it hits the seabed, taking a shallow 'bite' of the surficial sediment.	At Teignmouth the grab retrieved a range of fine sand, gravelly and shelly material, and stones. Stones can jam open the jaws of the grab, necessitating a repeat deployment.
Roxann System	An acoustic system used to produce a map of the nearshore and offshore zones of the study area, showing the size classification of the seabed sediments. The system works by comparing acoustic returns from the seabed at a number of different frequencies.	The system is mounted on a survey vessel, and needs an experienced specialist operator, both to make the survey and to interpret the results. An adequate number of bed samples must be taken to visually calibrate the system for site-specific conditions. Based on these point samples, the seabed composition can then be mapped in detail over the entire survey area.	Not used at Egmond. At Teignmouth, the system was installed on the survey vessel *Sir Claude Inglis* (HR). One survey was carried out, before the pilot campaign. For calibration purposes, 50 grab samples of the seabed were taken on a grid covering the survey area.

Instrument	Description	Practical considerations	COAST3D experience
Digital Side-Scan Sonar	An acoustic system designed to map the bedforms in the offshore and nearshore zones. Mounted on a survey platform (e.g. survey vessel or the WESP), can be used to build up a geo-corrected mosaic image displaying the bedforms throughout a study area.	Very portable, and can be mounted on a survey platform in a few hours. Tracklines need to be spaced at a small interval (typically 50 m) to enable an accurate mosaic to be assembled. Residual sediment transport directions can be deduced in locations where asymmetric bedforms occur. The accuracy of positioning during recording was a few centimetres in Egmond (due to mounting on WESP and very accurate DGPS system) and a few metres at Teignmouth (system was towed and less accurate DGPS system).	For the COAST3D project, the system was supplied/operated by Mag. At Egmond, the system was mounted on the WESP, which provided an ideal platform to work in very shallow waters. The system was able to resolve bed features over the entire recording width (2 × 45 m). A complete mapping of the experimental area was carried out towards the end of the main measurement campaign. At Teignmouth, the system was installed on the survey vessel *Sir Claude Inglis*. A complete mapping of the experimental area was carried out approximately midway through the main campaign.

Water levels/Waves

Instrument	Description	Practical considerations	COAST3D experience
Pressure Transducer (PT)	A device for measuring total pressure. When installed underwater, analysis of instantaneous pressures gives measure of wave height/period. Analysis of time-averaged pressure gives measure of water depth, provided that atmospheric pressure is known.	In use for many years, and now a standard, commercially available instrument. Needs to be mounted on a suitable platform, e.g. structure, frame, tethered buoy. Accuracy of wave measurement depends on water depth and wave period. Attenuation corrections must be applied.	Used extensively at both Egmond and Teignmouth. Not suitable for use at depths > 20 m, due to excessive attenuation effects. Data recovery rate was 90–100%. Accuracy for depth is 1–2%. Transducers on free-standing frames may not represent the water depth accurately if the frame settles into the seabed due to scour action. Accuracy for wave-height is 10–15% for waves of period 6 s in depth of 5 m. Wave energy at frequencies higher than 0.4 Hz is not detected by most commercially available instruments.

Wave Pole	A pole or pile driven into the seabed, and extending above the highest water level. Used as a mounting for a pressure transducer to measure water level, wave height and wave period	Five wave poles were installed by RWS at Egmond (independently of the COAST3D project) at water depths varying from about 4.7–0.1 m below mean sea level. The results indicated the cross-shore variation of wave heights. No wave poles were installed at Teignmouth.
Directional Wave Buoy	A surface buoy for measuring offshore wave conditions, including wave height, period and direction. Uses several accelerometers to measure the vertical motion of the buoy, and its tilt angle and direction. The resulting signals are radioed to a shore receiving station for processing and storage. Has been in use for many years, and is now a 'standard' instrument, typically deployed for a period of at least 12 months, with measurements every 3 h. Needs a minimum water depth of about 8 m in which to operate, and also needs a ship with fairly heavy lifting gear to deploy and recover. Not accurate for wave periods shorter than about 2 s, or longer than about 30 s. Being deployed some distance offshore, has disadvantage that also records waves that propagate from the coast, usually not relevant for coastal problems.	At Egmond, a directional waverider buoy was deployed by RWS offshore in a water depth of 15.7 m (below mean tide level), remaining on location for several months, from the start of the pilot campaign until after the end of the main campaign. Measurements were taken every hour. At Teignmouth, a directional waverider buoy was deployed offshore at a depth of about 9 m below mean tide level. Deployment was for the duration of the main measurement campaign (about 6 weeks), with wave measurements taken every hour.

| Wave Recording System (WRS) | The Wave Recording System is an array of six pressure transducers used to derive the wave height, period and directional spectra in the nearshore zone. It is especially valuable at sites where reflected waves are expected (e.g. from a seawall or a steep beach). | The six pressure transducers are deployed by divers in a triangular configuration on the seabed. Signals from the transducers are carried up to a surface buoy, and transmitted to base by GSM radio. Alternatively the data can be stored at the buoy, and downloaded from a boat. This is a research-level system developed by UPI that has had trials at several UK coastal sites. | At Egmond, the Wave Recording System was deployed in the nearshore zone, at a location where the water depth was 6 m. Unfortunately the onshore migration of the bar system during the main measurement campaign covered the WRS in 1.5 m of sand, and no data was obtained. At Teignmouth, the WRS was deployed in about 8 m of water, at the offshore boundary of the experimental area. Data was obtained at hourly intervals throughout the main measurement campaign |
| Inshore Wave Climate Monitor (IWCM) | The IWCM is an array of five electrical resistance wave staffs used to derive wave height, period and directional spectra in the inter-tidal zone. | The five wave staffs are driven into the beach in a triangular array, and are connected to a central data storage/battery power unit. Data recovery can be carried out weekly using a notebook PC. Wave staffs are fairly widely available, but the overall system has been specially developed by UPI. | At Egmond, the system was deployed during the main measurement campaign, at a location close to mean low water springs. However movement of the nearshore sand bars meant that the staffs were now in a deep trough, and could not be accessed for data retrieval. At Teignmouth, the system was deployed at East Pole Sand near the low water mark, and data was obtained at hourly intervals throughout the main campaign. |

X-band Radar	A system for the remote sensing of wave direction and wavelength. Based on interpretation of the 'noise' in the radar return signal, generated by reflections off ripples on the sea surface.	Uses a conventional maritime radar antenna, which needs to be mounted about 10 m above sea level. Requires a wind speed of a couple of metres per second to generate the wave field required for the system to operate. General patterns of wave direction, wavelength, wave refraction/diffraction etc. can be viewed on a conventional display screen. However quantitative information requires digitisation and storage of the images, and analysis using special software only recently developed by POL. Further developments may include the derivation of wave height from the signal, and the use of linear wave theory in an inverse mode to obtain bathymetry.	At Egmond, the X-band radar system operated for about a month during the main measurement campaign, recording a sequence of 64 images at 2.25 s intervals every hour. The antenna was mounted on top of the Relay Station at the back of the beach, and data was obtained over a semi-circular area of the nearshore zone, with a radius of about 1.8 km. The locations of the bars were easily discerned, as indicated by wave breaking. At Teignmouth, the antenna was mounted on the roof of Teignmouth Pier, and the system was operational for most of the main measurement campaign.
Velocities			
Electro-Magnetic Flow Meter	The EMF meter measures current strengths in two dimensions, by measuring the voltage generated by the conductor (water) flowing past two pairs of electro-magnetic poles.	The EMF meter has been in use for many years, and has become a 'standard' instrument for measuring instantaneous current strengths in two directions, usually in the nearshore/intertidal zones. Usually employed to give velocities in the horizontal plane, it can also be aligned to give velocities in the vertical plane. The meter needs careful calibration, preferably both before and after an experimental campaign. Problems can sometimes occur with 'zero-drift', although these can often be overcome at data analysis stage.	Used extensively at both Egmond and Teignmouth, usually in association with other instrumentation, for example tripods, CRIS etc. Practical working range for velocity is about 0.03–2.0 m/s. Inaccuracy is a maximum of about 15% for time-averaged velocities greater than about 0.5 m/s, with significantly larger errors at lesser velocities. Errors in wave orbital velocities may also be about 15%.

Instrument	Description	Deployment notes
S4 Current Meter	A submerged, spherical instrument package available commercially, and used principally to measure instantaneous velocities in the nearshore and offshore zones. Similar in concept to an EMF meter, but with self-contained power source, data storage etc. Some versions of the S4 incorporate other instrumentation, e.g. pressure transducer (to measure wave heights/periods), salinometer, thermistor.	The S4 current meter has been in use for many years now, and has become a standard instrument for measuring current strengths and directions in the nearshore and offshore zones. Because of its size (300 mm diameter) the S4 data may not be reliable in water depths of less than about 2 m, and must be at least 1 m below the water surface. The sphere and its sinker weights (or mounting frame) need to be deployed by boat with fairly heavy lifting gear (approx 400 kg). It is necessary to retrieve the sphere in order to download the data. For example, using 10 min burst sampling of average current speed at Teignmouth, S4s were deployed for a 27-day period. Calibration of the S4 is advisable before and after a measurement campaign.	A total of nine S4 instruments was deployed at Egmond (by UCa and HR) in water depths varying between about 1.1 and 5.5 m below mean sea level, mostly giving good data recovery. Nine S4s were deployed at Teignmouth, five in the nearshore zone, and three in the intertidal zone. Those in the intertidal zone were mounted on specially built frames (UCa), measuring at a distance of about 0.65 m above the seabed.
Acoustic Doppler Velocity Meter (ADV)	A device for measuring instantaneous currents, based on the Doppler shift in frequency of an acoustic signal due to the moving water.	A relatively new technique, but increasingly being used in coastal research. Gives instantaneous velocities and directions at the position of the sensor. Needs to be mounted on a suitable platform, for example structure, frame or tethered buoy.	In the COAST3D experiments, ADV meters were attached to one or two maxi-tripods (e.g. by UU).

Acoustic Doppler Current Profiler (ADCP)	An acoustic system used to measure the vertical profile of horizontal velocities.	A relatively new technique, but increasingly being used in coastal and estuary research. Usually mounted on a survey vessel with the transducer (e.g. 1500 kHz) pointing downwards, allowing instantaneous measurements of the vertical profile at many locations in the study area. But can also be mounted on the seabed, with the transducer pointing upwards, to give a time series of vertical profiles at a fixed location. Current speeds can be measured at vertical intervals of typically 0.5 m through the water column, but measurements cannot be made within the top 2 m and the bottom 1.5 m of the water column. The accuracy depends on height of cells and on averaging time.	At Egmond, mounted in a frame on the seabed, in a water depth of about 5 m. For various reasons, reliable data was only obtained for a period of about 9 days during the main campaign. At Teignmouth, two ADCPs were deployed, by HR. One was bottom mounted in a water depth of 8 m. The other was installed on a survey vessel; on two occasions during the main campaign, continuous profiling of current speed and direction was undertaken whilst travelling all along the boundary of the study area, including across the harbour entrance. Each survey was repeated at approximately hourly intervals throughout the tidal period.
Float tracking	A standard technique for observing (tidal) flow patterns over an area. Analysis of float positions at given times enables near-surface tidal velocity and direction to be determined	HRW standard drogues (1 × 1 m cruciform terylene panels) were deployed in the estuary mouth at Teignmouth. The centre of the drogues was set at either 1 m or 3 m below water surface. Surface floats attached to the panels were tracked by boat.	Three releases at Teignmouth during an ebbing spring tide showed how the flow from the estuary developed and interacted with the tidal flow outside the estuary.

Suspended Sediment Concentrations

Pumped Sampling	A technique for obtaining samples of the water and suspended sediment, which are then analysed to determine suspended sediment concentration and size grading, and other relevant parameters.	A standard technique which has been used by many organisations for very many years. At Teignmouth, 16 mm nozzles were positioned at known heights above the bed (usually on a frame), connected by flexible tubing to a pump and filtration unit mounted in an inflatable boat. Samples were typically taken every 1–2 min, comprising 20–40 l depending on sediment concentration. Each sample was pumped through a pre-weighed 40 micron nylon filter, which, together with a sample of the filtrate, was retained for analysis.	Direct samples for a flood tide on Spratt Sands gave maximum concentrations of suspended sand of 45 mg/l at 0.1 m above bed, and 20 mg/l at 0.5 m above bed. The median size of the suspended material was 0.10–0.12 mm.
Optical Backscatter System (OBS)	An instrument to measure suspended sediment concentrations, based on the proportion of light scattered by particles in suspension.	This system has been used for many years, and has become a 'standard' method of determining instantaneous suspended sediment concentrations. Unlike the similar acoustic system, the optical system gives only a point measurement at the position of the sensor. The measured concentration is very sensitive to particle size, and detailed calibration is necessary against *in situ* samples. If the background concentration (e.g. of silt) to be subtracted from the record is of the same order of magnitude as the sand concentration, the OBS sand concentrations will be rather inaccurate.	Used extensively at both Egmond and Teignmouth, in various combinations with other instrumentation, for example tripods, CRIS.

Acoustic Backscatter System (ABS)	A multi-frequency acoustic backscattering system to provide high-resolution vertical profiles of the concentration and median grain-size of suspended sediments along a (short) transect.	For the COAST3D project the system used three acoustic frequencies, and hence three transducers. The transducers were fixed to a bedframe, deployed either in the inter-tidal or nearshore zones. The system requires detailed acoustic and electronic calibration, and preferably some in situ suspended sediment measurements. Typically, measurements can be made at a vertical spacing of 1 cm throughout a 1 m transect above the bed.	The equipment was deployed throughout the Egmond pilot and main campaigns, and returned good data. Unfortunately in situ pumped samples of the suspended sediment were not obtained, so the data could not be calibrated accurately. At Teignmouth the system was deployed by POL on the maxi-tripod on the inter-tidal portion of Spratt Sand, where it recorded suspended sediment concentrations simultaneously with an optical backscatter system. The system was calibrated against in situ pumped samples over one ebb tide.

Morphodynamics/Sediment transport

Fluorescent tracers	Sand dyed with a fluorescent paint is injected at a fixed location on the beach, and its movement is tracked at each low water to give an indication of the direction and rate of sand movement.	A fairly standard technique that has been in use for many years now. The sand to be dyed is taken from the beach being studied. Different colour dyes are available. The amount of sand injected is typically about 100 kg (for each site/experiment), depending on the wave conditions. Generally, sand movement cannot be detected for periods longer than about 3–4 tidal cycles (again depending on wave/tide conditions)	Used at both Egmond and Teignmouth by UCa. During the main campaign in Teignmouth, several injections were carried out to quantify sediment transport within the swash zone. Due to high energy in this area, the dispersion of the dyed sand grains was very fast, and the moving layer was very thick. In these hydrodynamic conditions it is impossible to identify the edge of the fluorescent cloud to quantify a sediment transport rate. Four injections were carried out on the Egmond beach with two or three low tide detections by night to locate the fluorescent sand grains with ultra-violet lights. Sediment movements were quantified (direction and rate) showing the longshore and cross-shore sand transport components integrated over several tidal cycles.

Swash Morphodynamics	Rapid surveys employing a system of graduated rods are used to measure the evolution of the sea bed during the swash processes.	Graduated rods are placed in line along the cross-shore profile. A cylindrical ruler with a flat base is dropped over each rod in turn to measure the local bed level. The rods need to be well fixed on the beach, and their position accurately known. The method is operated by two people wearing wet suits. The complete measurement campaign lasts several hours, every rod in the swash zone typically being measured every 5 min, with an accuracy of about 5–10 mm. For safety reasons, measurement are not possible with wave breaker heights greater than about 1 m. Essentially a research technique developed by UCa.	Used at both Egmond and Teignmouth by UCa. Fifteen experiments were realised on the two different sites. Nine experiments were conducted on the lower beach at Egmond, lasting for 2–9 hours, and six were conducted on the upper beach at Teignmouth, lasting 5–8 hours. This experiment is manual, and requires the permanent presence of two experimenters for several consecutive hours, in all conditions of waves and weather.
Autonomous Sand Ripple Profiler	An acoustic system to scan ripple profiles, and to provide detailed data on bedform evolution. Measurements are made over a (short) transect.	The system is deployed in a bedframe, located either in the inter-tidal or nearshore zones. The ripple scanner requires software to extract and track the bed elevation over time. Minimum measurable variation in bed height is about 5 mm A research technique being developed by POL.	This experimental equipment was deployed by POL on two occasions during the main measurement campaign at Egmond, for 5 h and 29 h, respectively. The equipment was fixed to one of the Utrecht University maxi-tripods, in a water depth of 4.3 m. Bed profiles were measured every few minutes along a transect length of 3.5 m. The system was also deployed in the inter-tidal zone at Teignmouth, on Spratt Sand which has very pronounced bed features. Data was obtained throughout the main measurement campaign, with measurements from a few hours before to a few hours after each high tide.

| Tell-Tail Scour Monitor | An instrument to monitor the maximum depth of scour at a given location during a particular event, for example at a bridge pier during a major flood, or adjacent to a seawall during a major storm. Records movement of a vertical stack of 'tails' that waggle when water flows past but not when buried in the beach or sea bed. | Four instruments placed on the beach using a mechanical digger. Initial level set to cover range of interest. Recorded bed level with 100 mm resolution every 10 min over a vertical range of 0.8 m. Instruments must be levelled to a local datum, and bed elevation measured at low tides. | The instruments at Teignmouth showed that the bed elevation changed during a single tidal immersion by up to 0.3 m. Changes in bed elevation during the main campaign were at least 0.6 m. |
| ARGUS Video System | The ARGUS system consists of several digital video cameras set up to view the beach and surf zone. The system is programmed to record images for a few minutes every hour during daylight. | For best results the cameras must be mounted high above the beach, at a location with a power supply and telephone line (for transmitting images to the office base). The images give very valuable information on daily changes in intertidal and beach morphology, and can also indicate long-term changes in near-shore bathymetry, e.g. bar location. The information primarily provides a qualitative view of the changing bed morphology, although techniques to process this into a quantitative measure of the bathymetry are being developed. ARGUS systems are only available through Oregon State University, USA. | At Egmond, the ARGUS system was mounted on a very tall mast (height 40 m) located to the south of the experimental site. It was used primarily to monitor the position of the bar crests: comparison with the WESP surveys indicated that the ARGUS system gave a bar location accuracy of about 40 m, depending on local wave heights and water depths. Therefore the system should typically be used to detect changes in bar position and patterns over several weeks, months and years: the system should not be used to monitor daily changes in bar position. Accurate information on offshore waves and water levels is essential for quantifying bar location. At Teignmouth, the ARGUS system was mounted on the top of The Ness, the headland at the southern end of the experimental site. It was used primarily to monitor changes in the bars, shoals and channels seaward of the entrance to Teignmouth Harbour. Five cameras were used at each site. |

Instrument Carriers/Frames/Platforms

| Maxi-Tripod | A large frame placed on the seabed, providing a mounting platform for various measurement equipment. Designed for deployment mainly in the nearshore and offshore zones. Typical instrumentation includes: pressure transducer to measure water levels, wave heights/periods; one or more EMFs (or ADVs) for instantaneous current strength and direction at fixed elevation(s) above the seabed; one or more OBSs or an ABS for instantaneous suspended concentration at fixed elevation(s); a compass; a tilt-meter. Also includes power supply and data storage equipment. | Different versions of the tripod/bedframe have been developed by different organisations, but have essentially the same purpose. The tripods have to be deployed and recovered either by ship or by an amphibious vehicle such as the WESP. Settling of the tripods, in combination with migrating and changing bedforms (and therefore variable bed levels) causes some problems for analysis of the data, since the exact height of the sensors above the bed is not monitored. Tripods typically work in a burst-sampling scheme: every hour a series of measurements starts for a period of 20–40 min. Servicing requirements vary according to design, but typically the tripod has to be recovered every 20–30 days for data downloading. | For the COAST3D project, maxi-tripods/bedframes were deployed by UU, UCa, HR. At Egmond, deployment of the maxi-tripods was carried out smoothly and rapidly by the WESP. Seven tripods were deployed to measure cross-shore gradients in wave- and flow-parameters. The tripods were located at seabed elevations ranging from about 1.4–5.8 m below mean sea level. Measurements were taken every hour during the five weeks of the main experimental campaign. At Teignmouth, six bedframes/maxi tripods were deployed by the survey vessel *Sir Claude Inglis*. Three were deployed in the nearshore zone at water depths of about 2–6 m below mean tide level, while a fourth was deployed in the intertidal zone, on Spratt Sand. Two were used to provide long-term monitoring of waves, tides, currents, suspended concentrations and temperature at the boundaries of the study area. |

Mini-Tripods	A small tripod for deploying in the inter-tidal zone, equipped with basically the same instrumentation as the maxi-tripod, typically: pressure transducer, for waves, water levels; one EMF (or ADV) for currents; one OBS for suspended sediment concentration. Also includes power supply and (limited) data storage.	Different versions have been developed by different organisations. Small enough to be deployed by hand, usually around the time of low water. Same problems as maxi-tripod in determining exact height of sensors, but can be checked visually at each low water.	For the COAST3D project, mini-tripods/beach frames were deployed by UU, UCa and UPI. At Egmond, ten mini-tripods were used in the inter-tidal zone. These tripods were deployed over nearly every high water, in a configuration and at locations which depended on the morphology at the time of deployment. At Teignmouth, eight mini-tripods were deployed in the inter-tidal zone. Again the exact locations depended on the local morphology at the time of deployment.
WESP	The WESP, a Dutch acronym for Water and Strand Profiler, is an approximately 15 m high amphibious 3-wheel vehicle used primarily to carry out bathymetric surveys of the nearshore, intertidal and beach zones. It can also be used to deploy and recover instrumented tripods, and to tow the CRIS instrumented sledge (q.v.).	The WESP is generally used to measure cross-shore transects from the top of the beach seawards to a water depth of about 6–7 m (depending on wave conditions). A kinematic DGPS on the vehicle measures its position. The accuracy of the combined WESP/DGPS system is about 50–100mm for a flat or gradually sloping bed: for steeper profiles the accuracy is somewhat less. Developed and built by RWS, this is the only device of its kind in Europe. The COAST3D and KUST*2000 campaigns at Egmond were its first field trials, and it is still being evaluated.	At Egmond, cross-shore transects were measured with a spacing of 50 m, in wave conditions up to 2 m significant. Maximum bed slopes were about 6°, in which case the accuracy in the vertical was estimated to be about 100–200mm. Complete surveys of the experimental area were taken roughly every 2 days during the main measurement campaign. The WESP was not used at Teignmouth, partly because of the very considerable expense involved in transporting it from Egmond, and partly because of the very steep bed slopes near the harbour entrance.

| CRIS | The CRIS (Coastal Research Instrumented Sledge) is designed to make detailed sediment transport measurements in the nearshore and intertidal zones. It is used in combination with the WESP (towing, power supply, data handling, water sampling). The sledge can be equipped with various instrumentation, including electro-magnetic flowmeters, optical backscatter sensors etc. | The CRIS was originally intended to be used to give quasi-synoptic measurements along a cross-shore transect. In reality though, the time interval between two consecutive measurements is too long, resulting in non-steady wave and tidal conditions. By mounting the instrumentation on the relatively open structure of the CRIS instead of the rather substantial WESP, 'undisturbed' measurements of sediment transport can be made. The CRIS was developed by UU/RWS specifically for the KUST*2000 campaign. | At Egmond, the total time at each location was about 40 min – about 20 min to settle into the sediment, and then about 20 min of measurements. Not used at Teignmouth. |

Figure 8.1 Three views of the Inshore Wave Climate Monitor, deployed at the COAST3D field site at Teignmouth UK, to measure shoreline directional wave spectra. Photographs by courtesy of Tony Tapp and Dr David Simmonds, School of Civil and Structural Engineering, University of Plymouth, England.

8.3 Theory of physical models

8.3.1 Generic model types

A physical model may be defined as a physical system reproduced (at reduced size) so that the major dominant forces acting on the system are reproduced in the model in correct proportion to the actual physical system. To determine if a model can reproduce these dominant forces in correct proportion requires the application of the theory of similitude. This is introduced later in this section.

Traditionally, scaled physical models have been used extensively in the design of major hydraulic engineering works, notably river engineering schemes, estuary schemes, hydraulic structures, coastal engineering schemes and port and harbour developments. More recently, physical models have been used for two other purposes, namely as process models and validation models. Process models comprise experimental investigations of physical processes to improve our knowledge of the underlying physics. Validation models are used to provide test data against which numerical models may be compared, validated and calibrated. Design, process and validation models are also subdivided into two classes, fixed bed and mobile bed. A fixed bed model is rigid, with a moulded bathymetry, whereas a mobile bed model has a bed of mobile material. Fixed bed models are the most common and are often used even when the prototype has an erodible boundary (e.g. rivers, coastlines, estuaries) where the principal interest lies in water levels, velocities etc. Mobile bed models are needed when the principal interest lies in sediment deposition and erosion. In such models a choice of model sediment has to be made. This is not a straightforward matter and is discussed later in this section. Finally most models are constructed as (smaller) scale models of the prototype and are geometrically undistorted. However, some models are constructed as geometrically distorted models in which the vertical scale is smaller than the horizontal scale. This enables models of large areas with small depths (e.g. long sections of rivers or estuaries) to be built in available laboratory space.

8.3.2 Similitude

If a scale model is constructed such that all lengths in the model are in the same ratio to those in the prototype, then *geometric similarity* is achieved. The geometric scale is defined as the ratio of any length in the prototype (L_p) to that in the model (L_m), thus the length scale ratio (N_L) is defined as:

$$N_L = L_p/L_m \tag{8.1}$$

Scale ratios for area (N_A) and volume (N_V) follow directly from the length scale ratio, as area and volume are proportional to length squared and cubed, respectively; for example

$$N_A = \frac{A_p}{A_m} = \left(\frac{L_p^2}{L_m^2}\right) = \left(\frac{L_p}{L_m}\right)^2 = N_L^2 \tag{8.2}$$

$$N_V = \frac{V_p}{V_m} = N_L^3 \tag{8.3}$$

However, to achieve complete similarity between model and prototype also requires similarity of motions (e.g. similarity of velocities and accelerations), known as *kinematic* similarity and similarity of forces, known as *dynamic* similarity. Geometric similarity also provides similarity of velocities, but to achieve similarity of accelerations requires similarity of forces.

The simplest way to understand the implications of requiring dynamic similarity is to start from Newton's second law of motion, which states that the sum of the forces acting on a particle is equal to its mass times its acceleration (the inertial force F_i). In fluid mechanics problems the forces acting can include gravity, viscosity, surface tension, elastic compression and pressure forces. Restricting our attention to typical coastal engineering situations, the principal forces acting are those due to gravity (F_g) and viscosity (F_μ). Hence we may write:

$$m\frac{dV}{dt} = F_i = F_g + F_\mu$$

Diving by F_i yields

$$1 = \frac{F_g}{F_i} + \frac{F_\mu}{F_i} \tag{8.4}$$

For perfect similitude, these force ratios must each be equal between model and prototype. In practice, it turns out that this requirement cannot be met unless the scale is one. This will now be illustrated.

The ratio of inertial force to gravity force is equal to the Froude number (Fr) squared e.g.

$$\frac{F_i}{F_g} = \frac{mass \times acceleration}{mass \times gravity} = \frac{(\rho L^3)V^2/L}{(\rho L^3)g} = \frac{V^2}{Lg} = Fr^2 \tag{8.5}$$

The ratio of inertial force to viscous force is the Reynolds number (Re), for example

$$\frac{F_i}{F_\mu} = \frac{(\rho L^3)V^2/L}{\mu(V/L)L^2} = \frac{\rho L V}{\mu} = Re \tag{8.6}$$

These are both familiar dimensionless numbers of fundamental importance in free surface flows. Hence, for similitude it is necessary that the Froude and Reynolds numbers are the same in the model and prototype, for example

$$Fr_p = Fr_m \quad \text{and} \quad Re_p = Re_m$$

These relationships provide the similitude criterion by which model velocities and times may be related to the prototype values. Starting with the Froude criterion we have

$$\frac{V_p}{\sqrt{gL_p}} = \frac{V_m}{\sqrt{gL_m}}$$

or

$$N_V = \frac{V_p}{V_m} = \sqrt{\frac{g_p L_p}{g_m L_m}} = N_g^{1/2} N_L^{1/2} \qquad (8.7a)$$

as $N_g = 1$ this reduces to

$$N_V = N_L^{1/2} \qquad (8.7b)$$

Hence model velocities should be scaled up to prototype velocities by the square root of the length scale, according to the Froude criterion. The time scale can be found by noting that the velocity is distance/time, hence

$$\frac{t_p}{t_m} = \frac{L_p V_m}{L_m V_p} = \frac{L_p}{L_m}\sqrt{\frac{L_m}{L_p}} = \sqrt{\frac{L_p}{L_m}} = N_L^{1/2} \qquad (8.8)$$

Applying the same logic to the Reynolds criterion results in a velocity criterion given by:

$$\frac{V_p}{V_m} = \frac{v_p}{v_m} N_L^{-1} \qquad (8.9)$$

Combining Equations (8.7b) and (8.9) to find a common criterion gives:

$$\frac{v_p}{v_m} = N_L^{3/2} \qquad (8.10)$$

For a scaled model this implies that a model fluid must be used with a much smaller (scale dependent) viscosity to that of the prototype fluid (e.g. water).

In practice, this is a requirement that cannot be met. The result is that we must choose to use either the Froude or Reynolds scaling criterion. For typical free surface flows the Froude scaling criterion is predominantly used. However, this is at the expense of relaxing the conditions for perfect similitude. Thus the model will not perfectly reproduce all the phenomenon present in the prototype. This is referred to as the *scale effect*. Typical free surface flows normally operate in the rough turbulent region, in which frictional resistance is independent of Reynolds number. Thus, provided that the model also operates in the rough turbulent region, non-conformance with the Reynolds scaling criterion is not normally significant. Other scale effects will include those of surface tension. Here again, this will normally not be significant, provided the model is of sufficient size (e.g. water depths > 20 mm, wave periods > 0.35 s).

8.4 Short-wave hydrodynamic models

Many physical models used in coastal engineering require investigation of wind and swell waves (e.g. short waves) and associated effects. For this purpose a Froudian scaled model is used. Although the similarity arguments, developed above, may be used to justify the use of such a model, it can also be rigorously shown that the use of an undistorted Froudian scaled model satisfies all terms in the Navier–Stokes equations except viscous shear stress (see Hughes 1993). Hence, we can model refraction, shoaling, diffraction and reflection, surf zone processes (including turbulent energy dissipation), wave-induced currents and tidal currents. *This is more than can be achieved with any currently available numerical model.* It should be noted, however, that a distorted scale model cannot be used for wave modelling, except for the special case of long-wave modelling.

Table 8.1 lists the main similitude ratios for Froudian scaling. It should be noted that these ratios include scales for both fluid density (N_ρ) and specific weight (N_γ), as model experiments are normally conducted in fresh water, whereas the prototype fluid is normally seawater. These extra scales can be introduced by noting that from Equation (8.7a)

$$N_V = N_g^{1/2} N_L^{1/2}$$

Hence $$\frac{N_L}{N_t} = \sqrt{N_g N_L}$$

Table 8.1 Similitude ratios for Froude similarity.

Characteristic	Dimension	Froude ratio
	Geometric	
Length	L	N_L
Area	L^2	N_L^2
Volume	L^3	N_L^3
	Kinematic	
Time	T	$N_L^{1/2} N_\rho^{1/2} N_\gamma^{-1/2}$
Velocity	LT^{-1}	$N_L^{1/2} N_\rho^{1/2} N_\gamma^{-1/2}$
Acceleration	LT^{-2}	$N_\gamma N_\rho^{-1}$
	Dynamic	
Mass	M	$N_L^3 N_\rho$
Force	MLT^{-2}	$N_L^3 N_\gamma$
Specific Weight	$ML^{-2}T^{-2}$	$N_\rho N_g$
Pressure	$ML^{-1}T^{-2}$	$N_L N_\gamma$
Momentum	MLT^{-1}	$N_L^{7/2} N_\rho^{1/2} N_\gamma^{1/2}$
Energy	ML^2T^{-2}	$N_L^4 N_\gamma$
Power	ML^2T^{-3}	$N_L^{7/2} N_\rho^{-1/2} N_\gamma^{3/2}$

or $\quad N_t = \sqrt{\dfrac{N_L}{N_g}}$

As $\quad \gamma = \rho g$

$\quad N\gamma = N_\rho N_g$

and $\quad N_t = \sqrt{\dfrac{N_\rho N_L}{N_\gamma}}$

Short-wave hydrodynamic model experiments can be conducted in either a wave flume or a wave tank. In a wave flume, two-dimensional effects can be studied including stability of breakwater armour units, overtopping rates at coastal structures, wave reflection and transmission, wave forces on coastal structures and wave energy extraction devices. In a wave basin, three-dimensional effects can be studied, including refraction, diffraction and oblique reflection, longshore currents and testing of port and harbour layouts. Modern wave generators are capable of simulating regular or random wave sequences with a pre-defined wave energy spectrum (two-dimensional case) and a directional spectrum (three dimensional case). Active absorption of reflected waves can also be incorporated to ensure that the generated incident waves are not contaminated by re-reflected waves from the generator (see Hughes 1993 for details).

However, such models are not free of difficulties. Scale effects will be present, including those associated with reflection and transmission (which may be increased or decreased compared to prototype), viscous and frictional effects (which will generally be increased compared to prototype), and wave impact and shock forces (which may not be properly represented due to the effects of air entrainment). Another consideration, known as *laboratory effects*, also needs to be considered. In two-dimensional models these are mainly related to the effects of the side walls and end conditions and non-linear effects spuriously generated by the mechanical wave generation system. In three-dimensional models, selection of model boundaries to correctly mimic those of the prototype becomes more significant. In particular, large-scale circulations may be induced by the model boundaries and boundary reflections may occur, both of which will, in general, not be present in the prototype. The reader is referred to Hughes (1993) for the wealth of detail provided on these effects.

8.5 Long-wave hydrodynamic models

Such models are typically used to study harbour seiching, forces on moored vessels and harbour circulations and flushing. In the past, large three-dimensional models of estuaries were commonly undertaken. More recently, numerical models have largely superseded these. Long-wave models are based on the same scaling laws as short-wave models, but they may also be distorted. In this case the scaling is based on the vertical scale for wave height and the horizontal scale for velocity, wavelength and period. Provided only long waves are present, refraction and diffraction are correctly reproduced, as the celerity of shallow-water waves is dependent only on depth. Significant scale effects in distorted models can be expected for wave reflection, transmission and bottom friction.

8.6 Coastal sediment transport models

Mobile bed scale model investigations of coastal erosion and sediment transport are probably the most difficult hydraulic models to conduct. The state of knowledge regarding mobile bed modelling of the nearshore zone is still the subject of debate and uncertainty. However, such models can offer knowledge and insights in predicting the effects of coastal structures on shoreline evolution, scour and erosion in front of coastal structures and long- and cross-shore beach response to wave action.

The initial approach to the scaling of coastal sediment transport models is to establish an understanding of the dominant response mechanisms of the sediment. The common assumptions for such models is to assume that the sediments are reacting primarily to waves with currents added. This allows hypothesising the necessary similitude requirements and to scale the sediment accordingly. Such similitude requirements for bedload transport include:

Grain sized Reynolds number: $\mathrm{Re}_* = \dfrac{u_* D}{\nu}$

The entrainment function or Densimetric Froude number: $\theta = \dfrac{\rho u_*^2}{\gamma_i D}$

Relative density: $= \dfrac{\rho_s}{\rho}$

Relative length: $= \dfrac{\lambda}{D}$

Relative fall speed: $= \dfrac{W_s}{u_*}$

It is physically impossible to simultaneously satisfy all these requirements. A choice of parameters has to be made, which will determine the characteristics of the model sediment and its ability to reproduce particular responses to the dominant forces.

For bedload dominated transport models, Kamphuis (1985) proposed four possible models, as given in Table 8.2. The 'best model' satisfies the densimetric Froude

Table 8.2 Scaling laws for Kamphius's bed load models.

Model	N_{Re}	N_θ	$N_{\rho s/\rho}$	$N_{\lambda/D}$	N_D	$N_{\gamma i}$
Best	$N_L^{3/2}$	1	1	1	N_L	1
Lightweight	1	1	**	$N_L^{12/11}$	$N_L^{-1/11}$	$N_L^{3/11}$
Densimetric Froude	$N_L^{1/8} N_D^{11/8}$	1	**	N_L/N_D	*	$(N_L N_D)^{1/4}$
Sand	$N_L^{1/8} N_D^{11/8}$	$(N_L/N_D)^{1/4}$	1	$N_L N_D$	*	1

* means free choice
** means determined from $N_{\lambda i}$, but restricted to $1.05 < \rho_s/\rho < 2.65$

number, relative density and length, resulting in a model sediment with the same density as prototype and grain size reduced in accordance with the geometric scale. This can only be achieved if the prototype grain size is relatively large (e.g. gravel beaches). The 'lightweight' model preserves the grain-sized Reynolds number and densimetic Froude number, resulting in a lightweight model sediment with a grain size larger than that given by the geometric scale. However, it may not be possible, in practice, to obtain a model sediment with the required combination of density and grain size. The Densimetric Froude model is similar to the lightweight model, but allows more flexibility in the choice of model sediment. Lightweight models suffer from significant scale effects, due to their lower densities and larger grain sizes. For this reason many modellers prefer to use the sand model, which at least preserves the density of the prototype sediment. However, the sediment transport scales are significantly affected. This difficulty can be overcome by conducting a series of experiments using different model sediment sizes. Kamphuis (1974, 1975, 1985) gives a very detailed analysis of each model and its scale effects. Hughes (1993) also provides a detailed summary.

For suspension-dominated transport models, preservation of the relative fall speed (defined separately for suspended sediment transport) and relative density are considered to be most important. These criteria are used in Dean's surf zone sediment transport model (1985) on the grounds that turbulence, not bed shear stress, is dominant in the surf zone (for sand). Here, relative fall speed is defined as $H/(w_sT)$ and is also known as the *Dean number*. Physically, it represents the ratio of the time taken for sediment to fall a distance of one wave height to that of the wave period. If this ratio is larger than one, then suspended sediment is likely to predominate over bedload transport. The resulting fall speed scale relationship is given by:

$$N_{FS}\sqrt{N_L}$$

This allows calculation of the implied geometric scale, from the chosen model and prototype sediment fall speeds. Again, an excellent discussion of scale effects and other matters relating to suspension dominated transport models may be found in Hughes (1993); see also Dean and Dalrymple (2002) for a shorter summary.

By way of an example, Figure 8.2 illustrates a three-dimensional validation/process model for offshore breakwaters, conducted in the UK Coastal Research Facility at HR Wallingford at a model scale of 1 to 28. Initially, this model was constructed with a fixed bed. The model was used to generate random directional seas and measurements of wave and currents in the lee of the breakwaters taken to compare with numerical and field measurements. Subsequently, a mobile bed model was constructed. Two model sediments were employed – sand and anthracite – and their performance compared to each other and with field measurements (see Ilic *et al.* (1997) for further details).

Figure 8.2(a) Mobile bed sand model.

Figure 8.2(b) Mobile bed anthracite model.

Example 8.1   Three-dimensional coastal sediment transport model
It is proposed to build a 1:28 scale model, to study 3-D coastal sediment transport on a shingle beach, using the Kamphuis best model. Prototype parameters are $H_s = 2.0$ m, $T_s = 6.0$ s, $D_{50} = 15$ mm, $\rho_s = 2650$ kg/m³, $\rho = 1027$ kg/m³, $\nu = 1.36 \times 10^{-6}$ m²/s.

1   Determine the model wave height, period and grain size, assuming fresh water for the model.
2   Determine the prototype longshore current velocity, if the measured model longshore current velocity is 0.3 m/s.

3   Derive the scaling factor for grain sized Reynolds' number (Re_*) and discuss its implications.
4   Show that the scale factor for bed shear stress under wave action is correctly scaled.
5   Derive the scaling factor for longshore transport, using Kamphuis's equation and discuss its implications.

Solution
1

$$N_L = 28, N_\rho = 1027/1000 = 1.027, N_g = 1, N_\gamma = N_g N_\rho = N_\rho, N_D = N_L$$

Using Table 8.1

$$H_{sm} = H_{sp}/N_L = 2/28 = 71.4 \text{ mm}$$

$$T_{sm} = \frac{T_{sp} N_\gamma}{N_L^{1/2} N_\rho^{1/2}} = \frac{T_{sp} N_\rho^{1/2}}{N_L^{1/2}} = \frac{6x1.027^{1/2}}{28^{1/2}} = 1.15s$$

$D_m = D_p/28 = 0.54$ mm (Note, as the model grain size is a coarse sand, the model sediment will still act as a cohensionless material.)

2
Using Table 8.1

$$N_V = N_L^{1/2} \therefore V_{lscp} = V_{lscm} N_L^{1/2} = 0.3x\sqrt{28} = 1.59 m/s$$

3

$$Re_* = \frac{u_* D}{v}$$

Hence

$$N_{Re_*} = \frac{N_{u_*} N_D}{N_v}$$

For the Kamphuis best model $N_\theta = 1$, hence a scale factor for u_* can be derived from this:

$$N_\theta = \frac{N_{u_*}^2 N_\rho}{N_{\gamma_i} N_D} = 1$$

$$\therefore N_{u_*} = \left(\frac{N_{\gamma_i} N_D}{N_\rho} \right)^{1/2}$$

This is then substituted into the Reynolds scale equation to give:

$$N_{\mathrm{Re_*}} = \left(\frac{N_{\gamma_i} N_D}{N_\rho} \right)^{1/2} \frac{N_D}{N_v}$$

For the Kamphuis best model:

$$N_\rho \approx 1 \;\; \therefore \; N_{\gamma i} = N_\rho = N_v = 1 \text{ and } N_D = N_L, \text{ hence}$$

$$N_{\mathrm{Re_*}} = N_L^{3/2} \text{ (note, this is as given in Table 8.1)}$$

This implies that the prototype grain-sized Reynolds numbers will be 128 times greater than in the model for a scale of 28. By reference to the Shields diagram, it can be seen that this could alter the value of the critical entrainment function, if the model value enters the transition zone. This can be checked by calculating the dimensionless grain size for both model and prototype. In this case:

$$D_{*m} = \left[\frac{g(s-1)}{v^2} \right]^{1/3} D_m = 11.1$$

$$D_{*p} = \left[\frac{g(s-1)}{v^2} \right]^{1/3} D_p = 308$$

By reference to Figure 5.5, it can be seen that this is the case. The *critical* entrainment function for the model is less than for the prototype. As the value of the actual entrainment function in the prototype has been preserved in the model, then a *scale effect* has been introduced. This implies that the model sediment will be more mobile than the prototype sediment, giving relatively disproportionate transport rates.

4

From (3) we already have:

$$\therefore N_{u_*} = \left(\frac{N_{\gamma_i} N_D}{N_\rho} \right)^{1/2}$$

and $N_\rho \approx 1 \;\; \therefore N_{\gamma i} = N_\rho = N_v = 1$ and $N_D = N_L$, hence:

$$N_{u_*} = N_L^{1/2}$$

as

$$u_* = \left(\frac{\tau_0}{\rho} \right)^{1/2}$$

$$N_{u_*} = N_{\tau_0}^{1/2}$$

Hence

$$N_{\tau 0} = N_L$$

We can now check to see if the same scale applies to the bottom shear stress induced under wave action.

From Equations (5.11), (5.12), (5.6) and (5.7)

$$\tau_{ws} = \frac{1}{2} \rho f_w u_b^2 : f_{wr} = 1.39 \left(\frac{A}{z_0} \right)^{-0.52} ; \; z_0 = D_{50}/12 : \; A = u_b T / 2\pi$$

Hence

$$N_{\tau_{ws}} = N_\rho N_{fw} N_{u_b}^2$$

$$N_{fw} = (N_A / N_D)^{-0.52} = \left(\frac{N_{u_b} N_T}{N_D} \right)^{-0.52}$$

$$\therefore N_{\tau_{ws}} = N_\rho N_{u_b}^{1.48} N_T^{-0.52} / N_D^{-0.52}$$

For the Kamphuis best model

$$N_{u_b} = N_L^{1/2}, \; N_\rho = 1, \; N_T = N_L^{1/2}, \; N_D = N_1$$

$$\therefore N_{\tau_{ws}} = \frac{N_L^{0.74} N_L^{-0.26}}{N_L^{-0.52}} = N_L$$

This demonstrates that the bottom shear scale under wave action is the same as that produced by preserving the entrainment function between model and prototype. However, it should be noted that bedform roughness has not been accounted for and only bedload transport considered.

5

Starting from Equation (5.42)

$$Q_K = 2.27 H_{sb}^2 T_p^{1.5} (\tan \beta)^{0.75} D_{50}^{-0.25} (\sin 2\theta_b)^{0.6}$$

Hence

$$N_Q = N_H^2 N_T^{3/2} N_m^{3/4} N_D^{-1/4}$$

where m is beach slope

For a Froudian-scaled model

$$N_H = N_I : N_T = N_I^{1/2}$$

Hence

$$N_Q = N_L^2 \, N_L^{3/4} \, N_m^{3/4} \, N_D^{-1/4}$$

or

$$N_Q = N_L^{5/2} \, N_m^{3/4} \left(\frac{N_L}{N_D} \right)^{1/4}$$

For Kamphuis' best model

$$N_D = N_L, \text{ hence}$$

$$N_Q = N_L^{5/2} \, N_m^{3/4}$$

For a Froudian-scaled model, the expected scale for discharge is $N_L^{5/2}$. Hence, for the sediment transport scale, there is a scale effect which is expected to be proportional to $N_m^{3/4}$. This implies that model longshore transport rates will be larger than expected, as N_m will, in general, be greater than one. However, other scale effects are likely to be present, as the above treatment only considers bedload transport scaling and assumes the Kamphuis equation to hold true over all scales.

Chapter 9

Conceptual and detailed design

9.1 The wider context of design

9.1.1 Pressures on the coast

Coastal areas have always been a popular place for recreation, habitation and commerce. Typical features include ports, marinas, fishing harbours, roads, railways, power stations, agriculture, recreational resorts, residential property, agriculture and a wide variety of natural habitats. In many parts of the world, land immediately adjacent to the sea is significantly more valuable than elsewhere. It is therefore not surprising that that the coastal boundary is has been subject to both reclamation and protection in response to economic pressures. In many areas of the world there are soft eroding cliffs that experience erosion as well as low-lying coastal plains that are vulnerable to both erosion and flooding due to the action of the sea. This coupled with gradually rising sea levels due to global warming has resulted in an increase in the shorelines around the world suffering from erosion. Moreover, the prospect of accelerating sea level rise and possible changes in the frequency and direction of storms presents a high degree of risk and uncertainty when it comes to considering the most appropriate design scenarios for coastal structures.

Local approaches

Some design practices in the past, and in some places the present, might be classified as the 'brute force' approach. This is the principle that if a structure is big enough and strong enough it can withstand almost any of the conditions that it can be subject to with the exception of the most extreme events. However, this often takes no account of the morphological context in which the structure might exist. A further problem that seems to have persisted is that, whilst there have been significant advances in the appreciation of the interactions involved in regional coastal processes the physical areas of responsibility, and hence parochial interest, have been constrained to sub-areas of the coastal cell. In the event new works or repairs would be initiated as site-specific problems arose, sometimes as a result of ad hoc monitoring. Interaction with adjacent sections of coastlines, and the constraint that they might impose would often only be considered in relation to the specific problem at the site. The result of this process would be that, whilst a particular problem might be solved with capital works, the wider implications of this action would not be addressed. Thus regional strategy, social planning and environmental management would not have been fully considered so that all the possible options could be explored. These would generally include measures to mitigate any potential downdrift erosion problems, preservation of natural habitats or recognition of alternative uses and enjoyment of the coastal environment.

The ad hoc nature of this approach is unsatisfactory as it makes it extremely difficult to ensure that not only are schemes developed to be efficient and cost effective, but also natural process and natural resources are used to best effect in tandem with anthropological uses of the coastal area. The benefits of a more strategic approach to shoreline management should thus be easily appreciated.

Strategic approaches

In many parts of the world, the idea of 'integrated coastal management' (ICM) is proposed as being a more satisfactory way forward. This is a process that goes beyond the traditional approach of planning and managing activities on an individual scheme basis. Instead, the aim is to focus on the combined effects of all activities taking place at the coast to seek suitable environmental and socio-economic outcomes. Sustainable use, with environmental considerations underlying decision making in all sectors of activity, provides the basis for this type of management. It is geared to dealing with the coastal environment as a whole – coastal land, the foreshore, and inshore waters – and is forward looking, as well as trying to resolve the problems of present day use of the coast.

Integrated coastal management involves the comprehensive assessment, setting of objectives, planning and management of coastal systems and resources, taking into account traditional cultural and historical perspectives, cumulative impacts, and conflicting interests and uses. It is a continuous and evolutionary process for achieving sustainable development through participation of the public and private sectors and with the support and interest of local communities.

Global, regional and local issues – such as sea level rise, the concentration of populations and tourism on the coast, and depletion and damage to valuable natural resources such as fisheries and wildlife – are making coastlines one of the most pressured and threatened environments in the world. Most of the world's major cities are at the coast, and more than 50 per cent of the world's estimated 5.5 billion people live in coastal areas. It has been predicted that by 2020, 75 per cent of the world's projected population of 8.0 billion could be living within 60 kilometres of the shoreline, the majority in developing nations.

This concentration of population at the coast is a result of a number of factors including:

- the diverse and productive renewable resources base in coastal areas which include fisheries, forests and fertile soils;
- accessibility to maritime trade and transport routes through the construction of ports and harbours;
- abundant and attractive recreational and tourism opportunities;
- industrial investments such as power stations and oil/gas terminals;
- increasing demands for residential property on or close to the coastal strip.

The demands made by this population concentration have caused problems such as:

- over-exploitation of renewable resources like coastal fisheries, beyond sustainable yields;
- degradation of coastal water and marine ecosystems from land-based pollution including sediment run-off, fertilisers and untreated sewage;
- destruction of natural coastal habitats for construction or coastal aquaculture.

Coastal locations are also susceptible to a range of natural hazards such as storm surges, erosion and sea level change that can cause loss of life and property and damage to infrastructure, livestock and crops. Damage to the coastal infrastructure is often considered to be politically and economically unacceptable. However, in many circumstances it can no longer be assumed that defences should be maintained where they have previously existed. Through taking a strategic approach, there has been a significant change in the way that the design of coastal defences should be developed with the emphasis shifting from the provision of protection to managing the coastline in spatial scales that recognise the interactive nature of the processes that take place as well as over longer-term temporal scales. In doing so it is necessary to recognise that there is considerable uncertainty in defining all of the relevant parameters that can impact on the eventual outcome of adopting various policies so that management practices require a strategic approach that is largely based on risk analysis and continuous performance monitoring.

9.1.2   Early strategic approach in the UK (case study)

Formal shoreline management practices have been developing in the United Kingdom over the past twenty years or so. The Anglian Sea Defence Management Study (Fleming 1989; Townend *et al.* 1990) was the forerunner to the development of shoreline management plans (SMPs) around the coastline of England and Wales. The Anglian coast is some 750 kilometres in length and the initial analysis of the coastline was based on the collation of a number of factors considered to influence the choice of management policy as listed in Table 9.1. These were selected on the basis that they either:

- provided information on the direct influences and responses of the coast such as waves, coastal morphology and rate of retreat; or
- provided information on their implications with respect to the impact of accretion/erosion and any defence strategy that might be implemented.

This list is not exhaustive and other influences might be found to occur in particular circumstances.

The Anglian Sea Defence Management Project adopted an approach that involved collection of some spatial data such as shoreline position. Once collected some spatial and temporal data such as shoreline position was analysed using data-mining techniques and provide insights into the behavioural trends of the coastline. The data was also supplemented by the use of numerical modelling of various coastal processes as described in Chapters 5 and 6. The coastlines were then divided into management

Table 9.1 Factors influencing choice of management policy.

Main variable	Significance
Agriculture	• changes in habitat
	• drainage patterns and runoff
Birds	• assessment of environmental impact
Coastal movement	• indicate areas of high/low activity
	• assist with forecasting future movements
	• relationship to sediment budget
Coastal works	• interaction with coastal processes
Conservation sites	• special consideration to prevent undesirable changes
Currents	• influences sediment movement on offshore zones
	• links nearshore processes with far field effects
Ecology	• a measure of shoreline (cliff, dune, saltmarsh) stability
	• shelter, relationship with rivers and estuaries
	• assessment of environmental impact
Fisheries	• changes in habitat
	• potential environmental impact
Industry	• coastal impact on processes and environment
	• threat to habitats
Infrastructure	• constraint on the coastline
	• impact on local shoreline processes
Jurisdiction	• key to development of management strategy
Morphology	• basic description of coastline
	• physical significance (e.g. offshore banks dissipate wave energy, cliffs can provide a sediment supply)
	• width of foreshore indicates plan effects
	• slopes control form of incoming waves
	• indicates nature of sediment transport
	• represents sediment sources and sinks
	• intertidal features indicate beach cycles and onshore movement
Rainfall	• influences groundwater levels and river discharges
	• impact on sediment load in rivers
	• impact on cliff stability
Sediments	• determines mobility of material
	• forensic evidence for sources of materials
	• basis for sediment budget
Temperature	• seasonal variations may contribute to erosion
Water levels	• major effect on coastal processes
	• controls extent of wave influence on shoreline
	• relates to potential for land flooding
Water quality	• influences vegetation and hence shoreline stability
	• impact on marine life and alteration of habitats
	• density effects and transport regime
Waves	• fundamental to potential for shoreline erosion and accretion
	• influences height and movement of offshore banks
	• primary cause of infrastructure damage
	• linked to climate change
Wind	• generates waves and storm surges
	• governs sub aerial erosion and deposition

units, which are sections of coastline that exhibit coherent characteristics in terms of baseline geology, natural processes, existing defences, foreshore type and land use. It was then attempted to link the coastal management strategy to the objectives that needed to be satisfied through consultation with a wide range of stakeholders. In the case of shoreline management policy options for the management units identified they were simply described as:

- maintain existing line;
- set back defence;
- retreat the defence line;
- advance or reclaim.

On the face of it these options appear to be quite obvious and simplistic. However, it must be appreciated that a policy option selected on one section of coastline will invariably have an impact on the adjacent coastline and beyond. The first option applies to any existing line which is being defended and will generally be preferred whenever there is a substantial investment in infrastructure on the coast. However, this option can be linked to a change in the standard of service of the defence. On an eroding coast, set-back would be used to provide defences on the hinterland so that it is only necessary to defend against tidal inundation. The option could also be used to provide natural features 'room to move' (such as barrier beach and salt marsh systems), whilst retaining a level of defence against flooding. The retreat option is a managed withdrawal, allowing the coast to return to its natural state and can be an attractive option where the tidal floodplain is relatively narrow. It would also apply where no defence was to be provided on a naturally eroding coastline, but soft engineering expedients such as dune management, cliff drainage or beach management might be considered. Finally, the advance option allows for the possibility of limiting low-lying exposure by suitable reclamation or the use of tidal barriers. The option chosen is largely dependent on the existing infrastructure and erosion areas for any given length of coastline.

In order to implement any policy options it was proposed that various management options could be considered providing they were appropriate to the coastal classification. Those options were described as:

- do nothing – let nature take its course;
- reinstate – beach renourishment, saltings regeneration, structural reconstruction etc.;
- modify – remove features or structures, structural alterations, stabilisation (cliffs/ dunes/saltings) etc.;
- create – embayments, linear protection, intervention such as dredging, sand by-passing etc.

By defining policy options and management options for an entire coastline, the basis for a strategic management plan was established. It was recognised that, as actions based on this plan were to be undertaken so aspects of the coastal characteristics would be modified and this, in time, was likely to alter the coastal classification.

The foregoing describes some of the formative basic principles behind the development of 'shoreline management' in the UK and is differentiated from 'integrated coastal management' which includes a very much wider range of considerations with respect to the use and sustainable development of the wider coastal zone. It is beyond the scope of this book to cover these wider issues.

9.1.3 Current UK SMP approach (2011)

The strategic approach to shoreline management in the United Kingdom has been driven and sponsored by the Department for the Environment, Food and Rural Affairs (DEFRA). The overall objective can be stated as (Burgess 2002):

To reduce risks to people and the developed and natural environment from flooding and coastal erosion by encouraging the provision of technically, environmentally and economically sound and sustainable defence measures.

In this context sustainable management approaches are those which:

> take into account the relationships with other defences, developments and proc-
> esses . . . and which avoid as far as possible tying future generations into inflexible
> and expensive options for defence.

In order to assist in this process the coastline of England and Wales was initially divided into a number of primary cells and sub-cells which were defined as relatively self-contained units with respect to the movement of beach material. These are managed by groups that include representation from all of the authorities that have any statutory responsibility for coastline in the cell. The first round of SMPs to be developed provide the framework for defining the policy options that should be adopted in order to minimise the occurrence of flooding and coastal erosion in the context of sustainable development, whether related to the continuity of sediment transport processes or environmental conservation. At the same time the requirements of whatever legislation exists must be satisfied. The overall objectives of the shoreline management plan process are (Brampton 2002):

- to define, in general terms, the risks to people and the developed, historic and natural environment within the shoreline management plan area;
- to define the natural processes taking place in terms of forcing functions (e.g. waves, tides, currents) and responses (e.g. sediment movement, shoreline movement);
- to define the potential retreat or advance of the shoreline within the statutory planning horizon of 70 years;
- to consult and conciliate with all of the users of the coastline in the area;
- to identify the preferred policies for managing these risks over the next 50 years;
- to identify the consequences of implementing the preferred policies;
- to set out procedures for monitoring the effectiveness of the shoreline management plan policies;
- to ensure that future land use and development of the shoreline takes due account of the risks and preferred shoreline management plan policies.

SMP guidance has evolved quite rapidly over the past decade since the first edition of this book was published in 2004. In 2006 DEFRA published a three-volume document titled 'Shoreline management plan guidance'. This describes an SMP as a large-scale assessment of the risks associated with coastal processes and helps to reduce these risks to people and the developed, historic and natural environment. The strategy should aim to manage risks by using a range of methods which reflect both national and local priorities to:

- reduce the threat of flooding and erosion to people and their property;
- benefit the environment, society and the economy as far as possible, in line with the Government's 'sustainable development principles'.

Thus, the SMP itself is intended to define the policy option, but not the precise physical form of the defence option. There will almost certainly be a number of generic solutions that will satisfy the requirements that have been identified in the SMP. There are further stages of study required to reach a final definitive scheme for implementation. Generic solutions will be identified through a strategy study for sections of coastline that have been identified as requiring remedial or new works. The final stage concerns a specific scheme for which a scheme-specific study will compare alternative options and define the optimum scheme that best satisfies all of the technical, financial and socio-economic criteria that have been agreed. The overall framework is described in Table 9.2

Policy options were also modified in the 2006 DEFRA guidance providing the following four SMP policies available to shoreline managers:

- *Hold the existing defence line* by maintaining or changing the standard of protection. This policy should cover those situations where work or operations are carried out in front of the existing defences (such as beach recharge, rebuilding of the toe of a structure, building offshore breakwaters etc.) to improve or maintain the standard of protection provided by the existing defence line.

Table 9.2 Stages in assessing the risks of flood and erosion (DEFRA 2006).

Stage	SMP	Strategy	Scheme
Aim	To identify policies to manage risks	To identify appropriate Schemes to put the policies into practice	To identify the type of work to put the preferred scheme into practice
Delivers	A wide ranging assessment of risks, opportunities, limits and areas of uncertainty	Preferred approach, including economic and environmental decisions	Compare different options for putting the preferred scheme into practice
Output	Policies	Type of scheme (such as a sea wall)	Design of works
Outcome	Improve management for the coast over the long-term	Management measures that will provide the best approach to managing floods and the coast for a specified area	Reduce risks from floods and coastal erosion to people and assets

- *Advance the existing defence line* by building new defences on the seaward side of the original defences. This applies only to policy units where significant land reclamation is being considered.
- *Managed realignment* by allowing the shoreline to move backwards or forwards, with management to control or limit the movement (such as reducing erosion or building new defences on the landward side of the original defences).
- No *active intervention* where there is no investment in coastal defences or operations.

DEFRA published a series of Flood and Coastal Defence Project Appraisal Guidance in five volumes as follows:

- FCDPAG1 Overview (published March2001)
- FCDPAG2 Strategic planning and appraisal (published March2001)
- FCDPAG3 Economic appraisal (published December 1999)
- FCDPAG4 Approaches to risk (published March 2000)
- FCDPAG5 Environmental appraisal (published March 2000).

The selection of the most appropriate policy option requires a clear focus on the assessment and management of coastal flooding and erosion risks over a one hundred year period beyond the initial appraisal so that there is a strong need for awareness of the longer-term implication of coastal evolution. There is also a clear need for a better appreciation of the uncertainties associated with predicting future shoreline management requirements coupled with a recognition that current defence policies may no longer be feasible or acceptable in the future.

This series of FCDPAG documents did provide an invaluable guide to developing appropriate solutions to flood defence and coastal erosion problems within the UK legislative framework. However, there have been further developments and a Policy Statement for Flood and Coastal Erosion Risk Management (FCERM) published by DEFRA in 2009 replaces the previous policy guidance set out set out in the foregoing volumes. This policy statement refers to the adoption a of risk-based approach, giving more consideration to 'risk management' and 'adaptation' as opposed to only 'protection' and 'defence' as well as considering impacts within the whole of a catchment or shoreline process area.

DEFRA has encouraged the development of a strategic framework for flood and coastal erosion risk management based on Shoreline Management Plans (SMPs) and Catchment Flood Management Plans (CFMPs). These are high-level plans, which have various relationships with other high level plans, strategies and schemes and other planning initiatives as shown schematically in Figure 9.1.

More specific best-practice guidance on how to undertake appraisals has now been published by the Environment Agency (2010). A review of coastal risk management in the UK is provided by Pontee and Parsons (2010). Whilst these frameworks have been largely developed in this form in the UK they are equally applicable, in principle, to any region of the world. The outcome of the final stage focuses on the scheme appraisal process and also defines the type of structure or management strategy that should be adopted.

Legislation and required procedures will vary from country to country, so it is not possible to cover all of the possibilities in this book. However, a common theme of

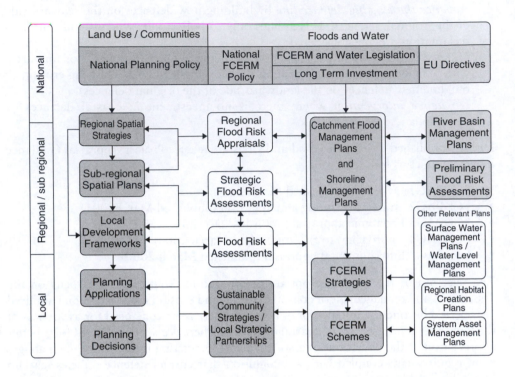

Figure 9.1 Relationship between high level plans, strategies, schemes and other planning initiatives (after DEFRA 2009).

any modern-day practice does focus on the need to properly carry out the appropriate planning steps, which include appropriate risk analyses leading to project optimisation. The new Coastal Engineering Manual Part V, Coastal Project Planning and Design (download from http://chl.erdc.usace.army.mil/cem) provides both a general framework as well as information that relates to procedures in the USA.

9.2 Coastal structures

There is a wide range of coastal works that might be employed to tackle a particular situation, each of which may perform a number of different functions. They will also have differing engineering lifespans as well as different capital and maintenance cost streams. The potential economic benefits will also have a strong influence on the final solution that might be adopted whilst still conforming to the objectives and policies developed through the shoreline management plan and strategy study. Figure 9.2 shows some of the more common types of coastal works that are often used and includes artificial headlands, groynes, offshore breakwaters, beach nourishment and sea walls. The basic advantages and disadvantages are also listed. CIRIA Report 153 also provides a useful summary guideline for the application of control works as given in Table 9.3. These comments provided assume that structures are

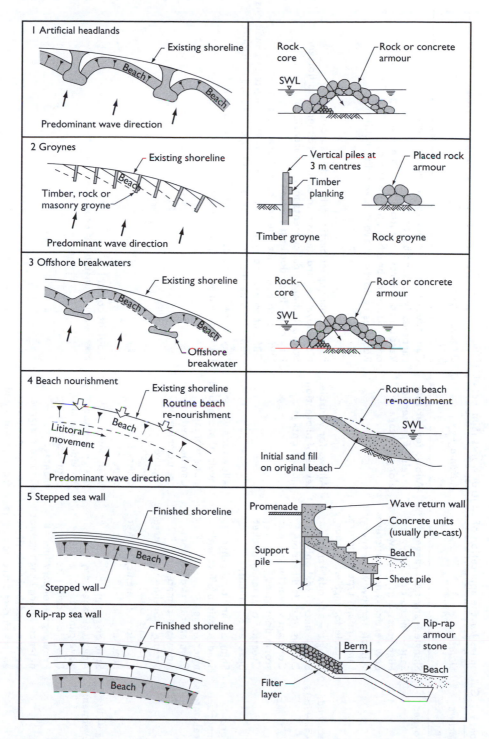

Figure 9.2 Common types of coastal works.

Table 9.3 Guidelines for the application of control works (Simm et al. 1996).

Structure type	Situation	Advantages	Disadvantages
Groynes	Shingle – any tidal range Sand – micro-tidal only High gross drift, but low net Low vertical-sided structures suitable for low wave energy Large mound type structures suitable for high wave energy	Allows for variable levels of protection along frontage	Can induce local currents which increase erosion, particularly on sand beaches Vertical structures potentially unstable with large cross-structure beach profile differences Requires recharge to avoid downdrift problems
Detached breakwaters	Shingle – any tidal range Sand – micro-tidal only Dominant drift direction Constant wave climate, not storm dominated Creation of amenity pocket beaches or salients	Allows for variable levels of protection along frontage	Large visual impact particularly with macro-tides May cause leeward deposition of fine sediment and flotsam Strong inshore tidal currents may be intensified May cause hazardous rip currents Difficult to construct due to cross-shore location Difficult to balance impact under storms and long-term conditions Difficult to balance impact on both shingle and sand transport
Shore connected breakwaters	Shingle – any tidal range Sand – limited effect with macro-tides Dominant drift direction Any wave climate Strong shoreline tidal currents ('fishtails' only) Creation of amenity pocket beaches	Allows for variable levels of protection along frontage Can be used to create amenity features Longshore and cross-shore control	May cause leeward deposition of fines and flotsam Little design guidance at present
Sea wall/Revetments	Sand or shingle Any tidal range, any wave climate Low gross drift rate Provides secondary line of defence where beach cannot be designed to absorb all wave energy during extreme events	Well developed design methods Provides equal protection along frontage Can be designed to support a sea front development	No drift control May become unstable if erosion continues

Sills	Shingle or sand Low wave energy Low and variable drift Submerged with micro-tides, regularly exposed with macro-tides	Creates perched beach Reduces shoreline wave climate	Storms may remove beach irreversibly Level of protection reduces during storm surge events
Beach drainage systems	Sand beaches, normally up to the high water line Any tidal range Any wave climate or drift rate	Responds to beach developments	Limited experience of use Long-term maintenance may be expensive

built and note that there are a number of wider environmental considerations that need to be considered in the context of a full appraisal.

Modern design practice places much emphasis on attempting to hold a healthy beach on the shoreline as the primary means of protection. A sufficiently substantial beach can accommodate the dynamic changes that are the result of differing climatic conditions. These so-called 'soft' solutions are generally considered to be more environmentally friendly than traditional 'hard' protection works. However, where human life may be at risk and high-density, high-value conurbations exist, the use of hard elements of a defence may be unavoidable.

There are a number of publications and standards that deal with general facets of coastal structure design and include some excellent information and detailed guidance. These are:

- A guide to managing coastal erosion in beach/dune systems (Scottish Natural Heritage, 2000).
- Beach Management Manual (Simm *et al*. 1996).
- Beach Management Manual (2nd edition, Rogers *et al*. 2010).
- BS6349 (1991). Maritime Structures – 1. General Criteria. British Standards.
- BS6349 (1991). Maritime Structures – 7. Guide to the design and construction of breakwaters. British Standards.
- Coastal, Estuarial and Harbour Engineers Reference Book (eds. Abbott M.B. and Price W.A. 1994).
- Coastal Protection (ed. Pilarczyk 1990).
- Concrete in coastal structures (ed. Allen 1998).
- Guide to the use of groynes in coastal engineering (Fleming 1990a).
- Guidelines for the Design and Construction of Flexible Revetments Incorporating Geotextiles in Marine Environment (PIANC 1992).
- Eurotop. Wave Overtopping of Sea Defences and Related Structures: Assessment Manual (EA: ENW: KFKI, www.overtopping manual.com, August 2007).
- ICE Design and Practice Guides, Coastal defences (ed. Brampton 2002).
- Manual on artificial beach nourishment (Delft Hydraulics Laboratory 1987).
- Manual on the use of rock in coastline and shoreline engineering (CIRIA/CUR 1991).
- The Rock Manual: The use of rock in hydraulic engineering (2nd Edn, CIRIA: CUR:CETMEF 2007).
- Overtopping of Sea walls – Design and assessment manual (Besley, Environment Agency 1999).
- Port Engineering (Bruun 1989).
- Revetment systems against wave attack. (McConnell 1998).
- Sea wall Design (Thomas and Hall 1992).
- Wave Run-up and Wave Overtopping at Dikes (TAW, Delft May 2002).

The US Army Corps of Engineers 'Shore Protection Manual' (1984) was once considered to be a standard reference document. However, it has been redrafted over the past decade in order to incorporate the wealth of developments that have taken place, but is still incomplete. It has been renamed as the 'Coastal Engineering Manual' and can be found on the internet at http://chl.erdc.usace.army.mil/cem.

A good understanding of the coastal environment at a site under consideration is an essential prerequisite to assessing the ability of a coastal defence option to perform as it is intended. A complex interaction exists between the various elements defining the coastal environment as discussed in the preceding chapters. The introduction of coastal protection works will invariably modify nearshore processes in some way and it is important to account for that feedback effect. Coastal features at any location for different erosion and accretion patterns which, in turn, are caused by the interaction of geological variations, wave climate, winds, currents and tides specific to a section of coastline. The causes and effects of these features must always be considered when dealing with works which affect littoral movement. The origin of beach material can be from inland sources bought to the coast by rivers or from the erosion of cliffs in the immediate or adjacent coastlines. Sometimes there can be shorewards pathways of sediment from offshore sources. In some cases those processes may no longer be active and the beach is comprised of relic features of material.

Knowledge of the geology underlying the nearshore zone is important because a stratum that is different from the surface material can affect the way in which a beach behaves. A thin veneer of loose material on an erodible platform can act as an abrasive and accelerate erosion, whilst its existence on an impermeable base will be inherently unstable and more mobile than an equivalent deep beach. These factors are also material to the design of foundations of coastal structures.

The following sections are intended to provide the reader with sufficient information on which to gain an appreciation of the principles involved in engineering design of coastal structures. Each section will consider, if appropriate, design guidance on determining the parameters that govern the basic geometry of the elements discussed. Thereafter, guidance with respect to appropriate material will be given. Design of the fabric of structures that can apply generically to many different types of structure will also be given in Section 9.4.

9.2.1 Groynes

Groynes, nearshore breakwaters and artificial headlands are all types of structure that are used to have a sufficient impact on nearshore sediment transport processes to modify beach response to the dominant wave and tidal conditions. All three generic types of structure are usually used with the objective of increasing the volume of beach material in both the backshore and the nearshore regions in recognition that a natural beach of either sand or shingle is the most efficient means of absorbing the wave energy from breaking waves. It is also commonly the most economic and environmentally friendly design approach.

Groynes are shore protection structures that are generally spaced at equal intervals along the shoreline and cross all or part of the intertidal zone, close to normal to the shoreline. Figure 9.3 provides some of the basic definitions of groyne dimensions. There are also a number of variations to simple straight groynes such as zig-zag (Figure 9.4b), 'T' head (Figure 9.4c and d) and 'Y' head. The first of these is intended to dissipate destructive flows from wave induced currents or wave breaking. The second is to create local wave height reduction through wave diffraction and the third is a variation that could be considered to have evolved into the fishtail groyne, which acts as an artificial headland control structure (see Section 9.2.4).

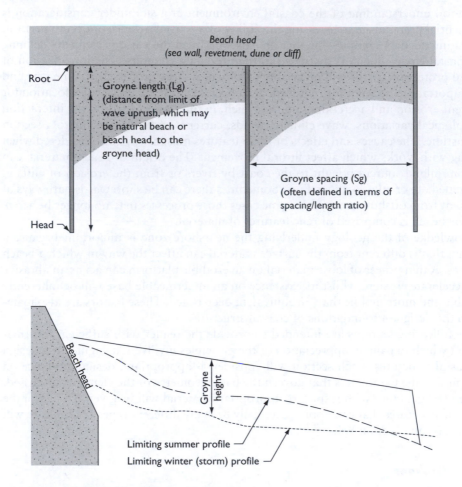

Figure 9.3 Definition of groyne dimensions.

They can be constructed of a variety of materials including for example timber, rock, concrete units and steel sheet piling, examples of which are shown in Figure 9.4 (a) to (h). They may be designed to be either permeable or impermeable to both fluid and sediment. Groynes have commonly been used with varying degrees of success on UK coastlines (Fleming 1990a and b). In general terms this element of a defence system is most appropriate to coastlines where the existing 'line' must be maintained and where there is a low net, but high gross alongshore drift (see Sections 5.2 and 6.3) Given this basic condition, a well-designed groyne system can:

- arrest or slow down the alongshore drift of material on a coastline and, by building up the volume of material in the groyne bays, stabilise the foreshore and protect the coastline;
- reduceng the impact of changes in shoreline orientation (Figure 9.5a and b);
- deflect strong tidal currents away from the shoreline;

Figure 9.4 Groynes: (a) typical groyne field, (b) zig-zag groynes, (c) 'T' head timber groynes, (d) 'T' head rock groyne, (e) massive timber groynes, (f) rock groynes, (g) concrete armour terminal groyne and (h) timber piled (deteriorated).

- help to hold material on a beach that has no natural supply and has been artificially nourished (Figure 9.5d);
- control seasonal shifts of material alongshore within a bay (Figure 9.5c);
- reduce the long-term erosive effect of wave activity in an area of coastal defence by accumulating beach material in front of hard beachheads such as sea walls, revetments and cliffs. This requires an adequate supply of material moving alongshore;
- improve the extent and quality of an amenity beach;
- increase the depth of beach material cover to an otherwise erodible seabed.

A major study of groyne systems in the UK was carried out by CIRIA (Fleming1990a and b) and resulted in the compilation of a large volume of data covering a wide range of beach types, as well as a guide to the uses of groynes in coastal engineering. Beaches were classified into four types and statistics were collected on groyne geometry and performance as summarised in Table 9.4. These parameters represent averages of the main parameters, but it will be noted that there remains a wide range of possibilities.

Median diameters in the above table are also nominal and a general relationship between grain diameter and beach slope is shown in Figure 9.6.

From the viewpoint of coastal defence the principle function of a groyne system is to retain a sufficient reservoir of beach material to withstand beach drawdown

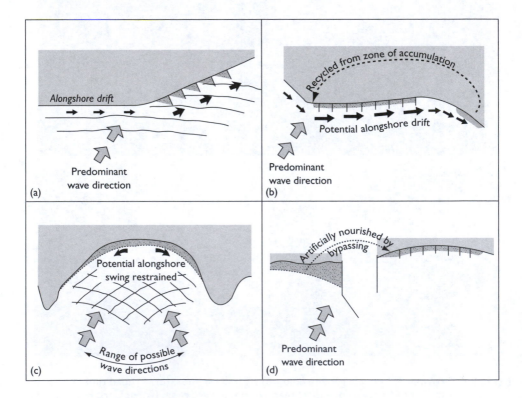

Figure 9.5 Some uses of groynes.

Table 9.4 Summary of groyne geometry by beach type (Fleming 1990a).

Beach type	Beach slope	Median diameter (mm)	Average lengths and ratios		
			Length	Spacing	Range of spacing/length
Shingle	1:6–1:10	10–40	60	60	0.5–1.7
Shingle upper/ sand lower	1:10 shingle 1:40 sand	10–40 0.3	50	50	0.5–1.5
Shingle/ sand mixed	1:30	2.0	70	85	0.6–2.4
Sand	1:100+	0.3	95	130	0.8–2.7

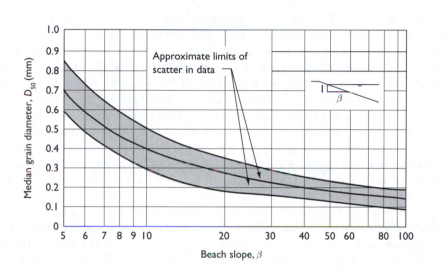

Figure 9.6 Relationship between beach slope and mean grain diameter.

during storms and hence maintain adequate protection to the beach head. In simple terms this is achieved by changing the orientation of the beach line within each of the groyne bays to become more closely aligned with the prevailing wave direction and thus reduce the rate of alongshore movement of material (see Chapters 5 and 6). The length and spacing must consider all possible combinations of wave height period and direction that might occur. This determines the theoretical plan geometry of the system whilst recognising that the vertical profile must take into account the possible variations in the cross-shore profile of the beach. The top level of a groyne will determine the maximum potential beach depth updrift of the groyne so that the structure should be designed for any combination of beach levels on either side of it between the local scour level and the desired maximum beach depth. These extremes will usually be determined by the natural limiting winter and summer beach profiles (see Figure 9.2). The significance of groyne height on the effectiveness of the system differs according to different beach types because the extent of scour also depends on the size of beach material. It must also be appreciated that an adjustment to increase groyne height to improve beach levels could lead to rip currents and erosion gullies if wave-induced currents are particularly strong. Consequently, on sand beaches (which are most sensitive to the height of groyne protruding above the beach) one of the traditional management practices has been to limit groyne height to protrude only 0.5–1.0 m above the seasonal beach profile. It follows from the above that, on shingle beaches, greater groyne heights are permissible and practical. Also, where beach recharge is contemplated, groyne heights must be determined to suit the nature of the recharge material.

When considering the length of a groyne, the cost of construction is largely influenced by the period of accessibility of the foreshore between tides. Without special provision in construction, an economic limit on length is reached inland of mean low water mark of spring tides (MLWS) or mean lower low water (MLLW). Therefore, in practice the length is often determined by tidal range and beach slope. The required length is, however, also related to the desired trapping effectiveness of the groyne system. In order to control sufficient alongshore drift, it may be necessary to go beyond this practical limit. To avoid outflanking, the landward end of a groyne should either abut a non-erodible longitudinal defence such as a cliff, sea wall or revetment, or, with an erodible beach head, it should be taken landward of the swash line thus allowing for beach drawdown in the most unfavourable combination of circumstances. Failure to recognise this requirement has caused outflanking of groynes by the sea, with consequent failure of the system. Thus the design of a groyne system should not be carried out in isolation from the type of beach head. Wave energy reflected by a wall, cliff or over-steep beach head is likely to move material offshore. Such conditions would not encourage a beach to improve or recover naturally, even under a favourable wave climate.

In order to provide a first-level estimate of the change in alongshore drift rate that a groyne system can *potentially* induce, it may be assumed that the volumetric transport rate of alongshore drift is directly proportional to the sine of twice the angle between the wave crest and the beach contour at the breaker line; Figure 9.7 illustrates how the ratio of the drift with and without the groynes (Q_g and Q_o respectively) is associated by the ratio

$$Q_g / Q_o = \sin 2(\alpha_o - \alpha_g) / \sin 2\alpha_o \approx (\alpha_o - \alpha_g) / \alpha_o \qquad (9.1)$$

where α_o represents the angle of incidence of waves to the ungroyned beach and α_g represents the change in angle due to groyning. The approximation relates to small angles of incidence.

For a given groyne length, maximum groyne spacing should take into account the resulting variation in beach level each side of the groyne as illustrated in Figure 9.7. Given the assumed lines of beach crest, the horizontal distance between points of corresponding height on the updrift and downdrift sides will be $S_g \tan(\alpha_o - \alpha_g)$. For a beach of slope β the difference in level across the groyne will be $S_g \tan(\alpha_o - \alpha_g) \tan \beta$, but this may assumed to be a maximum, as local sheltering will reduce this. Thus, in a bay where the direction of wave attack is confined, groynes may be more widely spaced than on an exposed promontory. It also follows that steeper beaches require more closely spaced groynes.

In some locations, situations can arise where wave attack during the beach-building summer period is at a relatively acute angle to the coastline (albeit that wave heights are

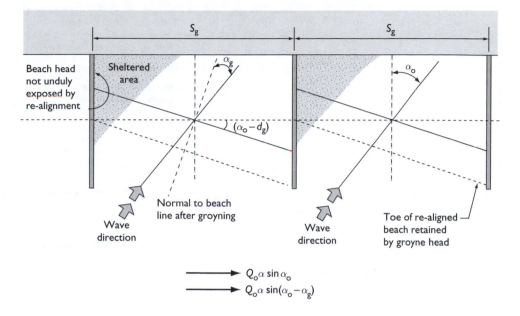

Figure 9.7 Definition of groyne design parameters.

moderate). This could require the groyne spacing/length ratio to be reduced to avoid large variations in beach-crest level. Thus, the rational determination of groyne spacing involves estimating the possible variation in beach shape that may take place within each groyne bay, while at the same time ensuring that an adequate reservoir of beach material is allowed to accumulate. In addition, the beach crest must be sufficiently far seaward to ensure that any sea wall or revetment is provided with reasonable protection by the beach at all times. The latter design consideration requires a good estimate of the equilibrium beach profile geometry under storm conditions (see Chapter 6).

Rather more sophisticated numerical beach plan shape models that can be linked to a combined refraction and diffraction wave model can provide methods of optimising groyne field geometry, an example of which is shown in Figure 9.8. The primary difficulty that arises in applying such a technique is that groynes do not usually pierce the water surface over their entire length so that it is necessary to make some basic assumption about the equivalent length of the groynes as represented in the model. This is not a simple matter and requires some good prototype calibration data. In the right circumstances a physical model might also be used to optimise the effects of beach shape, groyne length and spacing. However, there are considerable difficulties in creating a littoral environment of alongshore drift within a model as well as problems of rationalising scale effects. Such modelling of shingle-sized material is most likely to be successful, but it is not generally practical to model more than a few wave conditions from a limited number of directions.

From the practical viewpoint, groynes are generally constructed transverse to the general direction of the coastline. In order to minimise structural damage during

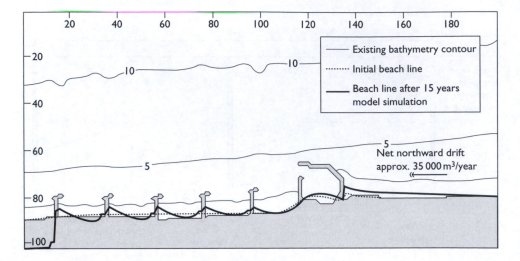

Figure 9.8 Beach plan shape model simulation.

storms, groynes should ideally be aligned directly into the direction of the maximum storm waves. In practice, this is not usually possible. At many sites, there is, in fact, a substantial drift in both directions due to the multi-directional nature of the wave climate. Groynes inclined slightly away from perpendicular to the coastline and in the downdrift direction (i.e. the direction of alongshore drift) are considered to provide the most effective control of littoral movement. However, where wave direction can vary and cause reverse drift, inclined groynes become angled updrift. This can lead to scour on the new downdrift side of the groyne. Downdrift angling should, therefore, only be considered for conditions of predominantly unidirectional drift.

Special considerations are required when dealing with the last downdrift groyne in a system. The importance of considering a length of coastline as a geographical cell has already been mentioned. Often the beach will terminate at an inlet to a tidal estuary or creek. Terminal groynes or training walls are sometime constructed in these cases to perform two functions. Firstly, they arrest as much of the alongshore drift as possible to prevent siltation of the inlet and secondly they preserve and improve a beach on the updrift side. It can be appreciated that a terminal groyne in a system might deliberately be made longer and higher in order to create a reservoir of material that can be mechanically transported updrift to nourish depleted beaches. Alternatively, in order to reduce the immediate impact of downdrift erosion, the groynes may be made progressively shorter in the downdrift direction. Beach nourishment should always be considered in addition to the use of groynes as a means of restoring or increasing the amount of beach material on a particular beach where either beach erosion has resulted from the starvation of an updrift supply or downdrift erosion must not be allowed to take place. Shingle-beach nourishment schemes usually present fewer problems than sand beach nourishment where careful selection of the particle size of material used is particularly important (see Section 9.4.10)

Finally, there are design considerations that relate specifically to the type of beach on which they are to be applied coupled with the materials, and hence three-dimensional geometry, of the groynes themselves. Table 9.5 (Simm *et al.* 1996) provides some commentary on the use of different types of material that have been used in groyne construction. The most common form of construction today is that of a rock mound, due to its inherent hydraulic efficiency. Some examples of differing groyne construction are illustrated in Figures 9.5 (a) to (h). Figure 9.5 (e) illustrates the intrusive nature of massive timber groynes on a shingle beach, Figure 9.5(f) shows a pair of rock groynes, Figure 9.5(g) shows a concrete armour unit groyne and downdrift erosion set-back at the boundary of a coastal defence scheme and Figure 9.5(h) shows the remnants of a dilapidated timber piled groyne system that constituted a hazard on the beach.

In conclusion, a well-designed groyne system can be effective in controlling beach movements, but the degree of success will, to a large extent, be dependent on the sediment supply whether natural or artificial. Groynes are simplest to design and most effective on shingle beaches. The corollary of this is that the adequacy of performance is less susceptible to poor design than for other types of beach. Finally, studies in both the Netherlands and the UK have considered the impact of offshore sand waves and sand banks. These can have a profound impact on the beach levels which response slowly over time to the movement of these features. Groynes can have little influence on such macro-scale movements.

Table 9.5 Impact of groyne construction materials (Simm *et al.* 1996).

Type/Material	Advantages	Disadvantages	Suggested applications
Vertical timber	Possible post construction adjustment	Cost and availability of hardwoods Environmental restrictions on hardwood sources Susceptible to physical abrasion and biological attack Vertical construction does not absorb wave or current energy Current induced beach scour pits along face and around head Unstable if large cross-groyne differentials in beach elevation develop or if large crest heights are required Difficult to construct below MLW Require maintenance	Low to moderate energy shingle beaches with low net drift
Rock mound	Hydraulic efficiency due to energy absorption Re-usable material Simple construction methods	Availability and transport of suitable rock Structures may be hazardous to swimmers and other beach users Accumulation of debris within structure Bed layer required if substrate is mobile	Low to high energy sand or shingle beaches with low net drift in areas where suitable rock is available Good for terminal structures

	Underwater construction possible Post-construction adjustment easy Stable, durable No size limit		
Concrete units	Hydraulic efficiency due to energy absorption Stable, durable Availability of materials	Rigorous construction methods required May be hazardous to swimmers and scramblers Accumulation of debris within structure Bed layer required if substrate is mobile	Low to high energy sand or shingle beaches with low net drift, in place of rock Good for terminal structures
Vertical concrete/ masonry	Availability of materials	No post-construction adjustment Expensive and complex construction particularly below MLW Near vertical construction does not absorb wave or current energy Maintenance required	Low to moderate energy beaches with low net drift Good for terminal structures
Steel sheet piles	Rapid construction Can be placed below low water	Vertical construction does not absorb energy No post-construction adjustment Suffer from abrasion; resulting jagged edges are a safety hazard Suffer from corrosion	Can be used to form foundation and sides of concrete structures, particularly below MLW
Gabions (rock-filled wire mesh baskets – see Fig 9.24e)	Low cost, rapid construction Hydraulically efficient	Not durable Particularly susceptible to vandalism Only suitable for small structures	Low energy sand or shingle beaches with low net drift
Rock-filled crib-work	Low cost due to smaller rock Hydraulically efficient	Movement of rocks can damage crib-work	Low to moderate energy sand or shingle beaches, with low net drift
Grouted stone or open stone asphalt	Low cost	Prone to settlement problems Susceptible to abrasion	Low to moderate energy sand or shingle beaches, with low net drift, on stable substrate
Rock apron around timber	Increase energy absorption of existing vertical structures	Interfaces subject to abrasion due to different interactions with waves	Refurbishment of old vertical groynes on low to high energy shingle or sand beaches with low net drift

9.2.2 Shore-connected breakwaters

Shore-connected breakwaters are differentiated from groynes by virtue of the fact that the former may be stand-alone structures and usually extend into deeper water than the latter and provide a rather more significant barrier to waves, wave-induced currents and hence alongshore sediment transport. As a category they include a variety of hybrid structures that do not conform to the design principles for groynes, but do require similar considerations to be made. They include both cross-shore elements and alongshore elements so that their primary influence on the beach geometry is to reduce the alongshore transport of material by generating dynamically stable formations between pairs of structures. At the same time a single structure can, in the right circumstances, be beneficial to the coastline. A basic ingredient is that the geometry uses wave diffraction as a means of holding the beach in the lee of the structure.

These structures might be generically described as bastions or artificial headlands, but an offshore breakwater that has become connected to the shoreline through a tombolo (see section 9.2.3) will also behave in the same way. Hence shore-connected breakwaters are structures which bridge the gap between groynes and detached breakwaters and, in some circumstances, the differences might be viewed as subtle. However, the fundamental mode of application of an artificial headland is to create stable beach formations between adjacent structures. If the structure extends sufficiently seaward the deflection of tidal currents off the shoreline may also be an important property. This implies less passage of alongshore drift of material that might be accommodated in a groyne or detached breakwater system, but does not rule it out completely. Figure 9.9(a) shows a naturally occurring tombolo, whilst Figure 9.9(b), (c) and (d) shows various applications of the principle.

A particularly effective form of artificial headland is known as the fishtail breakwater, which owes much of its development in the UK to Dr P.C. Barber. The concept of the fishtail breakwater is to combine the beneficial effects of the groyne and offshore breakwater and to eliminate the undesirable effects of the separate structures. The basic geometry of the fishtail breakwater is shown in Figure 9.10. The breakwater arms DA and DB act as wave energy dissipaters, whilst the arm DC intercepts the alongshore drift. Therefore the updrift beach is formed by the normal accretion process associated with any other alongshore barrier, whilst the downdrift beach is formed by the same diffractive processes associated with a detached, shore-parallel breakwater.

The arm AC is curved in plan so that the axial alignment at A is normal to the streamline of the diverted alongshore and tidal currents and the shoreward end C is normal to the beach line. An important feature of the outer section of the primary limb DA is that the seaward alignment does not allow the nearshore refracted waves from running inside the updrift shadow zone formed by the arm itself as this could result in a damaging 'mach stem' wave (see Chapter 2). Thus the curvature of CDA is designed to minimise wave reflection effects on the updrift side of the breakwater so that the area bounded by A and C should form a minor updrift diffraction zone in which the accretion is dominated by wave-induced currents. The arm DB is located in plan to allow waves sufficient distance to transform out of the current field and its length is dependent on achieving the desired diffraction effects that result in the downdrift beach remaining attached to the structure. This is partly dependent on the length of

Figure 9.9(a) Natural tombolo; (*b*), (*c*) and (*d*) shore-connected breakwater schemes.

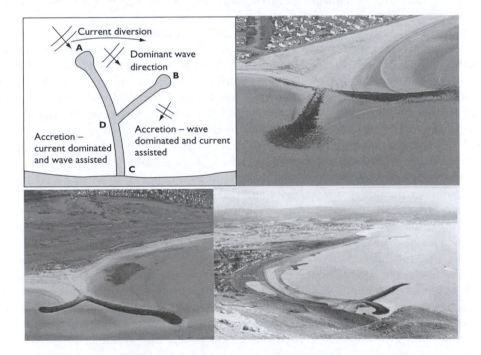

Figure 9.10 Basic geometry of fishtail breakwater and examples of various schemes.

DC. The overall dimensions of the breakwater are thus interdependent and a function of the incident wave height, direction and period, tidal range, beach morphology and the extent of required influence. In general terms the distance of the primary limb's outer roundhead A depends on the length of coast the breakwater is intended to influence, but should be greater than three inshore wave lengths as well as less than half the width of the active littoral zone. The relationship with an adjacent companion structure in creating a dynamically stable beach formation in the intermediate cell is also an important consideration, which is addressed further in the text. The crest levels vary throughout a fishtail breakwater and are dependent on the frequency of water levels and wave exposure along the length ADB. The crest between D and C should follow the 'equilibrium' beach profile. This type of structure can influence the beach in a number of ways. There is usually a steepening of the beach gradient in the immediate vicinity of the structure due to current and wave height steepness changes caused by the breakwater itself. Figure 9.10 also shows examples of fishtail groynes that have been constructed in the UK.

There are many examples of naturally occurring crenulate bays in nature as shown in Figure 9.11(b). When in perfect equilibrium wave refraction and diffraction results in the wave crests being parallel to the beach contours throughout the bay so that the theoretical alongshore movement of material is zero due to simultaneous breaking of waves along the shoreline. This principle has been developed by Silvester (1976) over many years since the early seventies. Figure 9.11(a) shows a definition sketch for a static equilibrium bay as defined by Hsu et al. (1989). The theory dictates that the beach between two headlands will erode an originally straight shoreline to form an equilibrium bay whose downcoast tangent is parallel to the inshore wave crest line at the point where the shadow line from the downdrift headland intercepts the original straight beach line. This shown as the transition point and determines the baseline length parameter R_0 together with a reference angle θ. At this point the tangent to the beach line is deemed to be parallel to the nearshore wave crests at the point at which they start diffracting on the updrift headland. The shape of the bay is thereafter defined through a relationship between a variable radius R radiating from the updrift diffraction point as a function of the angle β between the radius

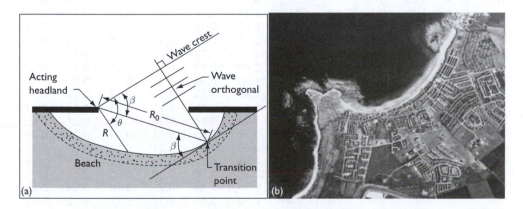

Figure 9.11 (a) Static equilibrium bay definitions and (b) example of natural bays.

and the incident wave crest at the diffraction point. The function is given as through the non-dimensional ratio R/R_0 versus the angle θ for a given β. Table 9.6 provides look-up values for increments of 5° between which linear interpolation is satisfactory. The method has also been applied to accretion behind a single offshore breakwater (Hsu and Silvester 1990) as well beaches downdrift of harbours (Hsu *et al.* 1993).

Both numerical and physical models may also be used to determine the variability of the beach line formation together with the cross-shore characteristics as described in Chapters 5 and 6. However, 1-line beach response models break down when the projected beach line deviates significantly from the original baseline, so that prediction of the beach line close to the headlands becomes difficult, depending on the degree of diffraction induced by the structure.

Artificial headlands and shore-connected breakwaters can take a number of different forms other than fishtail breakwaters and shore-parallel breakwaters that are attached to the shoreline by a tombolo (see section 9.2.3). As a general principle they must have a geometric shape that induces some degree of wave diffraction around the structure. It follows that the head of the structure will therefore have to be significantly wider that its root. The form of construction for this class of structure is generally rock or randomly placed armour units, the design principles of which are outlined in Section 9.4. There are, however, some examples of the use of steel pile crib work to contain smaller-sized rock as well as the use of pattern-placed armour units. Similar considerations with respect to the practicalities of construction to those outlined in section 9.2.1 apply.

The stable bay principle can be used very effectively in low to moderate tidal environments to create an interesting edge to reclamation. Figure 9.12 is an illustration of a conceptual master plan of a coastal development that employs the artificial headland principle in a number of ways in order to generate stable beaches in static equilibrium in positions where beaches did not previously exist. It should also be noted that the

Table 9.6 Radius ratios (R/R0) as a function of approach angle (θ) and local angle (β)

β/θ	30	45	60	75	90	120	150	180	210	240	270
10	0.37	0.26	0.20	0.17	0.15	0.12	0.10	0.10	0.08	0.08	0.08
15	0.53	0.38	0.30	0.25	0.21	0.17	0.15	0.14	0.12	0.11	0.11
20	0.70	0.50	0.40	0.33	0.28	0.23	0.20	0.17	0.13	0.13	0.13
25	0.85	0.61	0.48	0.41	0.34	0.27	0.24	0.21	0.18	0.16	0.16
30	1.00	0.72	0.57	0.48	0.40	0.32	0.28	0.23	0.20	0.18	0.17
35	—	0.82	0.65	0.55	0.47	0.37	0.31	0.26	0.22	0.19	0.19
40	—	0.91	0.73	0.62	0.42	0.41	0.34	0.28	0.23	0.20	0.20
45	—	1.00	0.80	0.68	0.58	0.46	0.38	0.29	0.24	0.21	0.20
50	—	—	0.87	0.74	0.64	0.50	0.40	0.31	0.24	0.21	0.21
55	—	—	0.94	0.80	0.69	0.54	0.43	0.32	0.24	0.21	0.20
60	—	—	1.00	0.87	0.74	0.58	0.45	0.32	0.24	0.21	0.20
65	—	—	—	0.91	0.79	0.62	0.46	0.31`	0.23	0.20	0.19
70	—	—	—	0.96	0.84	0.66	0.48	0.30	0.22	0.18	0.17
75	—	—	—	1.00	0.88	0.70	0.48	0.30	0.20	0.16	0.15
80	—	—	—	—	0.92	0.74	0.49	0.27	0.18	0.14	0.13
85	—	—	—	—	0.97	0.78	0.49	0.25	0.15	0.12	0.10
90	—	—	—	—	1.00	0.81	0.49	0.23	0.12	0.09	0.07

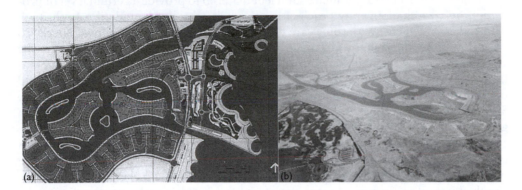

Figure 9.12 Doha West Bay Lagoon using artificial headlands: (a) concept plan and (b) post construction.

equilibrium bay principle has also been used to design the internal beaches in the lagoon system.

An example of the successful application of these principles is shown in Figure 9.13. Here a large area of seabed was reclaimed seaward of the natural coastline and a recreational amenity was required. As suitable beach was scarce it was necessary to use dredged coral fill to form the underlying reclamation geometry including the equilibrium bays, which were defined using the foregoing methodology. The beaches were then created by relaying a 1–2 m thick covering of sand that had been scraped off the original shoreline. In this location a 1 in 1-year wave height is of the order of 3.5 m with a period of 7 s and a maximum tidal range of 2 m. This has now been performing most satisfactorily for over 10 years.

When designing artificial headlands, similar principles to those mentioned in Section 9.2.1 apply with respect to the impact of the structure(s) on an existing beach.

Figure 9.13 Artificial beaches created using the static equilibrium bay methodology.

The objective will be to accumulate sufficient material to provide a certain level of protection or a width of amenity beach either through natural accumulation or through artificial nourishment. With some care the system can be designed to allow material to pass alongshore through the system once the beaches have stabilised. As with groyne systems the design methodology must include consideration of all possible combinations of wave attack, some of which might temporarily destabilise the 'equilibrium' of the formation. In addition allowance for possible downdrift deficits must be made. Shore-connected breakwaters can function on both shingle and sand beaches and are generally more satisfactory that groynes for the latter.

It should be mentioned that a different type of bay will form when the gap between two nearshore parallel breakwaters or natural features exists is small relative to the wavelength of the shallow-water wave. As depicted in classical wave diffraction theory the internal or shoreward wave pattern adopts a near circular geometry with the centre of the circle at the midpoint between the headlands. This can arise when the sea breaks through a stable durable coastline over a sill or breaches a parallel revetment. The formation is known as a 'pocket beach' which, geology permitting, will not only be symmetrical, but also have a depth to length ratio that is much greater than the so-called 'equilibrium bay' previously described. The formation is also virtually independent of the direction of wave approach (Dean 1977). A natural example of this type of feature exists at Lulworth Cove as shown in Figure 9.14.

9.2.3 *Detached breakwaters*

Detached breakwaters are simply that. They have no connection to the shoreline so that currents and sediment can pass between the structure and the waterline. In some texts they may be referred to as 'offshore breakwaters', 'nearshore breakwaters' or 'artificial reefs'. The latter infers a significant degree of overtopping can occur over the body of the structure so potentially there is a measurable element of wave transmission through and over the upper layers of the armouring. The commonest form of construction is parallel to the shoreline.

Figure 9.14 Natural pocket beach at Lulworth Cove.

Detached or nearshore breakwaters have been used extensively for coast protection or the creation of crescentic beaches with considerable success, particularly on coastlines where the tidal range is negligible or small. Detached breakwaters create a zone of reduced wave energy behind the breakwater as well as local patterns of wave-induced currents that, in turn, create a zone of sand deposition in the lee of the structure. In the absence of other influences, beach material will be transported into the area to form a tombolo or salient. Detached breakwaters can be used in much the same way as groynes to build up the volume of material that is capable of accommodating the drawdown that occurs under storm conditions. The shape of the beach that forms between adjacent breakwaters is that of a crenulate bay, which is inherently more stable and less volatile than the abrupt discontinuities caused by groynes. However, if attachment of tombolos is permitted, downdrift erosion problems will still occur. Detached breakwaters may also be used to deliberately create an area of deposition, for example updrift of a harbour entrance so that it can conveniently be dredged and deposited on the downdrift beach.

The fundamental difference between a groyne and an artificial headland is that the latter is a more massive structure designed to eliminate problems of downdrift erosion and promote the formation of beaches. While these structures may take a number of different forms, their geometry is such that, as with the offshore breakwater, wave diffraction is used to assist in holding the beach in the lee of the structure.

There have been two major detached breakwater/reef schemes in the UK situated at Elmer on the south coast and at Happisburgh/Winterton on the east coast. The former has been subject to much detailed research on shingle sediment transport processes around these types of structure as reported by Chadwick *et al.* (1994). The latter coastline is one that has experienced a long history of beach volatility and flooding (see Hamer *et al.* 1998). The sea defence strategy adopted was for the phased construction of a series of rock armour breakwaters coupled with long-term beach recharge and management. The strategy allowed for a review of the performance of each phase of the scheme in between each major construction phase. The first stage design allowed for extreme storm surge levels and resulted in the formation of mid-tide tombolos with some undesirable cutting back in the bays. Improvements in the second stage resulted in slighter shorter structures positioned at the same distance from the shoreline. The following text, based on a paper by Fleming and Hamer (2000) compares some of the design guidance found in literature with the measured performance of the two stages of this scheme. Much, if not all, of the outline design guidance prior to the early 1990s had been developed from analysis and observations of beaches in relatively sheltered and micro-tidal situations. In the case of Happisburgh to Winterton, the validity of applying this guidance has been considered for a site with a tidal range of 3 m and exposure to significant wave heights with an annual average value of close to 2 m. In Stage One of the construction programme, four reefs were built to a length of approximately 230 m at an offshore distance of 200 m from the shoreline. In Stage Two a further five reefs were constructed with a length of 160 m, at the same offshore distance. The reefs relating to the two stages can be seen in Figure 9.15(a) with the first stage in the foreground. The differences can easily be observed.

The principal terms used to describe offshore reef geometry are presented in Figure 9.16. Design guidance has focused on the relationships between these parameters and most notably on the ratio of structure length (L_s) to offshore distance (X)

(a)

(b)

Figure 9.15 Detached breakwater schemes at (*a*) Happisburgh to Winterton (Stage Two in foreground and Stage One in distance) and (*b*) Elmer.

and gap length (L_g) to offshore distance (X). It should be noted that values for L_s, L_g and X all vary with tidal height and the comparisons presented here relate to mid-tide values.

The normal practice when developing outline geometry, to determine the feasibility of an offshore reef scheme, is firstly to fix the offshore distance by reference to the sediment transport pathways. For example, if it is not desirable for the reef system to have a major impact on nearshore, as opposed to beach face, alongshore sediment transport, then it should be located inshore of any nearshore features that may be primary sediment pathways.

Having decided upon an optimum offshore distance, the standard relationships presented in Figure 9.17 might then be used to determine the length of reef that would result in different forms of beach response. Again, depending on the desired result, decisions may be taken to allow the beach shape to develop to form either salients or tombolos. Clearly, tombolos will be more disruptive than salients to the alongshore movement of sediment, but will offer more protection during severe storms and will offer greater amenity area.

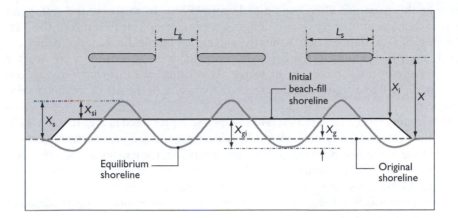

Figure 9.16 Design parameters for offshore breakwaters.

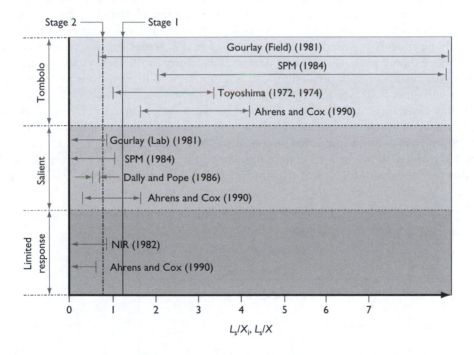

Figure 9.17 Compilation of design guidance in literature.

The annotations on Figure 9.17 demonstrate the actual ratios between offshore distance and structure length for the two reef designs that have been constructed in two stages of the Happisburgh to Winterton Sea Defence Strategy. It is evident that, whilst both reef designs are close to the boundary between shoreline response of salients and tombolos, the guidance was found to be most relevant, despite being developed for different prevailing conditions.

Another relationship investigated was the ratio of the distance offshore to the depth of water at the structure after Pope and Dean (1986). In this case the differences between each stage of the reef system design were imperceptible in terms of this ratio and the empirical guidance would similarly suggest negligible difference in behaviour. However, the observations in the field demonstrated a great sensitivity to structure length for reefs in the same depth of water. This difference in behaviour can only really be attributed to a significant difference in exposure of the site of the macro tidal environment

Another comparison relating to shoreline response is presented in Figure 9.18, which relates to the potential for erosion on the beaches opposite the gaps between reefs.

In Stage One of the Happisburgh to Winterton scheme, the gap length between reefs was approximately 230 m, which resulted in cutting back of the exposed beaches to the sea wall following storms. Reducing the gap length to 160 m in Stage Two reef design resulted in much reduced beach response and the formation of gentle crenulate bays between the structures.

Whilst some of the outline design guidance can be demonstrated to be adequate for the purposes of outline design and feasibility study, detailed design still requires a detailed understanding of the impact of any scheme on the adjacent beaches. A variety of tools as discussed elsewhere in this book may be used. For example 1-line beach plan shape models for predicting the effects of reef schemes can be quite effective. With more than 9 years of measured data in the vicinity of the reef system at Happisburgh, the results of such predictions can be compared to observations in the field as shown in Figure 9.19 (see also Hamer et al. 1998).

A very close agreement has been achieved between the predicted shoreline response and measured beach movements. Whilst it was noted that the initial phase of salient development was under-predicted, unless model bathymetry and hence wave field are regularly updated, the long-term development of the shoreline was well represented. In this case the downdrift beach erosion was initially predicted to be 150 000 m³ per

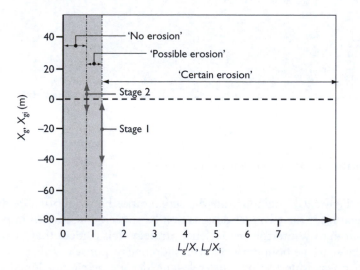

Figure 9.18 Comparison of observations to relationship after Rosati (1990).

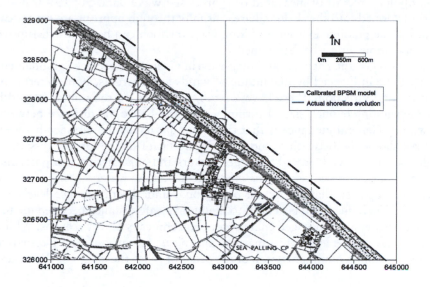

Figure 9.19 Beach plan shape predictions compared to measured response.

year on average and the latest validated version of the alongshore model suggests a value of closer to 130 000 m³ per year, which is considered to give strong evidence to support the predictions made in the past and those for the future.

9.2.4   *Port and harbour breakwaters*

Port and harbour breakwaters are, in principle, not different in design terms to other forms of breakwaters except that their functionality usually requires access along or behind the crest by both people and vehicles. By their nature they will also tend to be in deeper water and be very much more massive structures in order to withstand the forces of very large extreme waves.

In considering the layout of any area between harbour breakwaters, the following points should be considered:

- The entrance needs to be laid out such that wave penetration is minimised to acceptable design standards at the proposed berth positions. It is not necessary, or indeed often feasible, to achieve minimal wave activity in the outer reaches of a harbour. Different levels of protection are relevant to large ports, fishing harbours and small boat marinas. General guidance in given in BS6349, Part 1(1984) and maximum wave heights of between 0.3 m and 2.0 m may be acceptable depending on the size of vessel and the method of loading/unloading. The overlap between breakwaters will usually be such that the outer breakwater faces the direction of greatest wave exposure. Swan-neck style entrances should be generally be avoided.
- The acceptable downtime for operations related to the tolerable frequency of exceedence of the above conditions should be established. Evaluation requires a

reasonably long record of measured or synthesised wave data, say 10–20 years, to be transformed into the sheltered area, usually through appropriate numerical modelling. The possible existence of long waves, which can be highly disruptive to port operations, must be determined.

- Safe navigation of the entrance and manoeuvring inside the facility is fundamental and will be influenced by magnitude of winds and currents, both in terms of magnitude and direction. These have a strong influence on the minimum width of the navigation channel, size of turning circle and hence the distance between breakwaters. General guidance is that the minimum navigable distance between structures should be between 4 and 6 times the beam of the largest vessel (although 5 to 7 times is advocated for larger tankers and bulk carriers) for ports and fishing harbours. Allowance must be made for the below water extension of any breakwater slopes so that the lowest operational underkeel clearance levels are maintained. There may also be a need to maintain a safe distance between the toe of the breakwater and a dredged channel for geotechnical stability reasons. Whilst the same rules can be applied to marinas and pleasure harbours, an alternative criteria is to adopt a minimum width of the order of twice the length of the largest vessel.

- Generally, low-crested overtopped breakwaters are considerably cheaper and quicker to build than high-crested non-overtopped structures. The choice depends upon a number of issues, not least the use of space within the enclosed water area.

- Sediment transport patterns both on the beach and offshore are important with respect to potential siltation and hence maintenance dredging requirements. The geometry of breakwaters can have a significant influence on such movements. Mitigation measures for updrift accretion and downdrift erosion of the beach may also be necessary.

- Environmental issues are also of considerable importance and potential impacts may need to be considered early on when locating structures. Particular aspects may be specific areas of interest (spawning grounds, coral, seagrass etc.), the impacts of construction activities (suspended sediments) and effects on shoreline evolution.

- Level of tolerable maintenance and ease of operations/availability of material/ plant. The inherent damage allowances within designs should be clearly identified and minimised if this will be an issue.

Correct layout of the breakwater has major implications for both the functional design and costs of a facility. It is therefore recommended that numerical wave modelling is undertaken even at preliminary stages to determine and optimise appropriate layouts such that operational requirements are met. This should be used to investigate potential problems such as reflectivity and resonance within the harbour basin. It should also be recognised that a dredged channel, depending on its relative depth and orientation with respect to the wave climate, can have a significant impact on the wave propagation and the amount of wave energy that may be directed into or excluded from a harbour entrance. The cost of these types of studies should on average be equivalent to the construction of only 2 or 3 m of breakwater and usually have the potential to save considerably more.

There are many different types of breakwater when considering the detailed components. Burcharth (1994) provides an excellent summary of design principles for different breakwater types. In generic terms they may be classified as:

- rubble mound breakwaters;
- caisson breakwaters;
- composite breakwaters.

Rubble mound breakwaters consist of a core of relatively small-sized material (quarry run) covered by one or more filter layers of rock, finally protected on the exposed side by larger armour rock or concrete armour units. These are the most common type of breakwater and will generally be made of rock armour in water depths of 5–6 m. This is on the basis that median rock sizes of 6–8 tonnes are not commonly available from many quarries. There are, of course, exceptions and rock armour of the order of 15 tonnes has been produced in special circumstances where there is particularly competent rock. However, these sizes present their own handling problems. Depending on the size of armour there will be a point at which concrete armour units become more economical to produce and place. There have been many types of armour units developed over the years, some of which are described in Section 9.4.3. Figure 9.20 shows some of the more popular types such as the Stabit, Dolos, Tetrapod, Core-loc, Accropode and Modified Cube. The Core-loc and Accropode are recognised as being the most efficient armour units in terms of volume of concrete required per unit area of breakwater and it is not uncommon for claims of 20–30 per cent savings being possible when compared to other armour units. However, these assertions relate only to the unit cost of concrete in the armour layer alone and it should be appreciated that it is often more economical to use larger armour units than are theoretically required as the number of castings, unit movements and unit placements are reduced due to the greater coverage by the larger units. Hence the unit cost of concrete can easily become irrelevant when considering all of the elements of construction cost.

Where rock sizes cannot be produced to satisfy the statically stable design criteria described in Section 9.4.3, it is possible to design a so called 'berm breakwater'. This is a breakwater, which has an intermediate berm, usually at about mid tide to mean high water, and is designed to be dynamically stable. That is, under extreme conditions the breakwater armouring may move in a similar, but much more subdued, way than pebbles move on a beach. The face may therefore form an 'S' shape profile that is familiar with a coarse shingle or cobble beach. Berm breakwaters require rather larger volumes of material than conventional breakwaters, as sufficient armour must be placed to allow deformation to take place without threatening the integrity of the structure. At the same time armour stone sizes may be significantly smaller for the equivalent design wave height, by a factor of five or more. It is also possible and normal to allow a broader range of gradation of stone. This type of design is therefore well suited to situations where there is a suitable quarry within a short haulage distance to the structure location. Guidance on the design of berm breakwaters can be found in Van der Meer and Koster (1988), CIRIA/CUR (1991) and most recently a comprehensive state-of-the-art guide by PIANC (2003b).

When the depth of water becomes greater than about 15 m the volume of rock required for a rubble mound breakwater becomes extremely large. In addition

Figure 9.20 Concrete breakwater armour units (a) Stabit, (b) Doles, (c) Tetrapod, (d) Core-loc, (e) Accropode and (f) Modified Cube.

the footprint will be very large which can have implications where space might be restricted and wave penetration may be difficult to reduce at the entrance, as the distance between the navigation channel and the water line must increase with depth. Navigation itself may be compromised due to the large expanse of breakwater slope

that is below the water line. In these circumstances a vertical wall breakwater, usually in the form of a caisson will become a viable option. An example of this type of
breakwater is shown in Figure 9.21. A caisson may occupy all of the water depth
being founded on a rubble foundation or form part of a composite breakwater. The
latter allows the height of the caisson to be constant whilst varying the height of the
berm foundation accommodates variations in depth. The berm foundation also has
the effect of distributing the load over a larger area, thus reducing settlements and
allowing construction on weaker soils. Figure 9.22 shows the contrast between the
cross-sectional area of a composite caisson breakwater compared to the equivalent
rubble mound.

Benefits of the caisson design include reduced environmental impact due to significantly lower quarried rock and transport requirements as well as reduced construction risk as the caissons can be positioned quickly in selected weather conditions. The
disadvantages include the necessary use of reinforced concrete in the marine environ-

Figure 9.21 Caisson breakwater under construction

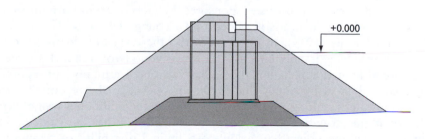

Figure 9.22 Comparison between rubble mound and equivalent caisson design.

ment, which should be avoided as far as possible and the structure's susceptibility to damage due to differential settlement, potentially high wave forces or seismic conditions. Therefore, the primary design issue for a composite breakwater is the resistance to the vertical component to sliding or overturning, which is resisted by the mass of the structure.

It is beyond the scope of this book to go beyond this brief description of breakwater types. There are many other types of breakwater, details of which can be found in Bruun (1989) or Burcharth (1994).

9.2.5 Floating breakwaters

Floating breakwaters are perceived to be lower-cost structures that are extremely versatile in that their position can be varied and their cost is not dependent on the depth of water or the tidal range. For maximum efficiency such breakwaters must have a high effective mass (represented by the sum of the mass of the structure and that of the body of water that moves with it), high damping characteristics, and possess natural frequencies of vertical and angular oscillation appreciably lower than those of the longest design wave. Lochner *et al.* (1948) explain how the above reasoning led to the design of the Bombardon, which was used during the Second World War to create temporary harbours at St. Laurent and Arromanches. It was a structure of cruciform cross-section 7.6 m × 7.6 m with a buoyancy tank in the upper arm that penetrated the water surface and had a beam of only 1.5 m. Units 61 m long were moored 15 m apart; with two such parallel lines of units separated by 244 m, with staggered centres the ratio of the transmitted to the incident wave was found to be approximately 0.3 for waves of 46 m wavelength and 3 m height. This corresponds to a wave period of approximately 3 s assuming 5 m depth of water.

A number of alternative laminar types of floating breakwater have been tested in the laboratory and at sea. These may be subdivided into semi-rigid and flexible types. The latter include floating plastic rafts containing compartments filled, or partially filled, with water or other liquids. The movement of the contained fluid provided a certain degree of damping, but these floating breakwaters present considerable mooring problems. Such mattresses only provide appreciable protection if they are of the order of half a wavelength in width.

Significant attention has been paid to the use of used car tyres in the construction of floating breakwaters of the type shown in Figure 9.23 for areas that are partially sheltered where the natural exposure limits the range of wave conditions. It is also a fact that the basic raw materials for such breakwaters are available in abundance. Much of the scientific research has been sponsored by Goodyear and a number of case histories and design guidance is provided by DeYoung (1978), McGregor and Miller (1978) and Harms (1979). In common with other types of floating breakwater, the tyre breakwater is only effective for wave periods of about 4 s or less and are therefore of limited application. There have also been a number of reported problems related to loss of buoyancy due to the air pocket at the top of the tyre being replaced by water resulting from wave agitation or the additional weight generated by heavy marine growth. For these reasons this solution has not gained a lot of credibility for anything other than a low-cost, limited durability or temporary solution.

Figure 9.23 Floating tyre breakwater.

Designers of marina pontoons have developed the floating breakwater concept by incorporating a skirt into the standard floating pontoon. This type of pontoon would typically be positioned strategically to deal with small locally generated wind waves.

Floating breakwaters are also being used in conjunction with wave energy devices and can therefore have a very significant dual purpose. Some of the theory is given by Count (1978) and more recent design practice is described by Bruun (1989), Tsinker (1986) and PIANC (1995).

9.2.6 Sea walls

Sea walls may be considered to be the last line of defence in a coastal protection scheme. In many circumstances they be the only line of defence, but this would be considered to be a last resort where no other more natural option is available. There are many potential impacts of a sea wall on a coastline, as it has no capacity to respond to natural events. If a coastline is naturally eroding the wall may hold the upper section of the profile, but will not prevent erosion of a vulnerable foreshore. As a general rule a sea wall should be positioned as far landward as feasible so as to allow the natural coastline as much freedom as possible. The alignment of the wall should be as smooth as possible and follow the natural contours rather than have severe changes in alignment to suit landward features. The consequences of terminating a sea wall on eroding coastline are clearly demonstrated in Figure 9.3(g).

The final and arguably most important characteristic is that the wall should be designed to dissipate as much wave energy as possible. Wave reflections from sea walls on erodible shorelines will definitely cause a redistribution of sediment and cause toe scour unless there is a very high net positive supply of sediment to the area. The evidence of beach steepening and foreshore lowering in areas where there are sea walls in the UK is irrefutable. The feedback effect results in deeper water closer to the sea wall to allow larger waves to break on the wall and hence accelerate the process. Figure 9.24 shows examples of a number of sea walls of differing construction. Figure 9.24(a) is a massive, near-vertical wall topped with a wave return wall together with large Tetrapod armour units placed at the toe sitting on a rock berm. It can be surmised that when the wall was built initially, reflections from the wall caused a

Figure 9.24 Examples of sea wall types: *(a)* Tetrapod toe, *(b)* stepped, *(c)* rock toe, *(d)* recurved wall, *(e)* gabions and *(f)* open stone asphalt with rock toe.

lowering of the foreshore to the extent that either the foundation was under threat or larger waves impacting the wall caused unacceptable overtopping volumes during storm events. The inclusion of concrete armour units, normally used for large break-water designs is unusual and indicates the severity of wave attack in this particular location. Figure 9.24(b) is a large concrete stepped sea wall. The use of a stepped profile is sometimes thought to create high wave energy dissipation and correspond-ing reduced wave reflection. However, experience suggests that except at particular water levels, wave reflection is not reduced that much, but the steps are user friendly for pedestrians. Figure 9.24(c) is the combination of a concrete wall providing a mas-sive barrier with pedestrian access combined with a significant rock toe to act as a significant wave energy dissipation and toe scour prevention feature. Figure 9.24(d) is a large smooth concrete sea wall incorporating a substantial wave return wall, which will be most efficient at a particular limited range of water levels. Figure 9.24(e) shows a gabion wall, which consists of relatively small stone encased in heavy-duty wire-mesh containers. As a general rule gabions are not recommended for use in the coastal zone as permanent works because they have a very limited life due to corrosion of the mesh despite measures such as plastic coating to prevent this. Finally, Figure 9.24(f) is an open stone asphalt revetment that has been reinforced with a rock toe some time after the initial construction. This demonstrates the use of a material that was insuf-ficiently robust for the environment in which it was used and subsequently required remedial works to reduce the wave impact.

There are many different types of revetment system that have been developed over the years. Many of these involve pattern placing of individual units. In many cases the underlying rock blanket is an integral part of the design and play s major role in the wave energy dissipation process. Figure 9.25 shows a number of examples. Figure 9.25(a) is a SeaBee unit slope topped by a wave return wall. These hexagonal units can be produced in a wide range of sizes and have even been used as breakwater armour units. In this example some of the units are manufactured to be deeper so that they protrude above the general slope in order to increase surface roughness and reduce wave run-up. Figure 9.25(b) are SHED units that are highly porous units with voids on all axes and are therefore very effective in absorbing wave energy. They are usually manufactured with fibre-reinforced concrete and are have also been used as breakwater armour units. They have been considered by some authorities to be dan-gerous for use where there is public access due to the possibility of someone becoming trapped within the voids. However, the same could be said for ordinary rock armour and other types of unit. Figure 9.25(c) are heavy interlocking precast blocks produc-ing a fairly rough surface whereas (d) are smaller precast concrete elements, known as Basalton blocks, that are light enough to be placed by hand. Figure 9.25(e) are porous interlocking blocks that can also be placed by hand and in the right situation will allow vegetation to grown up through the blocks. Figure 9.25(f) is a grout-filled mat-tress that is placed on a slope and filled *in situ*. Clearly these systems are suitable for varying degrees of exposure and it is necessary to obtain manufacturers catalogues for design information. Some systems incorporate single cable or dual cables so that large mats can be placed in a single operation. At the same time the cables play a key role in maintaining stability of the slope. Interlocking block systems may also interlock in both one and two dimensions. Some general guidance on designing different types of block system is given in section 9.4.3.

Figure 9.25 Examples of revetment systems: (*a*) SeaBees, (*b*) SHEDs, (*c*) interlocking blocks, (*d*) Basalton, (*e*) porous interlocking blocks and (*f*) grout-filled mattress.

9.2.7 Sills

Beach sills are not commonly used as coastal defence structures, but deserve a mention for both successful and unsuccessful applications. They may be described as lateral structures that are designed to be overtopped by wave run-up. If that run-up is laden with sediment this may become trapped behind the structure and create an artificial perched beach. Sills are also frequently used to create a man-made beach for recreational purposes in situations where there is insufficient room to create a full-depth beach profile. Most commonly this will be in sheltered situations such as the inside of a marina or dredged lagoon development. In these circumstances it is imperative that the toe of the beach should be no less than about 2 m below lowest low water to avoid any safety issues related to swimmers. Also, even in sheltered situations it is necessary to consider the possible variation of the beach profile that is being supported by an artificial toe.

Figure 9.26 shows a permeable sloping timber structure that was constructed along large lengths of the north Norfolk coast following the 1953 storm surge that caused so much damage to the east coast of England. It seems that, once found to be successful on one section of coastline, it was replicated along long lengths of coastline without any real understanding of why it appeared to work in the first place or taking into account any differences in physical setting. Figure 9.26(a) shows an area where the structure has apparently been quite successful in accumulating beach material behind it. This in turn should have reduced the rate of erosion of the cliffs at the beach head. However, both actions will have reduced the downdrift sediment supply through retention of existing material and reduction of new material production through cliff erosion. Indeed, cliff erosion can be a significant source of beach material in many circumstances. Figure 9.26(b) shows the identical structure on another section of the coastline, which has been clearly unsuccessful in trapping material behind it. The only apparent difference between the two areas would seem to be the size grading of the material on the beach. Another consequence of this type of scheme is the seriously negative impact it has on any beach recreation activity through the creation of an unsightly barrier to the natural beach.

Figure 9.27 shows an unusual application of the beach sill principle to a section of cliff fronting Fairlight village on the south coast of England. As can be seen a number of houses were under threat due to slow erosion of the near-vertical sandstone cliff. The rate of erosion was enhanced by a soft clay layer at the base of the cliff that was subject to wave action on every high tide. The cliffs were also a designated site of geological interest. The solution that was designed was a linear rock bund positioned seaward of the base of the cliff. This served a number of purposes as follows:

Figure 9.26 Sloping timber beach sill *(a)* filled and *(b)* empty.

Figure 9.27 Fairlight Cove linear bund.

- Cliff falls were episodic and unpredictable so that it was extremely dangerous to contemplate working immediately below the cliff itself.
- The bund protected the vulnerable clay layer from direct wave attack whilst maintaining exposure of the interesting geological strata.
- The bund would generally retain the fallen cliff material thus increasing the level of protection with time.
- The bund would also trap the sparse volumes of shingle drifting from the west whilst also allowing material to pass seaward of the bund.

Figure 9.27 shows an aerial view of the scheme together with a low-level oblique. The latter shows the retention of falling cliff material, the accumulation of alongshore drift behind the bund as well as the formation of a shingle beach in front of the bund that would have been facilitated by the reduction on wave reflection from the structure. When designed it was anticipated that cliff falls would continue to occur for at least 10 years until such time that the cliff attained its own natural stable slope. The figure shows some signs of relative stability by the vegetation that is establishing itself on the cliff face.

There have been a number of beach sill type structures proposed in the form of nearshore precast concrete reef blocks. Whilst great claims of success in building up beach volumes have been claimed, possibly due to a period of natural accretion during the monitoring period, there is little evidence to suggest that these measures are beneficial in the long term. Indeed, some data suggests that they may be detrimental and, like other beach sill structures, form an obstruction to the natural enjoyment of the beach.

9.3 Natural coastal structures

As discussed at the beginning of this chapter, the best types of coastal defence structures are those that occur naturally. The next best thing is to emulate those natural systems as closely as possible or to create conditions that encourage reinstatement to take place. Just as with structures these measures require a large element of design and a significant understanding of the coastal processes that are taking place.

9.3.1   Beach nourishment

Beach nourishment is also known as 'beach replenishment',' beach feeding' or 'beach recharge'. It entails finding a suitable source of material that is compatible with, but not necessarily identical to the material that occurs on the beach to be nourished. It is often the most satisfactory means of protecting a shoreline as it provided the necessary reservoir of material that allows a beach to respond normally to differing levels of wave attack. Interference with natural processes is reduced to a minimum and, where the size of beach has been enlarged there will be significant recreational and environmental benefits. A fundamental consideration of implementing a beach nourishment scheme will be the economic argument. For most schemes there will be an ongoing maintenance requirement to periodically place additional material following an initial campaign. Nevertheless some 30 years of experience in both the UK and the USA suggest that it is frequently a viable option, either as the sole method of increasing the level of service of coastal defence or in conjunction with beach control structures such as groynes, artificial headlands or detached breakwaters. Even greater benefits can be realised if the source of the borrow material is from maintenance dredging of a maritime facility such as the navigation channel of a port. Fowler (1998) describes such a scheme that has been implemented at Lee-on-Solent in the UK using dredged material from the access channel into the Port of Southampton.

There are a number of issues related to the planning and design of a beach nourishment scheme that are beyond the scope of this book. These include:

- identification of a suitable borrow area that will not have any impact on coastal processes following its exploitation;
- possible combining of materials from more than one source in order to provide the desired grading characteristics;
- different strategies for delivery of material to the beach and its initial profiling, as shown in Figure 9.28;
- selection of suitable plant for both dredging of beach nourishment material and distribution bearing in mind that land-based sources are rarely suitable;
- possible changes in grain size characteristics during handling;

Figure 9.28 Pumped delivery of beach nourishment.

- environmental impact of winning material and placing it;
- strategies for periodic maintenance including additional nourishment requirements.

There will be a number of choices related to each of these facets, but the rate of nourishment and position of placement in particular presents a number of possible alternatives. For example a shingle beach may be restored simply be feeding entirely at the updrift end of the system where there is a strong net drift in one direction and beach control structures such as groynes are involved, However, this may take some time to work through the system. If the material is placed at selected points a promontory may form which itself may act as a natural temporary groyne/headland causing short-term accretion and erosion trends associated with such structures (Dette 1977). Other strategies include placing the material within the active beach profile under water rather than on the upper beach. This is based on having the confidence that the size grading produced from the borrow area is such that the material will naturally migrate to the upper beach. This, in turn, suggests a coarser borrow material than that occurring naturally. Further guidance on these design principles is given in Section 9.4.10. More detailed information on beach nourishment can be found in a number of key texts that include Delft Hydraulics Laboratory (1987), Stauble and Kraus (1993), National Research Council (1995) and Dean (2002). For an overall detailed appreciation of beach management for beach nourishment is just one of the aspects the CIRIA 'Beach Management Manual' (2010) is highly recommended.

Two further aspects of beach management that involve the movement of beach material as a means of coastal protection are sand bypassing and beach recycling. The former involves moving material from an updrift area of accumulation to a downdrift area of erosion, both of which would normally have been created by construction of structures that interfere with the littoral drift process. Sand bypassing can be implemented by any mechanical means including land-based as well as marine-based plant. However, a particularly effective method of sand bypassing involves the use of a jet pump (Prestedge and Bosman 1994). A jet pump can be buried below a beach and, by a combination of jetting water with simultaneous suction, is self-priming and can be activated without the danger of becoming congested by excess sediment in the flow as can happen with a conventional dredge pump. There is the added attraction that the installation can be fixed, computer controlled and operated remotely. There has been a number of successful sand bypassing systems using this technique at tidal inlets and marinas.

Sand recycling involves moving material from a downdrift area of accretion back to an updrift point to act as a source. This strategy has been used for shingle beaches on the south coast of the UK for many years. For example, Dungeness foreland shown in Figure 9.29 accommodates a nuclear power station that has been protected by beach recycling over the past 30 years. The exposed coastline runs left to right corresponding to the west to east axis. Here there is a predominantly eastwards drift of shingle which migrates around the ness to a point where it is sheltered from the south-west waves and there is very little shingle drift northwards along the east-facing shore of the feature. The shingle ridges can be plainly seen, whereby the history of erosion and accretion over many hundreds of years can be appreciated. The beach recycling takes place by mechanical excavation of material from the accreted area to the north

Figure 9.29 Dungeness foreland.

transported by road to the updrift beach where it is deposited at a number of specific beach feeding points. This process has been monitored by analysis of annual photogrametric and topographic surveys through which the previous year's losses have been assessed and the next year's nourishment requirements have been determined. A recent innovation has been to implement a programme of managed retreat at a key point on the ness in order to realise significant savings in beach recycling quantities (Maddrell 1996). Periodically the question is raised as to whether it would be cheaper to construct a permanent coastal defence structure. The latest assessment showed that the beach recycling strategy remains less than half the NPV (net present value) of a beach retention system such as groynes or headlands and less than 20 per cent of the cost of a hard edge solution such as a revetment system with armour units.

9.3.2   Dune management

Dunes are accumulations of sand blown from the foreshore to the backshore by the wind, as shown in Figure 9.30; the sediment accumulates above the mean high water mark where it becomes vegetated. Further sediment is trapped by the presence of vegetation and deposition accelerates. These features should be viewed as a tremendously valuable resource in terms of providing a backshore reservoir of material to feed a beach during a period of extreme wave conditions. In the Netherlands this principle is used as the cornerstone to many of the lengths of vulnerable coastline to the extent that the 'system' is designed to withstand erosion associated with a 1 in 1000-year event. This reflects the extent of area of very low-lying hinterland that is protected by these features.

The formation of sand dunes is dependent upon two main factors. The first is an abundant supply of sand-sized sediment and the second a strong onshore wind to enable entrainment and transportation of sand from the beach to the dunes. Backshore dune development can be facilitated by a low-gradient sandy beach, which provides

Figure 9.30 Sand dunes.

a large expanse of beach sand exposed at low tide. Whilst establishment of colonising vegetation can influence dune morphology, it is not essential for their formation (Pye and Tsoar 1990). Dunes can move through migration that occurs through a mechanism of wind-driven saltation of the sand grains resulting in erosion of the front (exposed) face and deposition on the back (sheltered) face. It follows that any such movement is in the same direction as the prevailing wind. There are examples where sand dune migration has been part of a natural mechanism for transferring sand from one beach to another across a headland. As a general rule sand dune mobility is controlled by the rate of sand supply, the magnitude and frequency of wind and vegetation cover (Pye 1983). The marine erosion of dunes is more complex than that of cliffs because of the close interaction between the beach and the dune. The outcome of this is that dunes can both accrete and retreat. Erosion rates of dunes can be very high and rapid because they are composed of unconsolidated sands. The type of dune failure varies due to exposure, dune morphology and vegetation cover (Carter and Stone 1989).

The characteristic behaviour of a dune will depend upon its stage of evolution. A number of broad types can be recognised (Jay *et al.* 2003)

- *Embryonic dunes* – represent the first stage in the development of dune ridges and are formed by the deposition of sand along the high tide mark. They are low-lying mounds of sand and are often vegetated by salt-tolerant species; they are easily overwashed and removed during storms, releasing sand back to the beach.
- *Foredunes* – continuous or semi-continuous ridges of sand, often vegetated, which lie at the back of the beach and parallel to the shoreline. Parallel dunes can be modified during storm surges when wave overwash and breach may occur resulting in sand being swept landward in the form of fans or sheets. The height of foredunes is dependent upon the wind strength and sediment supply. Foredunes may become cliffed at their seaward margin during storms and undercutting at the dune toe can cause collapse and failure of the dune cliffs. Foredunes are also vulnerable to overwash (depending upon height) and breaching, particularly where the ridge is narrow and/or characterised by a series of blow-outs (see below).

- *Climbing dunes* – occur on some cliffed coasts there are sand dunes either piled against the cliff, forming climbing dunes, or at the top of the cliff. Where dunes have spilled inland and become separated from any source of sand, they have become relict cliff-top dunes, such as observed at locations along the Cornish coast.
- *Relict dunes* – also present where there are no contemporary sources of sand or where the link between the beach and dunes has been broken, for example on shingle beach ridges such as at Blakeney Point, Norfolk (Steers 1960).
- *Blow-outs and parabolic dunes* – generally form where dunes are unstable, possibly due to a lack of stabilising vegetation cover. There are two main ways in which they form: where natural gaps or storm-damaged cliffs in the foredune ridge are exploited by winds; and by erosion processes, for example the deflation of a poorly vegetated terrain. The movement of blow-outs and parabolic dunes is dependent upon the direction, frequency and strength of the onshore winds.
- *Transgressive dunes* – mobile dune forms which develop where sand blown inland from a beach has been retained by vegetation or where previously vegetated dunes become unstable and the numerous blow-outs merge to form an elongate dune (Bird 2000).

Dunes perform two functions in terms of defence: they provide a temporary store of sediment to allow short-term adjustment of the beach during storms; and they provide a protective barrier to the hinterland. It is also recognised that vegetation on dunes is an essential feature in maintaining stability of the dune system. Damage to that vegetation caused by beach users treading a common path is sufficient to cause extensive instability over a large area due to the creation of a vulnerable erosion route. In recognition of the foregoing pedestrian walkways and access bridges to the beach now commonly protect dune fields in areas of human activity.

Dunes systems have to be treated with extreme care particularly with respect to the introduction of structures. In the past the default response to shoreline erosion has been to construct coastal defence structures irrespective of the hinterland morphology. Consequently it has not been uncommon for sea walls to have been constructed in the front of sand dunes, thus inhibiting the natural exchange of material between the beach and the dune system. Figure 9.31(a) shows the relative effectiveness of protect-

Figure 9.31 Interaction of structures with dunes.

ing a dune system with a hard defence coupled with the undesirable consequences of terminating that protection. Figure 9.31(b) shows the folly of building a permanent structure on the top of a sand dune system, which has been all but destroyed, despite its steel sheet piling protection, in a single storm. An excellent publication covering the practical treatment of dune systems is from Scottish Natural Heritage (2000).

9.3.3 Tidal flats and marshes

Tidal flats and marshes are formed by an accumulation of fine sediments, such as sands, silts and clays, at the shoreline. They are usually formed in areas with a relatively large tidal range and a degree of shelter against direct action from ocean-generated waves. They are therefore generally found in estuaries. Tidal flats are often characterised by sandflats and/or mudflats and vegetated saltmarshes. Deposition of sediment flocs (mass of mud particles) occurs when the shear velocity of the water flow is at, or close to a minimum; therefore an important factor in determining the rate of accumulation of cohesive sediments is the duration of slack water periods. A second factor is the level of suspended sediment concentration in the tidal flows providing a further control on the rate of accumulation

Saltmarshes have developed in many of the relatively sheltered areas, associated with outer estuaries, around the UK. Pioneer saltmarshes develop when the tidal flat is high enough to result in a decrease in the frequency and duration of tidal inundations of the upper sections of the profile; vegetation that is tolerant to high salinity levels begins to colonise the surface. This results in reduced tidal velocities and increased sediment deposition. As the increased sedimentation within pioneer saltmarshes occurs, so the frequency and duration of tidal inundation decreases further, leading to the colonisation of the sediment by many varied salt-tolerant plant species. Thereafter, erosion of a salt marsh may occur due to further variations in sea level, reduced supply of fine sediments or deterioration in the health of vegetation due to chemical pollution. A typical formation is shown in Figure 9.32 that exhibits the normal mode of highly complex meandering channels of differing size and significance.

Figure 9.32 Typical natural saltmarsh.

As described in the NRA R&D Note 324 (1999) the development of saltmarsh is usually dependent on the existence of a mudflat to seaward that is capable of reducing tide and wave energy sufficiently to allow the above processes to occur. The relationship is a complex one with the potential for there to be exchanges of material between the two during periods of high wave energy. This interdependency has the result that each feature has a much better chance of survival if the other is present. However, mudflats do exist without saltmarshes, but they are prone to erosion especially at the upper levels. Likewise marshes do exist without mudflats, but are usually protected by some other means whether artificial or natural. Thus the combined marsh/mudflat landform is an efficient unit that should be considered to be inseparable by coastal managers.

In the past many of these area have been reclaimed and used for cultivation or rearing livestock. However, this has resulted in providing front-line coastal defence for the saltmarshes which themselves have previously been natural forms of primary defence. Even when there is a hard line of coastal defences to the landward boundary of the saltmarsh, this can prevent the natural readjustment of the marsh in response to changes in sea level. Saltmarshes have been subject to particular attention in the UK in recent years due to recognition of their rich ecological value as well as the realisation that they can play a key role in the coastal defences in the areas in which they exist. Modern practice has sought to reinstate the inventory of saltmarsh in the country. That reinstatement has been realised through the removal of sections of sea defence, thus allowing inundation of the previously protected area on every high tide. Tidal flats and saltmarshes are extremely efficient dissipaters of wave and tidal energy (Möller *et al.*, 1996) and as such are vitally important for reducing the risk of flooding to low-lying hinterland. Apart from their value in protecting flood defences direct attack by waves and currents, saltmarshes have other economic values such as their productivity with respect to fish and wild life, the protection of other resources inland and the amenity value including recreational uses.

9.4 Design guidance notes

The following sections represent a series of design guidance notes rather than prescriptive rules. These have been developed through decades of design experience and the contribution of Kevin Burgess (Technical Director of Coastal Engineering, Halcrow Group Ltd., UK) to these is fully acknowledged. Any design process must take in to account a multitude of parameters ranging from the context of a scheme to details of form and materials. For example, the economic form and profile of a sea wall are closely related to the materials and method of construction. The scope of this book precludes detailed discussion of each and every aspect, which would take many volumes to incorporate all of the relevant scientific research that has supported this subject area in contemporary times. Reference is made to a number of first-class publications, which can provide some of the detail required for a more rigorous treatment of the subject matter.

Factors that impact on design include local geology, the tidal range and position of the structure in relation to high and low water, wave climate both ambient and extreme, limitations on access and availability of structural components including natural durable rock. The use of concrete in the marine environment is common and can present its own problems (Allen, 1998). Other materials such as bitumen, open stone asphalt and steel can be used when appropriate. The durability of different materials is an important consideration; apart from corrosion, erosion of a marine structure caused by the abrasion and impact of mobile beach material may be the most significant factor determining the life of the structure.

For breakwaters and other barriers founded in deep water in exposed situations, the economic solution is to be found in a permeable structure designed to dissipate wave energy harmlessly and the same principles apply to the design of coast protection works for which the minimisation of reflected wave energy is vitally important to the preservation of protective beaches.

The following two sub-sections of these design guidance notes deal with wave run-up and overtopping. These topics are vitally important with respect to optimising structure geometry of coastal protection or breakwaters. In particular the economic viability of a project may be heavily dependent on the elevation of the structural elements through the determination of the quantum of run-up and overtopping coupled to the selection of tolerable limits. The importance of these topics has led to a number of significant research projects both within the European Community and elsewhere that have been carried out since the publication of the first edition of this book. The combined knowledge gained from researchers in the UK, the Netherlands and Germany has been distilled into a publication known as the EurOtop Manual (2007). This manual replaces and/or incorporates earlier guidance published by the Environment Agency (EA, UK), Technical Advisory Committee on Flood Defences (TAW, Netherlands) and Archive for Research and Technology on the North Sea and Baltic Coast (EAK, Germany). Significant new information was obtained through the EC CLASH project by collecting data from several nations, and further advances in national research projects. In doing so the manual extends or revises advice on wave run-up and overtopping predictions given in the CIRIA /CUR Rock Manual (1991), the Revetment Manual (McConnell 1998), British Standard BS6349, US Coastal Engineering Manual and ISO TC98. The EurOtop manual presents a comprehensive set of design methods and equations for wave run-up, mean overtopping and individual overtopping for a range of coastal structures including coastal dykes and embankment sea walls, armoured rubble slopes and mounds, and vertical and steep sea walls. Additionally, it introduces for the first time some elements of probabilistic design methods, including the relevant equations and statistical parameter values. To complement the EurOtop Manual a set of web-based calculation tools were also developed for ease of use. These are available on the overtopping manual web site (www.overtopping-manual.com). The EurOtop Manual also incorporates other significant works such as Besley(1999) which is extensively referred to in Section 9.4.2.

Before proceeding with discussing the latest formulations presented in the EurOtop Manual (2007) it is necessary to expand on some of the wave spectrum relationships given in Sections 3.2.2–3.2.4. This because the manual departs from the use of traditional parameters such as significant wave height and peak wave period to use definitions that can be applied in a generalised way to different types of wave spectra.

Wave height

The wave height used is the incident significant wave height at the toe of the structure, called the spectral wave height $H_{mo}=4(m_0)^{1/2}$. The usual definition of significant wave height, being the average of the highest one-third of the waves is not used on the grounds that, whilst both definitions produce almost the same value in deep water, situations in shallow water can lead to differences of 10–15 per cent. These differences arise due to the presence of the foreshore causing larger waves in the spectrum to break so that the significant wave height is reduced. It is reasonably simple to calculate a wave height distribution and associated significant wave height $H_{1/3}$ using the method of Battjes and Groenendijk (2000).

Wave period

As explained in Section 3.2.2 the common definitions of wave period are the peak period T_p, the average wave period T_m and the significant period $T_{1/3}$. The ratio T_p/T_m usually lies between 1.1 and 1.25 and T_p, and $T_{1/3}$ are almost identical. Generally the EurOtop Manual uses the spectral wave period $T_{m-1,0}$ $(=m_{-1}/m_0)$. It is pointed out that this wave period gives more weight to the longer periods in the spectrum than the average period and gives similar values of wave run-up or overtopping for the same values of spectral wave period and wave height, independently of the type of wave spectrum. Thus the wave run-up and overtopping can be evaluated for say double-peaked or 'flattened' spectra in a straightforward manner. For a single-peaked spectrum a fixed relationship of $T_p = 1.1 \, T_{m-1,0}$ is applied.

Wave steepness and breaker parameter

Wave steepness is defined as $s_o = H_{m0}/L_o$. Values of the order of 0.01 characterise a swell sea, whilst values in the range 0.04 to 0.06 relate to typical wind-driven seas. The breaker parameter, also known as the 'surf similarity parameter' or 'Iribarren number' is central to most wave run-up and overtopping formulae. It is defined as

$$\xi_{m-1,0} = \tan\alpha_{wall}/(H_{m0}/L_{m-1,0})^{1/2} \qquad (9.2)$$

where α_{wall} is the angle of the front face of the structure relative to the horizontal and $L_{m-1,0}$ is the deep-water wave length corresponding to $gT^2_{m-1,0}/2\pi$. This parameter defines the type of wave breaking whereby for $\xi_{m-1,0} > 2$–3 waves are considered to be surging (generally not breaking), for $\xi_{m-1,0} = 2$–3 collapsing breakers, for $\xi_{m-1,0} < 2$–3 $<$ 0.2 plunging breakers and for $\xi_{m-1,0} < 0.2$ spilling breakers.

Probabilistic or deterministic design

The EurOtop Manual (2007) provides formulations for both probabilistic and deterministic (or safety assessment) design. This allows either method of assessment to be performed. The deterministic prediction returns the mean measured results plus one standard deviation. The probabilistic method is based upon 50 per cent of measured results exceeding the prediction; this is generally similar in value, but not identical, to the arithmetic mean. The results for the deterministic overtopping analysis are therefore higher for the majority of design cases, unless additional factors of safety are being introduced elsewhere in the analysis. For preliminary designs the deterministic approach is recommended. However, as a design is refined there may be instances where a probabilistic approach will provide more insight into the overall performance of a structure, but one must not lose sight of the other uncertainties that exist such as the variability of the foreshore morphology and the uncertainty of the parameters used such as wave height and wave period. Only the deterministic relationships are given here in the relevant sections. The probabilistic relationships generally only differ in the numeric values of coefficients and corresponding maxima and the reader is referred to the original manual for further detailed information.

Principal types of structure

The manual is primarily concerned with three principal types of structure: sloping sea dikes and embankment sea walls; armoured rubble slopes and mounds; and vertical battered or steep walls. The primary difference between dikes or embankment sea walls and rubble mound armour structures concerns the permeability if the core and, whilst the manual treats these two defence types differently, this distinction is not strictly necessary for the normal range of common coastal applications. This will be explained further in Section 9.4.2.

9.4.1   Wave run-up

An important objective of the design of most types of coastal structure founded on a soft or erodible seabed is to maximise the destruction of wave energy by causing the wave to break on the wall. Also by a suitable selection of wall profile and degree of surface roughness the intention is to promote maximum turbulence of the swash over the surface of the wall. Where the wall is also providing protection against flooding by the sea, it is important to ensure that, in the worst combination of circumstances, the run-up of breaking waves does not cause unacceptable overtopping (see Section 9.4.2). Hunt (1959) summarises useful analytical and experimental data concerning these factors in relation to walls with inclined seaward faces of simple or composite form, with and without berms, encountered by unbroken waves. The most important conclusion was that the slope of the face of the sea wall to the horizontal to ensure breaking of the wave is given by

$$\tan \alpha_{wall} = 8/T \, (H_i / 2g)^{1/2} \tag{9.3}$$

Such a slope will result in the reflected wave being approximately 50 per cent of the incident wave height. Hence the minimum slope of the face of the wall, or at least the apron up to the point of breaking, may be determined in relation to the longest wave of critical height. Another conclusion was that the run-up, R, of a breaking wave, measured vertically above the mean surface level of the sea at the time, may be related to the incident wave height H_i by consideration of a number of factors set out by Hunt (1959) in the non-dimensional form so that

$$R/ H_i = K \tan \alpha_{wall}/8/T \, (H_i/2g)^{1/2} \tag{9.4}$$

where K is a constant for a smooth plane surface with a value of about 2.3. For a surging wave this ratio will be no greater than 3, and it has been shown theoretically that in the absence of friction

$$R/ H_i = (\pi / 2\ \alpha_{wall})^{1/2} \qquad \text{for } \pi/4 < \alpha < \pi/2 \qquad (9.5)$$

It is common practice to express the effect of surface texture of the slope's surface to a coefficient of roughness defined as the ratio of the rough to smooth run-up. A selection of values is given in Table 9.7. It is clear from this that the roughness coefficient is a function of both the permeability of the face of the wall as well as its roughness.

The EurOtop Manual (2007) introduces a roughness factor γ_f which is not necessarily the same as the roughness coefficient. Values for both coastal dikes/embankment sea walls and armoured rubble slopes are given in Table 9.7. Further details on the influence of dimensions of artificial roughness are given in the original publication and the values for armoured rubble slopes are stated to be for a structure slope of 1:1.5. It is also advocated that for significant wave heights of less than 0.75 m, grass influences the run-up process and lower roughness influence factors are recommended by TAW(1977). This is because of the relatively greater roughness that grass has on thin run-up depths. The recommended relationship is

Table 9.7 Roughness coefficients and factors for different surface textures (various sources).

Type of slope protection	Roughness coefficient, r	Roughness factor from EurOtop, γ_f
Smooth concrete or asphalt	1.0	1.0
Smooth concrete blocks with little or no drainage	1.0	1.0
Stone blocks pitched or mortared	0.95	
¼ stone set 10 cm higher	0.9	
Stepped	0.9	
Turf	0.85–0.9	1.0
Basalt		0.9
Rough concrete	0.85	
Small blocks over 1/25 surface		0.85
Small blocks over 1/9 surface		0.8
One layer of rock armour on an impermeable base.	0.8	0.60
Open stone asphalt	0.8	
Stones set in cement, ragstone etc.	0.75–0.8	
Ribs (spacing/width = 7, height/width = 5–8)		0.75
Fully grouted stone	0.6–0.8	
Partially grouted stone	0.6–0.7	
Rounded stones	0.6–0.7	
One layer of rock armour on a permeable base	0.55–0.60	0.45
Two layers of rock armour	0.5–0.6	0.55 (impermeable core) 0.40 (permeable core)
Hollow cube armour units, one layer	0.5	
Antifers		0.47
Dolos and Accropode™ armour units	0.4	0.46
Xbloc®		0.45
CORE-LOC®		0.44
Stabit armour units	0.35–0.4	
Tetrapods, two layers	0.3	0.38

$$\gamma_f = 1.15\, H_s^{0.5} \qquad H_s < 0.75 \text{ m}$$

It should also be noted that the roughness coefficients used in neural network calculations (see later sub-section) are different to those used in the empirical equations.

Historically in the Netherlands and Germany in particular, the definitive run-up parameter has been the 2 per cent run-up level. This is the level above still-water level exceeded by only 2 per cent of the waves and its origins are considered to be probably arbitrary in terms of being a sufficiently small number to determine a satisfactory design with limited overtopping. However, some formulae (such as the calculation of wave forces on crown walls – see Section 9.4.7) require the use of other percentage run-up values in their formulations.

For both coastal dikes/embankment sea walls and armoured rubble slopes/mounds the 2 per cent wave run-up level for deterministic design is similar and given as

$$R_{u2\%}/H_{m0} = 1.75\, \gamma_b\, \gamma_f\, \gamma_\beta\, \xi_{m-1,0} \quad \text{with a maximum of}$$

$$R_{u2\%}/H_{m0} = 1.00\, \gamma_f\, \gamma_\beta [4.3 - 1.6/\xi_{m-1,0}^{0.5}] \tag{9.6}$$

where γ_b is the influence factor for a berm (see Equation 9.22), γ_f is the influence factor for roughness elements on a slope (see Table 9.7) and γ_β is the influence factor for oblique wave attack given as $\gamma_\beta = 1 - 0.0022\, |\beta|$ in the range $0° \leq |\beta| \leq 80°$ and $\gamma_\beta = 0.824$ for $|\beta| \geq 80°$. For armoured rubble slopes/mounds with surging waves the influence of roughness reduces so that the roughness influence factor increases linearly to 1.0 from $\xi_{m-1,0} = 1.8$ to $\xi_{m-1,0} = 10$ and is unity thereafter. For a permeable core a maximum is reached for $R_{u2\%}/H_{m0} = 2.11$

For composite slopes Saville (1958) suggested that a reasonable estimate of run-up could be related to an equivalent plane slope that intersects the actual slope at the position of the breaker point and the extreme run-up. This requires an iterative solution to resolve and will under-estimate the run-up for concave slopes. The introduction of a berm into a slope can provide a very effective means of reducing run-up provided the width of the berm represents a significant part, say 20 per cent, of the wavelength. This is typically of the order of 10 m for shallow-water coastal defence structures. A berm is also generally most effective when positioned at or above the still-water level. In a high tidal range environment the definitive level will often be taken as mean high water springs or that determined from a joint wave and water level probability analysis. (see Chapter 7). Further details on the impact of a berm on wave overtopping according to the EurOtop Manual (2007) are given in the following Section where it can be seen that the recommended procedure does not differ significantly from Saville's original suggestion.

9.4.2 Wave overtopping and crest elevation

For the majority of coastal structures, quantification of overtopping, that is, the discharge of water over the crest, dictates the crest elevation required. In this sense it is one of the parameters that has the greatest potential to have an impact on the cost of a structure. Curiously it is not something that has attracted a significant amount of research until relatively recently. However, modern design practice is based on using the rate of overtopping discharge as a criteria rather than wave run-up, which does not quantify discharge over a structure. (The run-up approach was historically used as a consequence of lack of sufficient design data.)

Conventional design practices require a structure to defend against those conditions that would damage the structure itself as well as the nature of the land use or development behind it. The latter will often make it necessary to provide higher standards for safety reasons or to prevent damage to property than would be required for protection of the structure alone. Such features could be other structures, such as buildings, roadways or working areas as in the case of a port structure. It is necessary to consider the safety of people or vehicles behind the structure. However, designing to meet these criteria can lead to the development of extremely large structures. Consequently, it is only usual to design for these conditions in exceptional circumstances, for example where a highway lies directly behind the sea wall, or on a downtime principle, allowing the condition to only be exceeded a certain number of times per year. In most cases it will usually be much cheaper to have a warning/restricted access commitment than to build the larger defence structure. It is also worth noting that tolerable discharges, although appearing small, can result in considerable localised flooding and depth/duration of flooding could be the controlling factor in determining the appropriate level of overtopping discharge. Tolerable discharges can be calculated through knowledge of the drainage capacity, or determination of the size of the flood area and limiting acceptable depth, converting this into a total acceptable volume per linear metre of defence. In the latter case, actual discharge would then also be calculated as a total volume, rather than a mean rate, calculating incremental volumes with water level variation across the peak of the tide. These are the primary reasons for limiting overtopping discharge and it is therefore important to establish the design criteria that relate to all facets of the structure and its intended performance.

Overtopping discharges are usually calculated and quoted as mean discharges (litres/sec/m run) and can appear to be relatively small values. However, the actual discharge occurs as a random series of large single impact events (i.e. every wave crest) with a frequency equal to the wave period. It should also be realised that overtopping calculation methods have limitations to their accuracy and the physical model data from which the methods are derived generally exhibit considerable scatter. It is generally accepted that even the most reliable methods cannot provide absolute discharges, and they can only be assumed to produce overtopping rates that are accurate to within one order of magnitude. Likewise the tolerable discharges defined in various publications should not be taken as absolute values. They represent an order of magnitude for which damage or unsafe conditions may exist.

Calculation of overtopping rates

Some of the earliest information on calculating overtopping rates was undertaken in the 1950s, the results of which are presented in the Shore Protection Manual (1984). This was superseded by work carried out by a number of investigators, most notably Owen (1980) who established the formulation framework that continues to be used today. A more recent definitive and comprehensive work, which addresses overtopping for different structural forms, has been carried out and published by HR Wallingford (Besley 1999). It also reviews work undertaken elsewhere. This has since been extended and revised in the EurOtop manual (2007) which is meant to replace the 1999 manual. Some of the principal elements from the HR and EurOtop manuals are now presented.

Following Besley (1999), the mean overtopping discharge for a plain rough armoured slope as defined in Figure 9.33 may be calculated from the following equations:

$$R_* = R_c/(T_m(gH_s)^{0.5}) \qquad (9.7)$$

Where R_c is the freeboard defined as the height of the crest above the still-water level, H_s is the significant wave height, g is acceleration due to gravity and T_m the mean period of the wave at the toe of the structure. Equation (9.7) is valid between the limits $0.05 < R_* < 0.30$. A second parameter is defined as

$$Q_* = A \exp(-BR_*/r) \qquad (9.8)$$

Where A,B are empirical coefficients dependent on the slope of the structure (see Table 9.8) and r is the roughness coefficient as given in Table 9.7. This equation is valid in the range $0.05 < R_* < 0.30$. The mean overtopping discharge rate per metre length of structure in m³/s/m is

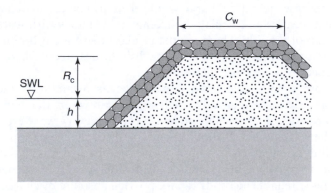

Figure 9.33 Definition sketch for wave overtopping rough plane slope.

Table 9.8 Empirical coefficients – simply sloping sea walls (after Besley 1999).

Sea wall slope	A	B
1:1	7.94×10^{-3}	20.1
1:1.5	8.84×10^{-3}	19.9
1:2	9.39×10^{-3}	21.6
1:2.5	1.03×10^{-2}	24.5
1:3	1.09×10^{-2}	28.7
1:3.5	1.12×10^{-2}	34.1
1.4	1.16×10^{-2}	41.0
1:4.5	1.20×10^{-2}	47.7
1:5	1.31×10^{-2}	55.6

$$Q_m = Q_* T_m g H_s \tag{9.9}$$

If the structure has a permeable crest berm a reduction factor C_r may be applied and this is

$$C_r = 3.06 \exp(-1.5\, C_w / H_s) \tag{9.10}$$

where C_w is the crest width as indicated in Figure 9.33. If C_w / H_s is less than 0.75 it can be assumed that there is no reduction and $C_r = 1$

For waves approaching at an angle to the slope a further reduction factor may be applied based on investigations by Banyard and Herbert (1995). For a simple slope this is

$$O_r = 1 - 0.000152\, \beta^2 \tag{9.11}$$

where β is the angle between the normal to the slope and the direction of wave propagation. Also see recommendations of the EurOtop project in further sub-sections.

As indicated in the previous section the introduction of a berm into a slope can be a very effective means of reducing the crest level to a lower elevation than would be required for a simple plane slope for the same overtopping discharge. This can often be important for aesthetic reasons or facilities where a line of sight over the structure is a key feature. Besley (1999) proposes that for a slope with a berm that is below the still-water level, Equations (9.6)–(9.9) can be used together with modified empirical coefficients given in Table 9.9 together with the slope of the upper section of the structure as indicated in Figure 9.34. For berms above still-water level it is suggested that an equivalent slope based on the plane that joins the intersection of the lower slope with the still-water level and the top of the seaward slope of the upper section as shown in Figure 9.34 should be used. Then Equations (9.6)–(9.9) may be used in conjunction with the empirical coefficients given in Table 9.8 that most closely fit the equivalent slope. Also see sub-section on 'Complex Slopes'.

Another method of reducing the height of a structure is to include a wave return wall. Coastal defence structures can therefore sometimes incorporate a wave return wall either directly at the top of the slope or at some distance retired from the crest

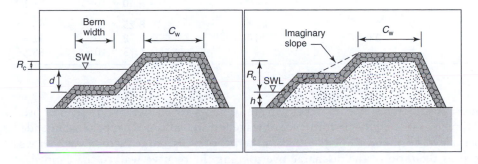

Figure 9.34 Definition sketch for bermed sea wall.

Table 9.9 Empirical coefficients – bermed sea walls – berm at or below SWL (after Besley, 1998)

Sea wall slope	Berm elevation m	Berm width m	A	B
1:1			6.40×10^{-3}	19.50
1:2	−4.0	10	9.11×10^{-3}	21.50
1:4			1.45×10^{-2}	41.10
1:1			3.40×10^{-3}	16.52
1:2	−2.0	5	9.80×10^{-3}	23.98
1:4			1.59×10^{-2}	46.63
1:1			1.63×10^{-3}	14.85
1:2	−2.0	10	2.14×10^{-3}	18.03
1:4			3.93×10^{-3}	41.92
1:1			8.80×10^{-4}	14.76
1:2	−2.0	20	2.00×10^{-3}	24.81
1:4			8.50×10^{-3}	50.40
1:1			3.80×10^{-4}	22.65
1:2	−2.0	40	5.00×10^{-4}	25.93
1:4			4.70×10^{-3}	51.23
1:1			2.40×10^{-4}	25.90
1:2	−2.0	80	3.80×10^{-4}	25.76
1:4			8.80×10^{-4}	58.24
1:1			1.55×10^{-2}	32.68
1:2	−1.0	5	1.90×10^{-2}	37.27
1:4			5.00×10^{-2}	70.32
1:1			9.25×10^{-3}	38.90
1:2	−1.0	10	3.39×10^{-2}	53.30
1:4			3.03×10^{-2}	79.60
1:1			7.50×10^{-3}	45.61
1:2	−1.0	20	3.40×10^{-3}	49.97
1:4			3.90×10^{-3}	61.57
1:1			1.20×10^{-3}	49.30
1:2	−1.0	40	2.35×10^{-3}	56.18
1:4			1.45×10^{-4}	63.43
1:1			4.10×10^{-5}	51.41
1:2	−1.0	80	6.60×10^{-5}	66.54
1:4			5.40×10^{-5}	71.59
1:1			8.25×10^{-3}	40.94
1:2	0.0	10	1.78×10^{-2}	52.80
1:4			1.13×10^{-2}	68.66

of the seaward slope. A comprehensive study was carried out by Owen and Steel (1991) who evaluated the performance of wave return walls of the type recommended by Berkeley-Thorn and Roberts (1981), in terms of a discharge factor D_f which was defined as the ratio of the discharge overtopping the recurve wall to the equivalent discharge in the absence of the wall. Referring to the reference Figure 9.35, Equa-

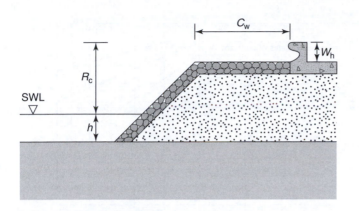

Figure 9.35 Definition sketch for wave return wall.

tions (9.6)–(9.8) apply in order to determine Q_m which is the discharge per metre run (m³/s/m) at the base of the return wall and is the same as that at the crest of the slope for a smooth impermeable crest. A and B are the same empirical coefficients given in Table 9.8. A dimensionless wall height is defined as

$$W_* = W_h / R_c \tag{9.12}$$

where W_h is the height of the wave return wall and R_c is the freeboard to the top of the wall as previously defined. From hereon, the procedure for impermeable and impermeable structures differs. For an impermeable structure, given the dimensionless wall height, the seaward slope of the sea wall and the set-back distance of the wave return wall, Table 9.10 provides values of an adjustment factor A_f which in turn is used to define an 'adjusted slope freeboard' given by:

$$X_* = A_f R_* \tag{9.13}$$

Figure 9.36(a) is then used to determine a discharge factor (D_f) for the given conditions so that the mean discharge over the wall is:

$$Q_w = Q_m D_f \tag{9.14}$$

Further analysis of wave approach angle has identified that the large increases only occur for very small discharge factors and decreases only occur for discharge factors greater than about 0.3. The recommended approach is to use the 'worst case' combination of D_f and O_r. This translates into using D_f only when the $D_f \geq 0.3$, and using ($D_f \times O_r$) when $D_f < 0.3$.

For roughened slopes or those incorporating a berm, Besley (1999) recommends the determination of a smooth slope that gives the same overtopping discharge at the top of the slope, for the same wave conditions. That 'equivalent slope' is then used to obtain the adjustment factor from Table 9.10. However, this may well produce slopes

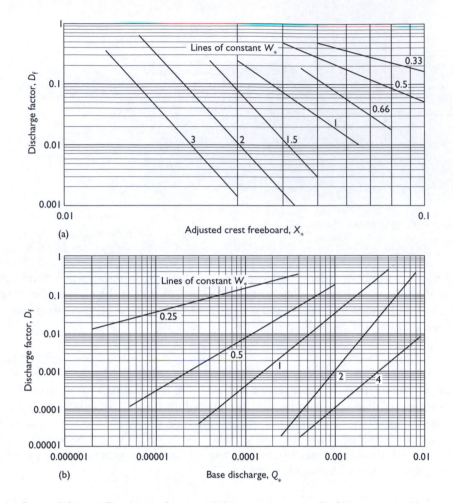

Figure 9.36 (a) Discharge factors with wave return walls for impermeable slopes and (b) permeable slopes (Besley1999).

Table 9.10(a) Adjustment factors – wave return walls on impermeable sea walls (after Besley 1999), $W^* = W_h/R_c \geq 0.6$

Sea wall slope	Crest berm width (C_w) m	A_f
1:2	0	1.00
1:2	4	1.07
1:2	8	1.10
1:4	0	1.27
1:4	4	1.22
1:4	8	1.33

Table 9.10(b) Adjustment factors – wave return walls on permeable sea walls (after Besley 1999), $W^* = W_h/R_c < 0.6$

Sea wall slope	Crest berm width (C_w) m	A_f
1:2	0	1.00
1:2	4	1.34
1:2	8	1.38
1:4	0	1.27
1:4	4	1.53
1:4	8	1.67

that lie outside the range of available data. The alternative is to calculate the overtopping using the method of Van der Meer *et al.* (1998) described later.

For wave walls on permeable slopes Besley (1999) re-analysed the data from Bradbury and Allsop (1998) to produce Figure 9.36(b). The base discharge is calculated in the same way as that described for permeable crests in Equations (9.8) and (9.9). Given W^* as defined in Equation (9.12) the discharge factor is obtained directly from Figure 9.36(b) so that the mean overtopping discharge becomes:

$$Q_w = Q_m \, C_r \, D_f \qquad (9.15)$$

For plain vertical walls Besley (1999) summarises the work of Allsop *et al.* (1996). A parameter h_* is defined as

$$h_* = (h / H_s)(2 \, \pi h / (g \, T_m^2)) \qquad (9.16)$$

where h is the water depth at the toe of the structure for which reflecting waves dominate when $h_* > 0.3$ and impacting waves when $h_* < 0.3$. For the former the mean overtopping discharge per metre run of wall is given as:

$$Q_m = 0.05 \exp(-2.78 \, R_c / H_s) \, (g \, H_s^3)^{0.5} \qquad (9.17)$$

where R_c is the total freeboard to the crest of the structure and is valid in the range $0.03 < R_c / H_s < 3.2$. For angled wave attack the reduction factor is:

$$O_r = 1 - 0.006 \, \beta \qquad \text{for } 0° < \beta < 45°$$

$$O_r = 0.72 \qquad \text{for } \beta > 45° \qquad (9.18)$$

For impacting waves:

$$Q_m = 0.000137 \, ((R_c / H_s) \, h_*)^{-3.24}) \, h_*^2 \, (g \, h^3)^{0.5} \qquad (9.19)$$

Which is valid in the range $0.05 < ((R_c / H_s) \, h_* < 1.00$. There is no equivalent expression for different angles wave attack. Besley (1999) also provides empirical expressions for composite vertical walls fronted by a mound that may be submerged or emergent.

EurOtop methods

Much of the foregoing has been incorporated into the EurOtop project and formulae adjusted accordingly. The EurOtop deterministic formula for mean overtopping discharge on simple slopes for dykes and sea walls (for $\xi_{m-1,0} < 5$)is given by.

$$\frac{Q_m}{\sqrt{gH_{mo}^3}} = \frac{0.067}{\sqrt{\tan\alpha}}\gamma_b\xi_{m-1,0}\cdot\exp(-4.3\frac{R_C}{\xi_{m-1,0}H_{mo}\,\gamma_b\gamma_f\gamma_\beta\gamma_v}) \qquad (9.20)$$

where $\xi_{m-1,0}$ has been defined in the introduction to Section 9.4. This expression corresponds to breaking waves for which $\gamma_b\,\xi_{m-1,0} < \cong 2$ .For non-breaking waves where $\gamma_b\,\xi_{m-1,0} > \cong 2$, there is a maximum value of

$$\frac{Q_m}{\sqrt{gH_{m0}^3}} = 0.2\exp\left\{-2.3\frac{R_c}{H_{m0}\gamma_f\gamma_\beta}\right\} \qquad (9.21)$$

where as before γ_b is the influence factor for a berm (see Equation 9.22), γ_f is the influence factor for roughness elements on a slope (see Table 9.7) and γ_β is the influence factor for oblique wave attack and γ_v is the wave wall influence factor. The influence factor for oblique wave attack is given as

$\gamma_\beta = 1-0.0033\,|\beta|$ for dikes and sea walls in the range $0° \le |\beta| \le 80°$

and

$\gamma_\beta = 0.736$ for $|\beta| \ge 80°$

Thus, for breaking waves the structure influences the incident wave before it reaches the crest and reduces the resulting overtopping discharge. However, the structure would need to have very flat slopes or the waves have very short periods for the wave breaking on the structure condition to occur.

For armoured slopes and mounds overtopping discharge is given by the same equation as (9.21), that is, the equation relating to non-breaking waves. This arises because in the majority of cases rubble mound structures have steeper side slopes and are located in deeper water so that it is highly unlikely that waves would break on the slope of the structure. Thus, the difference on overtopping discharge for the two classes of structure come about due to the different properties of the fabric of the structure, noting that influence factors for roughness are generally rather lower for rock or concrete armour cover layers. In addition for armoured slopes and mounds, the reduction in overtopping due to wave obliquity is much faster so that the influence factor becomes

$\gamma_\beta = 1-0.0063\,|\beta|$ in the range $0° \le |\beta| \le 80°$

and

$\gamma_\beta = 0.496$ for $|\beta| \ge 80°$

The coefficients, 4.3 and 2.3, in the EurOtop overtopping formulae given in Equations (9.20 and 9.21) are one standard deviation below the mean value, thus giving a conservative estimate of the overtopping discharge. For probabilistic design, these coefficients are changed to 4.75 and 2.6 respectively, which are the mean values (μ), with standard deviations σ of 0.5 and 0.35. Thus for reliability estimates, 95.5 per cent of all possible values can be found by setting the coefficients to $\mu \pm 2\sigma$. It should also be noted that the EurOtop overtopping formula maximum value does not contain a slope parameter. According to the manual, overtopping of armoured rubble slopes steeper than about 1 in 2 is very little influenced by the slope and thus this maximum value applies for all such structures.

Further formulations are given for conditions where $\xi_{m-1,0} > 7$. This case represents very heavy wave breaking on a shallow foreshore when the shape of the wave spectrum can change (flatten) and long waves can be present. Under these shallow foreshore conditions Equation (9.20) can significantly underestimate the wave overtopping discharge and an alternative should be used. Interpolation should be performed for conditions lying in the range $5 < \xi_{m-1,0} < 7$.

Complex slopes

In the EurOtop manual, the procedure for determining overtopping on composite slopes is to first establish a so-called 'characteristic slope'. This is defined as the average slope from the point of wave breaking to the maximum wave run-up height. This has to be determined using an iterative procedure. Thereafter the EurOtop recommended formulae for dykes and sea walls is applied.

Referring to the definition diagram in Figure 9.37 the breaking limit is chosen as $1.5H_{mo}$ below the still-water line and the limit of wave run-up estimated as $1.5H_{mo}$ above the still-water line. The initial estimate of the characteristic slope is then given by:

$$\tan\alpha = \frac{3H_{mo}}{L_{slope} - B}$$

As a second estimate, the wave run-up height, determined using the initial characteristic slope, is then used to find a second estimate:

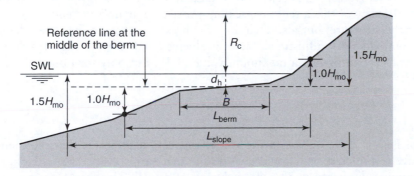

Figure 9.37 Definition sketch for complex slopes.

$$\tan\alpha = \frac{1.5 H_{m0} + R_{u2\%}}{L_{slope} - B}$$

where the deterministic value of run-up is given by Equation (9.5).

The berm width is defined as the flat part of the profile that has a slope of less than 1:15. The effectiveness of a berm is dependent on its level in relation to the still-water level at the mid-point of the berm which defines the depth of water d_h as defined in Figure 9.37. The berm reduction coefficient is given as:

$$\gamma_b = 1 - r_b (1 - r_{hb}) \qquad \text{for } 0.6 \leq \gamma_b \leq 1.0 \tag{9.22}$$

where

$r_b = B/L_{berm}$
$r_{hb} = 0.5 - 0.5\cos(\pi d_h / R_{u2\%})$ for a berm above the still-water line
$r_{hb} = 0.5 - 0.5\cos(\pi d_h / 2H_{mo})$ for a berm below the still-water level.

The berm is most effective when it is at still-water level and an optimum berm width is achieved if the reduction factor reaches a value of 0.6 so that for a berm at still-water level the optimum width becomes:

$$B = 0.4 L_{berm} \tag{9.23}$$

For the reduction factor for the presence of a small vertical wall or very steep slope at the top of the wall an expression is provided in the EurOtop manual, but it is strictly limited to a restricted range of geometry and does not have general applicability. Other facets of wave overtopping of coastal dikes and embankments addressed in the EurOtop manual include individual overtopping volumes, flow depths and flow velocities on the seaward slope, the crest and the landward slope. Similarly for armoured rubble slopes and mounds additional topics include wave run-up height for shingle beaches, percentage of overtopping waves, individual overtopping volumes, overtopping velocities and scale and model uncertainties. Special mention should be made of the last of these. Results from the CLASH project (see further sub-section) suggested significant differences in wave overtopping between field data and equivalent physical hydraulic model results. In view of this a tentative scaling procedure, mainly dependent on the roughness of the structure, but also including the presence of wind, has been developed. For smooth dikes the adjustment factor is always 1.0. For very rough 1:4 slopes with wind the factor can reach a maximum value of 30, but only for upscaled rates less than $10^{-5} \mathrm{m^3/s/m}$.

Wave walls

The work of Besley (1999) given in the foregoing section indicates that the inclusion of a wave return/crest wall could give results that saw considerable reductions in the overtopping discharge due to the inclusion of a wall. The recent second edition of *The Rock Manual* (CIRIA/CUR/CETMEF 2007) also makes reference to this method. However, EurOtop, which is intended to supersede earlier guidance, suggests a rather less significant effect on overtopping volumes through the addition of a wave wall.

For coastal dikes and embankment sea walls the method proposed for determining the wave wall coefficient (γ_v) considers the wall slope between vertical (maximum reduction) and 45° (zero reduction) and linearly interpolated between the two limits so that $\gamma_v = 1.35 - 0.0078\,\alpha_{wall}$. It is acknowledged that data is limited and there is a list of limiting geometry given in the manual. Outside this range it might be possible to use the Neural Network approach (see further sub-section).

For armoured rubble slopes and mounds the method assumes that a wave wall crest has the same impact as a revetted structure continuing to the same height. Therefore the crest elevation used within the overtopping discharge, Equation (9.21), is taken as the level of the top of the wave wall.

Vertical and steep sea walls

The EurOtop manual contains considerably more information concerning vertical and steep sea walls than previous guides, based on a number of research programmes undertaken between the publication of the Besley manual in 1999 and the publication of the EurOtop manual in 2007. The reader is therefore recommended to read the EurOtop manual to gain a fuller understanding of current knowledge. Here, attention is restricted to noting the revisions to the Besley manual equations already described.

Equation (9.16) is slightly revised to incorporate the new wave parameters which differentiates between non-impulsive (pulsating) conditions when $h_* > 3$ and impulsive conditions $h_* < 0.2$. It becomes:

$$h_* = 1.35\frac{h}{H_{m0}}\frac{2\pi h}{gT_{m-1,0}^2} \tag{9.24}$$

Equation (9.17), for deterministic design with non-impulsive conditions, becomes

$$Q_m = 0.4\exp\left(-1.8\frac{R_c}{H_{m0}}\right)\sqrt{gH_{m0}^3} \quad \text{valid in the range } 0.1 < R_c/H_{m0} < 3.5 \tag{9.25}$$

Equation (9.19), for deterministic design for impulsive conditions, becomes

$$Q_m = 0.00028(h_*\frac{R_c}{H_{m0}})^{-3.1}h_*^2\sqrt{gh}$$

valid in the range $0.03 < h_* R_c/H_{m0} < 1.0$ \hfill (9.26)

Near vertical walls with 10:1 and 5:1 batters are found commonly for older UK sea walls and breakwaters. Mean overtopping discharges for walls under impulsive conditions are slightly in excess of those for a vertical wall given in Equation (9.26) over a wide range of dimensionless freeboards. Suggested multiplying factors are:

10:1 battered wall
5:1 battered wall

$$Q_{m\ 10:1\ batter} = 1.3\ Q_{m\ vertical}$$
$$Q_{m\ 5:1\ batter} = 1.9\ Q_{m\ vertical}$$

For oblique waves and non-impulsive conditions the coefficient 1.8 in Equation (9.25) is replaced by $1.8/\gamma_\beta$ where

$$\gamma_\beta = 1 - 0.0062\ |\beta| \qquad \text{for } 0° \le |\beta| \le 45°$$

and

$$\gamma_\beta = 0.72 \qquad \text{for } |\beta| \ge 45°$$

For oblique waves combined with impulsive conditions the picture is significantly more complex and the reader is referred to the EurOtop Manual for further details.

Other facets of wave overtopping of vertical or near vertical walls that are dealt with in the EurOtop Manual and include broken wave conditions with a submerged toe, broken wave conditions with an emergent toe, composite vertical walls and vertical walls with a bullnose wave return wall or parapet, effects of wind, scale and model effect corrections, percentage of overtopping waves, individual overtopping volumes, overtopping velocities, spatial extent of overtopping and downfall pressures due to overtopping discharge. Of special mention is the effect of wind, which intuitively might be considered to be a major influence on overtopping of vertically projecting water jets. However, it was found that, whilst a scaling factor of up to four times the model scale measurements for upscaled values less than $10^{-5} m^3/s/m$. It is consequently stated that, in many practical cases, the influence of wind can be disregarded.

Neural Network and the CLASH database

The CLASH database is an EXCEL document compiled in 2005 and contains all the measurements and results collected during the production and development of the EurOtop Manual (2007). The database lists 31 hydraulic and structural parameters for each test. The database is of limited use in isolation, but forms the basis for the Neural Network analysis.

The Neural Network (NN) is an intelligent database tool that considers a list of 15 hydraulic and structural input parameters, and through interrogation of the CLASH database returns an overtopping result based upon the 'best fitting' empirical measurements. The hydraulic parameters are wave height, wave period, wave angle of approach and water depth just in front of the structure. The structural parameters describe almost every possible structural configuration by a toe (2 parameters), two structure slopes (including vertical and wave return walls), a berm (2 parameters) and a crest configuration (3 parameters). The tenth structural parameter is the roughness factor for the structure and describes the 'average roughness for the whole structure'. Guidance is given, but it is acknowledged that estimation is not easy if the structure has different roughness on various parts of the structure. Common sense dictates that only that part of the structure over which waves run up should be used for this purpose. It should also be noted that the roughness coefficients used for the NN (Table 1 of downloadable instructions) are not the same as those used in the foregoing empirical equations.

The NN is potentially an extremely powerful tool if the database on which it is based is sufficiently large and covers a wide range of structure types. It also allows conditions to be varied so that a design can be optimised with respect to limiting overtopping discharges. On the other hand it is something of a 'black box' approach and does not assist the user to understand how the output was determined and hence assess the validity of the results. Also the CLASH network was only developed using data sets with recorded overtopping data (tests with 'no overtopping' were not considered) and therefore the network will always return a quantitative prediction of overtopping, even in the range where no overtopping would be expected. However, the NN is proposed as a useful tool where there is a lack of appropriate empirical formulation due to unusual overtopping scenarios, but it must be appreciated that success is dependent on there being sufficient data for the scenario being investigated. Alternatives to this include bespoke physical modelling of the structure or computational modelling of the wave propagation in the surf zone using the non-linear modelling techniques mentioned in Section 3.9. The NN outputs a mean overtopping discharge for quantiles 2.5%, 5%, 25%, 50%, 75%, 95% and 97.5%. Comparison with empirical methods should be based on the mean result. The CLASH database is freely available for download from www.clash.ugent.be/results/Database_20050101.xls and the Neural Network for on-line use or download from www.deltares.nl/en/software/630304/overtopping-neural-network.

Designing for overtopping

The generally accepted values for limiting values of mean overtopping rates have been completely reviewed and values recommended in the EurOtop Manual are considered to supersede all earlier guidance on tolerable discharges. These are summarised in Figure 9.38.

The following comments relate to the application of Figure 9.38:

- The definition of 'protected' and 'unprotected' is derived from reference to a concrete revetment/pavement, the later referring to compacted soil, grass or clay.
- The Dutch have different criteria that can be adopted which are more stringent (Van der Meer *et al.*1998). However, it is not suggested these should be universally applied without more detailed analysis of acceptable risks at any location.
- A common misconception is that overtopping discharges reduce by an order of magnitude for every 10 m behind the crest or wave wall of a structure. This is a misinterpretation of a statement by Owen (1980). He stated that the *tolerable limits for damage* (and therefore safety) could be increased by a factor of 10, at a distance 10 m behind the crest or wave wall. This is important with respect to flooding aspects as clearly the volume of water overtopping the crest or wave wall is a function of the wave conditions and the geometry of the structure.
- In terms of safety (and *only* this), Besley proposes the calculation of maximum or peak discharges. This is logical and it is recommended that this approach is adopted. This concludes that all structures become dangerous for pedestrians when the largest overtopping event exceeds $0.04\ \mathrm{m^3/m}$ and that all structures become dangerous for vehicles driven at any speed when the largest overtopping event exceeds $0.06\ \mathrm{m^3/m}$.

Hazard type and reason	Mean discharge (l/s/m)
For pedestrians	
Trained staff, well shod and protected, expecting to get wet, overtopping flows at lower levels only, no falling jet, low danger of fall from walkway.	1–10
Aware pedestrian, clear view of the sea, not easily upset or frightened, able to tolerate getting wet, wider walkway.	0.1
For vehicles	
Driving at low speed, overtopping by pulsating flows at low flow depths, no falling jets, vehicle not immersed.	10–50
Driving at moderate or high speed, impulsive overtopping giving falling or high velocity jets 0.01–0.05	0.0–0.05
For property behind the defence	
Significant damage or sinking of larger yachts.	50
Sinking small boats set 5–10 m from wall.	10
Damage to larger yachts.	
Building structure elements.	1
Damage to equipment set back 5–10m.	0.4
For damage to defence crest or rear slope	
Embankment sea walls/sea dikes	
No damage if crest and rear slope are well protected.	50–200
No damage to crest and rear face of grass covered embankment of clay.	1–10
No damage to crest and rear face of embankment if not protected.	
Promenade or revetment sea walls	0.1
Damage to paved or armoured promenade behind sea wall.	
Damage to grassed or lightly protected promenade	200
or reclamation cover.	50

Figure 9.38 Overtopping limits recommended in the EurOtop manual (2007).

- Breakwaters are not usually designed for the above tolerable overtopping discharges unless they have facilities located on or directly behind them, or require frequent access, in which cases the above considerations apply. This is generally because the size of armour on the crest and rear slope is much larger than that considered as 'protection' by the critical limits specified above.

Whilst work has been conducted to investigate tolerable overtopping discharges, little has been carried out to establish the sizing of protective cover layers to avoid overtopping damage. Guidance is given by Pilarczyk (1990) and reproduced in CUR/CIRIA (1990) (see page 272 box 54), although the original reference provides more background. This provides a method for calculating both the size of rock required and the width of protection, although the former requires calculation of run-up levels and the latter includes a factor related to the importance of the structure without specific guidance on what this value should be. Consequently the methods require some degree of interpretation.

A minimum practical width of protection of three primary armour stone widths, i.e. $3D_{n50}$ (see Equation 9.26), is suggested, although as a conservative rule of thumb, Pilarcyck also suggests that the crest and lee slope may be protected over a width equal to the projected extent of run-up beyond the crest of the structure.

Earlier work by Knauss in 1979 related protection stone size directly to maximum permissible overtopping discharge rate, Q_{max} so that

$$Q_{max} = 0.625 \ g^{-1} \ ((\rho_r / \rho - 1) \ D_{n50})^{1.5} \ (1.9 + 0.6 \ \phi_p - 3 \sin \alpha_i) \tag{9.27}$$

where α_i is the back slope of the structure and ϕ_p is a stone arrangement packing factor and which may vary between 0.6 for natural dumped rock fill, through 1.1 for manually placed rock fill and 1.25 for manually placed blocks. This would appear to offer a useful comparison with the Pilarczyk (1990) approach. Note that this should be based upon momentary discharge (i.e. per wave), not time-averaged values.

In order to avoid ambiguity, the general descriptions used in setting the tolerable discharges need to be broadly applied so that there are distinctions made, for example, between turfed/compacted gravel and formal protection such as armour stone or concrete blocks.

9.4.3   Armour slope stability

Armour slopes, whether on the seaward face of a coastal defence structure or a breakwater, take the same general generic form. Most commonly this consists of a primary armour of rock, concrete blocks or mass concrete units overlaying one or more filter layers that in turn overlay a core. The primary armour resists wave forces by mass (gravity) and interlock, to varying degrees depending upon the characteristics of the rock or the armour unit. Common practice is to extend primary armouring to a depth of at least 2 times the significant wave height (H_s) below the lowest water level. Numerous variations can be incorporated into a design depending on the need or otherwise for a crown wall and roadway, the quality of the available materials and the geotechnical properties of the foundation. Figure 9.39 adapted from BS6349: Part 7 (1991) shows many of the potential components of a breakwater together with the possible causes of failure due to wave action. This design standard also shows a number of different examples of breakwater configuration. There are many other types of designs using various types of materials examples of which are shown in Section 9.2.6 and discussed later.

The ultimate choice between using rock or manufactured armour units depends upon a number of factors. A key one is often the availability of rock of sufficient mass to withstand extreme wave conditions at a particular location. Another may be the technical resources that are available for construction. Some concrete units, for example, require placement to precise pattern arrangements, whilst others can be placed randomly. Maximum lifting capacity of plant may be another restriction that influences choice.

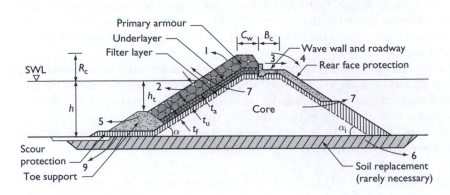

Figure 9.39 Causes of failure due to wave action and some dimensional definitions (adapted from BS6349: Part 7). Key: I Loss or damage to armour units 6 Foundation failure 2 Movement of armour 7 Loss of core material 3 Cap movement 8 Slumping due to excess pore pressure 4 Overtopping causing lee scour 9 Sea bed scour/liquefaction 5 Toe erosion

Rock armour slopes

Various methods for the prediction of the size of armour units designed for wave attack, particularly rock, have been proposed in the past few decades. The decision over which formulae to use has been the subject of much debate but most practitioners are now generally agreed that the Van der Meer (1988a) method is most appropriate. This is based upon an extensive series of model tests conducted at Delft Hydraulics, which included a wide range of core/underlayer permeabilities and wave conditions.

For many years the formula of Hudson (1959) was used fairly universally. This may be expressed as the required median mass of armour rock or concrete armour unit

$$M_{50} = \rho_r H^3 / (K_D \Delta^3 \cot \alpha_{\text{wall}}) \qquad (9.28)$$

where K_D represents a non-dimensional factor, $\Delta = (\rho_r / \rho - 1)$ and is the relative buoyant density of the armour stone and ρ_r and ρ are the density of rock and water respectively, H is the chosen design wave height and α_{wall} the angle of the slope to the horizontal. (Note that the notation W_{50} is also used to describe the median mass.) The corresponding nominal diameter of rock is determined as

$$D_{n50} = (M_{50} / \rho_r)^{1/3} \qquad (9.29)$$

Values of K_D have been obtained from model tests for different types of armour stone and various concrete armour units. Whilst the values obtained for any particular form of rock armouring vary to a certain degree, depending on the wave steepness and breaking characteristics, the mass of the more efficient concrete units will be a relatively small fraction of that of natural stone needed to withstand comparable conditions. This is illustrated by the recommended values of K_D given in Table 9.12.

The Hudson equation has a number of limitations, including:

- There are potential scale effects due to the small scales at which most of the tests were conducted.
- The original values of K_D were based on regular waves only, whereas irregular wave conditions are essential.
- The formula takes no account of wave period or storm duration (and thus the amount of wave energy).
- There is no description of the amount of damage sustained although it is generally accepted that the original formula represents up to 5 per cent damage. However, later model tests enabled values of K_D to be related to a given damage level.
- The formula applies to non-overtopped and permeable-core structures only.
- One of the issues of contention in the use of the Hudson equation for structure design centres on the advice given in the *Shore Protection Manual* (1984) with respect to use of the value $H_{1/10}$ for the design wave. This presented a considerable increase over the earlier 1973 edition of the *Shore Protection Manual* and all previous work where H_s had been used. This alone has the effect of increasing armour size by a factor of 2 and has subsequently been reproduced by BS6349 (1991), but without any additional supporting data. However, it is now accepted that for armour stone values for K_D derived using $H_{1/10}$ are $K_D = 1.0$ for structures with an impermeable core, $K_D = 2.0$ for permeable core structures and breaking waves and $K_D = 4.0$ for permeable core structures with non-breaking waves.

Despite the foregoing limitations the Hudson formula is often applied where a better alternative is not currently available, for example, for certain mass concrete armour units. The formula in Equation (9.28) can be re-written using $H_{1/10} = 1.27 H_s$ together with a stability parameter defined as $N_s = H_s /(\Delta D_{n50})$ so that

$$N_s = H_s /(\Delta D_{n50}) = (K_D \cot \alpha_{wall})^{1/3}/1.27 \tag{9.30}$$

This stability parameter can be used to classify various structures as follows:

- $N_s < 1$ for caissons and sea walls for which no damage is allowed, using the height or width of the structure as the characteristic length.
- $N_s = 1\text{--}4$ for statically stable structures. This generally applies to uniform slopes of rock or concrete armour units for which only limited damage is permissible under extreme loading conditions. Here the characteristic length is the median nominal diameter of the stones or armour units.
- $N_s = 3\text{--}6$ for dynamic or reshaping breakwaters. These structures, generally comprising of armour stone, are characterised by a steep seaward upper and lower slope separated by a near horizontal berm around still-water level known as a 'berm breakwater'. The primary objective is for the breakwater face to deform during initial wave attack to create a profile that is inherently more stable in response to further wave attack.
- $N_s = 6\text{--}20$ for dynamic rock slopes for which the size of armour rock is such that it will always move subject to severe wave attack. This type of structure can be described as a 'rock beach' which trends towards a new equilibrium in response to each different severe wave condition.

In the case of rock armoured slopes it is recommended that the method and formulae of Van der Meer (1988a, b) should be used, unless a better approach can be proved for any particular application, such as through model testing for example. These methods are presented in a wide range of publications, although it is suggested that The Rock Manual (2007), which is effectively the second edition of CIRIA/CUR (1991), provides the most comprehensive and reliable source of information for the design of rock armoured structures and should be used as the primary reference.

The formulae of Van der Meer (1998a) are quite straightforward. The formulae have been derived initially for deep-water conditions with one for plunging waves (waves that are breaking) and one for surging waves (non-breaking waves). Minimum stability is found at the transition between these two wave states. This transition can be determined from comparison of the armour slope surf similarity parameter, ξ_m, similar to Equation (9.2), but using T_m which may expressed as

$$\xi_m = \tan\alpha_{wall} / s_{om}^{0.5}$$

in which s_{om} is the deep-water wave steepness corresponding to the mean wave period and α_b is the slope of the beach. This may be compared to a critical value,

$$\xi_{mc} = ((c_{pl}/c_s) \, P^{0.31} \, (\tan\alpha_{wall})^{0.5})^{1/(P+0.5)} \tag{9.31}$$

where P is a notional permeability coefficient given in Figure 9.40. Values of P range between 0.1 and 0.6, but for any structure with a geotextile as part of the filter layer a value of 0.1 is recommended. Empirical coefficients c_{pl}, c_s have values of 6.2 and 1.0 respectively with a standard deviation of 0.4 and 0.08 respectively.

The formula for plunging waves is used where $\xi_m < \xi_{mc}$ is:

$$H_s / (\Delta D_{n50}) = c_{pl} \, P^{0.18} \, (S_d \, N^{-0.5})^{0.2} \, \xi_m^{-0.5} \tag{9.32}$$

The parameter S_d describes the damage level, which is related to the percentage of displaced rocks or armour units related to a certain area and N is the number of waves during the design storm.

For surging waves where $\xi_m > \xi_{mc}$ the relationship becomes:

$$H_s / (\Delta D_{n50}) = c_s \, P^{-0.13} \, (S_d \, N^{-0.5})^{0.2} \, (\cot\alpha)^{0.5} \, \xi_m^P \tag{9.33}$$

It should be noted that for structure slopes of 1 in 4 or flatter, only the Equation (9.32) for plunging waves should be used irrespective of the surf similarity parameter. Limits of validity of Equations (9.32) and (9.33) are given in Table 5.24 of *The Rock Manual* (2007).

Thus the Van der Meer formulae are used to determine the median nominal diameter rock size (D_{n50}), which is the equivalent cube size, and thus suitable to obtain an accurate conversion to block mass from Equation (9.29) as well as armour layer thickness. This D_{n50} parameter is also used in assessing other characteristics of the construction. As with any deterministic approach a sensitivity analysis should be carried out. *The Rock Manual* (2007) recommends that the values of the coefficients c_{pl} and c_s in Equations (9.32) and (9.33) can be adjusted by 1.64 times the standard deviation

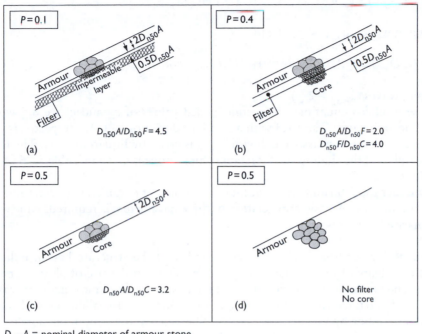

$D_{n50}A$ = nominal diameter of armour stone
$D_{n50}F$ = nominal diameter of filter material
$D_{n50}C$ = nominal diameter of core material

Figure 9.40 Notional permeability factor P (after Van der Meer 1990).

which is equivalent to the 5 per cent exceedance limit. These values are c_{pl}= 5.5 and c_s = 0.87.

The foregoing description of the Van der Meer formulae only applies to deep-water conditions. It is therefore necessary to define 'shallow water' with respect to the limit of application of these formulae. In shallow water, the distribution of wave height deviates from that offshore. In a further development of the Van der Meer equations, further work has been conducted into the application of these in shallow water. It is advocated that $H_{2\%}$ provides a better description of the wave conditions than the significant wave height for reasons previously discussed in the introduction to Section 9.4. Given the known relationship of $H_{2\%}/H_s$ = 1.4 it is possible to simply modify the coefficients in Equations (9.32) and (9.33) to apply to the shallow-water wave distribution so that c_{pl} = 8.7 and c_s = 1.4. As previously noted the methodology developed by Battjes and Groenendijk (2000) can be used for the purpose of evaluating $H_{2\%}$. Using data sets related to shallow-water conditions it was proposed by Van Gent et al. (2004) to modify the Van der Meer equations (1988a) by using a different characteristic wave period. They used $T_{m-1,0}$ to take into account the influence of the shape of the wave energy spectrum rather than simply the mean wave period T_m derived from time-domain analysis. This is the same approach adopted by the EurOtop Manual (2007). The empirical coefficients c_{pl} and c_s were determined by regression analysis so that the modified equations become

$$H_s / (\Delta D_{n50}) = c_{pl} P^{0.18} (S_d N^{-0.5})^{0.2} (H_s / H_{2\%}) \xi_{m\text{-}1,0}^{-0.5}$$ (9.34)

for plunging waves $(\xi_{m\text{-}1,0} < \xi_{mc})$ and

$$H_s / (\Delta D_{n50}) = c_s P^{-0.13} (S_d N^{-0.5})^{0.2} (H_s / H_{2\%})(\cot\alpha)^{0.5} \xi_{m\text{-}1,0}^{P}$$ (9.35)

for surging waves $(\xi_{m\text{-}1,0} \geq \xi_{mc})$.

In these modified equations the recommended values of c_{pl} and c_s are 8.4 with a standard deviation of 0.7 and 1.3 with a standard deviation of 0.15, respectively. The surf similarity parameter associated with $T_{m\text{-}1,0}$ is given by Equation (9.2). The range of applicability of the shallow-water relationships above is given in Table 5.26 of *The Rock Manual* (2007).

The Van der Meer formulae take account of a range of parameters to determine the size of armour rock. A basic appreciation of these parameters is required, which may be summarised as follows:

- *Permeability coefficient* (P) – The permeability of the structure has an influence on the stability of the armour layer and depends upon the size of filter layers and core. The permeability coefficient used here has no physical meaning, but was introduced into the formulae to ensure that the structure permeability was taken into account. Testing by Van der Meer (1988a) showed that the armour on more permeable structures has greater stability, with an increase in stability of 35 per cent as P shifts from 0.1 to 0.6, equating to a difference by a factor of 2.5 in the mass of stone for the same wave height. In breakwater design, P will usually be around 0.4. Only in the case of thin revetments, armour layers incorporating geotextiles and shallow slopes less than 1:4 are P values approaching 0.1 likely to be applicable. It should be noted that the D_{n50} relationships quoted in the diagrams indicate the basis for the model testing to which the P values relate. These provide guidance upon selection of appropriate P value; but they do not form the basis for design of layer thicknesses.
- *Damage level* (S_d) – The damage level coefficient is defined as $S_d = A_e / D_{n50}^2$. The physical interpretation of this is the number of cubic stones with dimension D_{n50} eroded within a D_{n50} wide strip of the structure. Structure stability can be described by the development of damage, which in this case is the displacement of armour stone under design conditions. It is not common practice to design rock/rubble mound structures for no damage. The nature of this type of structure, the range of material sizes within the grading and the variability in wave energy in the wave train, mean that some shifting and displacement of units may reasonably be expected Without going into the details Van der Meer has suggested that the limits of S_d mainly depend on the slope of the structure. He suggests that an initial damage value of between 2 and 3 is comparable to the Hudson formula given in Equation (9.25), which gives 0–5 per cent damage. Table 9.11 gives design values for S for a two diameter thick armour layer. For S values higher than 15–20, deformation of the slope will occur and the structure will develop an S-shape profile and must be analysed as a dynamically stable structure where some profile development is acceptable, such as riprap slopes.
- *Damage number* (N_{od}). The damage parameter (S_d) is less useful in the case of complex concrete armour units or for the evaluation of movement of rock in

physical model tests. In these cases it is more satisfactory to express damage in terms of the number of displaced units within a strip width D_{n50}. The damage number is

N_{od} = (no of units displaced out of armour layer) D_{n50} /width of tested section

which may also be expressed as a percentage, whence

N_d = no of units displaced out of armour layer/total no of units within the reference area.

The reference area can be the complete armour face, but this can be misleading and it is better to define the slope between two levels within which a tolerable extent of damage can be sustained.

- *Number of waves* (N) – The number of waves (i.e. the duration of the storm) will affect exposure of a structure and thus the degree of damage potentially suffered. This parameter is perhaps more important for dynamically stable rather than statically stable structures, but should be considered in all cases. It can be calculated quite simply by assessing likely storm duration and having knowledge of wave period. Development of the formulae was based upon values of N between 1000 and 7000. For values of N > 7000 the damage tends to be overestimated and is not recommended. Where uncertainty exists over storm duration, a commonly adopted value for N is 3000 waves, or the number occurring over a 3-hour period where wave conditions are depth limited.

- Further work by HR Wallingford (McConnell 1998) has explored the effect of rock shape and thus layer thickness upon damage, producing revised stability coefficients for use in the Van der Meer equations, which may increase or decrease rock size. Ignoring tabular shapes (usually inappropriate for armour), the suggested increase in rock weight for other shapes is up to 95 per cent. However, it is often uncertain at the time of design exactly what the shape of the rock will be. Consequently, it is not recommended that such modification factors are used without supporting information such as physical model testing.

- In reality, rocks are not cubic, and the actual 'equivalent sieve size' (D_{50}) will depend upon its shape. The most commonly accepted and recommended relationship is $D_{50} = D_{n50}/0.84$ and in which case the thickness of a double-layer rock armour slope may nominally be taken to be $2D_{n50}$. The CIRIA/CUR Rock Manual (1991) provides comprehensive details on rock shape/size characteristics (see pages 87 to 94) and a more general expression for the thickness of a layer or layers of rock is

$$t_A = n \, k_\Delta \, D_{n50} \tag{9.36}$$

where n is the number of layers and k_Δ is a layer thickness coefficient. The number of units per unit area is then

$$N_u = n \, k_\Delta \, (1 - n_v) \, D_{n50}^{-2} \tag{9.37}$$

where n_v is the volumetric porosity. Although lower values are sometimes quoted a practical minimum values for k_Δ is 1.0 and can be as high as 1.2 for specially placed round or semi-round rock shapes. BS6349: Part7 (1991) suggests a more limited

Table 9.11 Design values for S_d for two-layer armouring (after Van der Meer 1990).

Slope	Initial damage	Intermediate damage	Failure
1:1.5	2	3–5	8
1:2	2	4–6	8
1:3	2	6–9	12
1:4	3	8–12	17
1:6	3	8–12	17

range of between 1.02 for randomly placed smooth quarry stone to 1.15 for randomly placed rough quarry stone. For rock the porosity may vary between 35 per cent for very round stones to 40 per cent for rough quarry stone. Values for concrete armour units are given later.

Single-layer rock armouring is not generally advocated except for the most sheltered situations. This is for two reasons. First, a single layer will perform differently from a double layer, with reduced interlock, greater internal reflectivity, lesser wave energy dissipation and hence reduced stability. This makes calculation of rock armour sizing difficult, with all formulae derived from model testing of double-thickness layers. Secondly, the filtering characteristics are also lost, with potentially large voids between individual blocks. There may be scope to form a single layer with a graduated reduction in size for secondary layers (i.e. slightly smaller rock), although such proposals require physical model testing to develop an acceptable design. The usual practice is to provide a double layer of rock armour, with a thickness equivalent to $2D_{n50}$.

Dynamically stable armouring

Structures designed using the above methods are 'statically stable' structures. Whilst they are not rigid and do have potential to adjust their profile or 'settle' into place, their design is based upon zero to minor damage with the mass of individual units large enough to withstand the wave forces. 'Dynamically stable' structures are ones where subsequent profile development is acceptable and incorporated into the design. Typically such structures are constructed from rock and would include rip-rap revetment slopes and berm breakwaters.

The principle behind this type of design is that the materials can move until an equilibrium profile results, in much the same as a beach responds to wave activity, but to a far lesser extent. The benefits of this approach are that a much wider grading and potentially smaller size of material can be used. There is also a lesser requirement for individual placement of units, although due to the greater mobility, a much larger quantity of material will normally be required. The key design consideration is the determination of the expected extent of mobility of the material and ensuring that a minimum thickness of protection is obtained at all points such that the underlying materials are not exposed. The design of dynamically stable slopes can be based upon Van der Meer (1988a, b), although an additional design consideration for such structures is the potential for transport of materials along the structure. *The Rock Manual* (2007) and PIANC (1992) provide the necessary information for design.

Mass concrete armour unit slopes

Mass concrete armour units are generally used where rock of sufficient size cannot be obtained in the required quantities. Whilst rocks of up to 20 tonnes may be sourced a practical median weight limit is 10–12 tonnes in most cases, depending on the type of rock. Concrete armour units have been developed considerably to provide a high degree of interlock and hence stability, whilst at the same time being robust enough to withstand breakage. Some of the more popular units are shown in Figure 9.41. Concrete armour units can be broadly categorised under three headings; gravity bocks, interlocking units and energy dissipators. Their primary characteristics can be described as follows:

- Gravity blocks. These primarily provide stability due to their own self weight in a similar way to rock. However, their shape can play an important part in enhancing stability and hence influences the size of unit required. Gravity blocks such as the Cube, Antifer and Tripod are in general bulkier and more robust than other types of unit. A tightly packed armour layer is produced which gives less porosity and can result in higher run-up and wave reflection than other types of unit. Interaction between units plays an important part in their stability. For example, Tripods will interlock to some degree and can be expected to resist movement better than Cubes as reflected in the K_D values given in Table 9.12. Likewise the

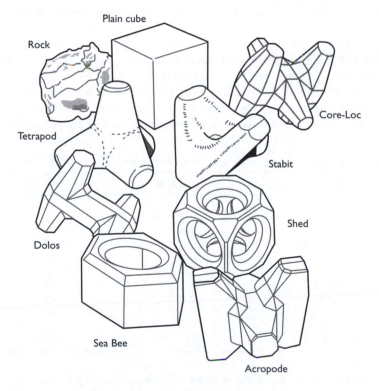

Figure 9.41 Selection of concrete armour units.

Antifer (grooved cube) will also result in improved stability. These units generally offer the simplest construction, both in casting and placing and may therefore offer an economical solution despite the relatively smaller stability coefficients.

- Interlocking units. These also depend a great deal on self weight for stability although there is relatively greater interaction between units. Units in this category include Akmon, Tetrapod, Stabit, Dolosse, Accropode®, Core-Loc™ and Xbloc®. They provide a greater stability to weight ratio because of their interlocking ability produced by their geometric design. This requires more complicated casting, but is not unduly difficult. Placing generally requires more care, but they are generally considered to be robust units.
- Energy dissipators. These are dealt with in a separate section following.

With the exception of energy dissipators, mass concrete armour units are sometimes described as randomly placed armour units and this really is an exaggeration as the units are usually placed to a general pattern. However, the satisfactory packing and interlock are more important than the pattern itself. The main variable in the Hudson formula that differentiates between various armour unit types, and hence design options, is the stability coefficient K_D. Recommended values can be found in a number of publications such as SPM (1984), BS6349 (1991) or *The Rock Manual* (2007). When adopting appropriate K_D values the definition of 'breaking' waves is those waves, which break as a result of the foreshore directly onto the structure. This and 'non-breaking' waves are not the same as 'plunging' and 'surging' waves as used by Van der Meer, nor is the 'breaking' wave the same as a wave which breaks some distance offshore.

General practice is to use the latest K_D values based on physical modelling testing coupled with defining the representative wave height $H_{1/10}$ as the design wave. Table 9.12 gives indicative values for K_D based on information from a variety of sources. It should be emphasised that these values should only be used for preliminary design purposes and new information may become available as further physical modelling is carried out, particularly more recently developed units such as the Accropode®,

Table 9.12 Nominal values of K_D and porosity for initial design only.

Unit	K_D 0% damage	K_D 5% damage	Porosity (n_v)
Rock	1.1–4.0[1]		37
Cube	6.5	5.0	45
Antifer	6.0	7.0	46
Tetrapod	6.0	7.0	50
Tripod	—	8.0	50
Akmon	8.0	9.5	52
Accropode®	9.5–15[1]	—	50
Stabit	10.0	12.0	55
Dolosse	8.0	12.0	56
Core-loc™	13–16[1,2]	—	
Xbloc®	13–16		59–63

Notes:

1 Varies between structure trunk, structure head, or shape and number of layers for rock.

2 $K_D = 9$ for trunk on steep seabed slope.

the Core-Loc™ and the Xbloc®. It should be acknowledged that for certain units the K_D values include high factors of safety (between 1.5 and 2.0), although modifications to published values should only be considered if supported by physical model testing. Also note that K_D values for non-breaking waves are higher than for breaking waves and values for the head of a structure can be around 25 per cent lower in both cases.

Some studies have been undertaken by Van der Meer (1988b) to develop methods for determining stability, based upon his work on rock slopes, considering one- and two-layer Cubes, Dolosse, Tetrapods and the Accropode®. Whilst these formulations provide the benefits of considering a wider range of factors than the Hudson formula, they have not yet been widely adopted. Part of the reason for this may be the general perception that the Hudson approach provides a 'safer' design, which is preferable for breakwaters. Also, as discussed below, use of the smallest size possible of armour unit does not necessarily result in the most economical design. However, *The Rock Manual* (2007) provides full coverage of these expressions which allow greater insight into levels of damage and potential failure than the Hudson equation.

For ocean conditions it has been suggested that for very large units the mass becomes more important and the shape/interlock effect reduces. For example, many ocean-facing structures have been built with massive concrete cubes/blocks of the order of 100 tonnes plus. There is some evidence to suggest that alternative units which rely more on interlock, and possibly offer a 50–60 per cent reduction in size, may not be as stable in these conditions. This is because the extreme swell wave conditions that can occur, have the potential to completely lift units out of place, in which case mass weight alone becomes the critical factor.

Whilst a wide variety of mass concrete armour units exist, there are only a few that are likely to be considered in most applications. There is currently an increase in the use of Accropode®, Core-Loc™ and Xbloc®. The reason for this is twofold: they offer a high level of stability (K_D values well in excess of 12) and they are single-layer armour systems. These units are also well supported by extensive physical model testing and do have inherent factors of safety. A drawback with these units is the potential complexity of their manufacture (although this has not prevented widespread use of the Accropode®) and they also carry royalty charges.

An alternative developed by Halcrow is the Stabit, used extensively over the past 40 years, particularly in the Middle East. This no longer carries royalties but does have an even more complex shape and placing arrangement, although again it has not prevented its use. Of the other units developed, it is recommended that it is generally only worth considering Cubes (or modified cubes along the lines of the Antifer), or in coastal defence applications a smaller simple unit, the Tripod. These generally offer the simplest construction, both in casting and placing, and may therefore offer an economical solution.

Some units such as the Core-Loc™, Accropode®, Xbloc® and Stabit are considered to have increased stability with steeper slopes due to the manner in which they interlock so that 1:1.33 slopes have been advocated. Whilst this is counter to the stability formulations the data is not sufficiently broad to parameterise this. However, there is a practical issue that the steeper slope makes placement and control of the core and any underlayers that much more difficult due to consideration of temporary stability during construction. Consequently, slopes of 1:1.5 are frequently the steepest adopted. An obvious corollary to this is that stability of these units does not increase with slopes flatter than 1:2. *The Rock Manual* (2007) also suggests that a further reduction in hydraulic stability numbers is warranted for situations involving depth-limited wave heights in combination with steep forshore slopes. A reduction of the order of 10 per cent is recommended.

For concrete armour units, consideration should always be given to the use of standard sizes as not only can previous model testing and design information be used, but it is likely that casting forms will be more readily available from previous projects. The designer should also consider the overall potential construction costs rather than simply the volume of concrete used in armour unit production. Smaller units require greater numbers to cover the same slope area and therefore need a greater number of units in production, transport and placing within the works. Consequently, use of particular-sized units may be more economically advantageous than smaller theoretical requirements. Also in situations where placing might be particularly difficult there might be some practical advantage in the use of simpler units such as Cubes or Antifers which are also offer simpler manufacture. Thus the choice of armour unit is not simply about the highest stability to weight ratio and a number of other issues need to be addressed. Clearly, the range of choices will be greater for less-exposed situations.

Detailed documentation is available on design of armour units such as the Accropode® Core-Loc™ and Xbloc® from the original developers of the units who are Sogreah (Grenoble, France), CERC (Vicksburg, USA) and Delta Marine Consultants (The Netherlands), respectively. These should be used in developing solutions with these units.

Energy-dissipating armour units

This group are single-layer, pattern-placed units that generally produce a flush face to the structure above an underlayer of rock which itself plays a relatively more significant role in the performance of the unit. Energy is dissipated through both the voids in the cover layer of units, but also the underlying rock and this can in some cases force wave breaking. Resistance to uplift is achieved by very accurate placing such that the sides of adjacent units are flush or interlocked in such a way that there is very high friction between individual units. Units in this category include the SeaBee, SHED and COB of which the SeaBee can be considered to be the most robust and has been used in a number of successful coastal defence schemes around the UK (see Figure 9.25). They can withstand relatively large waves for a lightweight unit, but generally require a very stable toe and capping beam to maintain integrity of the slope. Thus construction in anything other that shallow water can be difficult and slow, particularly with respect to achieving accurate placement under water. It also follows that the displacement of one or two units can lead to rapid unravelling of the slope so that construction risks can be relatively high.

The unit for which the most design information is available is the SeaBee (Brown 1979). There is very limited design information available for the SHED, which is a slender cubic frame, reinforced by glass fibres. Model testing has indicated that a unit weighing only 2 tonnes can be stable even when exposed to waves up to 7 m high. There is no known available design information for the COB and there is no obvious reason to use this in favour of the other units mentioned.

Other revetment protection systems

A number of alternative forms of armouring exist, primarily for use in revetment systems. These are generally appropriate for more moderate wave conditions, up to about 2 m, although this varies with type (see for example CIRIA/CUR (1991) pages 290–293). They may also be used as part of composite systems, for example as erosion protection above a main sea wall or revetment (it is unlikely that these solutions would be considered for breakwaters). These systems, some of which are shown in Figure 9.25 include:

- concrete block/slab revetments;
- concrete block mats (generally proprietary systems);
- grouted or pitched stone;
- bituminous systems, including open stone asphalt;
- gabion baskets and mattresses;
- fabric and other (e.g. grout) filled containers;
- reinforced grass slopes.

It is not within the scope of this publication to fully expand upon the use or design of these systems in any detail, but to refer the reader to appropriate references and highlight any key points of note. Essential reading on this subject includes:

- 'Guidelines for the Design and Construction of Flexible Revetments Incorporating Geotextiles in Marine Environment', PIANC (1992);
- *Coastal Protection* pages 197–367, Pilarczyk (1990), although some of the later is incorporated into the PIANC guidelines;
- *Revetment Systems Against Wave Attack – a Design Manual*, McConnell (1998).

Flexible revetments are designed on a different basis to rock and concrete unit armouring. They are much more sensitive to the degree of permeability/impermeability of the primary cover layer, the drainage and hence pore water pressure within the sublayer, uplift pressures, current/flow velocities, sliding and settlements. There are a variety of methods for calculating stability and determining size requirements, which are described within the key literature referenced above.

Good information on failure mechanisms is reproduced in all of the cited references. McConnell (1998) also provides good simple-to-follow guidance on how to produce a design of layer thickness and underlayer requirements for the different types of revetment system. This includes worked examples as well as typical information for inclusion in specifications. The PIANC Guidelines (1992) result from inputs of extensive international experience in the design and construction of revetment systems and should be referred to. In addition Pilarczyk (1990) provides comprehensive information on the design and use of asphalt systems. He also provides design information for a number of different types of revetment system. The basis for this is a general empirical and stated as 'approximate' formula which is:

$$H_s/\Delta_u D = \Psi_u \phi \cos\alpha/\xi^b \tag{9.38}$$

$$\xi^b = \tan\alpha \, (H_s / L_o)^{-0.5} = 1.25 \, T_z H_s^{-0.5} \tan\alpha \tag{9.39}$$

in which Ψ_u is a system-determined (empirical) stability upgrading factor based on a value of unity for rip-rap, ϕ is a stability function for incipient motion at $\xi = 1$, D is the thickness of the protection unit, α is the slope angle, Δ_u is the relative density of the system unit and b is an exponent related to the interaction between waves and the revetment type incorporating factors such as friction and porosity and has values in the range between 0.5 and 1.0 corresponding to rough permeable slopes through to smooth impermeable slopes. A value of 2/3 can be considered to be a common representative value. D and Δ_u are defined for specific systems as:

Rock; $D = D_{n50} = (M_{50} / \rho_s)^{1/3}$ and $\Delta_u = \Delta = (\rho_s - \rho) / \rho_w$
Blocks; $D =$ thickness of block and $\Delta_u = \Delta$
Mattress; $D =$ average thickness and $\Delta_u = (1 - n) \Delta$ where $n =$ bulk porosity of fill
   material, ϕ varies between 2.25 for incipient motion and 3.0 for maximum
   tolerable damage.

Given the foregoing, Table 9.13 gives the various empirical values for the parameters, particularly the stability upgrading factor.

When considering proprietary systems, it is recommended that the manufacturer is contacted and provided with relevant information regarding the site. They will provide design details themselves, although these should always be checked at detailed design stage. Further detailed design guidance on flexible revetments incorporating geotextiles can also be found in PIANC (1992).

Port and harbour breakwaters

The plan layout of a breakwater will be established by a number of factors including water depth, size of water area to be impounded/location of assets to be protected, manoeuvrability of vessels, wave climate, sediment transport, sea bed bathymetry, local geology, dredging requirements and occasionally aesthetics (for example coastal developments). The largest cost savings can usually be made through minimising the length of breakwaters. The second major cost saving arises through reducing the height of the breakwater noting that an increase in height adds width at the base so that seeking shallowest seabed levels is also advantageous. As a rule of thumb, a 10 per cent increase in breakwater height will produce a 15–20 per cent increase in volume due to the increased width at the base. Likewise a slight flattening of the side slopes, for example from 1 in 1.5 to 1 in 1.75 will increase volume by approximately 10 per cent due to the increased volume at the base.

The height of a breakwater should ideally be the lowest that provides the protection required and meets the service requirements. A small reduction in height will usually bring greater savings in material volumes and costs than a small reduction in width. Inclusion of a crown wall can be an effective means of providing a lower crest. Whilst this can be more expensive and difficult to construct, consideration must also be given to permanent access along a completed breakwater either for operational reasons or for access by maintenance plant. Typically, widths of 2 m and 4 m may be adopted as a minimum for pedestrian and permanent vehicular (single lane) access, respectively.

Table 9.13 Stability parameters for various revetment systems (after Pilarczyk 1990).

Criterion $H_s/\Delta_m D = \Psi_u \phi_s \cos\alpha/\xi^b_m = \Psi_u 2{,}25 \cos\alpha/\xi^b_m$
Limits/notes: $\phi_{s(rock)} = 2.25$; $\cot\alpha \geq 2$

System	D	Δ =	Ψ_u	Description	Subl.	Limits/
	b	$(1 - n)\Delta$				
Rock	D_n	$\Delta_m = \Delta$	1.0	Riprap (2 layers)	Gr	remarks damage
(reference)	b = 0.5	= 1.65	1.33	Riprap (tolerable damage)	Gr	1–3 stones damage < D_n
	D	Δ	1.00	Poor quality (irregular) stone	Gr	
Pitched stone	Average thickn.	stone = 1.65	1.33	Good quality (regular) stone	Gr	
	b = 2/3		1.50	Natural basalt	Gr	
	b = 2/3			Loose closed blocks (on	G + S	H_s < 1.5m
Blocks/	to 1	Δ	1.50	sand)	Gr	
block-	D	concrete	1.50	Loose (closed) blocks	Gr	
mats		1.2–1.9		Blocks connected to	G + C	Open area
		= 1.4	1.50	geotextile	Gr	blocks > 10%
			2.00	Loose closed blocks	Gr	grout =
	b = 2/3		2.00	Cabled blocks/Open blocks		crushed stone
				Grouted (cabled)blocks/		
			2.50	Interlocked blocks adeq. Dsgnd		
Grout	D_n	Δ	1.05	Surface grout (30% voids)	Gr	Avoid imperm.
	b = 0.5	stone				
	to 2/3	= 1.65	1.50	Pattern grout (60% voids)	Gr	H_s < 3 + 4
Open Stone Asphalt/	t thickn.	Δ asphalt	2.00	Open stone asphalt	G + S	U_p < 7 m/s
	cover layer	(= 1.15)	2.50	Open stoneasphalt/concrete	SA	H_s < 3 + 4 m
Open concr.	b = 2/3					
	t	Δ_m	2 + 3.0	Gabion/mattress as a unit	G + S	H_s < 1.5m
Gabions	b = 0.5 D_n	mattress				(max 2m)
		Δ stone	2 + 2.5	Stone-fill in a basket	(G) + C	min = 1.8D_n
			1.00	P_m << 1 less perm mattr.	S/C	H_s < 1.5 m
Fabric	t	Δ_m	1.50			(max. 2m)
Containers	thickn.			$P_m \approx 1$		
	Mattr.	Mattress	2.00	P_m > 2 permeable mattress		
	b = 2/3			of special design		
Grass	clay	Δ	—	Grass-mat on poor clay	C	U_p < 2 m/s
	= 0.5m	Clay		Grass-mat on proper clay		U_p < 3 m/s

Notes:

Gr = granular, G + S = geotextile on sand, G + C = geotextile on clay

SA = sand asphalt, S = sand, C = clay, U = permissible velocity

P_m = permeability ratio of cover layer and sublayer/subsoil (k_c/k_{fa})

During construction a safe working level will be chosen, often on the crest of the core or secondary underlayer, but usually about 2–3 m above MHWS in exposed situations. A minimum running surface width of 7 m is recommended to allow for two trucks to pass comfortably and for a large crawler crane to advance along the structure.

Width and height will also be determined from hydraulic performance characteristics whereby the structure needs to be sufficiently high to limit wave overtopping and wide or impervious enough to limit wave transmission. Width at the base may also be kept to a minimum through adopting steep side slopes, with particular scope in many cases to achieve this on the lee side of the breakwater. Flatter side slopes reduce overtopping and height, but will often have little influence upon transmission at the water line. Slopes flatter than 1 in 2.0 also become progressively more difficult to construct due to limitations on the reach of plant when constructing from the crest as well as the extent of work required profiling underlayers from the natural tipped slope. Steeper slopes are also preferable with some armour units, providing an increase in stability.

Structure roundheads and transitions

The foregoing design principles relate to the general stability of armour cover layers. Particular considerations need to be made for changes in the structure, such as at structure terminations and roundheads on the ends of breakwaters or groynes. These can experience particular stability problems. Waves breaking over a roundhead can concentrate and significantly increase instability due to very high velocity and complex flows, particularly on the lee side of the head. To deal with this and provide the same stability as for the main trunk section, it is usual to flatten the slope, increase the armour weight, or both. Jensen (1984) reports that there is a tendency for the most complicated units, such as Dolos or Tetrapods, to require the greatest weight increase as they depend more on interlock than on gravity.

Wave energy dissipation on roundheads is complicated and it is these elements of structures that feature most in breakwater failures. One way of dealing with this is through the definition different stability coefficient K_D values for use in the Hudson formula as suggested in the *Shore Protection Manual* (SPM 1984, pp. 7–206). Values between trunk and roundhead sections vary differently depending on the armouring being considered. For Core-Loc™ and Accropode®, the stability is reduced by about 20 per cent, whereas for rock armour the reduction is up to 50 per cent. A rule of thumb from laboratory testing experience (Sogreah) shows that a minimum roundhead radius of between 2.5 to 3 times H_s measured at highest water level can be adopted in most cases.

Whilst alternative K_D values are also published for rock armour this method is not recommended for rock slopes. As an alternative Allsop (1983) developed the Van der Meer (1988a) formulae for the sizing of rock on roundheads. The equations remain exactly the same except for the initial constants. Resolving these show that the relationship between the nominal rock diameter for the roundhead is $1.30D_{n50}$ relative to the trunk for both equations. This equates approximately to an increase in weight by a factor of 2.2 or alternatively a flattening of the slope with the same weight by the same factor (e.g. from a slope of 1:2.0 to 1:4.4). This is a somewhat larger increase than the 25–75 per cent suggested by the ratio of K_D values quoted in SPM (1984). Further reading includes *The Rock Manual* (2007).

Hydrodynamic forces exerted by waves dissipating their energy can be extreme and, where possible, abrupt changes in armour slope geometry must be avoided. As a rule of thumb, if the radius of the corner is more than 20 times H_s, then the corner can be considered as part of the trunk and dimensioned in the same way. If it is less than 20 times H_s, the corner should be dimensioned as if it were a roundhead. It is not recommended to have corners with radius tighter than 3 times H_s.

Construction can be difficult when there are changes in slope, especially going round corners. It may therefore be preferable to maintain the same armour slope throughout and simply increase the armour size locally as needed. This design strategy can deal with the problem of armour tending to 'peel off' when there are abrupt changes in slope.

The same rules apply to convex corners with special attention needing to be paid to the crest height as increased run-up can result in increased overtopping. Other design features to note are:

*   Transitions between different sizes of units should be on a diagonal with the smaller armouring size overlying the larger.
*   Where changes of sizes of unit occur, these should take place at a minimum distance of five to six armour units clear of changes in breakwater direction or slope.
*   Transitions for change of slope should normally occur over a distance of at least 10 armour units.

Whilst energy dissipaters such as the SHED and COB have been used on breakwaters, no information is known to be currently available on the design of roundheads using these units. Reference should therefore be made to existing constructions and supported by physical model testing.

9.4.4   Crest and lee slope armour

For breakwaters the width of the crest may be determined by a number of factors including for example the need or otherwise for any superstructure, ease of construction trafficking, or minimising wave transmissions. In the absence of any other controlling factors, a minimum requirement is for the crest to be protected by a continuation of the primary armour, to a width of at least three units, which in the case of rock is $3D_{n50}$.

Stability considerations on the lee slope of breakwaters include direct wave attack, wave overtopping damage and, to a lesser extent, wave transmission uplift forces. Lee slope armouring is also dependent upon other factors such as the configuration at the crest and geometry of any buttress wall. These are somewhat different from those for overtopping of coastal defences as discussed in Section 9.4.2. When there is significant overtopping, the traditional approach has been to continue the primary armour units on the seaward side over the crest and down the leeward slope to minimum sea level. However, in shallow-water cases where high overtopping discharges may be expected, this should be extended to the toe (SPM 1984). It is also possible to sometimes provide

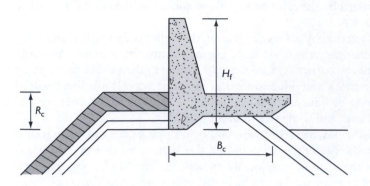

Figure 9.42 Buttress wall and roadway.

a steeper slope to the lee side of the structure without having to increase the armour size.

Unfortunately, reliable and consistent guidance on reducing lee-side armour is currently unavailable. However, physical model testing can be used to demonstrate the scope to reduce armour weight or steeper slopes on the lee side of breakwaters, which will often allow rock armour on the lee side of a concrete unit armoured breakwater. Features such as the incorporation of a buttress or wave wall, the width of the crest and slope of the lee armour will all influence protection requirements. In some instances overtopping water can be deflected over and beyond the lee slope by designing a crest slab behind the buttress wall to act like a spillway. Figure 9.42 shows a crest configuration that has been designed in this way. When carrying out tests on this type of arrangement it is necessary to consider a range of design conditions, as the most extreme events, when overtopping is high, do not necessarily represent the worst-case scenario.

In all cases the possibility of damage due to direct attack by internal waves whether diffracted into the lee side or locally generated must be considered. In these circumstances, the basic methods for primary armour stability apply. Uplift forces created by wave transmission and differential hydrostatic head across the breakwater may also need to be checked. This is unlikely to be an issue, however, unless the breakwater is very narrow and highly porous, or there is a substantial reduction in the size of the lee-side armour.

9.4.5 Rock grading

The Rock Manual (2007) provides a significant amount of detail on various parameters that have been derived to describe the geometric properties of a rock samples. The rock mass distribution is expressed in terms of the percentage lighter by mass cumulative curve and is usually plotted on log-linear scale. Thereafter the median mass for which 50 per cent of the rocks are lighter is designated as M_{50}. Thus, the steepness of a grading curve represented by the M_{85}/M_{15} ratio expresses the width of the grading. Grading widths may be described as in Table 9.14.

The log-linear equation is commonly used for both narrow and wide gradations and can usefully be expressed as

Table 9.14 Range values for rock grading description.

Descriptor	(D_{85}/D_{15})	(M_{85}/M_{15})
Narrow or 'single size' gradation	< 1.5	1.7–2.7
Wide gradation	1.5–2.5	2.7–16.0
Very wide or 'quarry run'	2.5–5.0+	16.0–125+

$$M_{p} = M_{50} (M_{85}/M_{15})^{((p-50)/70)} \qquad (9.40)$$

where p is the percentile value.

Graded rock is divided into three classes:

- 'heavy gradings' for larger sizes that are used in armour layers and placed into the works individually;
- 'light gradings' which may be used for armour layers in mild wave conditions, underlayers or filter layers. These are both produced and placed into the works in bulk;
- 'fine gradings' which are produced by square mesh screening and therefore less than 200 mm maximum dimension.

For practical reasons standard gradings are always used for both light and fine gradings. For heavy gradings it is usually relatively easy to define non-standard gradings as stones are selected and handled individually. In both cases four parameters are used to specify the grading. These are the 'extreme upper class limit' (EUCL), the 'upper class limit' (UCL), the 'lower class limit' (LCL) and the 'extreme lower class limit' (ELCL). A further parameter is defined as the arithmetic average weight of all the blocks in a consignment (M_{cm}).

Conventional gradings for shoreline and coastal armour layers as well as berm breakwaters are generally narrow and classed as non-standard. In these cases the CIRIA/CUR Manual (1991, Box 26, p101) recommends values given in Table 9.15 for two different weight ranges. These guidelines allow a range for M_{50}, effectively allowing a 5–10 per cent reduction in size. It is the recommendation that the calculated W_{50} is adopted as the lower bound of that range.

The rock grading for underlayers is usually described by the standard gradings for heavy and light gradings. In these cases the CIRIA/CUR Manual (1991, Table 19, p97) provides a detailed table of requirements for various weight ranges. The underlying principle is that the percentage by weight lighter on a cumulative plot should be

Table 9.15 Definition of non-standard specification for narrow heavy gradings, after CIRIA/CUR (1991).

Weight range (tonne)	EUCL	UCL	LCL	ELCL	M_{cm} range	M_{85}/M_{50}	M_{50} range
0.5–3	2.25 M_{50}	1.5 M_{50}	0.45 M_{50}	0.30 M_{50}	0.8–1.0 M_{50}	2.0–4.0	0.9–1.1 M_{50}
> 3	2.10 M_{50}	1.4 M_{50}	0.7 M_{50}	0.47 M_{50}	0.95–1.1 M_{50}	1.5–2.5	0.95–1.1 M_{50}

less than 2 per cent, between 0 and 10 per cent for LCL, between 70 and 100 per cent for UCL and greater that 97 per cent for EUCL. The Manual also provided a derivation for non-standard specification for wide light and light/heavy gradings (Box 27, p103), which will generally be more applicable to dynamically stable slope protection and to underlayers and filter layers.

The more recent *Rock Manual* (2007) summarises the requirements of the new European Standard EN 13383 that has been devised for armour stone. It includes a system for grading materials suitable for armouring and filtering, but not for core materials. The manual includes guidance on the following:

- grading widths;
- standard grading system of EN 13383 for armour stone;
- Rosin–Rammler curves;
- graphical illustration of grading curves;
- fragments and effective mean mass;
- requirements for compliance with EN13383 Standard gradings;
- additional useful information on EN 13383 Standard gradings;
- the relationship between median mass and the effective mean mass;
- non-standard gradings;
- core materials.

The definition of grading widths given in Table 9.14 remains the same, but whilst the general principal of defining weight ranges is similar, the weight ranges given in Table 9.15 are redefined for heavy gradings, typical of armour layers as:

- ELL (Extreme Lower Limit) – the mass below which no more than 5 per cent passing by mass is permitted;
- NLL (Nominal Lower Limit) – the mass below which no more than 10 per cent passing by mass is permitted;
- NUL (Nominal Upper Limit) – the mass above which no less than 70 per cent passing by mass is permitted;
- EUL (Extreme Upper Limit) – the mass below which no less than 97 per cent passing by mass is permitted.

Heavy, light and coarse European EN 13383 standard grading requirements are given in Table 3.5 of *The Rock Manual* as specific numerical values for each class designation rather than general ratios as given in Table 9.15 above. An important parameter is the effective mean mass M_{em} which is the average mass of a sample of stones without fragments below the ELL value of the grading. By excluding fragments from the total mass of a sample of stones it is possible to obtain a meaningful average mass simply by bulk weighing and counting all the stones, thus providing a rapid method of grading control. For cover layer applications the range of M_{em} will normally be specified. Approximate relationships for any grading width are given based on measurements of 'numerous projects' identical to those in EN 13383 across a range of heavy standard and light gradings. These are

$$M_{50}/M_{em} = 0.860 \ (M_{85}/M_{15})^{0.296} \qquad (9.41)$$

is based on field project data.

$$M_{50}/M_{em} = 0.860 \text{ (NUL/NLL)}^{0.201} \tag{9.42}$$

is based on a direct theoretical relationship.

$$M_{50}/M_{em} = 1.61 M_{50}^{-0.05} \tag{9.43}$$

is only applicable to standard gradings and reflects a systematically wider grading with decreasing M_{50} based on empirical data combined with theoretically derived results. It is pointed out that Equation (9.41) will give a better prediction than Equation (9.43) for gradings that are uncharacteristically wide or narrow for a given M_{50} as might arise with certain non-standard gradings. There is considerably more information in *The Rock Manual* on the specification and limits of certain classes of both standard and non-standard gradings and the reader is referred accordingly.

9.4.6 Underlayers and internal stability

The design of the internal elements of a breakwater can be as important as the external armouring. The underlayers in particular are part of the wave energy dissipation system and their nature will have an influence upon armour stability. It is also necessary to ensure that the internal layers will not be lost through washout, resulting in settlements, deformations and failure.

Where possible, it is advantageous to match requirements to quarry production because use of all the available grading is easier and cheaper to produce. This is not always possible because of the uncertainty over which quarry will be used, but measures to attempt to accommodate this can be taken by provision of overlapping gradings for different layers. Costs may also be strongly influenced by the armour rock specification, the overall volume of material required and the placement techniques to be used.

There are unusual design cases where an internal layer or core needs to provide an alternative function, such as restricting internal flows or, preventing wave transmission. Examples from various projects include providing a barrier for a cooling water intake, producing differential water levels to promote circulatory flows for water quality, and protecting against oil spillage. Techniques that have been used include incorporation of sand, sand-asphalt, geomembranes and geotextiles within the core of the structure. In providing such designs, particular care needs to be given to the influence of the internal structure upon wave pressures and internal set up of pore pressures, which can act as additional destabilising forces upon both the armouring and the superstructure.

Traditionally breakwater and revetment design has been based upon secondary layers/underlayers being sized by mass, relative to the mass of the armour layer. Whilst having some value in terms of armour stability, stone dimension characteristics can be more important than mass in many applications. Common practice now is to use filter design rules based upon stone dimensions, although mass still plays a part in determining primary underlayers, particularly when concrete armour units are used.

Filter layers may be provided for a number of reasons: to prevent washing out of finer material, provide drainage, protect sub-layers from erosion due to flows, and to regulate an uneven formation layer. A brief overview of filter design is provided by McConnell (1998, pp. 111–114), and a more technical but very useful discussion within Pilarczyk (1990, pp. 260–264). Underlayers, cores and filters are usually made up of granular material, generally quarried rock. River gravel may occasionally be used as a filter, although attention should be given to the potential lesser internal stability of such material given its rounder shape.

Goetechnical stability is a fundamental requirement. An extremely comprehensive description of internal stability issues and their consideration during design is provided in Section 5.4 of *The Rock Manual* (2007). In many cases, application of simple rules as described below will be adequate and a detailed analysis of internal failure mechanisms will not be required. However, a sound appreciation of the potential geotechnical problems and design requirements is recommended to enable that decision to be taken. In particular any seismic activity must be carefully investigated with respect to the possibility of potential liquefaction of soils beneath the base of the foundation.

Primary underlayer

As a general rule, use of a median underlayer mass (M_{50U}) can be related to the median mass of the armour layer (M_{50A}). A range expressed as a fraction of the armour layer is considered appropriate for underlayers in structures such as breakwaters and exposed revetments that are subject to severe wave attack. Table 9.16 gives values that have been used for rock (SPM, 1984) and concrete armour units (BS6349:Part7, 1991). The upper limit not generally so important, but the lower limits should be treated as an absolute minimum to prevent losses through the armour layer.

A relatively large underlayer produces an irregular surface, providing more interlock between armour and underlayer, and this also produces a more permeable layer, improving wave dissipation and armour layer stability. Where design information is not available the basic filter rules can be applied as a cross-check.

The underlayer in a revetment often doubles up as a filter layer, sitting above a fine material such as clay or sand with or without an intervening geotextile as shown in Figure 9.43. It is important that small particles beneath the filter are not washed out through this layer: it is also important that the filter/underlayer itself is also not lost through the armour layer. For these reasons, the design of internal layers needs to be appropriately sized to suit the dimensional characteristics of the materials both above

Table 9.16 Mass range of rock in underlayers.

Armour unit with mass M_{50A}	Mass of underlayer rock
Rock	$M_{50U}/0$ to $M_{50U}/15$
Tetrapod	$M_{50U}/10$ to $M_{50U}/20$
Stabit	$M_{50U}/5$ to $M_{50U}/10$
Dolos	$M_{50U}/5$ to $M_{50U}/10$
Accropode®	$M_{50U}/7$ to $M_{50U}/15$
Core-loc™	$M_{50U}/7$ to $M_{50U}/15$
Xbloc®	$M_{50U}/7$ to $M_{50U}/15$

Figure 9.43 Primary armour and underlayer under construction.

and below. To achieve this, a multi-layer system may develop, or it may be preferable to incorporate a geotextile as a substitute for a layer of material where dimensions need to be reduced or suitable material is not available.

Filter rules

There are various filter rules that have been used in the design of breakwaters and revetments. An important step is to understand what the criteria are, and why they are important, such that only those of relevance are applied and they are used appropriately. The basic considerations are as follows:

- Internal stability (uniformity) criterion – the grain size distribution within each layer should be approximately uniform to reduce the potential for internal migration of particles through the absence of intermediate grain sizes.
- Interface stability (piping) criterion – prevent finer particles of an underlayer from being washed out through the layer above.
- Permeability criterion – permeability should be sufficient for the hydraulic gradient through it to be negligible compared with that through the underlying material to prevent local build up of hydraulic gradient concentrations.
- Segregation (uniformity) criterion – the grading of each layer should be approximately parallel and not to far apart, to minimise segregation.

There is general agreement between different publications on the filter rules to be used and those below are recommended for adoption. A good description of designing with filter rules is provided in *The Rock Manual* (2007). These have been developed to take a more detailed account of the gradation of the layer and are summarised in Table 9.17. Here the subscripts refer to filter (F) and base (B). However, the F to B relationships may also be applied to rock armour and filter layer respectively. The

Table 9.17 Filter rules from various sources.

Criterion	Filter Rule	Comments
Internal Stability – geometrically tight	$D_{60}/D_{10} < 10$ $D_{10} < 3D_5$ $D_{20} < 3D_{10}$ $D_{30} < 3D_{15}$ $D_{40} < 3D_{20}$	Pilarczyk (1998) The Rock Manual
Interface Stability – geometrically tight	$D_{15F}/D_{85B} < 5$	The Rock Manual Both materials uniformally graded
Permeability	$D_{15F}/D_{15B} > 5$ $D_{15F}/D_{15B} > 1$	General purpose Relaxed – wide-graded material
Segregation	$D_{50F}/D_{50B} < 20$ to 25 $D_{50F}/D_{50B} < 5$	Filters – Pilarczyk (1984) Underlayers – CIRIA/CUR

term 'underlayer' refers to the layer underneath the primary armour and is synonymous with 'filter layer' in a two-layer armour system.

In addition to the general guidance given in Table 9.17, *The Rock Manual* (2007, Figure 5.135) provides a design nomogram for the interface stability of granular filters based on a geometrically open approach, which is rather less rigorous than the geometrically tight criteria given in Table 9.17 and more appropriate to situations where hydraulic gradients are not severe.

A simplified form of relationship was developed by Bakker *et al.* (1994) for the design of geometrically open filters in bed protection. This is

$$D_{15F}/D_{50B} = 15.3 \, R/(C_o \, D_{50t}) \tag{9.44}$$

where R is the hydraulic radius (m), C_o is a coefficient accounting for the hydraulic gradient at the filter interface and the average hydraulic gradient in the bed layer with a conservative value of 30 and D_{50t} is the median sieve size stone diameter of the top layer. This relationship is valid for either a top layer placed directly on an existing non-cohesive subsoil or a top layer and one, two or more filter layers on an existing non-cohesive subsoil.

The thickness for any rock layer will nearly always be a minimum of at least 2 stones calculated as $2D_{n50}$, although filters may require considerably greater thickness to be effective and practical. As the nominal diameter becomes smaller this number may increase as the thickness needs to be a practical minimum for placement and deal with irregularities and placement tolerances.

Considerable detail on the effects of rock shape and placement techniques upon layer thickness is provided in *The Rock Manual* (2007). Calculations are based on a variable layer coefficient. However, in many circumstances when the shape of rock is equant or irregular and the placement of the material is well controlled, it is appropriate to use a layer coefficient (k_t) of 1.0 so that the layer thickness is simply a multiple of the nominal diameter D_{n50}. For cores and layers of multiple stone thicknesses, the layer coefficient becomes irrelevant. For other materials, recommended minimum layer thicknesses depend upon the nature of material, likely deformation and placement conditions for which McConnell (1998) provides some useful guidance.

Geotextiles

A geotextile or geomembrane is a synthetic permeable textile manufactured in sheets and used to prevent the migration of soil or filter material. It may be fabricated as woven, non-woven or composite material. The first of these is a single layer geotextile formed by an interlaced thread system whereas the second is formed by fibre fleeces which may be bonded by needle punching, adhesion or melting. A composite material is a multilayer system, each of differing structure.

Currently published guidance on the design and specification of geotextiles includes:

- 'Code of Practice, Use of Geotextile Filters on Waterways' (BAW 1993);
- PIANC (1992);
- *The Rock Manual* (2007).

The preceding section provides criteria to determine whether two adjacent materials have satisfactory filter characteristics. The BAW Code of Practice also includes a very useful so-called 'CISTIN/ZIEMS diagram' to check the need for an additional filter layer which may be provided by a suitable geotextile. This uses the relative gradings using uniformity coefficients (D_{60}/D_{10}) for both the base material and the filter layer. In general it can be said that the more widely graded the materials under consideration, there is greater margin for difference in median grain size. The BAW method provides significant potential refinement of a design together with a lot of very useful guidance on most aspects of geomembrane selection.

Having made the preliminary selection of geotextile, the specification needs to be based on manufacturers' data sheets, bearing in mind that the construction cost will invariably be less if the final specification may be met by using a range of products from different suppliers. The following aspects should also be considered:

- Long-term performance as a filter – The BAW Code provides empirical guidance on the thickness of armour layers required to cover and protect geotextiles from long-term damage. The thicknesses quoted are all less than 700 mm, which is generally less that the thickness required for cover armour stability. It also provides guidance on minimum strengths for geotextiles due to tensile loads and abrasion, which tend to be less than or equal to 12 kN/m. Such loads tend to be low compared to the strength required to resist damage due to rock placement. Long-term damage due to UV weathering, shipping and chemical composition of groundwater are also relevant factors.
- Short-term damage during construction – The strength and density of geotextiles for use in coastal structures is often determined by the need to minimise damage during construction, rather than long-term strength requirements. However, there is conflicting advice from manufacturers regarding the response of woven and non-woven products to rock placement. The longitudinal and cross-direction threads of a woven product may be separated as rock is placed. This may affect the filter characteristics, and allow greater loss of fines from the underlying material. Equally, non-woven geotextiles are compressed differentially by the placement of rock, which affects the pore size and permeability performance. Guidance

on damage caused by rock placement is provided in a technical note 'Geotextile Filters in Revetment Systems' by Naue Fasertecknik, with reference to their own (non-woven) products.

In general, the fabric of non-woven needle-punched geotextiles tends to be more robust than woven materials under irregular, punching loadings. Indeed, rough handling may puncture some woven products that have a reasonably high strength rating. However, the use of woven fabrics underwater can be very difficult due to their buoyancy and increased weight when wet. The designer should also be aware that the placing of geotextile in any depth of water is difficult and that a natural granular filter will often enable greater quality control during construction.

Given the wide range of products available, the most reliable guidance for placement is to follow the manufacturers' instructions. There are a few points that require additional emphasis:

- Storage – Regardless of the type of product, it will be safer to specify that the material should be kept out of the light and in manufacturer's wrappings until the time at which it is to be placed in the works. This should provide better protection against mechanical damage as well as UV damage.
- Lap length – When geotextile sheet width is not large enough to avoid overlaps, manufacturers often state that lap lengths may be as little as 200–300 mm. Such recommendations are usually based on horizontal placement onto fine materials. The placement of a geotextile onto a rock filter, on a slope will require a greater overlap to allow for (a) difficulties in placement on uneven or inclined surfaces, (b) movement of the geotextile as the overlying rock is placed and (c) the lap to be held in place by adjacent stones if the armour size is large. The BAW Code specifies 0.5 m in the dry and 1.0 m in the wet. It also recommends that all overlaps should run parallel to the slope and, given that overlaps across the slope are unavoidable, the lower lap should be placed over the upper lap. Construction experience suggests that a minimum lap length of 1000 mm is the practical minimum allowing for sensible construction tolerances. If the armour size is larger than this, then the lap should be equal to the stone size to ensure that the lap is held in place by adjacent stones. If the placement is expected to be particularly difficult, or in a substantial depth of water then the lap length may also be increased for example by a factor of 1.5 to 2. Given the generally low rates for supply and placement of geotextiles, compared with rock armour, the additional cost of providing greater confidence in the overlap is minimal, but note the foregoing comments on working under water.

9.4.7   Crown walls

Crown and wave return walls are often used on revetments and breakwaters to reduce wave overtopping without raising the crest of the structure as discussed in Section 9.4.2. Frequently pedestrian or vehicular access will also be incorporated into the feature. Wave forces on the wall will not only depend on the incident wave conditions, but also on the detailed geometry of the armour in relation to the wall. Depending on the degree of protection afforded by the primary armour the primary loading is on the face of the structure coupled with an uplift force on the underside of the element.

There are no generally applicable methods for predicting forces on crown walls independent of the crest geometry, and physical model testing is often required to provide the necessary design data. Data from Jensen (1984) and Bradbury and Allsop (1998) has been fitted to an empirical equation, which serves as reasonable framework for further model studies. The maximum horizontal force is described as

$$F_{H} / (\rho \, g \, H_{f} \, L_{p}) = a \, H_{s} / (R_{c} - b) \qquad (9.45)$$

where H_{f} is the total height of the crown wall face that can be impacted by waves either directly or through the voids in the armour (see Figure 9.42), R_{c} is the freeboard between crest of the armour and still-water level (sometimes notated as A_{c}) and L_{p} is the wavelength corresponding to the peak period. The coefficients have been derived from available data and vary between $0.025 < a < 0.54$ and $0.011 < b < 0.032$, their magnitude being largely dependent on the degree of exposure for the various cross section given in Allsop (1998) or *The Rock Manual* (2007, Table 5.49). The equivalent expression for the uplift force is

$$F_{U} = (\rho \, g \, B_{c} \, L_{p}/2) \, (a \, H_{s} / (R_{c} - b)) \qquad (9.46)$$

where B_{c} is the width of the crown wall element. These force values can be used to design the stability of the crown wall element. The vertical uplift must be resisted by the weight of the element whilst the horizontal force must be resisted by friction. A friction coefficient of 0.5 may be assumed when the crown wall element is cast *in situ* onto the underlayer. This may be increased to as much as 0.8 to 1.0 if a significant key into the underlayer can be assured. A corollary is that precast units will be less resistive.

Other methods for calculating wave pressures on crown walls due to Pedersen (1996) and Martin (1999) are given in *The Rock Manual* (2007).

9.4.8 Scour and toe stability

Wave and current velocities are often increased by the presence of a coastal structure due to factors such as wave reflections and wave downrush. Structures are also usually required in areas of high shoreline volatility, or coastal instability and erosion. This can result in localised scour around and in front of a structure, which needs to be considered in design. Toe stability is essential because failure of the toe will often lead to failure throughout the entire structure. Past work by CIRIA (1986) determined that approximately 12 per cent of sea wall failures arise directly from erosion of the beach or foundation material, and that scour is at least partially responsible for a further 5 per cent of failures. This is a problem that is not always fully appreciated but needs to be understood and considered fully in the design of coastal structures.

Whilst a distinction needs to be made between natural shoreline movements and structure-induced scouring, design must accommodate both. Natural movements may be considered in two broad categories, which are long-term change and short-term volatility.

The first, a retreat of the whole coastal system, will continue to occur regardless of any shoreline structure, with beach and seabed levels decreasing as the natural shoreline position seeks to move landward. The extent of this can best be determined from an understanding of historic evolution on a site-specific basis. This is usually best appreciated by analysis of the whole nearshore profile to the seaward depth of closure. This information may not be available to enable comparison and it will be necessary to make best use of whatever information can be obtained. However, it should be appreciated that extrapolating rates of change from historic maps can be misleading as map publication dates are often different from actual survey dates and mapping of high and low water lines may be inaccurate, depending upon tidal states at time of surveys. There is also the possibility of seasonal volatility.

Short-term volatility is a change in beach levels that take place seasonally or in response to individual storms, and may result from both cross-shore and alongshore movements of material (see Chapters 5 and 6). In the UK, average differences in beach levels of in excess of 1 m directly in front of the structure between summer and winter are not uncommon, whilst lowering in excess of 2 m on the same beaches may occur during a single storm. The extent of such changes requires assessment on a site by site basis, from knowledge of waves, water levels, beach material and volume reserves. Assessment needs to be made from experience in understanding beach evolutionary processes, to provide an estimate of the extent of changes that need to be taken into account by the design. Account also needs to be taken of sea level rise that will accelerate change.

The magnitude of any scouring as a result of structural influences is difficult to predict. It may sometimes be unobserved because maximum scour occurs during the height of a storm, with some recovery before the waves have abated and water levels lowered. Further research since the mid-1980s has helped to improve upon the SPM (1984) rule that the maximum depth of scour under wave action (d_s) is approximately equal to the height of the maximum unbroken wave that can be supported by the depth of water (H_{max}). Research by Powell (1987) reproduced in CIRIA/CUR (1991, Figure 187) goes some way to addressing this, showing depth of scour is variable with both wave steepness and water depth. It also suggests that scour is not predicted to occur for water depths greater than $3H_s$. This data is also limited to shingle beaches and vertical walls. Further work by Powell and Whitehouse (1998) has made contributions to the subject through examination such factors as reflection coefficients, sea wall slopes, wave steepness and water depth on both shingle and sand beaches. Kraus and McDougal (1996) present a wide-ranging literature review of research mainly in the USA. Whilst this does not provide quantitative information, it is a comprehensive compilation of views and may aid understanding of wider processes as well as highlighting some of the research being conducted in this field. Table 9.18 presents a number of 'design rules' from a variety of sources that have been suggested together with some commentary.

Selection of toe protection

Toe protection provides insurance against scouring and undermining of a structure, and it provides support against sliding to the structure armour/face. It therefore needs to be provided to an adequate depth and be of sufficient size/stability to prevent the occurrence of these two possible failure modes. Important considerations in establishing the nature of toe protection required are location of the structure in relation to

Table 9.18 Design 'rules' for scour design.

Rule	Comment
Scour depth is equal to H_{max}	Only under certain circumstances, this is a very limiting statement which may either under or overestimate scour
Maximum scour over the range $0.02 < H_s/L_m < 0.04$ is approximately equal to wave height	Partially true. Powell's graphs suggest that this is also a function of water depth
Maximum scour occurs when the structure is located around the plunge point of breaking waves where d_{sw}/H_s^* is approximately 1.5	Generally supported, although Powell's (1998) graphs suggest d_w/H_s^* relationship is closer to 2.0
The depth of scour is directly proportional to structure reflection coefficient	Whilst reflectivity is an issue, this is a generalisation which does not appear to hold true in all cases.
Scour can be minimised for structures with a smooth impermeable face, by adopting a slope flatter than about 1:3	Possibly reduced , but not minimised
Impermeable slopes of 1:1.5 to 2 offer no significant reduction in scour depth compared to that at a vertical wall	Evidence only relates to shingle beaches
Impermeable slopes of 1:3 reduce scour typically by 25%, and up to 50%, compared to that at a vertical wall	Evidence only relates to shingle beaches
Scour is only significantly reduced for slopes of 1 in 4 or less on sand beaches	Needs further evidence. The concept of using sloping sea walls to reduce scour is in doubt
Rock slopes do not tend to cause scour and and can encourage localised accretion	A very general statement but the first part is probably true
Depth of scour is *directly proportional* to structure reflection coefficients	Not substantiated, but there is some influence
The maximum scour is expected to occur when the water level is highest when larger waves exist	Not necessarily as breaking waves may be experienced at smaller water depths. Depends on tidal range
Reflection is probably not a significant contributor to beach profile change or to scour in front of sea walls, at least for the duration of a storm	Not true or supported by experimental evidence
If the beach is close to equilibrium shape the arrival of a storm may not cause a significant change in profile and hence erosion	Probably true in most cases
Scour potential is greatest where the water depth at the toe is less than twice the height of the maximum unbroken wave	True

Note
* d_{sw} is the depth of scour at the wall.

the wave break point, form of structure with respect to reflectivity and nature of the seabed

Special attention should also be given to areas where scour may be intensified, such as changes in alignment, structure roundheads, channels and downdrift of groynes.

The basic principle behind flexible toe protection for revetments is to provide an extension of the face such that the foundation material is kept in place beneath the structure to the bottom of the maximum depth of scour. This can be achieved in four ways:

- Buried toe – where construction conditions permit the cover layer is extended by burying the toe in an excavated trench to the depth of predicted scour. It may be appropriate to backfill the trench with granular fill or rock, depending on natural conditions.
- Extension of cover layer on the bed – laying a 'falling' apron on the bed which will drop into any scour hole that develops.
- A combination of both – trenching in a falling apron reduces the undermining risk and possible erosion of toe protection but avoids full depth excavation.
- Static toe restraint – examples are sheet piling, timber, a concrete toe beam or anchor bolts to prevent sliding but driven/founded deep enough to prevent undermining. This form of toe may be preferable where a more static defence structure is in place such as a sea wall or concrete revetment, or where space is constrained and/or deep water is required such as for the edge protection within a marina.

The choice of design strategy is strongly related to the nature of the seabed as follows:

- Erosion-resistant strata at foundation level. CIRIA/CUR (1991) proposes that the armour layer should be keyed into the stratum at a minimum depth of $0.5D_{n50}$ to ensure that sliding of the armour layer does not occur (see Figure 302). Alternatively, in the case of a very hard rocky foreshore a toe beam such as concrete or anchor bolts could be dowelled into the rock. Adoption of this advice is recommended – although a further acceptable alternative is to provide a sufficiently wide toe of sufficiently large units to prevent sliding.
- Limited-resistant strata at foundation level. Some types of clay beach have low undermining scour potential. In these cases continuation of the armour slope down to the predicted depth of scour is recommended.
- Beach or other mobile material at foundation level. The founding level should be based upon the predictions of beach level variation and scour depth with the addition of an allowance for the risk of undermining. The potential for scour may be classified in relation to the ratio of the predicted scour depth to the incident wave height such as 'low' ($d_s < H$), 'low to moderate' ($H < d_s < 1.25H$) and 'moderate to severe' ($1.25H < d_s < 1.5H$).

Typical sea wall toe designs where scour is foreseen are shown in Figure 9.44 after McConnell (1998). These cover the majority of conditions described above. For more massive structures such as revetments in highly exposed situations or breakwaters it is normal practice to provide a toe bund to support the primary armour layer coupled

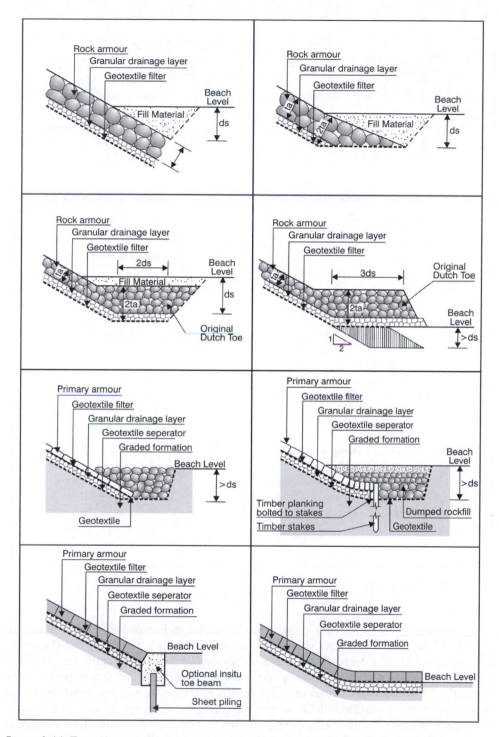

Figure 9.44 Typical sea wall toe designs to combat scour (after McConnell, 1998).

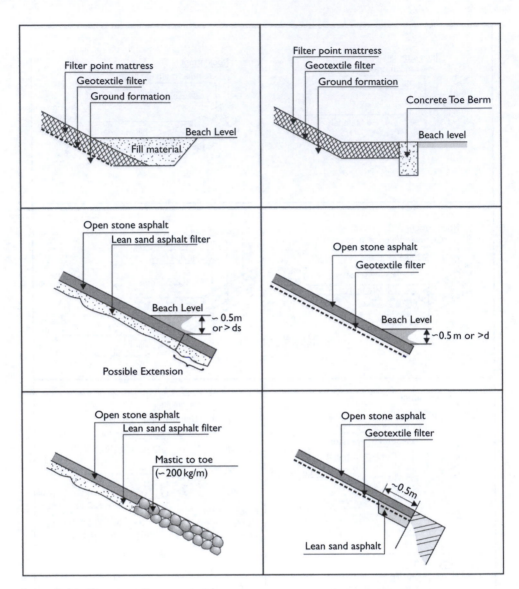

Figure 9.44 (Continued)

with anti-scour bed protection as generically shown in Figure 9.39. Basic guidance in BS6349:Part 7 (1991) states that if the water depth is less than $2H_s$ and the structure slope is less than 1 in 3, a toe bund is required. Typical toe details are given for different foundation cases. However, difficulties arise in shallow water because the theoretical size of toe protection required results in the surface of the bund becoming closer to the water surface, which in turn leads to greater exposure and thus heavier stone requirements. In these cases the alternatives described in Figure 9.44 should be adapted.

Armoured toe design

As a basic rule of thumb, if the rock or concrete units in the toe of a structure have the same dimensions as the armour cover layer, the toe will be stable. However, in most cases there is a strong cost advantage in reducing the size of material.

The Rock Manual (2007) discusses issues surrounding the relationship between the stability parameter $N_s = H_s / (\Delta D_{n50})$ and the relative water depth at the toe of the structure (H_s/h) for values less than 0.5. At this point the waves become depth limited and the relationships apply accordingly. For breakwaters in very large water depths (>20 m) other approaches are appropriate. The various levels of damage criteria (S_d) are described as:

- 0–3 per cent – little or no movement of stones at the toe;
- 3–10 per cent – toe flattened out but still functioning and acceptable;
- > 20 per cent represents failure.

The Rock Manual (2007) suggests a relationship between the ratio of the depth of water above the toe bund to the total depth at the toe (h_t/h) and the stability parameter $H_s / (\Delta D_{n50})$. Design values for low damage levels in near depth limited conditions are given in Table 9.19. For lower ratios of h_t/h the stability formula for armoured slope as given in Section 9.4.3 should be adopted.

For most coastal structures, wave forces (downrush and breaking) present the critical conditions when determining stability of the toe. However, currents can become important, particularly in deeper water or more sheltered sites where wave activity is restricted. CIRIA/CUR (1991) recommends that where currents exceed 1 m/s, the armour layer of the toe protection nominal diameter is increased by a factor of at least 1.3. In all cases a minimum thickness of $2D_{n50}$ is recommended as a minimum.

Van der Meer *et al.* (1995) developed an expression for the stability of a toe berm formed by two layers of stone of 2.68 t/m³ density for sloping structures. The formulae for the stability of the toe gives the relative toe depth in two ways:

$$H_s / (\Delta D_{n50}) = (1.6 + 0.24 / (h_t / D_{n50})) N_{od}^{-0.15} \qquad (9.47)$$

and

$$H_s / (\Delta D_{n50}) = (2.0 + 6.2 / (h_t / h)^{2.7}) N_{od}^{-0.15} \qquad (9.48)$$

It is pointed out that a toe with a relatively high level of $h_t / h < 0.4$ comes close to a berm and, therefore, close to the stability of the armour layer on the sloping front face of the structure. These Equations (9.47) and (9.48) may be applied within the range $0.4 < h_t / h < 0.9$ and $3 < h_t / D_{n50} < 25$.

Table 9.19 Toe stability as a function of h_t / h.

Depth ratio (h_t / h)	$H_s / (\Delta D_{n50})$
0.5	3.3
0.6	4.5
0.7	5.4
0.8	6.5

The research results of Van der Meer *et al.* (1995) were modified by Burcharth *et al.* (1995) so that it could be applied to stones having other densities or to parallelepiped concrete blocks. The stability parameter, which defines the nominal stone diameter, was given as

$$H_s / (\Delta D_{n50}) = 1.6 / (N_{od}^{-0.15} - (0.4\, h_t / H_s)) \tag{9.49}$$

For Equations (9.47)–(9.49), N_{od} is the number of units displaced out of the armour layer within a strip width of D_{n50} and a value of 0.5 represents almost no damage, 2 corresponds to acceptable damage and a value of 5 corresponds to failure. All other parameters have been previously defined.

Defining the width of scour protection is a largely a matter of engineering judgement and in the case of a falling apron design must be wide enough or contain sufficient material to collapse safely into the anticipated depression in the seabed. Whilst scour can be assumed to be greatest within one-quarter wavelength of the front of the armour slope, protection over this area will often be well in excess of actual requirements. For breakwater or revetment toe bund design, BS6349 recommends a minimum width $4D_{n50}$, which is slightly more than $3.3D_{n50}$ recommended by CIRIA/CUR (1991). Further, any shoulder of smaller layers on the seabed should have a width of at least 2 m. For revetments, the most common guidance is that a toe apron should extend to a width of at least three times the depth of scour, although Pilarczyk (1990) recommends a minimum width requirement of between one and two times the incident wave height.

Toe protection against currents may require smaller protective stone, but a wider apron, although little definitive guidance is currently available. Pilarczyk (1990) does, however, provide a formula by Hales and Houston (1983) for the breaking wave stability of a rock blanket extending seaward from the toe of a permeable rubble slope on a 1:25 slope foreshore, which can be used as a first indication of decreasing stone size (D_{n50}) with distance from the structure (B_r). This is:

$$H_B / (\Delta D_{n50}) = 20\, (B_r / T_p)^{2/3}\, (gh_t)^{-1/3} \tag{9.50}$$

The Rock Manual (2007, Table 5.47) provides numeric values for armour stone grading in toe berms with shallow water and gently sloping foreshores and within specified wave height ranges.

Vertically faced structures

The preceding design guidance relates to sloping structures. Where the superstructure is vertically faced, such as in the case of a caisson structure or concrete sea wall, design guidance regarding the rock berm on which such a structure would sit is more limited. A complicated expression developed by Tanimoto *et al.* (1982) is reproduced in *The Rock Manual* (2007, p. 625). A simpler relationship is given by Magridal and Valdes (1995) for two layers of quarrystone where the stability criterion is

$$H_s / (\Delta D_{n50}) = (5.5\, h_t / h - 0.6)\, N_{od}^{0.19} \tag{9.51}$$

where h_t is the depth of the foundation of the structure and N_{od} is the number of units displaced from a strip width D_{n50}; a value of 0.5 represents almost no damage, 2 corresponds to acceptable damage and a value of 5 corresponds to failure. This equation is valid in the range $0.5 < h_t/h < 0.8$ or $7.5 < h_t/D_{n50} < 17.5$.

Finally rock is often the favoured material for toe protection because of its flexibility. However, other forms of toe protection are available such as various mattresses. Reference should be made to suppliers' literature with regard to the use, applicability and dimensioning of these systems.

General considerations

In addition to the design of the fabric of any structure there are further aspects that should be allowed for in relation to ground conditions, a knowledge of which is essential:

- settlements due to soft seabed material;
- rotational slip due to failure of seabed material;
- seismic activity (how will structure react, choice of core, foundation and armouring can be important);
- displacement of soft material during placement;
- long-term seasonal changes in bathymetry;
- scour potential of seabed materials (which may increase due to presence of structure);
- dredged side slopes on channels which may flatten in time and threaten the integrity of the toe or a structure.

Methods of construction and local resources can have a strong influence on design considerations and should always be considered. These aspects include:

- Availability of materials – in particular this may influence choice of armouring (for example concrete or rock) and shape of structure (adoption of less steep slopes or berm breakwater profile).
- Local construction resources – if the quality of construction is in doubt it is necessary to make due allowances in design sizing and tolerances.
- Best use of materials – the exact dimensions of a structure should ideally be proportioned to optimise the use of rock obtained by quarrying so that design should suit local material availability.
- Type of plant – this can influence both the method and duration of construction as well as limit the size/weight of individual components of the works.
- Trafficking of plant – provision of sufficient space at a construction level above water level to enable plant movement, material supply, crane manoeuvrability and inclusion of passing places as features in final construction.
- Design details – should always consider the practicalities of construction so that the number of different construction activities are kept to a reasonable minimum.
- Partial construction – evaluation of risks on partially completed sections of the works should be considered in the design process.
- Heath and safety – all aspects of health and safety during the lifetime of a structure whether during construction or thereafter and legal requirements will vary from place to place.

9.4.9 Design of sea walls

All of the foregoing design guidance is relevant to the design of sea walls that have a sloping seaward face and are protected with different types of cover layer. Modern design practice would normally dictate that any new structure should have as flat a slope as possible and be protected by a cover layer or layers that destroy as much energy as possible. However, there may be a number of reasons why this is not possible such as the availability of space, the desirability of public access to the wall or constraints of a legacy system. In these circumstances the designer may choose to use stepwork or some other form of near vertical structure. The depth of water at the wall will determine the potential significance of the wave forces on the structure due to breaking waves. A detailed analysis of the wave forces on vertical structures is due to Goda (1974, 2000) and Tanimoto *et al.* (1976). Goda's design method allows a consistent calculation procedure, regardless of whether wave breaking takes place or not. It is limited to predicting the pulsating, quasi-static wave loads and does not attempt to predict impulsive forces. The method determines the horizontal and uplift pressure forces and hence determines the resistance to sliding and the overturning moment. It is currently the most widely accepted design method and has been in use in Japan, for the design of caisson breakwaters, since 1979. It replaces the earlier formulae of Sainflou, for standing waves and those of Hiroi and Minikin for breaking waves. Allsop (2000) provides an excellent review of wave forces on vertical sea walls and provides a detailed summary of design procedures and the equations used. The CEM also provides a useful summary of current procedures and design equations, including Goda's method.

Design equations for impulsive forces are yet to be determined (see Section 2.5), but there is excellent advice (see Goda 2000 and Allsop 2000) on determining the conditions under which to expect impulsive wave loads and how to design to avoid inducing impulsive wave forces. The use of physical scaled models is also recommended for detailed design of major new structures of this type. However, it needs to be recognised that significant problems of scaling will be inevitable when considering impact forces, because of aeration effects. A new method has recently been proposed to address this problem, which may provide some of the answers needed (Cuomo, Allsop and Takahashi 2010b).

In any event it is not common for the design of a sea wall to be sensitive to the absolute value of the wave loading unless it is in the form of a thin vertical structure. More often the requirements of robustness and durability will provide construction elements that will withstand the wave forces. This is evidenced by the fact that there are very few failures of sea walls that have occurred due to a failure of the fabric of the structure. In contrast, the nature of the material on which the sea wall is constructed and the design of filter systems to contain relatively weak permeable material is vitally important.

One example of how things can go terribly wrong is demonstrated in Figure 9.45. Here, the sea wall has been constructed out of pre-cast concrete units sitting on a core of sand, the cross section of which can be realised from Figure 9.45(a). There was no attempt to contain the underlying sand that was vulnerable to being leached out by overtopping waves. The consequences of this are shown in Figure 9.45(b), which was the result of a fairly modest wave condition. However, any weakness such as this will soon be exploited by breaking waves.

A comprehensive publication on use and design of sea walls in the UK is by Thomas and Hall (1992). It covers a wide range of aspects, but in particular the functional requirements of a wall itself including:

- stability against wave attack;
- wave reflections;
- run-up and overtopping;
- spray;
- aesthetics;
- durability and likely life;
- ease of construction and requirements for construction;

(a)

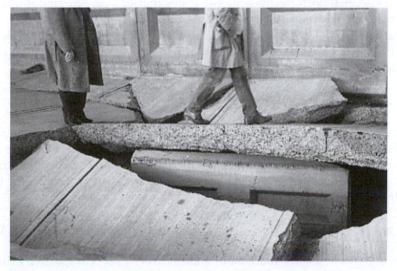

(b)

Figure 9.45 Failure of a pre-cast concrete sea wall.

- availability of materials;
- required level and ease of maintenance;
- flexibility with respect to scour and settlement;
- strength with respect to imposed loads;
- ease of access along and across the wall;
- safety;
- cost.

The design of breakwaters with vertical walls, most commonly caissons, is a specialised subject and beyond the scope of this book. PIANC (2003a) is a detailed and comprehensive guide covering all of the aspects of detailed design and should be considered as essential reading if such a design is to be undertaken.

9.4.10   Beach nourishment design

The significant benefits of beach nourishment have been discussed in Section 9.3.1. Selection of suitable borrow material is most important if a nourishment scheme is to be successful. Fine material tends to be unstable on a beach and moves offshore rapidly spreading itself over large areas. Coarse material tends to be more stable, but is not always economically available. In any event, coarser material will generate a steeper beach that might not be desirable for recreational purposes. There have also been examples where the change in beach material grading characteristics have generated undesirable features such as the increased intensity and frequency of rip currents that are dangerous to swimmers. Thus the objective of selecting suitable borrow material for a beach nourishment design should not be limited to estimating the proportion of material that will be lost after placing, but also the characteristics of the beach that will be generated in relation to its intended use.

One approximate method for estimating the losses that can be expected to occur from an area that has been nourished is based on the composite grain size distribution of both the borrow material and the native beach material. These methods (Dean 1974; James 1975) are based on the comparison between the respective grain size distributions described on the phi scale, which is:

$$\phi = -\log_2 (D) = -3.322 \log_{10} (D) \tag{9.52}$$

where D is the grain size diameter in mm. Grain size distributions on beaches generally exhibit a log-normal form and the borrow material is assumed to be similar. Grain size distributions are defined by two principal parameters. The first is the 'phi mean (μ)' that is a measure of the central tendency of the grain size distribution, which for a lognormal distribution is

$$\mu = (\phi_{84} + \phi_{16}) / 2 \tag{9.53}$$

where ϕ_{84} and ϕ_{16} are the eighty-fourth and sixteenth percentiles respectively. The second is the 'phi sorting' or 'phi standard deviation (σ)' that is a measure of the spread of the grain sizes about the phi mean and for log-normal distributions is approximated as

$$\sigma = (\phi_{84} - \phi_{16})/2 \qquad (9.54)$$

Comparison between the native material, subscripted 'n', and the borrow material, subscripted 'b', can be made by evaluating the phi mean difference

$$\delta = (\mu_b - \mu_n)/\sigma_n \qquad (9.55)$$

and the phi sorting ratio

$$\sigma_r = \sigma_b / \sigma_n \qquad (9.56)$$

These parameters can be used to derive an 'overfill ratio', R, in mathematical terms involving standard integrals. However, Figure 9.46 summarises the outcome and is sufficient to appreciate the indicators provided by the methodology. The figure is split into four quadrants for which quadrants 1 and 2 represent regions where the borrow material is more poorly sorted than the native material. Quadrants 1 and 4 represent regions where the borrow material has a finer phi mean than the native material. Points that lie in quadrants 2 or 3 will generally result in a more stable fill. Points lying in quadrant 1 will result in a more stable fill for some combinations, but losses could

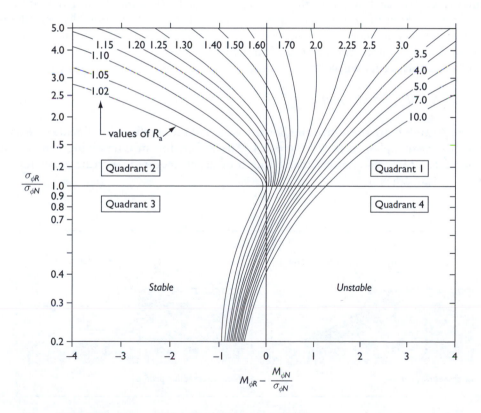

Figure 9.46 Adjusted overfill ratio (after James 1975).

be large. Points lying in quadrant 4 will generally be increasingly unstable. It should be emphasised that, whilst the figure indicates a fairly high level of precision, it should only be used as an indicative and relative descriptor of potential behaviour, so that the method should only be used in conjunction with other beach fill design techniques (Davison *et al.* 1992).

Dean (1991) developed a method based on his equilibrium profile model whereby volumes of fill would be estimated by comparing the equilibrium profile of the borrow material with that of the native material. A simplified version has been proposed by Houston (1996). This model is also consistent with the original Bruun's rule (1962) (see Equation 6.11), which simply stated suggests that the beach profile will always respond to sea level rise by adjusting the seaward profile by an equivalent amount by a balancing offshore movement of material from the upper beach, thus resulting in shoreline recession. More recently he has published a book on the subject, Dean (2002). His method is based on assuming that both the natural and nourished beach profiles conform to the characteristic polynomial equation given in Section 6.2.4.

As can be appreciated from Figure 9.46 there are compelling reasons to use a borrow material that is of similar size or coarser than the native material, in which case the post-nourished beach profile should also be similar or steeper. Referring to Figure 9.47, the nourished profile may intersect the native profile landward or seaward of the closure depth (d_c as defined in Section 6.2.4) depending on the relative slopes and the amount of dry beach width B_d that is being reclaimed. Here, the dry beach width is a free parameter in the succeeding equations. It needs to be determined by consideration of the expected cut back of the equilibrium profile under storm conditions, using either modelling predictions or historical measurements.

Intersection occurs when

$$B_d (A_n / d_c)^{3/2} \leq 1 - (A_n / A_b)^{3/2} \qquad (9.57)$$

where A_n and A_b are the native and borrow values of the coefficient in Equation (6.10), which is also explained in Section 6.2.4. Thereafter, for an intersecting profile, the volume of beach material per metre length of beach required to create an increased dry beach width is

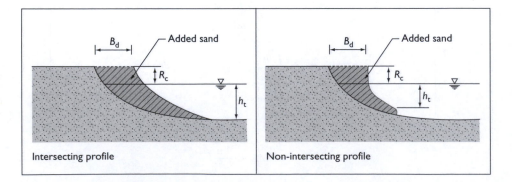

Figure 9.47 Profile definitions (after Dean 1991).

$$V_b = B_d R_c + (3/5)A_n B_d^{5/3} / (1 - (A_n / A_b)^{3/2})^{2/3} \tag{9.58}$$

and for a non-intersecting profile

$$V_b = B_d R_c + 3 d_c^{5/2}((B_d / d_c^{3/2} + (1/ A_b)^{3/2})^{5/3} A_n - (1/ A_b)^{3/2})/5 \tag{9.59}$$

In the less common case of non-intersecting profiles with a borrow material being finer than the native material, the volume of material that must be placed before there is any dry beach after the profile has adjusted to equilibrium, is

$$V_b = 3 d_c^{5/2}(1 / A_b)^{3/2} (A_n / A_b - 1)/5 \tag{9.60}$$

Dean (2002) also provides very much greater detail as well as a number of different relationships for less uniform borrow material conditions and methods of placement.

Another parameter which can be useful to estimate is the expected project half life (t_{50}). This is the time taken for the fill volume to reduce to 50 per cent of its original quantity, within the original project beach length (l). An approximate expression for this, accurate to about 15 per cent, taken from the *Coastal Engineering Manual* is

$$t_{50} = \frac{(1/2)^2 \pi}{4\varepsilon} \tag{9.61}$$

where

$$\varepsilon = \frac{KH_b^2 C_{gb}}{8} \left(\frac{\rho}{\rho_s - \rho} \right)\left(\frac{1}{1-p} \right)\left(\frac{1}{R_c + d_c} \right)$$

However, great care should be taken when making this estimate, as the result is very sensitive to the assumed breaking wave height. As the project half life may be several or many years for a large beach fill project of several to tens of kilometres, a morphologically averaged wave height is needed. This can be estimated as the long term average of, where suitable records exist.

Pilarczyk et al. (1986) have proposed a method based on a similar theme whereby the relationship between the native and the nourished profile is represented as a function of the relative fall velocity (w_s) of the respective mean grain sizes. The nourished profile is defined as

$$X_b = (w_{sn} / w_{sb})^{0.56} X_n \tag{9.62}$$

where X_n and X_b are the distance offshore of a given contour line from the intersection of mean sea level with the native and nourished profile respectively. Given the desired width of dry beach, the volume required can be readily calculated for the intersecting profile case. For the non-intersecting case it is assumed that closure is achieved by reducing the thickness at depth d_c linearly to zero at depth $3d_c$.

Profile methods have also been developed specifically for shingle beaches. Powell (1993) has proposed an equilibrium slope method for beaches to be nourished with sediment of a dissimilar grading. The method is described by Simm et al. (1996) together with some commentary on limits of application. Indeed, it should be recognised that, whilst all of the foregoing methods are useful indicators, it is normal practice to test a design against a variety of predictors. Obviously the value of detailed field monitoring data can never be undervalued in providing calibration and verification data. It can also be appreciated that predictive methods such as beach plan shape models and cross-shore beach profile models (see Chapter 6) can also play a significant complimentary role in beach nourishment design. This is particularly the case where beach control structures, which will change the natural alongshore drift rate, are to be introduced.

There are a number of other factors that need to be considered when designing a beach nourishment project as follows:

- Identification of a suitable borrow area can be a major task in itself. In the UK there are licensed areas and stringent statutory processes for developing new areas. These provide the necessary checks and balances with respect to impact of the potential borrow site on the wave regime as it might impact the coastlines as well as all of the other environmental issues. These are described in Brampton (2002).
- Environmental impact of the method of placing that may vary from spraying to pumping or bottom dumping (see Figure 9.28). Various methods are described in Simm *et al.* (1996) and Dean (2002).
- There may be handling losses during the dredging operation if the borrow material contains moderate to large fractions of fine sand. This can change the beach fill characteristics, sometimes for the better.
- There are likely to be initial profile losses as the placed profile is likely to be at variance with the natural equilibrium profile of the borrow material, although the methods described above are intended to account for this.
- There has been experience in the UK where two sources of borrow material have been mixed in the dredge hopper in order to achieved a target grain size distribution envelope. However, if the grain size distribution is plotted in the normal geotechnical format of a cumulative grain size distribution curve, the bi-modal distribution that can be created by mixing two sources can be easily masked. The result can be that there is, in fact, very little sediment in the mix at or around the target median grain size! It is therefore essential that grain size distributions are plotted as absolute percentage occurrence within chosen grain size intervals.
- When a beach fill is placed it is likely to be quite poorly sorted, especially if two sources of material have been mixed. During the sorting process, as the sediment is being worked by larger wave events, beach cliffing can take place. Such beach cliffs in excess of 1 m have been experienced in the UK for particularly poorly sorted sediments. This can pose a significant hazard to the beach user and has required expensive remedial measures of reprofiling the beach with mechanical plant to be carried out on more than one occasion until natural sorting has taken place.

Once a beach nourishment programme has been completed and the initial losses due to sorting have taken place, a maintenance programme involving periodic renourishment will usually be required. Thus, when a beach nourishment scheme is being evaluated all of the costs during the nominal lifetime of the scheme must be included. Methods of carrying out such evaluations are referred to in Section 9.1.

9.5 Design example

The following is a design example of a simple coastal defence protection revetment in a relatively sheltered location and a modest tidal range environment. The revetment is intended to protect a road from which a sea view is considered to be an important aspect. As in any design process, some experience is required in guessing the initial structure geometry. The first guess is shown in Figure 9.48. There is always a trade-off between employing a steeper slope requiring less material, but larger armour and higher crest as opposed to a flatter slope with smaller armour and a lower crest level which may be more aesthetically pleasing. In this case, given the relatively mild wave climate, a 1:2 seaward slope should be sufficient. The initial crest level should be based on an elevation that is a minimum of the MHHW plus the 1:100 year wave height.

DESIGN CRITERIA

Water levels

MHHW	+2.25 (CD)
MSL	+1.45 (CD)
MLLW	+0.45 (CD)

Road level

RL	+5.0 (CD)

Nearshore wave parameters (usually derived from a refraction/diffraction study)

1:1-year return period (RP) $H_s = 1.6\,\text{m}, T_z = 5\,\text{s}, T_p = 6.25\,\text{s}$
1.100-year return period (RP) $H_s = 3.2\,\text{m}, T_z = 7\,\text{s}, T_p = 8.75\,\text{s}$

Shoreline bed slope   1:100

Maximum allowed overtopping (see Figure 9.38)

Max limit for unsafe at all speeds   0.001 l/s/m

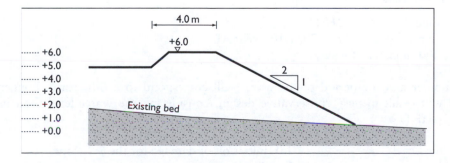

Figure 9.48 Initial structure geometry.

| Max limit for unsafe at high speeds | 0.01 l/s/m |
| Max limit for unsafe to park | 0.1 l/s/m |

Sea bed level at toe level

Assuming the crest level of +6.0 (CD) and existing bed profile, the toe level will be about +0.2 (CD).

Wave breaking according to Goda (2000) (see Section 2.6.2)

The full set of equations is given in Section 2.6.2. In this example it is assumed that the deep-water wave height (H_o) is the same as the nearshore wave height (H_s)

Examine depth limited wave at MHHW (1 Year RP, $H_s = 1.6$ m, $T_z = 5$ s, $T_p = 6.25$ s):

Wave length	L_o	= 39.1 m
Water depth at Toe	d	= 2.05 m
Relative water depth	d/H_o	= 1.281
Relative water depth	d/L_o	= 0.052
Wave steepness	H_s/L_o	= 0.041
	β_0	= 0.096
	β_1	= 0.542
	β_{max}	= min(0.92, $f(H_o, L_o)$ = 0.83) = 0.92
Shoaling coefficient	K_s	= 1.02
	H_{sb}	= min($\beta_0 H_o + \beta_1 d$, $\beta_{max} H_o$, $K_s H_o$)
		= 1.27 m unless $d/L_o > 0.2$
Design wave height at toe	H_{si}	= 1.27 m H_{sb} if $H_b < H_s$
Similarly for 1:100 RP	H_{si}	= 1.42 m H_{sb} if $H_b < H_s$

DESIGN OF THE ARMOUR ROCKS

Methodology

The Van der Meer formula will be used for stability criteria as described in Section 9.4.3.

Design parameters

Side Slope	2H:1V
$H_s = 3.2$ m,	$T_m = 7$ s, 1:100-year RP
$N = 3000$ (number of waves)	

This number and the period of the waves will correspond to a 6-h storm. Longer storms will result in very conservative design, considering the water level used in design is MHHW.

| Damage Level | = 2.0 (Initiation of damage for the 1:2 slope) |
| Bed Slope | = 1:100 |

Toe Level	= +0.20 (CD)
Water Level	= +2.25 (CD)
Permeability Coefficient (P)	0.2
Roughness coefficient	= 0.55 (to layers of rock)

Design procedure and results

The design parameters mentioned in the previous section has been used as base values and sensitivity analysis has been performed to check the effect of variation in number of waves, toe level, water level and wave period. The graphs can be seen in Figure 9.49. The resulting M_{50} for the base parameters is 800 kg. Based upon the results of sensitivity analysis and uncertainties in some of the design parameters (as P, N, .), the M_{50} equal to 1000 kg for the armour layer rocks is selected.

Derivation of non-standard rock grading (see Table 9.15).

M_{50} gradings	Tonnes	1 Narrow heavy 0.5–3 tonne
ELCL	$(y < 2)$	= 0.3
LCL	$(0 < y < 10)$	= 0.45
UCL	$(70 < y < 100)$	= 1.5
EUCL	$(97 < y)$	= 2.25
Min M_{em}	(Effective mean weight, i.e.	= 0.8
Max M_{em}	excluding pieces less than ELCL)	= 1
Min M_{50}	(Expected range of M_{50})	= 0.9
Max M_{50}		= 1.1
M_{85}/M_{15} range		2.0–4.0

Required rock grading for given M_{50} = 0.45 to 1.50 tonnes

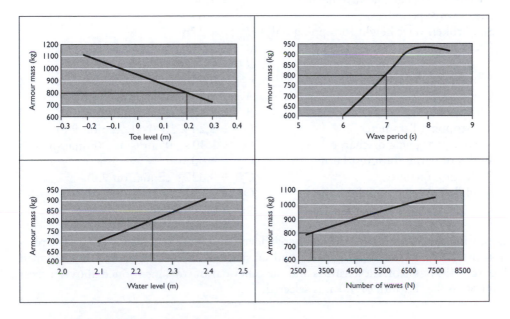

Figure 9.49 Sensitivity analysis of armour weight.

CREST LEVEL AND CONFIGURATION

The crest level is determined based upon allowable overtopping, existing road level and importance for keeping sea view from the road.

Overtopping

The overtopping rates calculated using methods in section 9.4.2. The design parameters were as follows:

Nearshore significant wave height	H_s	= 1.60–3.20 m
Mean wave period	T_m	= 5.00–7.00 s
Water level		+2.25 m (CD)
Sea bed level at toe structure		+0.2 m (CD)
Water depth	d	= 2.05 m
Sea bed slope (1 in . . .)		100
Goda sig. broken wave height	H_{sb}	1.27 (1-year RP), 1.42 (100-year RP)
Deep-water wave steepness (Broken wave)	S_{om}	= 0.02–0.03
Crest level		+5.80–6.6 (CD)
Roughness coefficient	r	= 0.55
Sea wall slope (1 in . . .)		= 2.0
Width of permeable crest berm	C_w	= 2, 3 and 4 m

Overtopping of rock slopes without crest walls

Nearshore significant wave height	H_s	= 1.60 m
Mean wave period	T_m	= 5.0 s
Sig. broken wave height (o/topping only)	H_{sb}	= 1.27 m
Crest elevation (SWL = MHHW)	R_c	= 3.75 m (Figure 9.33)
	R^*	= 0.2124 (Equation 9.5)
Roughness coefficient	r	= 0.55
Coefficient	A	= 9.39×10^{-3} (Table 9.8)
Coefficient	B	= 2.16×10^1 (Table 9.8)
Overtopping Parameter	Q^*	= $2,24 \times 10^{-6}$ (Equation 9.6)
Mean overtopping discharge	Q	= 1.40×10^{-4} m³/s/m (Equation 9.7)
Width of permeable crest berm	C_w	= 4.00 m
Reduction factor	C_r	= 0.027 (Equation 9.8)
Modified overtopping discharge	Q	= 3.80×10^{-6} m³/s/m
		= 0.004 l/s/m

Repeating the foregoing for different crest levels and wave conditions allows the relationship between overtopping rates and crest levels to be determined as shown in Figure 9.49. Based on these values and the other parameters mentioned above, a crest level of +6.0 and width of 4.0 m is selected.

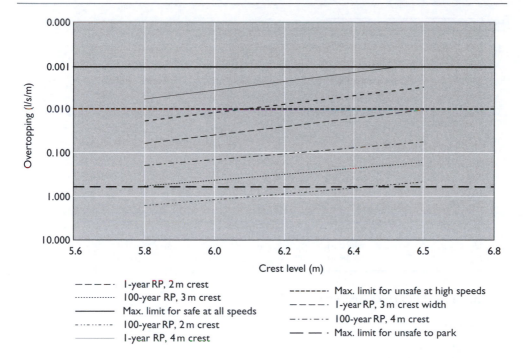

Figure 9.50 Relationship between overtopping and crest level for crest width of 4 m.

DESIGN OF UNDERLAYER

M_{50A} 1000 kg
ρ_r 2650 kg/m³
Water absorption   0.0 per cent
Narrow heavy grading for armour materials, so $M_{85}/M_{15} = 2\text{--}4$

For design purpose and because the rocks will be used as armour units $M_{85}/M_{15} = 2.0$ will be used.

$$D_{n50A} = (M_{50A}/\rho_r)^{1/3} = 0.72\,\text{m} \quad \text{(Equation 9.26)}$$
Assume $M_{50F}/M_{50A} = 1/15$, $M_{50F} = 66.7\,\text{kg}$ (Table 9.16)

which will be wide, light and light/heavy grading, according to non-standard rock grading (see Section 9.4.5).

The size of underlayer stones should be within 30 per cent of the nominal weight selected. Using a log-normal distribution for the grading non-standard rock:

$$M_{15} = M_{50}*(M_{85}/M_{15})^{-0.5} \quad \text{(Equation 9.35)}$$
$$M_{85} = M_{50}*(M_{85}/M_{15})^{0.5} \quad \text{(Equation 9.35)}$$

So for the armour stones:

M_{50}	1000 kg
D_{50}	0.72 m
M_{85}/M_{15}	2
M_{15}	707.1 kg
M_{85}	1414.2 kg
D_{15}	0.64 m
D_{85}	0.81 m

And for the underlayer:

M_{50}	66.67 kg
D_{50}	0.29 m
M_{85}/M_{15}	11
M_{15}	20.1 kg
M_{85}	221.1 kg
D_{15}	0.20 m
D_{85}	0.44 m

Checking for filter rules (see Table 9.17)

Stability

$D_{15A}/D_{85F} < 5$
$D_{15A}/D_{85F} = 1.47$ OK

Permeability

$4 < D_{15A}/D_{15F} < 20$
$D_{15A}/D_{15F} = 3.28$ Not OK

Segregation

$D_{50A}/D_{50F} < 25$
$D_{50A}/D_{50F} = 2.47$ OK

The design should be revised

Assume $M_{50F}/M_{50A} = 1/30$, $M_{50F} = 33.0$ kg

M_{50}	33.3 kg
D_{50}	0.23 m
M_{85}/M_{15}	11
M_{15}	10.1 kg
M_{85}	110.6 kg
D_{15}	0.16 m
D_{85}	0.35 m

and filter rules:

Stability
$D_{15A}/D_{85F} < 5$
$D_{15A}/D_{85F} = 1.86$ OK

Permeability

$$4 < D_{15A} / D_{15F} < 20$$
$$D_{15A} / D_{15F} = 4.13 \quad \text{OK}$$

Segregation

$$D_{50A} / D_{50F} < 25$$
$$D_{50A} / D_{50F} = 3.11 \quad \text{OK}$$

So the rock selected for the underlayer is appropriate

THICKNESS OF LAYERS

D_{n50A}	= 0.72 m
D_{n50F}	= 0.23 m
Rock type	Smooth quarrystone
K_Δ (for smooth)	= 1.05
K_Δ (for rough)	= 1.15
Number of armour layers (n)	= 2

Thickness of armour layer

$$t_A = n \times K_\Delta \times D_{n50A} = 2 \times 1.05 \times 0.72 = 1.512 \text{ m (say 1.5 m)}$$

Thickness of filter layer

$$t_F = n \times K_\Delta \times D_{n50F} = 2 \times 1.15 \times 0.23 = 0.53 \text{ m}$$

However, since the reclamation materials are probably sand and impermeable, it is better to increase the thickness of underlayer at least to the amounts mentioned in Figure 9.40.

$$t_F = 1.5 \times D_{n50A} = 1.5 \times 0.72 = 1.08 \text{ m (say 1.0 m)}$$

DESIGN OF TOE

The toe of the structure is in shallow water and exposed to the breaking wave. The maximum scour depth (d_s) can be estimated in a number of different ways as described in Table 9.18. This suggests that the scour depth could be about the maximum unbroken wave height that can be supported by the original depth (H_{max}), alternatively, the actual wave height at the toe. In either case, scour of this magnitude would be unacceptable and some toe protection would be required. Considering relative small armour size, the same material should be used in toe protection as for the primary armour. The width of toe protection should be a be minimum of 4 rocks, so that

$$W_d = 4 \times 0.72 = 2.9 \text{ m}$$

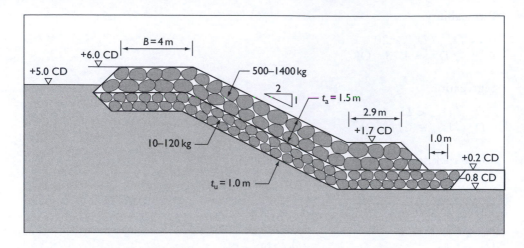

Figure 9.51 Cross section of revetment based on design calculations.

Also, due to soft material of the bed (sand), the underlayer will be embedded in the bed to act as a bed protection layer under the toe. In order to transfer the stress to bed materials, the underlayer should be extended from the end of the toe at least equal to its depth, which is 1 m. The final design section for the rock revetment is shown in Figure 9.51.

Appendix A

Summary of statistical concepts and terminology

Basic statistics

Statistical investigations may be descriptive or inferential. Generally the former type involves fairly simple techniques whilst the latter demands a higher level of critical judgement and mathematical methods. Suppose we are confronted with a set of measurements or observations obtained from past records. The task is to select a few procedures and measures by which the significant aspects of the data may be highlighted. This may be through graphing, averaging or classification. This type of analysis is descriptive as no information about theoretically related probability distributions is sought. If, on the other hand, we wish to draw conclusions about the population of the measured variable from the available sample of measurements, then certain assumptions must be made and any results interpreted accordingly. This type of analysis is inferential and is based on the mathematical theory of probability.

Averages

Many statistical inferences about a population must be made from a random sample. The first step consists of describing the numerical characteristics of the sample, usually through averages that indicate the tendency and variability of the sample.

An average is a typical or representative value, employed to replace a set of numbers. There are different kinds of averages including the mode or most frequently occurring value, the median or middle value of an ordered group, and the arithmetic mean.

The arithmetic mean, $\bar{x}$, of a set of values (or 'variates') $x_1, x_2, \ldots\ldots x_N$ is defined as

$$\bar{x} = \frac{1}{N} \sum_{i=1}^{N} x_i \tag{A.1}$$

and is often termed 'the average' in everyday discussion.

The deviation of a variate, x_i, from its mean, is defined as

$$d_i = x_i - \bar{x} \tag{A.2}$$

The sum of the deviations of a set of variates from its arithmetic mean is zero. The variability in a set of observations may also be described by averages. Common measures of variability include:

- the range (difference between the maximum and minimum values in the sample);
- the mean absolute derivation (or mad) defined by

$$\text{mad} = \frac{1}{N} \sum_{i=1}^{N} \left| x_i - \bar{x} \right| \tag{A.3}$$

which although a robust statistic is not readily used in algebraic manipulation;

- The variance,

$$\sigma^2 = \frac{1}{N} \sum_{i=1}^{N} (x_i - \bar{x})^2 \tag{A.4}$$

The variance is the average of the square of the deviations and is therefore non-negative. The positive square root of the variance, σ, is termed the standard deviation. When estimating the variance of a population from a sample many statisticians prefer to replace N by $N-1$ in the denominator of Equation (A.4). This provides an 'unbiased' estimate of the population variance. The larger N becomes the closer the two formulae agree.

If we have measurements of two variables (e.g. wind speed and wave height) we may wish to characterise the degree to which they are similar. That is, one may provide a good indication of the behaviour of the other. This can be of practical importance. For instance, it is generally easier and cheaper to obtain wind observations than wave measurements. Thus, if we can make good predictions of wave conditions from the wind measurements significant savings may be made. One measure of similarity is given by the covariance.

If we denote the two sets of variates by x_i and y_i and their respective means by and then the covariance is defined as

$$\begin{aligned}
\text{Cov}(x,y) \quad &= \frac{1}{N} \sum_{i=1}^{N} \left(x_i - \bar{x} \right) \left(y_i - \bar{y} \right) \\
&= \frac{1}{N} \sum_{i=1}^{N} x_i y_i - \frac{1}{N} \sum_{i=1}^{N} x_i \cdot \frac{1}{N} \sum_{i=1}^{N} y_i
\end{aligned} \tag{A.5}$$

The correlation coefficient r of x_i and y_i is defined by

$$R = \frac{\text{cov}(x_i, y_i)}{\sigma_x \sigma_y} \tag{A.6}$$

where σ_x and σ_y are the standard deviations of x_i and y_i respectively.

Distributions

A graphical means of obtaining an indication of the probability distribution of a set of N measurements is to construct a histogram of frequency distributions. Firstly we define a set of intervals. For example if we are considering wave heights we might choose the 1 m intervals 0–1 m, 1–2 m, 2– 3 m and so on. The number of intervals (sometimes termed 'bins') is determined by the range of the measurements and the choice of interval. Secondly we go through the set of measurements, noting into which interval each one falls, to calculate the number of measurements in each interval. Plotted as a histogram the results will take the form of Figure A.1.

If instead we plot the cumulative frequency (i.e. the number of observations with a value equal to or less than the maximum of the current interval), we obtain a plot like Figure A.2. The frequency plot provides an easy way of determining the mode

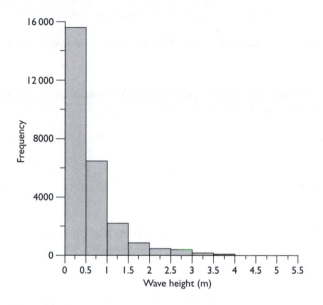

Figure A.1 Histogram (prototype pdf).

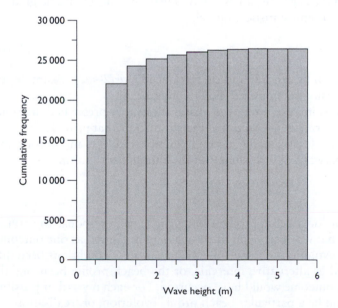

Figure A.2 Histogram (prototype cdf).

of the sample while the median may be found from the cumulative frequency plot by reading off the value on the x-axis corresponding to the intersection of the cumulative frequency 'curve' and the line $y = {}^N/_2$.

In the limit of a large number of observations we may reduce the size of the interval and the frequency and cumulative frequency histograms will more closely approximate a smooth continuous curve. Formally, if $f(x)$ is a non-negative continuous function of x over the interval $a \leq x \leq b$ the limit of the sum $\sum_{i=1}^{N} f(x_i') \cdot (x_i - x_{i-1})$ as N tends to infinity and $x_i - x_{i-1}$ tends to 0 exists and is designated as the definite integral of $f(x)$ from a to b; that is, $\int_a^b f(x)dx$

The mean (denoted by μ) and variance of a continuous random variable are defined in a manner analogous to that used in the discrete case:

$$\mu = E\{x\} = \int_a^b f(x) \cdot x \cdot dx \tag{A.7}$$

$$\sigma^2 = E\{(x - \mu)^2\} = \int_a^b f(x)(x - \mu)^2 dx \tag{A.8}$$

Here, $E\{\}$ denotes the mean or expected value.

The Normal or Gaussian density function is widely used to model observations and is defined by

$$f(x) = \frac{1}{\sqrt{2\pi}} \, e^{-\left(\frac{x-\mu}{4\sigma}\right)^{\frac{1}{2}}} \tag{A.9}$$

For variables that do not take negative values (such as wave height) the Rayleigh density function can provide a useful statistical model,

$$f(x) = \frac{x}{\alpha^2} \, e^{-x^2/2\alpha^2} \qquad x \geq 0 \tag{A.10}$$

In this case the mean is given by $\alpha\sqrt{(\pi/2)}$ and the variance by $\alpha^2(2-\pi/2)$. Examples of Gaussian and Rayleigh density fractions are shown in Figure A3.

Many inferential methods involve fitting the observations to a prescribed distribution. The fitting process determines the values of the parameters that provide an 'optimum' solution. Typically, a least squares method is employed, (i.e. minimising the square of the deviations between observations and the chosen distribution).

Stochastic processes

Consider the mean tide line on a beach profile evolving in time in response to varying environmental forcing. What we observe on the beach may be viewed as one outcome of an experiment (i.e. the evolution of the profile). Had the wave conditions been different, or storms occurred at alternative intervals, or the beach profile been slightly modified, then a different outcome would have occurred. For each myriad of possible conditions the result would be a particular beach profile evolution, or 'realisation'. A stochastic process is a rule for assigning to every outcome of an experiment a function

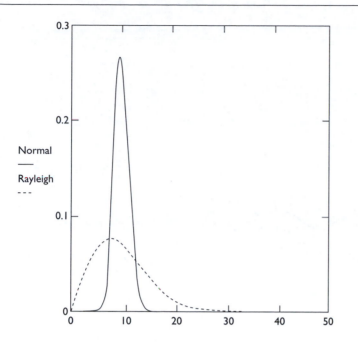

Figure A.3 Gaussian and Rayleigh probability density functions (pdfs), both with mean of 10. The Gaussian pdf has a standard deviation of 1.5. The standard deviation of the Rayleigh pdf follows directly from specifying the mean, and in this case is approximately equal to 5.2.

$x(t)$. In the example above $x(t)$ is the time-evolution of the mean tide line position, *for a particular realisation*. Illustrative realisations of $x(t)$ are shown in Figure A.4, together with the mean over all possible realisations or 'ensemble average' (denoted by $<x>$).

The statistics of a stochastic process maybe calculated in an analogous manner to continuous random variables. So, the mean of $x(t)$ is

$$E\{x(t)\} = \int_{-\infty}^{\infty} f(x,t)x\, dx \tag{A.11}$$

and the *autocorrelation* is

$$R(t_1,t_2) = E\{x(t_1)x(t_2)\} = \int_{-\infty}^{\infty}\int_{-\infty}^{\infty} f(x_1,x_2,t_1,t_2)x_1\, x_2\, dx_1 dx_2 \tag{A.12}$$

where t_1 and t_2 denote the time evolution in two different realisations. The *autocovariance* $C(t_1,t_2)$ of $x(t)$ is the covariance of the random variables $x(t_1)$ and $x(t_2)$;

$$C(t_1,t_2) = R(t_1,t_2) - \mu(t_1)\,\mu(t_2) \tag{A.13}$$

and its value for $t_1 = t_2$ equals the variance of $x(t)$.

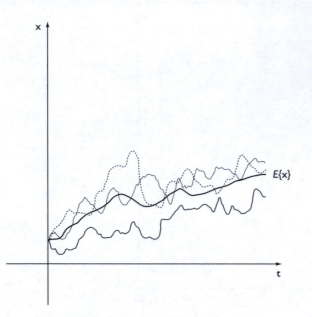

Figure A.4 Realisations and mean process.

Appendix B

Maximum likelihood estimation

Example 1

The principle may be illustrated for the binomial distribution. This distribution occurs in situations of repeated sampling or trials such as tossing a coin or rolling a die. The trials in a sequence of trials are said to be independent if the probabilities associated with each trial do not depend on the results of the preceding trials. For example, the probability of 'tails' on a given toss of a symmetric coin is ½, irrespective of what is known about the results of previous tosses. But if we try to get an ace by drawing cards one at a time without replacing them in the pack, the trials are dependent (the probability of drawing an ace at any particular turn will depend on how many previous cards have been taken and how many of these were aces!).

When an event has constant probability, p, of success, the probability of m successes in n independent trials can be computed as follows. A sequence of m successes and $n - m$ failures is represented by a sequence of m letters S and $n - m$ letters F:

SSFS.........FFS

Since the trials are independent, the probability of any one sequence is

$$ppqp.........qqp = p^m q^{n-m}$$

where $q = 1 - p$. However, the m successes can occur in any order within the n trials so the total number of possible sequences with m successes is $^nC_m = n!/\{m!(n - m)!\}$, where $n! = n.(n - 1).(n - 2) .... 2.1)$. The probability of m successes in n trials is $^nC_m p^m q^{n-m}$ which is the Binomial distribution.

To illustrate the use of this formula we find the probability that a 6 will occur exactly 4 times in the course of 10 throws of a die. Here, $p = 1/6$, $q = 5/6$, $n = 10$, $m = 4$. Hence the probability is

$$\frac{10!}{4!6!}\left(\frac{1}{6}\right)^4\left(\frac{5}{6}\right)^6 = 0.05427$$

The same ideas can be applied to determine the probability of the annual maximum water level exceeding, say, the 1 in 50-year level exactly m times in the next n years.

Now we apply these ideas to the problem of calculating a maximum likelihood estimate.

Example 2

We have been given a coin which is suspected may be biased. We must determine which of three hypothetical values of the probability of obtaining 'heads' is most likely: 0.4, 0.5 or 0.6. We are also told that in 15 tosses of the coin 9 heads and 6 tails were obtained.

If $p = 0.4$ the probability of a sample result such as that given would be:

$$^{15}C_9(0.4)^9(0.6)^6 = 0.061;$$

If $p = 0.5$ the probability becomes

$$^{15}C_9(0.5)^9(0.5)^6 = 0.153;$$

And if $p = 0.6$ the probability becomes

$$^{15}C_9(0.6)^9(0.4)^6 = 0.207.$$

(Note that $^{15}C_9 p^9 q^6$ is the likelihood function, where we specify values of p.)

The use of the principle of maximum likelihood to decide among the three possibilities leads to the choice $p = 0.6$ since this is the value of p that would have made the given sample the most likely result.

Example 3

Ten throws of a suspect die give the result 6, 6, 6, 1, 6, 6, 3, 6, 6, 4. For what values of p is the probability of the observed result a maximum?

The probability of getting seven 6s and three other numbers is

$$^{10}C_7 p^7 q^3$$

The probability is maximum when $p^7 q^3 = p^7(1-p)^3$ is a maximum. In turn this is maximum when the logarithms are a maximum. That is when

$$\log(p^7(1-p)^3) = 7\log(p) + 3\log(1-p)$$

is a maximum. Differentiating, and equating to zero gives,

$$7/p = 3/(1-p)$$

or $p = 0.7$ for the maximum. This estimate is the maximum likelihood estimate.

The principle of maximum likelihood is equally applicable to continuous distributions, and this is discussed in the next two examples.

Example 4

Suppose we are sampling from a normally distributed population with known variance σ^2 and that it is required to find the maximum likelihood estimator of the population mean μ on the basis of a sample of size N from the population X_1, X_2, X_3, X_4, X_5, ..., X_N.

The density of each X_i is

$$f(X_i \mid \mu) = \frac{1}{\sigma\sqrt{2\pi}} e^{-(X_i-\mu)^2/2\sigma^2}$$

Assuming the trials to be independent, the likelihood function is simply the product of the N density functions:

$$L(X_1, X_2, ...X_N \mid \mu) = (f(X_1 \mid \mu)f(X_2 \mid \mu).........f(X_i \mid \mu)$$

That is,

$$L(X_1, X_2, ...X_N \mid \mu) = \left(\frac{1}{\sigma\sqrt{2\pi}} e^{-(X_1-\mu)^2/2\sigma^2} \right)$$

$$\left(\frac{1}{\sigma\sqrt{2\pi}} e^{-(X_2-\mu)^2/2\sigma^2} \right).....\left(\frac{1}{\sigma\sqrt{2\pi}} e^{-(X_i-\mu)^2/2\sigma^2} \right)$$

or,

$$L(X_1, X_2, ...X_N \mid \mu) = \left(\frac{1}{\sigma\sqrt{2\pi}} \right)^N e^{-\sum_1^N (X_i-\mu)^2/2\sigma^2}$$

To minimise $L(X_1, X_2, ...X_N|\mu)$ we take the logarithm, differentiate and set the result equal to zero:

$$\log(L) = N\log\left(\frac{1}{\sigma\sqrt{2\pi}} \right) - \frac{\sum_1^N (X_i-\mu)^2}{2\sigma^2}$$

so

$$\frac{d\log(L)}{d\mu} = -\frac{\sum_1^N 2(X_i-\mu)(-1)}{2\sigma^2} = \frac{\sum_1^N (X_i-\mu)}{\sigma^2}$$

Setting this equal to zero we obtain:

$$\sum_{1}^{N}(X_i - \mu) = 0$$

or $\qquad \sum_{1}^{N} X_i - \sum_{1}^{N}\mu = 0$

or $\qquad \sum_{1}^{N} X_i - N\mu = 0$

which implies that $\mu = \dfrac{\sum_{1}^{N} X_i}{N}$

Thus, for populations having a normal distribution the sample mean is a maximum likelihood estimator of μ.

Example 5

We are given the following sequence of 10 independent wave height measurements: H_i = 3.2, 4.6, 2.9, 2.4, 5.6, 4.0, 2.5, 3.1, 2.0, 3.3, for i = 1, . .,10.

(a) Find the maximum likelihood estimate Gaussian density function given that the variance is 1.1.
(b) Find the maximum likelihood estimate for the Rayleigh density function parameter b where $f(H) = (H/b^2)exp\{-H^2/2b^2\}$ for $H \geq 0$, 0 otherwise.
(c) Plot the empirical distribution based on frequency of occurrence for wave height bins 0–1, 1–2, 2–3, 3–4, 4–5, and 5–6. Calculate the corresponding probability densities from the two maximum likelihood density functions in (a) and (b).
(d) Which of (a) and (b) would be your preferred choice and why?

(a) From Example 4 we know that the maximum likelihood estimate for μ is given by

$$\mu = \frac{\sum_{i} H_i}{10} = 3.33m$$

Therefore the density function is

$$f(H) = \frac{1}{\sigma\sqrt{2\pi}} e^{-(H-\mu)^2/2\sigma^2} =$$

$$\frac{1}{1.1\sqrt{2\pi}} e^{-(H-3.33)^2/2(1.1)^2} = 0.3627 e^{-(H-3.33)^2/2.42}$$

(b) Using the Rayleigh density function the likelihood function maybe written as

$$L = \frac{H_1 H_2 H_{3......} H_{10}}{b^{20}} e^{-\sum_i H_i^2 / 2b^2}$$

Taking logarithms

$$\log(L) = Log(H_1 H_2 H_{3......} H_{10}) - 20\log(b) - \frac{\sum_i H_i^2}{2b^2}$$

Differentiating with respect to the unknown parameter b we obtain

$$\frac{d}{db}\log(L) = -\frac{20}{b} + \frac{\sum_i H_i^2}{b^3}$$

Equating this to zero gives:

$$b = \sqrt{\frac{\sum_i H_i^2}{20}}$$

Squaring each of the 10 values of wave height and summing these, dividing by 20 and then taking the square root gives $b = 2.41$. The maximum likelihood Rayleigh density function is therefore

$$f(H) = \begin{cases} \dfrac{H}{5.81} e^{-H^2/11.61} & H \geq 0 \\ 0 & otherwise \end{cases}$$

(c) The empirical distribution is determined by sorting the given values:

Bin	0–1	1–2	2–3	3–4	4–5	5–6
Number of occurences	0	1	3	4	1	1
Empirical Probability	0	0.1	0.3	0.4	0.1	0.1
Gaussian	0.15	0.9	0.28	0.36	0.21	0.05
Rayleigh	0.08	0.22	0.25	0.22	0.13	0.07

The probabilities are determined by dividing the number of occurrences in each bin by the total number of observations, in this case 10. Values for the maximum likelihood Gaussian and Rayleigh functions are obtained by substituting the mid-bin value of wave height (e.g. 1.5 for the 1–2 m bin), into the respective density functions. The results are shown in the table above and are plotted in Figure B.1 below.

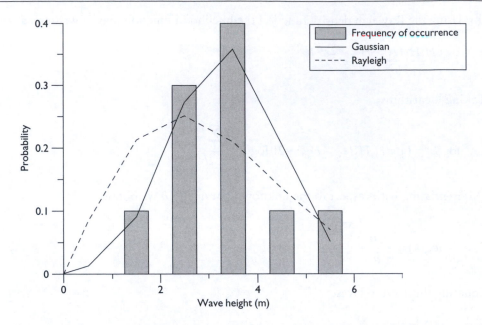

Figure B.1 Wave height distributions: empirical (histogram); best-fit Gaussian (full line); best-fit Rayleigh (broken line).

(d) The sample mean and variance are 3.33 and 1.20. The mean of the maximum likelihood Gaussian function is 3.33, and we are given the variance to be 1.1. The mean and variance of the maximum likelihood Rayleigh function are 3.02 and 2.49, respectively (see Appendix A). Thus on a comparison of the mean and variance of the sample and 'best fit' distributions the Gaussian density function appears to provide a better fit to the data. However, from a physical viewpoint this is not an ideal model because it gives a non-zero probability of negative wave heights (e.g. check that $f(-0.5) = 0.00085$ for the distribution found in (a)), and the Rayleigh density would be a better choice. In practice you would want many more than 10 observations to have confidence in the distribution obtained from maximum likelihood estimation of parameters.

Harmonic analysis results

Results of a harmonic analysis of water level recordings for a site near Cromer, UK (see Section 4.4).

Number	Harmonic name	Frequency (cycles/hour)	Amplitude (m)	Phase (degrees)
1	Z0	0	2.7783	0
2	SSA	0.00022816	0.0572	323.93
3	MSM	0.00130978	0.0329	306.63
4	MM	0.00151215	0.018	212.12
5	MSF	0.00282193	0.0173	245.18
6	MF	0.00305009	0.0345	320.57
7	ALP1	0.03439657	0.0094	24.03
8	2Q1	0.03570635	0.0105	355.49
9	SIG1	0.03590872	0.0065	274.12
10	Q1	0.0372185	0.0425	61.85
11	RHO1	0.03742087	0.013	84.18
12	O1	0.03873065	0.1604	115.68
13	TAU1	0.03895881	0.0095	51.35
14	BET1	0.04004044	0.0032	114.16
15	NO1	0.0402686	0.0224	120.05
16	CHI1	0.04047097	0.0015	133.64
17	P1	0.04155259	0.0565	268.64
18	K1	0.04178075	0.1477	285.17
19	PHI1	0.04200891	0.0033	329.99
20	THE1	0.04309053	0.0053	242.61
21	J1	0.0432929	0.0064	358.92
22	SO1	0.04460268	0.0062	88.31
23	OO1	0.04483084	0.0036	10.81
24	UPS1	0.04634299	0.001	61.26
25	OQ2	0.07597495	0.0062	138.12
26	EPS2	0.07617731	0.0076	221.5
27	2N2	0.0774871	0.0381	124.05
28	MU2	0.07768947	0.0163	173.65
29	N2	0.07899925	0.2969	135.83
30	NU2	0.07920162	0.062	136.98
31	M2	0.0805114	1.5589	159.56
32	MKS2	0.08073956	0.0065	326.48
33	LDA2	0.08182118	0.03	154.03
34	L2	0.08202355	0.0667	171.18
35	S2	0.08333334	0.5183	205.86
36	K2	0.08356149	0.1396	203.58

(Continued)

Number	Harmonic name	Frequency (cycles/hour)	Amplitude (m)	Phase (degrees)
37	MSN2	0.08484548	0.0168	15.09
38	ETA2	0.08507364	0.0044	271.95
39	MO3	0.11924206	0.0267	202.36
40	M3	0.1207671	0.0115	198.8
41	SO3	0.12206399	0.0124	287.03
42	MK3	0.12229215	0.0288	11.77
43	SK3	0.12511408	0.0112	79.62
44	MN4	0.15951064	0.0337	206.15
45	M4	0.1610228	0.0867	223.4
46	SN4	0.16233258	0.0109	284.18
47	MS4	0.16384473	0.0681	268.85
48	MK4	0.16407289	0.0197	270.93
49	S4	0.16666667	0.0098	337.86
50	SK4	0.16689482	0.0052	348.7
51	2MK5	0.20280355	0.016	60.38
52	2SK5	0.20844741	0.0005	184.25
53	2MN6	0.24002205	0.017	287.56
54	M6	0.2415342	0.0275	312.35
55	2MS6	0.24435614	0.0261	354.39
56	2MK6	0.24458429	0.0055	6.88
57	2SM6	0.24717806	0.0061	26.61
58	MSK6	0.24740623	0.0025	32.44
59	3MK7	0.28331494	0.0031	130.32
60	M8	0.32204559	0.0062	359.32

References

Abbott, M.B. and Price, W.A. (eds.), 1993. *Coastal, Estuarial and Harbour Engineer's Reference Book*, Chapman and Hall, London, UK

Abernethy, C.L. and Gilbert, G. 1975. *Refraction of Wave Spectra*, Hydraulics Research Station, Wallingford Report INT117, May 1975.

Abramowitz, M. and Stegun, I. 1964. *Handbook of Mathematical Functions*, Applied Mathematics Series, Vol. 55 (Washington: National, Bureau of Standards; reprinted by Dover Publications, New York).

Ackers, P., 1993. Sediment transport in open channels: Ackers and White update. *Proc. Instn. Civil Engrs; Water, Maritime and Energy*, 101, pp. 247–9.

Alexander L.V., Tett, S.F.B. and Jónsson, T., 2005. Recent observed changes in severe storms over the United Kingdom and Iceland. *Geophysical Research Letters*, 32, L13704.

Allen, J.R.L. (ed.), 1998. *Concrete in Coastal Structures*, Thomas Telford, London, UK.

Allen, J., 1985. Field evaluation of beach profile response to wave steepness as predicted by the Dean model, *Coastal Engineering*, 9, pp. 71–80.

Allsop, N.W.H., 1983. Low-crested breakwaters: Studies in random waves, *Proc. Coastal Structures '83*, ASCE, USA.

Allsop, N.W.H., 1998. Hydraulic performance and stability of coastal structures, in *Concrete in Coastal Structures*, Allen.

Allsop, N.W.H., McKenna, J.E., Vicinanza, D. and Whittaker, T.T.J., 1996. New design methods for wave impact loadings on vertical breakwaters and seawalls. *25th International Conference on Coastal Engineering*, Sept 96, Florida, ed. Billy L. Edge, American Society of Civil Engineers, pp. 2508–2521.

Allsop, N.W.H. 2000 Wave forces on vertical and composite walls. In Herbich, J. (ed.), *Handbook of Coastal Engineering*, McGraw-Hill, New York.

Al-Mashouk, M., Reeve, D.E., Li, B. and Fleming, C.A., 1992. ARMADA: An efficient spectral wave model, *Hydraulic and Environmental Modelling: Coastal Waters*, Ashgate, Proc. 2nd Int. Conf. On Hydr. and Env. Modelling of Coastal, Estuarine and River Waters, Vol. 1, pp. 433–444.

Ames, W.F., 1977. *Numerical Methods for Partial Differential Equations*, Thomas Nelson and Sons, London.

Ang, A.H-S. and Tang, W.H., 1984. *Probability Concepts in Engineering Planning and Design, Volume II – Decision, Risk and Reliability*, John Wiley & Sons, New York.

Aranuvachapum, S. and Johnson, J.A., 1979. Beach profiles at Gorleston and Great Yarmouth, *Coastal Engineering*, 2, pp. 201–213.

Arhens, J.P. and Cox J., 1990. Design and Performance of Reef Breakwaters, *Journal of Coastal Research*, pp. 61–75.

Arthur, R.S., Munk, W.H., and Isaacs, J.D., 1952. The direct construction of wave rays, *Trans. Am. Geophys. Union*, 33, pp. 855–865.

Baas, A.C.W., 2002. Chaos, fractals and self-organization in coastal geomorphology: simulating dune landscapes in vegetated environments, *Geomorphology*, 48(1–3), pp. 309–328

Bagnold, R.A., 1963. *Beach and nearshore processes* Part 1: Mechanics of marine sedimentation. The sea: Ideas and observations, Vol. 3, Interscience, New York/Hill M. N., pp. 507–528.

Bagnold, R.A., 1966. An approach to the sediment transport problem from general physics. Professional Paper, 422—I. US Geological Survey, Washington, DC.

Bailard, J.A., 1981. An energetics total load sediment transport model for a plane sloping beach. *Journal of Geophysical Research*, 86(CII), pp. 10938–10954.

Bailard, J.A., 1984. A simplified model for longshore sediment transport. *Proceedings of the 19th International Conference on Coastal Engineering*, Houston, American Society of Civil Engineers, pp. 1454–1470.

Bakker, W.T., 1968. The dynamics of a coast with a groin system, *Proc. 11th International Conference on Coastal Energy*, ASCE. London, pp. 492–517.

Bakker, W.T., Klein Breteler, E.H.J. and Roos, A., 1970. The dynamics of a coast with a groyne system, *Proc. 12th International Conference on Coastal Engineering*, ASCE, pp. 1001–1020.

Banyard, L. and Herbert D.M., 1995. The effect of wave angle on the overtopping of seawalls, HR Wallingford Report SR396, UK

Barnett, T.P., 1968. On the generation, dissipation and prediction of ocean wind waves, *J. Geophys. Res.*, 73, pp. 513–529.

Barnett, T.P. and Preisendorfer, R., 1987. Origins and levels of monthly and seasonal forecast still for United States surface air temperatures determined by canonical correlation analysis, *Monthly Weather Review*, 115, pp. 1825–1850.

Batchelor, G.K., 1967. *An Introduction to Fluid Dynamics*, Cambridge University Press, London and New York.

BAW, 1993. Code of practice: Use of geotextile filters on Waterway, Bundesanstalt fur Wasserbau, Karlsruhe, Germany.

Beavan, J., Wang, X., Holden, C., Wilson, K., Power, W., Prasetya, G. et al., 2010. Near-simultaneous great earthquakes at Tongan megathrust and outer rise in September 2009, *Nature* 466(7309), pp. 959–963.

Beji, S. and Battjes, J.A., 1994. Numerical simulation of nonlinear wave propagation over a bar, *Coastal Engineering*, 23, pp. 1–16.

Benjamin, J.R. and Cornell, C.A., 1970. *Probability, Statistics and Decision Theory for Civil Engineers*, McGraw-Hill Book Company.

Berkeley-Thorn, R. and Roberts, A.C., 1981. *Sea Defence and Coast Protection Works*, Thomas Telford, UK.

Berkhoff, J.C.W., 1972. Computation of combined refraction-diffraction, *Proc. 13th International Conference on Coastal Engineering*, Lisbon, pp. 55–69.

Bernabeu, A.M., Medina, R. and Vidal, C., 2003. A morphological model of the beach profile integrating wave and tidal influences, *Marine Geology*, 197, pp. 95–116.

Besley, P., 1999. Overtopping of Seawalls – Design and assessment manual, R&D Technical Report W178, Environment Agency, UK.

Bijker, E.W., 1971. Longshore transport computations. ASCE, *J. Waterways, Harbors and Coastal Engineering*, 97(WW4), pp. 687–701.

Bird, E.C.F., 2000. *Coastal Geomorphology: an Introduction*, John Wiley & Sons Ltd, Chichester, UK.

Birkemeier, W A. 1985. Field Data on Seaward Limit of Profile Change. *Journal of Waterways, Port, Coastal and Ocean Engineering*, ASCE, 111(3), pp. 598–602

Birolini, A., 1999. *Reliability Engineering; Theory and Practice*, Springer-Verlag, Berlin.

Boccotti, P., 2000. *Wave Mechanics for Ocean Engineering*, Elsevier Oceanography Series.

Bodge, K.R., 1992. Representing equilibrium beach profiles with an exponential expression, *J. Coastal Res.*, 8(1), pp. 47–55,

Booij, N., Holthuijsen, L.H. and Ris, R.C., 1996. The SWAN wave model for shallow water, *Proc. 25th International Conference on Coastal Engineering*, ASCE, Orlando, pp. 668–676.

Bouws, E., Günther, H., Rosenthal, W. and Vincent, C.L., 1985. Similarity of the wind wave spectrum in finite depth water, *J. Geophys. Res.*, 90, pp. 975–986.

Bosma, K.F. and Dalrymple, R.A., 1996. Beach profile analysis around Indian River inlet, Delaware, USA, *Proc. 25th International Conference on Coastal Engineering*, ASCE, Orlando, pp. 2829–2842.

Bracewell, R.N., 1986. *The Fourier Transform and its Applications*, 2nd Edn. McGraw-Hill, Singapore, 474p.

Bradbury, A.P., 2002. Predicting breaching of shingle barrier beaches- recent advances to aid beach management, *DEFRA Conference on Coastal and River Engineers*, 2002.

Bradbury, A.P. and Allsop, N.W.H., 1988. Hydraulic performance of breakwater crown walls, Report SR146, Hydraulics Research Ltd., Wallingford, UK

Brampton, A. (ed.), 2002. *ICE Design and Practice Guides: Coastal defences*, Thomas Telford, UK.

Brampton, A.H., 1999. Wave climate change: indication from simple GCM outputs, HR Wallingford Ltd, Report TR 80, Wallingford, UK.

Bretschneider, C.L., 1952. Revised wave forecasting relationships, *Proc. 2nd International Conference on Coastal Engineering*, ASCE, Council on Wave Research.

Bretschneider, C.L., 1958. Revisions in wave forecasting: Deep and shallow water, *Proc. 6th International Conference on Coastal Engineering*, ASCE, Council on Wave Research.

Bretschneider, C.L., 1959. Hurricane design – wave practices, *Trans. ASCE*, 124, pp. 39–62.

Bretschneider, C.L., 1977. *On the Determination of the Design Ocean Wave Spectrum*, James K.K. Look Laboratory of Oceanographic Engineering, University of Hawaii, 7(1), pp. 1–23.

Broch, J.T., 1981. *Principles of Analog and Digital Frequency Analysis*, Tapir, Norway.

Broker, I.B., Deigaard, R. and Fredsoe, J., 1991. Onshore/offshore sediment transport and morphological modelling of coastal profiles, *Proc. Coastal Sediments '91*, ASCE, Seattle, pp. 643–657.

Broomhead, D.S. and King, G.P., 1986. Extracting qualitative dynamics from experimental data, *Physica, 20D*, pp. 217–236.

Brown, C.T., 1979. Armour units – random mass or disciplined array, *Coastal Structures '79*, Vol. 1, ASCE, USA.

Bruun, P.M., 1954. *Coast Erosion and the Development of Beach Profiles*, Beach Erosion Board Tech. Memo No. 44, US Army Corps of Engineers.

Bruun, P.M., 1962. Sea-level rise as a cause of shore erosion, *J. Wtrway., Harb. and Coast. Engrg. Division* ASCE, 88 (WW1), pp. 117–130.

Bruun, P.M., 1983. Review of conditions for uses of the Bruun rule of erosion, *J. Coastal Engrg.*, 7(1), pp. 77–89

Bruun, P.M., 1989. *Port Engineering*, Vols. 1 and 2, Gulf Publishing, Houston, USA

BSI, BS6349, 1991. *Maritime Structures*, Part 1 – General Criteria, Part 2 – Design of Quay Walls, Jetties and Dolphins, Part 3 – Design of Dry Docks, Locks, Slipways and Shipbuilding Berths, Shiplifts and Dock and Lock Gates, Part 4 – Design of Fendering and Mooring Systems, Part 5 – Code of Practice for Dredging and Land Reclamation, Part 6 – Design of Inshore Mooring and Floating Structures, Part 7 – Guide to the Design and Construction of Breakwaters. British Standards Institute, London, UK.

Burcharth, H.F., 1994. The design of breakwaters, Ch. 29, in *Coastal, Estuarial and Harbour Engineer's Reference Book*, see Abbott and Price (1993).

Burcharth, H.F., Frigaard, P., Uzcanga, J., Berenguar, J.M., Madrigal, B.G. and Villanueva,

J., 1995. Design of Ciervana Breakwater, Bilbao. *Proc. Adv. Coast. Structures Breakwaters Conf.*, Institution of Civil Engineers, Thomas Telford, UK, pp. 26–43.

Burcharth, H.F. and Sorensen, J.D., 1998. Design of vertical wall caisson breakwaters using partial safety factors, *Proc, 26th International Conference on Coastal Engineering*, ASCE, Copenhagen, pp. 2138–2151.

Camfield, F.E., 1980. *Tsunami Engineering*, Special Report No. 6, US Army Coastal Engineering Center, Ft. Belvoir, VA.

Carter, D.J.T., 1982. *Estimation of Wave Spectra from Wave Height and Period*, MIAS Reference Publication No. 4.

Carter, D.J.T., Challenor, P.G., Tucker, M.J, Srokosz, M.A., Pitt, E.G. and Ewing, J.A., 1986. *Estimating Wave Climate Parameters for Engineering Applications*, Offshore Technology Report no. 86, HMSO, London.

Carter, R.W.G. and Stone, G.W., 1989. Mechanism associated with erosion of sand dune cliff, Malligan, Northern Ireland, *Earth Surface Processes and Landforms*, 14, pp. 1–10.

Carter, R.W.G., 1988. *Coastal Environments*, Academic Press, London.

Carter, R.W.G., Forbes, D.L., Jennings, S.C., Orford, J.D., Shaw, J. and Taylor, R.B., 2003, Barrier and lagoon coast evolution under differing relative sea level regimes-examples from Ireland and Nova-Scotia, *Marine Geology*, 88(3–4), pp. 221–242.

Cavaleri, L. and Rizzoli, P.M., 1981. Wind wave prediction in shallow water: Theory and application, *J. Geophys. Res.*, 86, pp. 10961–10973.

Chadwick, A.J., 1989a. Measurement and analysis of inshore wave climate. *Proc. Inst. Civ. Eng. Part 2*, 87, Mar, pp. 23–38.

Chadwick, A.J., 1989b. Field measurements and numerical model verification of coastal shingle transport, in *Advances in Water Modelling and Measurement BHRA 1989*, pp. 381–402.

Chadwick, A.J., 1991a. An unsteady flow bore model for sediment transport in broken waves. Part 1: the development of the numerical model. *Proceedings of the Institution of Civil Engineers*, Part 2(91), pp. 719–737.

Chadwick, A.J., 1991b. An unsteady flow bore model for sediment transport in broken waves. Part 2: the properties, calibration and testing of the numerical model. *Proceedings of the Institution of Civil Engineers*, Part 2(91), pp. 739–753.

Chadwick, A.J., Fleming, C.A., Simm, J. and Bullock, G., 1994. Performance Evaluation of offshore breakwaters. A field and Computational Study. *Proc. of Coastal Dynamics '94*, Barcelona, Spain, ASCE, February 1994.

Chadwick, A.J., Ilic, S. and, Helm-Petersen, J., 2000. An evaluation of directional analysis techniques for multidirectional, partially reflected waves: Part 2 application to field data. *Journal of Hydraulic Research*, 38(4), pp. 253–258.

Chadwick, A.J., Pope, D.J., Borges, J. and Ilic, S., 1995a. Shoreline directional wave spectra, Part 1: an investigation of spectral and directional analysis techniques. *Proc. Inst. Civ. Eng., Water, Maritime and Energy*, 112, Sept. pp. 198–208.

Chadwick, A.J., Pope, D.J., Borges, J. and Ilic, S., 1995b. Shoreline directional wave spectra, Part 2: instrumentation and field measurements. *Proc. Inst. Civ. Eng., Water, Maritime and Energy*, 112, Sept. pp. 209–214.

Chadwick, A. J., Karunarathna, H., Gehrels, R, Massey, A. C., O'Brien, D. and Dales, D., 2005. A new analysis of the slapton barrier beach system. *Maritime Engineering*, 158(4), pp. 147–161.

Chambers, J.M., Cleveland, W.S., Kleiner, B. and Tukey, P.A., 1983. *Graphical Methods for Data Analysis*, Duxbury, Boston.

Chang, H.H., 1988. *Fluvial Processes in River Engineering*, John Wiley & Sons, New York.

Cheong, H.F. and Shen, H.W., 1983. Statistical properties of sediment movement, *J. Hydraulic Engng.*, 109(12), pp. 1577–88.

CIAD, 1985. *Computer Aided Evaluation of the Reliability of a Breakwater Design*, Final Report, CIAD Project Group, July, 1985.

CIRIA and CUR, 1991. *Manual on the Use of Rock in Coastal and Shoreline Engineering*, CIRIA Special Publication 83/CUR Report 154, London.

CIRIA 1977. *Rationalisation of Safety and Serviceability Factors in Structural Design*, Report 63, London.

CIRIA 1986. *Sea Walls: Survey of Performance and Design Practice*, CIRIA Technical Note 125, London.

CIRIA, 1996. *Beach Management Manual*, Report 153, London.

CIRIA, 2010. *Beach Management Manual*, Second Edition, Publication no. RP787, London.

CIRIA/CUR/CETMEF, 2007. *The Rock Manual. The Use of Rock in Hydraulic Engineering (2nd edition)*, C683, CIRIA, London

COAST3D, 2001. *Final Volume of Summary Papers*. Report TR121 version 2. HR Wallingford, Wallingford, UK.

Coles, S.G. and Tawn, J.A., 1994. Statistical methods for multivariate extremes: an application to structural design (with discussion), *Appl. Statistics*, 43, pp. 1–48.

Copeland, G.J.M., 1985. A practical alternative to the 'mild-slope' wave equation, *Coastal Engineering*, 9, pp. 125–149.

Cornell, C.A., 1969. A probability-based structural code, *ACI-Journ.*, 66, pp. 974–985.

Cotton, P.D., Carter, D.J.T., Allan, T.D., Challenor, P.G., Woolf, D., Wolf, J., Hargreaves, J.C., Flather, R.A., Li, B., Holden, N. and Palmer, D., 1999. *JERICHO Final Report to the British National Space Centre*, 37 pp.

Count, B.M., 1978. On the dynamics of wave power devices, *Proc. R. Soc. A.*, 363, pp. 559–579.

Cox, D.R. and Hinkley, D.V., 1974. *Theoretical Statistics*, Chapman & Hall, London.

Cuomo, G., Allsop, W., Bruce, T. and Pearson, J. (2010a). Breaking wave loads at vertical seawalls and breakwaters. *Coastal Engineering* 57, pp. 424–439

Cuomo, G., Allsop, W. and Takahashi, S. (2010b). Scaling wave impact pressures on vertical walls. *Coastal Engineering* 57, pp. 604–609.

Cunningham, B., 1908. *A Treatise on the Principles and Practice of Harbour Engineering*, Charles Griffin & Co., London.

CUR/TAW, 1990. *Probabilistic Design of Flood Defences*, Centre for Civil Engineering Research and Codes, Gouda, The Netherlands, Report 141.

Dabees, M.A. and Kamphuis, J.W., 2000. N-line: efficient modelling of 3-D beach change, *Proc. 27th International Conference on Coastal Engineering*, ASCE, Sydney, pp. 2668–2681.

Dally, W.R. and Pope, J., 1986. *Detached Breakwaters for Shore Protection*, Tech Rep CERC-86–1, USACE, Waterways Exp Stn, Vicksburg, USA.

Dales, D.G. and Al-Mahouk, M., 1991. Application of a 3-D beach response model to the Norfolk coast, *Proc. MAFF Conf. of River and Coastal Engineers*, Loughborough.

Dalrymple, R.A. and Kirby, J.T., 1988. Models for very wide-angle water waves and wave diffraction, *J. Fluid Mech.*, 192, pp. 33–50.

Dalrymple, R.A., 1992. Prediction of storm/normal beach profiles. *Journal of Waterways, Port, Coastal and Ocean Engineering* 118, pp. 1993–2000.

Damgaard, J.S. and Soulsby, R.L., 1996. Longshore bed-load transport. *Proceedings of the 25th International Conference on Coastal Engineering*, Orlando, American Society of Civil Engineers, pp. 3614–3627.

Darbyshire, M. and Draper, L., 1963. Forecasting wind-generated sea waves. *Engineering*, 195 (April).

Darras, M., 1987. IAHR list of sea state parameters: a presentation. *IAHR Seminar Wave Analysis and Generation in Laboratory Basins*. XXII Congress, Lausanne, 1–4 Sept 1987, pp. 11–74.

Davidson, M.A., Bird, P.A.D., Bullock, G.N. and Huntley, D.A., 1996. A new non-dimensional number for the analysis of wave reflection from rubble mound breakwaters. *Coastal Engineering*, 28, pp. 93–120.

Davies, A.M. and Flather, R.A., 1978. Application of numerical models of the northwest European continental shelf and the North Sea to the computation of the storm surges of November-December 1973, *Dt. Hydrogr. Z. Erg. H.*, A, 14.

Davies, A.M. and Flather, R.A., 1987. Computing extreme meteorologically induced currents, with application to the northwest European continental shelf, *Continental Shelf Research*, 7, pp. 643–683.

Davison, A.T., Nicholls, R.J. and Leatherman, S.P., 1992. Beach nourishment as a coastal management tool; an annotated bibliography on development associated with the artificial nourishment of beaches, *J. Coastal Res.*, 8(4).

Davison, A.C. and Smith, R.L., 1990. Models for exceedances over high thresholds (with discussion), *J. Roy. Statist. Soc.*, B, 52, pp. 393–442.

De Vriend, H.J., 1997. Prediction of aggregated-scale coastal evolution, *Proc. Coastal Dynamics, 1997*, ASCE, Gdansk, pp. 644–653.

De Vriend, H.J., Zyserman, J., Nicholson, J., Roelvink, J.A., Pectron, P. and Southgate, H.N., 1993a. Medium term 2DH Coastal area modelling, *Coastal Engineering*, 21, pp. 193–224.

De Vriend, H.J., Copabianco, M., Chesher, T., de Swart, H.W., Latteux, B. and Stive, M.J.F., 1993b. Approaches to long-term modelling of coastal morphology: a review, *Coastal Engineering*, 21, pp. 225–269.

Deacon, M.B., 1971. *Scientists and the Sea 1650–1900: a Study of Marine Science*, Academic Press, New York.

Dean, R.G., 1973. Heuristic models of sand transport in the surf zone. *Proc. 1st Australian Conference on Coastal Engineering*, Engineering Dynamics in the Surf Zone, pp. 208–214.

Dean, R.G., 1974. Compatibility of beach materials for beach fills, *Proc. 14th International Conference on Coastal Engineering*, ASCE, USA.

Dean, R.G., 1977. *Equilibrium Beach Profiles: US Atlantic and Gulf Coasts*, Ocean Engineering Report No. 12, Dept. of Civil Engineering, University of Delaware.

Dean, R.G. 1985. Physical modeling of littoral processes, in *Physical Modelling in Coastal Engineering*, R. A. Dalrymple, ed., A.A. Balkema, Rotterdam, The Netherlands, pp. 119–139.

Dean, R.G., 1991. Equilibrium beach profiles: characteristics and applications, *J. Coastal Research*, 7(1), pp. 53–84

Dean, R.G., 2002. *Beach Nourishment; Theory and Practice, Advanced Series on Ocean Engineering – Volume 18*, World Scientific, Singapore.

Dean, R.G. and Dalrymple, R.A., 1991. *Water Wave Mechanics for Engineers and Scientists*, Advanced Series on Ocean Engineering, Vol. 2, World Scientific, Singapore.

Dean, R.G. and Dalrymple, R.A. 2002. *Coastal Processes with Engineering Applications*. Cambridge University Press, Cambridge, UK.

Dean, R.G. and Dalrymple, R.A., 2004. *Coastal Processes with Engineering Applications*, Cambridge University Press, 475 pp.

Debernard, J.B., Røed, L.P., 2008. Future wind, wave, and storm surge climate in the Northern Seas: a revisit. Tellus 60A: 427–438.

DEFRA, 2002. *The Futurecoast Project*, London.

DEFRA, 2009. Appraisal of Flood and Coastal Erosion Risk Management, June, published by DEFRA (www.defra.gov.uk/environment/flooding).

Defant, A., 1961. *Physical Oceanography*, Vols 1 and 2, Pergamon, Oxford.

Delft Hydraulics Laboratory, 1987. *Manual on Artificial Beach Nourishment*. DHL, Delft, The Netherlands

Der Kiureghian, A. and Liu, P-L., 1986. Structural reliability under incomplete probability information, *Journal of Engineering Mechanics*, 112(1), pp. 85–104.

DETR, 1999. *Indicators of Climate Change in the UK*, London.

Dette, H.H., 1977. Effectiveness of beach deposit nourishment, *Proc. ASCE Speciality Conference 'Sediments 77'*, pp. 211–227, ASCE, USA

Dewey, R.K. and Crawford, W.R., 1988. Bottom stress estimates from vertical dissipation rate profiles on the continental shelf, *J. Phys. Oceanogr.*, 18, pp. 1167–1177.

Dickson, M., Walkden, M. and Hall, J.W., 2007. Systematic impacts of climate change on an eroding coastal region over the twenty-first century. *Climatic Change*, 84(2), pp. 141–166.

Dingemans, M.W., 1997. *Water Wave Propagation over Uneven Bottoms*. Advanced Series on Ocean Engineering, Vol. 13. World Scientific, London.

Ditlevsen, O., 1979. Narrow reliability bounds for structural systems, *J. Struct. Mech.*, 7, pp. 435–451.

DoE, 1990. *Atlas of Tidal Elevations and Currents Around the British Isles*, prepared by Proudman Oceanographic Laboratories as Department of Energy – Offshore Technology Report OTH 89 293.

Dodd, N., 1998. A numerical model of wave run-up, overtopping and regeneration, *Journal of Waterways, Port, Coastal and Ocean Engineering*, ASCE 124(2), pp. 73–81.

Donelan, M.A., 1980. Similarity theory applied to the forecasting of wave heights, periods and directions, *Proc. Can. Coast. Conf., Nat. Res. Council*, Canada, pp. 47–61.

Dong, P. and Chen, H., 1999. Probabilistic predictions of time-dependent long term beach erosion risks, *Coastal Engineering*, 36(3), pp. 243–261.

Douglas, B.C., Kearney, M.S. and Leatherman S.P., 2001. *Sea Level Rise: History and Consequences*. Academic Press, San Diego, US, 232 pp.

Dyer, K.D., 1986. *Coastal and Estuarine Sediment Dynamics*, John Wiley & Sons, Chichester.

EEFIT (Earthquake Engineering Field Investigation Team), 2009. Preliminary field report on the Samoan earthquake and tsunami of 29 September 2009. The report can be downloaded at www.istructe.org/knowledge_expertise/EEFIT/Pages/reports.aspx.

Efron, B. and Hinkley, D.V., 1978. Assessing the accuracy of the maximum likelihood estimator: observed versus expected Fisher information (with discussion), *Biometrica*, 65, pp. 457–487.

Efron, B., 1982. *The Jacknife, the Bootstrap and Other Resampling Plans*, CBMS-NSF Regional Conference Series in Applied Mathematics, Monograph 38, Society for Industrial and Applied Mathematics, Philadelphia

Environment Agency, 2000. *Risk Assessment for Sea and Tidal Defence Schemes*, Final R & D Report, Report No. 459/9/Y.

Environment Agency, 2010. Flood and Coastal Risk Management Appraisal Guidance, FCERM-AG, March, published by Environment Agency (www.environment-agency.gov.uk).

Environmental Resources Management, 2000. *Potential UK Adaptation Strategies for Climate Change*, London.

EurOtop, 2007. Wave Overtopping of Sea Defences and Related Structures: Assessment Manual, EA: ENW: KFKI (www.overtopping-manual.com).

European Commission, 2000. Communication from the Commission to the Council and the European Parliament on Integrated Coastal Zone Management: A Strategy for Europe, COM/2000/547final, EC, Brussels.

Ewing, J.A., 1971. A numerical wave prediction method for the North Atlantic Ocean, *Dtsch. Hydrogr. Z.*, 24, pp. 241–261.

Falconer, R.A., 1986. A two-dimensional mathematical model study of the nitrate levels in an inland natural basin, *Proc. Int. Conf. On Water Quality Modelling in the Inland Natural Environment*, BHRA Fluid Engineering, Bournemouth, Paper J1, June, pp. 325–344.

Falconer, R.A. and Owens, P.H., 1990. Numerical modelling of suspended sediment fluxes in estuarine waters, *Estuarine, Coastal and Shelf Science*, XXX1, pp. 745–762.

Fang, G., Kwok, Y.-K., Yu, K. and Zhu, Y., 1999. Numerical simulation of principal tidal constituents in the South China Sea, Gulf of Tonkin and Gulf of Thailand, *Continental Shelf Res.*, 19, pp. 845–869.

Feller, W., 1957. *An Introduction to Probability Theory and Its Applications*, Vol. 1, John Wiley & Sons, New York, 509 pp.

Feyen, L., Dankers, R., Barredo, J.I., Kalas, M., Bódis, K., Roo, A., et al., 2006. *Flood Risk in Europe in a Changing Climate. Institute of Environment and Sustainability, PESETA Project*, Report EUR 22313 EN, 20 pp.

Fisher, R.A. and Tippett, L.H.C., 1978. Limiting forms of the frequency distributions of the largest or smallest member of a sample, *Proc. Camb. Phil. Soc.*, 24, pp. 180–190.

Flather, R.A., 1984. A numerical model investigation of the storm surge of 31 January and 1 February 1953 in the North Sea, *Q. J. Roy. Meteor. Soc.*, 110, pp. 591–612.

Flather, R.A., 1987. Estimates of extreme conditions of tide and surge using a numerical model of the northwest European continental shelf, *Estuarine, Coastal and Shelf Science*, 24, pp. 69–93.

Fleming, C.A., 1989. *The Anglian Sea Defence Management Study*, Coastal Management, Institution of Civil Engineers.

Fleming, C.A., 1990a. *Guide on the Uses of Groynes in Coastal Engineering*, Technical Report No. 119, CIRIA, London.

Fleming, C.A. (ed.), 1990b. Groynes in Coastal Engineering: Data on Performance of Existing Groyne Systems, Technical Note No 135, CIRIA, London.

Fleming, CA, 1994, Beach response modelling. In Abbolt, M.B. and Price, W.A. (eds) *Coastal, Estuarial and Harbour Engineer's Reference Book*, Chapman and Hall, London, pp. 337–344.

Fleming, C.A., and Hamer, B., 2000. Successful Implementation of an Offshore Reef Scheme on an Open Coastline, *27th International Conference on Coastal Engineering, Sydney*, ASCE.

Fleming, C.A. and Hamer, B., 2001. Successful implementation of an offshore reef scheme, *Proc. 27th International Conference on Coastal Engineering*, July 2000, Sydney, ASCE, pp. 1813–1820.

Fleming, C.A., Pinchin, B. M. and Nairn, R.B., 1986. *Evaluation of Coastal Sediment Transport Estimation Techniques, Phase 2: Comparison with Measured Data*, Canadian Coastal Sediment Study, National Research Council, Canada, Report C2S2–19.

Fowler, R.E., 1988. One man's rubbish is another man's resources: the beneficial use of dredged material in the Lee-on-the-Solent Coast Protection Scheme, *33rd MAFF Conference of River and Coastal Engineers*, UK

Fowler, R.E. and Allsop, N.W.H., 1999. Codes, standards and practice for coastal engineering in the UK, *Proc. Coastal Structures Conference*, ASCE, Santander, Spain.

Fredsoe, J. and Deigaard, R., 1992. *Mechanics of Coastal Sediment Transport*, Advanced Series on Ocean Engineering 3, World Scientific, Singapore.

Gardner, J.D., Hamer, B., and Runcie, R., 1997. Physical protection of the seabed and coasts by artificial reefs, *Proc. 1st European Artificial Reef Research Network Conference*, March 1996, Ancona, Italy, Southampton Oceanography Centre, pp. 17–36.

Gelci, R., Cazalé, J. and Vassal, J., 1957. Prévision de la houle. La méthode des densités spectroangulaires, *Bull. Inform. Comité Central Oceanogr. D'Etude Côtes*, 9, pp. 416–435.

Gill, A.E., 1982. *Atmosphere-ocean Dynamics*, International Geophysics Series, Vol. 30, Academic Press, New York.

Goda, Y., 1974. New wave pressure formulae for composite breakwaters, *Proc. 14th International Conference on Coastal Engineering*, ASCE, New York.

Goda, Y., 2000. *Random Seas and Design of Maritime Structures*, Advanced Series on Ocean Engineering, Vol. 15, World Scientific.

Godin, G., 1972. *The Analysis of Tides*, Liverpool University press, Liverpool.

Golding, B.W., 1983. A wave prediction system for real-time sea state forecasting, *Q. J. Roy. Meteor. Soc.*, 109, pp. 393–416.

Gonzalez, M. and Medina, R., 2001. On the application of static equilibrium bay formulations to natural and man-made beaches, *Coastal Engineering*, 43, pp. 209–225.

Gotoh, H., Hayashi, M and Sakai, T., 2003. Simulation of tsunami-induced flood in hinterland of seawall by using particle method, *Proc. 28th International Conference of Coastal Engineering*, Cardiff, ed. Jane McKee Smith, pp. 1155–1167.

Gourlay, M.R., 1981. Beach processes in the vicinity of offshore breakwaters, *Proc. 5th Australian Conf. on Coastal Engineering*, Perth, Australia.

Gradshteyn, I.S. and Ryzhik, I.M., 1980. *Table of Integrals, Series and Products. Corrected and Enlarged Edition*, Academic Press (London).

Gravens, M. B., Kraus, N. C., Hanson, H. (1991). *GENESIS: Generalized Model for Simulating Shoreline Change*, Report 2, Workbook and System User's Manual. U.S. Army Corps of Engineers.

Grijm, W., 1961. Theoretical forms of shoreline. *Proceedings of the 7th International Conference on Coastal Engineering*, ASCE, pp. 219–235.

Gulev, S.K. and Hasse, L., 1999. Changes in wind waves in the North Atlantic over the last 30 years. *International Journal of Climatology*, 19, pp. 1091–1117.

Günther, H., Rosenthal, W., Stawarz, M., Carretero, J.C.. Gomez, M., Lozano, L., Serrano, O. and Reistad, M., 1998. The wave climate of the Northeast Atlantic over the period 1955–1994: the WASA wave hindcast, *The Global Atmosphere and Ocean System*, 6, pp. 121–163.

Gunawardena, Y., Ilic, S., Southgate, H.N. and Pinkerton, H., 2008. Analysis of the spatio-temporal behaviour of beach morphology at Duck using fractal methods, *Marine Geology*, 252(1–2), pp. 38–49.

Hales, L.Z. and Houston, J.R., 1983. Erosion control of scour during construction, Tech Rep. HL-80-3, Report 4, USACE, Vicksburg, USA.

Hamer, B.A., Hayman, S.J., Elsdon, P.A. and Fleming, C.A., 1998. Happisburgh to Winterton Sea Defences: Stage Two, *Breakwaters and Coastal Structures '98*, Inst. of Civil Eng. UK.

Hammersley, J.M. and Hanscomb, D.C., 1964. *Monte Carlo Methods*, Methuen, London.

Hanson, H., Gravens, M.B. and Kraus, N.C., 1988. Prototype applications of a generalised shoreline change numerical model, *Proc. 21st International Conference of Coastal Engineering*, ASCE, pp. 1265–1279.

Hanson, H. (1987). *GENESIS – a Generalized Shoreline Change Numerical Model for Engineering Use*, Ph.D. Thesis, Dept. of Water Resources Eng., Lund Inst. of Tech./Univ. of Lund, Report No. 1007.

Hanson, H., Larson, M., Kraus, NC and Capobianco, M., 1997. Modeling of seasonal variations by cross-shore transport using one-line compatible methods, *Proc. Coastal Dynamics '97*, ASCE, pp. 893–902.

Hanson, H., Larson, M., Kraus, N.C. and Gravens, M.B., 2006. Shoreline response to detached breakwaters and tidal current: comparison of numerical and physical models. *Proc. 30th International Conference of Coastal Engineering*, ASCE.

Hardisty, J., 1990. *Beaches Form and Process*, Unwin Hyman, London.

Harms, V.W., 1979. Design criteria for floating tire breakwaters, *Proc. ASCE, J. of Waterway, Port and Ocean Div.*, USA

Hashimoto, H. and Uda, T., 1980. An application of an empirical prediction model of beach profile change to Ogawara Coast, *Coastal Engineering Japan*, 23, pp. 191–204.

Hasofer, A.M. and Lind, N.C., 1974. An exact and invariant first order reliability format, *Proc. ASCE, J. Eng. Mech. Div.*, pp. 111–121.

Hasselmann, K., 1988. PIPS and POPs: the reduction of complex dynamical systems using principal interaction and oscillation patterns, *J. Geophysics Res.* 93 (D3), pp. 11 015–11 021.

Hasselmann, K., Barnett, T.P., Bouws, E., Carlsen, H., Cartwright, D.E., Enkee, K., Ewing, et al., 1973. Measurements of wind-wave growth and swell decay during the joint North Sea wave project (JONSWAP). *Deutsches Hydrographisches Zeitschrift*, 8 (12), 95.

Hasselmann, S. and Hasselmann, K., 1985. The wave model EXACT-NL, in *Ocean wave Modelling*, Plenum Press, New York, pp. 249–75.

Hawkes, P.J., Bagenholm, C., Gouldby, B.P. and Ewing, J., 1997. *Swell and Bi-Modal Wave Climate around the Coast of England and Wales*, Report SR 409, HR Wallingford Ltd, Wallingford, UK.

Hedges, T.S., 1987. Combinations of waves and currents: an introduction. *Proc. Inst. Civ. Engrs.*, Part 1, 1987, June, pp. 567–585.

Hedges, T.S., 1995. Regions of validity of analytical wave theories. *Proc. Inst. Civ. Eng., Water, Maritime, and Energy*, 112, June, pp. 111–114.

Hedges, T.S. and Shareef, M., 2002. Predicting seawall overtopping by bimodal seas, *Proc. ICCE 2002, World Scientific*, 2, pp. 2153–2164.

Hendershott, M.C. and Munk, W.H., 1970. Tides, *Annu. Rev. Fluid Mech.*, 2, pp. 205–224.

Holland, K.T., Holman, R.A., Lippmann, T.C., Stanley, J. and Plant, N., 1997. Practical use of video imagery in nearshore oceanographic field studies. *Journal Oceanic Engineering*, 22(1), pp. 81–92.

Honda, K. and Mase, H., 2007. Application of nonlinear frequency-domain wave model to Mach stem evolution and wave transformation on reef, *Proc. Coastal Structures*, pp. 1113–1124.

Horel, J.D., 1984. Complex principal component analysis. Theory and examples, *Journal of Climate and Applied Meteorology*, 23, pp. 1660–1673.

Horikawa, K., 1978. *Coastal Engineering*, University of Tokyo Press, Tokyo.

Horikawa, K. (ed.), 1988. *Nearshore Dynamics and Coastal Processes, Theory Measurement and Predictive Models*, University of Tokyo Press, Tokyo.

Hosking, A. and McInnes, R., 2002. Preparing for the impacts of climate change on the Central South Coast of England: a framework for future risk. *Journal of Coastal Research* 36(SI), pp. 381–389.

Houston, J.R., 1996. Simplified Dean's method for beach-fill design, *Journal of Waterways, Ports, Coastal and Ocean Engineering*, 122(3), pp. 143–146.

Hsu, J.R.C. and Silvester, R., 1990. Accretion behind single offshore breakwater, *Journal of Waterways, Port, Coastal and Ocean Engineering*, 116(3), pp. 362–380.

Hsu, J.R.C., Silvester, R. and Xia, Y.M., 1989. Generalities on static equilibrium bays, *Journal of Coastal Engineering*, 12(4), pp. 353–369.

Hsu, J.R.C., Uda, T. and Silvester, R., 1993. Beaches downcoast of harbours in bays, *Coastal Engineering*, 19.

Hsu, T-W, On, S-H and Wang, S-K, 1994. On the prediction of beach changes by a new 2-D empirical eigenfunction model, *Coastal Engineering*, 23, pp. 255–270.

Hubbert, G.D., Leslie, L.M. and Manton, M.J., 1990. A storm surge model for the Australian region, *Q. J. Roy. Meteor. Soc.*, 116, pp. 1005–1020.

Hudson, R.Y., 1959. Laboratory investigations of rubble-mound breakwaters, *Proc. ASCE*, 85(3), pp. 93–121.

Hughes, S.A. 1993. *Physical Models and Loboratory Techniques in Coastal Engineering*. World Scientific, Singapore, New Jersey, London, Hong Kong.

Hulme, M. and Jenkins, G.J., 1998. *Climate Change Scenarios for the UK: Scientific Report. UKCIP Technical Report No.1*, Climate Research Unit, Norwich, UK, 50 pp.

Hulme, M., Jenkins, G.J., Lu, X., Turnpenny, J.R., Mitchell, T.D., Jones, R.G., et al., 2002. *Climate Change Scenarios for the United Kingdom: the UKCIP02 Scientific Report.* Tyndall Centre for Climate Change Research, School of Environmental Sciences, University of East Anglia, Norwich, UK, 120 pp.

Hunt, I.J., 1959. Design of seawalls and breakwaters, *Proc. Am. Soc. Civ. Engnrs.*, 85, WW3, pp. 123–152.

Hunt, J.N., 1979. Direction solution of wave dispersion equation. *Journal of Waterways, Ports, Coastal and Ocean Engineering*, 105, pp. 457–459.

Huntley, D.A., Davidson, M., Russell, P., Foote, Y. and Hardisty, J., 1993. Long waves and sediment movement on beaches: recent observations and implications for modelling. *Journal of Coastal Research*, Special Issue, 15, pp. 215–229.

Iida, K., 1959. Earthquake energy and earthquake fault, *J. Earth Sci.*, Nagoya University, 7, pp. 98–107.

Iida, K., 1969. *The Generation of Tsunamis and the Focal Mechanism of Earthquakes, Tsunamis in the Pacific Ocean* (ed. W.M.Adams), East-West Center Press, University of Hawaii, pp. 3–18.

Ilic, S. and Chadwick, A.J., 1995. Evaluation and validation of the mild slope evolution equation model for combined refraction–diffraction using field data, *Coastal Dynamics 95*, Gdansk, Poland, pp. 149–160.

Ilic, S., Chadwick, A.J., Helm-Petersen, J., 2000. An evaluation of directional analysis techniques for multidirectional, partially reflected waves: Part 1 numerical investigations. *Journal of Hydraulic Research*, 38, No 4, pp. 243–252.

Ilic, S., Chadwick, A.J., Li, B., Fleming, C.A., 1997. Composite modelling of an offshore breakwater scheme in the UK CRF. In Thornton, *Coastal Dynamics 97, International Conference on the Role of Large Scale Experiments in Coastal Research*, June 1997. In Thornton, E. (ed.), ASCE, pp. 684–693.

Inman, D.L., Elwany, M.H.S. and Jenkins, S.A., 1993. Shorerise and Bar–Berm Profiles on Ocean Beaches, *J. Geophys. Res.*, 98, C10, pp. 18 181–18 199.

Ioualalen, M., Pelinovsky, E., Asavanant, J., Lipikorn, R. and Deschamps, A., 2007. On the weak impact of the 26 December Indian Ocean tsunami on the Bangladesh coast, *Natural Hazaards and Earth System Science*, 7(1), pp. 141–147.

IPCC, 1990. *Strategies for Adaption to Sea Level Rise*, Ministry for Transport and Public Works, The Hague, The Netherlands.

IPCC, 1992. *Global Climate Change and the Challenge of the Rising Sea*, Ministry for Transport and Public Works, The Hague, The Netherlands.

IPCC SRES, 2000. *Special Report on Emissions Scenarios (SRES). A Special Report of Working Group III of the Intergovernmental Panel on Climate Change.* Cambridge University Press, Cambridge, UK, 559 pp.

IPCC WG1, 2001. *Climate Change 2001: the Scientific Basis. Contribution of the Working Group I to the Third Assessment Report of the Intergovernmental Panel on Climate Change.* Cambridge University Press, Cambridge, UK, 881 pp.

IPCC WG1, 2007. *Climate Change 2007: the Physical Science Basis. Contribution of the Working Group I to the Fourth Assessment Report of the Intergovernmental Panel on Climate Change*, Cambridge University Press, Cambridge, UK, 940 pp.

IPCC WG2, 2007. *Climate Change 2007: Impacts, Adaptation and Vulnerability. Contribution of the Working Group II to the Fourth Assessment Report of the Intergovernmental Panel on Climate Change.* Cambridge University Press, Cambridge, UK, 1000 pp.

Jamal, M.H., Simmonds, D.J., Magar, V. and Pan, S., 2010. Modelling infiltration on gravel beaches with an XBEACH variant, *Proc. ICCE 2010 (Shanghai)*.

James, W.R., 1975. *Techniques in Evaluating Suitability of Borrow Materials for Beach Nourishment*, Report TM60, USACE, CERC, Vicksburg, USA.

Jay, H., Burgess, K.A. and Hosking, A., 2003. An Awareness of Geomorphology for Coastal Defence Planning. *Proc. International Conference on Coastal Management 2003*, Brighton, pp. 327–341, Institution of Civil Engineers, London, UK

Jensen, O.J., 1984. *A Monograph on Rubble Mound Breakwaters*, Danish Hydraulic Institute, Denmark.

Johnson, H.K. and Kamphuis, J.W., 1988. N-line model for a large initially conical sand island, *Proc. Symposium Mathematical Modelling of Sediment Transport in the Coastal Zone*, IAHR, Copenhagen, pp. 275–289.

Kaas, E., Andersen, U., Flather, R.A., Williams, J.A., Blackman, D.L., Lionello, et al., 2001. *Synthesis of the STOWASUS-2100 Project: Regional Storm, Wave and Surge Scenarios for the 2100 Century*. Danish Climate Centre Report 01–3, 27 pp.

Kamphuis, J.W. 1974. Practical scaling of coastal models, *Proc. of the 14th Coastal Engineering Conference*, American Society of Civil Engineers, Vol. 3, pp. 2086–2101.

Kamphuis, J.W. 1975. Coastal mobile bed model – does it work? *Proc. of the 2nd Symposium on Modeling Techniques*, American Society of Civil Engineers, Vol. 2, pp. 993–1009.

Kamphuis, J.W. 1985. On understanding scale effect in coastal mobile bed models. In Dalrymple, R.A. (ed.) *Physical Modelling in Coastal Engineering*, A.A. Balkema, Rotterdam, The Netherlands, pp. 41–162.

Kamphuis, J.W., 1991. Alongshore sediment transport rate, *Journal of Waterways, Port, Coastal and Ocean Engineering*, 117, pp. 624–640.

Kamphuis, J. W. (1993). Effective modelling of coastal morphology. *Proceedings of the 11th Australian Conference on Coastal Engineering*, Institute of Engineers of Australia, Sydney, Australia, pp. 173–179.

Kamphuis, J.W., 2001. *Introduction to Coastal Engineering and Management*, Advanced series on Ocean Engineering, Vol. 16, World Scientific, Singapore.

Kamphuis, J.W., Davies, M.H., Nairn, R.B. and Sayao, O.J., 1986. Calculation of littoral sand transport rate. *Coastal Engineering*, 10, pp. 1–21.

Kay, R. and Alder, J., 1999. *Coastal Planning and Management*, E & FN Spon, London.

Kirby, J.T., 1986. Rational approximations in the parabolic equation method for water waves, *Coastal Engineering*, 10, pp. 355–378.

Kleinhans, M.G. and Van Rijn, L.C., 2002. Stochastic prediction of sediment transport in sand-gravel bed rivers. *J. Hydraul. Eng.*, 128, ASCE, pp. 412–425.

Knauss, J., 1979. Computation of maximum discharge at overflow rock fill dams, *13th Congress of Large Dams*, New Delhi, Q50, R.9, pp. 143–160.

Komar, P.D., 1976. *Beach Processes and Sedimentation*, Prentice-Hall, Englewood Cliffs, NJ.

Komar, P.D. and McDougal, W.G., 1994. The analysis of exponential beach profiles, *Journal of Coastal Research*, 10(1), pp. 58–69.

Koshizuka, S., Tamako, H. and Oka, Y., 1995. A particla method for incompressible viscous flow with fluid fragmentation, *Comp. Fluid Dynam. J.*, 4(1), pp. 29–46.

Koutitas, G.K., 1988. *Mathematical Models in Coastal Engineering*, Pentech Press, London.

Kraus, N. C. 1997. *History and Heritage of Coastal Engineering*. ASCE, New York

Kraus, N.C., Larson, M., Kniebel, D.C., 1991. Evaluation of beach erosion and accretion predictors. *Proc. Coastal Sediments 91*, ASCE, pp. 572–587.

Kraus, N.C. and McDougal, W.G., 1996. The effects of seawalls on the beach: Part 1: An updated literature review. *J. Coastal Res.*, The Coastal Education and Research Foundation, 12(3), 691–701.

Kreibel, D.L. and Dean, R.G., 1985. Numerical simulation of time dependent beach and dune erosion, *J. Coastal Engrg.*, 9, pp. 221–245

Lagrange, J.L., 1781. Mémoire sur le théorie du mouvements des fluides, *Nouv. Mem. Acad. R. Sci. Belleltt*. Berlin. {Reprinted in *Oeuvres*, Vol. IV, pp. 695–750, Gauthier-Villars, Paris, 1869}.

Laplace, P.S., 1778/79. Recherches sur plusiers points du système du monde, *Mem. Acad. R. Sci. Paris*, 1775; 75–182, 1776; 117–267, pp. 525–552.

Larson, M., Hanson, H., Kraus, N. C. (1987). *Analytical Solutions of the One-Line Model of Shoreline Change*, Technical Report CERC-87–15, USAE-WES, Coastal Engineering Research Center, Vicksburg, Missisippi.

Larson, M., Hanson, H. and Kraus, N.C., 1997. Analytical solutions of one-line model for shoreline change near coastal structures, *Journal of Waterways, Port, Coastal and Ocean Engineering, ASCE*, July/August, pp. 180–191.

Larson, M., Donnelly, C., Jimenez, J., and Hanson, H., 2009. Analytical model of beach erosion and overwash during storms. In *Proceedings of the Institution of Civil Engineers: Maritime Engineering 162 (Issue MA3)*, pp. 115–125.

Larson, M., Donnelly, C. and Hanson, H., 2005. Analytical modelling of dune response due to wave impact and overwash. In *Proceedings of Coastal Dynamics 2005*. American Society of Civil Engineers (on CD).

Larson, M., Erikson, L. and Hanson, H., 2004. An analytical model to predict dune erosion due to wave impact, *Coastal Engineering, 51*, pp. 675–696.

Lawrence, J., Chadwick, A.J. and Fleming, C.A., 2001. A phase resolving model of sediment transport on coarse grained beaches. *Proc. International Conference on Coastal Engineering Coastal Engineering 2000*, Billy L. Edge (ed.) Vol. 1 pp. 24–636. *American Society of Civil Engineers*, Reston, Virginia.

Lay, T., Ammon, C.J., Kanamori, H., Rivera, L., Koper, K.D. and Hutko, A.R., 2010. The 2009 Samoa-Tonga great earthquake triggered doublet, *Nature*, 466(7309), pp. 964–968.

Le Mehauté, D. and Soldate, M., 1978. Mathematical modelling of shoreline evolution, *Proc. 16th International Conference on Coastal Engineering*, ASCE, London, pp. 492–517.

Leadbetter, M.R., Lindgren, G. and Rootzen, H., 1983. *Extremes and Related Properties of Random Sequences and Series*, New York, Springer

Lee, T.T., 1972. Design of a filter system for rubble mound structure, *Proc. 13th International Conference on Coastal Engineering*, ASCE, USA.

Leendertse, J.J., 1964. *Aspects of a Computational Model for Long-Period Water-Wave Propagation*, The Rand Corporation, RM-4122-APRA.

Leggett, J., Pepper, W.J. and Swart, R.J., 1992. Emissions scenarios for IPCC: An update. In Houghton J.T., Callander B.A., Varney S.K., (eds), *Climate Change 1992. The Supplementary Report to the IPCC Scientific Assessment*, Cambridge University Press, Cambridge.

Li, B., 1994a. A generalised conjugate gradient model for the mild-slope equation, *Coastal Engineering*, 23, pp. 215–225.

Li, B., 1994b. An evolution equation for water waves. *Coastal Engineering*, 23, pp. 227–242.

Li, B. and Anastasiou, K., 1992. Efficient elliptic solvers for the mild-slope equation using the multi-grid technique, *Coastal Engineering*, 16, pp. 245–266.

Li, B., Reeve, D.E. and Fleming, C.A., 1993. Numerical solution of the elliptic mild-slope equation for irregular wave propagation, *Coastal Engineering*, 20, pp. 85–100.

Li, B., Reeve, D.E. and Fleming, C.A., 2002. An investigation of inshore wave conditions using satellite data, *Proc. I.C.E. Water and Maritime Engineering*, 154(4), pp. 275–284.

Li, K.S., 1992. Point-estimate method for calculating statistical moments, *J. Engrg. Mech.*, ASCE, 118(7), pp. 1506–1511.

Li, Y., Simmonds, D.J. and Reeve, D.E., 2008. Quantifying uncertainty in extreme values of design parameters with resampling techniques, *Ocean Engineering*, 35(10), pp. 1029–1038.

Liang, G., White, T.E. and Seymour, R.J. 1992. Complex principal component analysis or seasonal variation in nearshore bathymetry, *Proc. International Conference on Coastal Engineering*, ASCE, pp. 2242–2250.

Lippman, T.C., Holman, R.A. and Hathaway, K.K., 1993. Episodic, non-stationary behaviour of a two sand bar system at Duck, NC, USA, *Journal of Coastal Research*, S1(15), pp. 49–75.

Lin, P. and Liu, P.L.-F., 1998. A numerical study of breaking waves in the surf zone, *J. Fluid Mech.*, Vol. 359, pp. 239–264.

Liu, P.L.-F. and Tsay, T.-K., 1984. Refraction-diffraction model for weakly nonlinear water waves, *J. Fluid Mech.*, 141, pp. 265–274.

Lochner, R., Faber, O. and Penny, W.G., 1948. The 'Bombardon' floating breakwater, The Civil Engineer in War, 2, Docks and Harbours (Institution of Civil Engineers).

Longuet-Higgins, M.S. and Stewart, R.W., 1964. Radiation stresses in water waves: a physical discussion, with applications, *Deep Sea Res.*, 11, pp. 529–562.

Longuet-Higgins, M.S., 1970. Longshore currents generated by obliquely incident sea waves. *Journal of Geophysical Research*, 75, pp. 6778–6789.

Lopez de San Roman, B. and Southgate, H.N., 1998. *The Effects of Wave Chronology in Beach Plan-Shape Modelling*, HR Wallingford Ltd, Report TR 67.

López de San Román-Blanco, B., Coates, T., Holmes, P., Chadwick, A. J., Bradbury, A., Baldock, T., et al., 2006. Large scale experiments on gravel and mixed beaches: experimental procedure, data documentation and initial results. *Coastal Engineering*, 53(4), pp. 349–363.

Lorenz, E.N., 1956. Empirical Orthogonal Functions and Statistical Weather Prediction. Cambridge, Department of Meteorology, MIT, Rep.1, p. 49.

Lorenz, E.N., 1963. Deterministic nonperiodic flow, *Journal of Atmospheric Science*, 20, pp. 130–141.

Lorenz, E.N., 1993. *The Essence of Chaos*, UCL Press, London, 226p.

Maddrell, R.J., 1996. Managed coastal retreat, reducing flooding risks and protection costs, Dungeness Nuclear Power Station, UK, *Coastal Engineering*, 28, pp. 1–15, Elsevier, The Netherlands.

Madsen, P.A. and Larsen, J., 1987. An efficient finite-difference approach to the mild-slope equation, *Coastal Engineering*, 11, pp. 329–351.

Madsen, P.A., Sørensen, O.R. and Schäffer, H.A., 1997. Surf zone dynamics simulated by a Boussinesq type model, Part 1. Model description and cross-shore motion of regular waves, *Coastal Engineering*, 32, pp. 255–287.

MAFF, 1993a. Flood and coastal defence: Project appraisal guidance note, Ministry of Fisheries and Food (now DEFRA), London, UK.

MAFF, 1993b. Coastal defence and the environment: A guide to good practice, Ministry of Fisheries and Food (now DEFRA), London, UK.

MAFF, 1993c. Strategy for flood and coastal defence in England and Wales, Ministry of Fisheries and Food (now DEFRA), London, UK.

MAFF, 1995. Guidance on preparation of shoreline management plans, Ministry of Fisheries and Food (now DEFRA), London, UK.

MAFF, 1997. Interim guidance for the strategic planning and appraisal of flood and coastal defence schemes, Ministry of Fisheries and Food (now DEFRA), London, UK.

MAFF, 2000. Flood and coastal defence project appraisal guidance: Approaches to risk, Report FCDPAG4, UK Ministry of Agriculture, Fisheries and Food, London.

MAFF, 2001. Flood and Coastal Defence Project Appraisal Guidance: Overview (including general guidance), FCDPAG1, Ministry of Fisheries and Food (now DEFRA), London, UK.

Magridal, B.G. and Valdes, J.M., 1995. Study of rubble mound foundation stability. *Proc. Final Workshop*, MAST II, MCS-Project.

Martin, F.L., 1999. Experimental study of wave forces on rubble mound breakwaters crown walls. PIANC Bulletin no. 102, pp. 5–17.

Masselink, G. and Pattiaratchi, C.B., 2001. Seasonal changes in beach morphology along the sheltered coastline of Perth, Western Australia, *Marine Geology*, 172, pp. 243–263.

McConnell, K., 1998. *Revetment Systems Against Wave Attack*, Thomas Telford, UK.

McDaniel, S.T., 1975. Parabolic approximations for underwater sound propagation, *J. Acoust. Soc. Am.*, 58, pp. 1178–1185.

McDowell, D.M., 1989. A general formula for estimation of the rate of transport of bed load by water, *J. Hydraulic Res.*, 27(3), pp. 355–61.

McDowell, D.M. and O'Connor B.A., 1977. *Hydraulic Behaviour of Estuaries*, Macmillan, London.

McGregor, R.C. and Miller, N.S., 1978. Scrap tyre breakwaters in coastal engineering, *Proc. 16th International Conference on Coastal Engineering*, Hamburg, Vol. III, pp. 2191–2208.

Meadowcroft, I.C., Reeve, D.E., Allsop, N.W.H., Diment, R.P., and Cross, J., 1995a. Development of new risk assessment procedures for coastal structures, In J.E. Clifford, *Advances in Coastal Structures and Breakwaters*, Thomas Telford, London, pp. 6–46.

Meadowcroft, I.C., von Lany, P.H., Allsop, N.W.H. and Reeve, D.E., 1995b. Risk assessment for coastal and tidal defence schemes, *Proc. 24th International Conference on Coastal Engineering (ICCE '94)*, Kobe, Japan, 23–24 October 1994, pp. 3154–3166.

Melchers, R. E., 1999. *Structural Reliability Analysis and Prediction*, 2nd Edition, John Wiley & Sons, Chichester, 437 pp.

Melville, W.K., 1980. On the Mach reflexion of a solitary wave, *J. of Fluid Mech.*, 98, pp. 285–297.

Miles, J.W., 1957. On the generation of surface waves by shear flows, *J. Fluid Mech.*, 3, pp. 185–204.

Miles, J.W., 1980. Solitary waves, *Annual Review of Fluid Mechanics*, 12, pp. 11–43.

Mojfeld, H.O., 1988. Depth dependence of bottom stress and quadratic drag coefficient for barotropic pressure-driven currents, *J. Phys. Oceanogr.*, 18, pp. 1658–1669.

Möller, I., Spencer, T., French, J.R., Leggett, D.J. and Dixon, M., 1996. Wave transformation over salt marshes: a field and numerical modelling study from North Norfolk, England. *Estuarine, Coastal and Shelf Science*, 49, pp. 411–426.

Möller, I. and Southgate, H.N., 2000. Fractal properties of beach profile evolution at Duck, North Carolina, *J. Geophysical Research*, 105, C5, pp. 1489–1507.

Moore, T., Zhang, K., Close, G. and Moore, R., 2000. Real time river level monitoring using GPS heighting, *GPS Solutions*, 4(1), ISSN 1080–5370, June 2000.

Mulder J.P.M., Koningsfeld M. van, Owen M.W., Rawson J. 2001. Guidelines on the selection of CZM tools, *Report RIKZ/2001.020*, Rijkswaterstaat, April 2001.

Munk, W.H., 1949. Surf beats, *Trans. Am. Geophys. Union*, 30, pp. 849–854.

Nairn, R.B., Rodrick, J.A. and Southgate, H.N., 1990. Transition zone width and implications for modelling surface hydrodynamics. In *Proc. 22nd International Conference on Coastal Engineering*, Delft, ASCE.

Nairn, R.B. and Southgate, H.N., 1993. Deterministic profile modelling of nearshore processes. Part 2: Sediment transport and beach profile development, *Coastal Engineering*, 19, pp. 57–96.

National Research Council, 1995). *Beach Nourishment and Protection*, National Academy Press, Washington DC.

Neilsen, P., 1992. *Coastal Bottom Boundary Layers and Sediment Transport*. World Scientific Publishing. Singapore, Advanced Series on Ocean Engineering, vol. 4.

Newe, J., Peters, K. and Dette, H.H., 1999. Profile development under storm conditions as a function of the beach slope. *Proc. Coastal Sediments, 1999*, Hauppauge, New York, Vol. 3, pp. 2582–2596.

Nicholson, J., Broker, I.A., Roelvink, J.A., Price, D., Tanghy, J.M. and Moreno, L., 1997. Intercomparison of coastal area morphodynamic models, *Coastal Engineering*, 31, pp. 97–123.

Niedoroda, A.W., Reed, C.W., Swift, D.J.P., Arato, H., Hoyanagi, K., 1995. Modelling shore-normal large-scale coastal evolution, *Marine Geology*, 126, pp. 181–199.

Niedoroda, A.W., Reed, C.W., Stive, M.J.F. and Cowell, P., 2001. Numerical simulations of coastal-tract morphodynamics, *Proc. Coastal Dynamics 2001*, ASCE, Lund, Sweden, p 403–412,

Nir, Y., 1982. Offshore artificial structures and their influence on Israel and Sinai Mediterranean beaches, *Proc. 18th International Conference on Coastal Engineering*, ASCE, USA

NRA, 1999. A Guide to the understanding and management of saltmarshes, National Rivers Authority (now The Environment Agency), R&D Note 324, UK.

Ochi, M.K., 1982. Stochastic analysis and probabilistic prediction of random seas, *Advances in Hydroscience*, Vol. 13, pp. 218–375.

Ochi, M.K. and Hubble, E.N., 1976. On six-parameter wave spectra, *Proc. 15th International Conference on Coastal Engineering*, 1, pp. 301–328.

Orford, J.D., Carter, R.W.G., Jennings, S.C. and Hinton, A.C., 1995. Processes and time scales by which a coastal gravel dominated barrier respond geomorphologically to sea level rise-Story Head Barrier, Nova-Scotia, *Earth Surface process and Land Forms*, 20(1), pp. 1–37.

Oumeraci, H., Kortenhaus, A., Allsop, N.W.H., de Groot, M.B., Crouch, R.S., Vrijling, J.K. and Voortman, H.G., 2001. *Probabilistic Design Tools for Vertical Breakwaters*, A.A. Balkema, Lisse, pp. 73.

Overtopping manual web site (2007), www.overtoppingmanual.com/calculation_tool.html accessed on 23/08/10.

Owen, M.W., 1980. *Design of Seawalls Allowing for Wave Overtopping*, HR Wallingford Report EX924, UK

Owen, M.W., 1988. *Wave Prediction in Reservoirs: Comparison of Available Methods*, Report No. EX 1809, Hydraulics Research Ltd, Wallingford, UK.

Owen, M.W. and Steele. A.A.J., 1991. *Effectiveness of Recurved Wave Return Walls*, HR Wallingford Report SR 261, UK.

Owen, M.W., Hawkes, P., Tawn, J. and Bortot, P., 1997. The joint probability of waves and water levels: A rigorous but practical new approach, *Proc. 32nd MAFF Conf. of River and Coastal Engineers*, Keele, UK, B4.1–B4.10.

Ozasa, H. and Brampton, A.H., 1980. Mathematical modelling of beaches backed by seawalls, *J. Coastal Engrg.*, 4(1), pp. 47–63.

Palmer, T.N., 1999. A nonlinear dynamical perspective on climate prediction, *J. Climate*, 21, pp. 575–591.

Pedersen, J. (1996). Wave forces and overtopping on crown walls of rubble mound breakwaters. An experimental Study. PhD thesis, series paper no. 12, Department of Civil Engineering, Aalborg University.

Pedrozo, A., Simmonds, D.J., Otta, A.K., Chadwick, A.J., 2006. On the cross-shore profile change of gravel beaches. *Coastal Engineering*, 53(4), pp. 335–347.

Pelnard-Considere, R., 1956. Essai de theorie de l'evolution des forms de rivages en plage de sable et de galets. *Societe Hydrotechnique de France, Proc. 4th Journees de l'Hydraulique, les Energies de la Mer, Question III*, Rapprot No. 1, pp. 289–298.

Pender, G. and Li, Q., 1996. Numerical prediction of graded sediment transport. *Proc. Instn Civ. Engrs Wat., Marit. and Energy*. 118, Dec., pp. 237–245.

Penney, W. and Price, A., 1952. The diffraction of sea waves and shelter afforded by breakwaters. *Philos. Trans Royal Society (London) Series A*, 244, pp. 236–253.

Peregrine, D.H., 1967. Long waves on a beach, *J. Fluid Mech.*, 27, pp. 815–827.

Peregrine, D.H., 1983. Breaking waves on beaches, *Ann. Rev. Fluid Mech.*, 15, pp. 149–178.

Perlin, M. and Dean, R.G., 1983. *A Numerical Model to Simulate Sediment Transport in the Vicinity of Structures*, Report, MR-83–10, US Army Corps of Engineers.

Phillips, O.M., 1957. On the generation of waves by turbulent wind, *J. Fluid Mech.*, 2, pp. 417–445.

Phillips, O.M., 1958. The equilibrium range in the spectrum of wind-generated waves, *J. Fluid Mech.*, 4, pp. 426–434.

PIANC, 1992. *Analysis of Rubble Mound Breakwaters*, supplement to PIANC Bulletin No. 78/79, PIANC (Brussels).

PIANC, 1995. *Floating Breakwaters; a Practical Guide for Design and Construction*, Report of Working Group 13, Supplement to Bulletin No. 85

PIANC, 2003a. *Breakwaters with Vertical and Inclined Concrete Walls*. Report of Working Group 28, MarCom Report WG28.

PIANC, 2003b. *State-of-the-Art of Designing and Constructing Berm Breakwaters*, Report of Working Group 40, MarCom Report WG40.

Pickands, J., 1975. Statistical inference using extreme order statistics, *Ann. Statist.*, 3, pp. 119–131.

Pierson, W.J. and Moskowitz, L., 1964. A proposed spectral form for fully developed wind seas based on the similarity theory of S.A. Kitaigorodskii, *J. Geophys. Res.*, 69, pp. 5181–5190.

Pierson, W.J., Tick, L.G. and Baer, L., 1966. Computer-based procedures for predicting global wave forecasts and wind field analyses capable of using wave data obtained by spacecraft, *6th Naval Hydrodynamics Symposium*, Washington.

Pilarczyk, K.W., 1984. *The Closure of Tidal Basins*, Huis in't Veld (ed.), Delft University Press.

Pilarczyk, K.W., 1990. Design of seawalls and dykes, in *Coastal Protection*, Pilarczyk (ed.), A.A. Balkema, Rotterdam.

Pilarczyk, K.W., Van Overeem, J. and Bakker, W.T., 1986. Design of beach nourishment scheme, *Proc. 20th International Conference on Coastal Engineering*, ASCE, USA

Plimer, I., 2009. *Heaven and Earth – Global Warming: the Missing Science*, Quartet, London.

Pontee, N.I. and Parsons, A., 2010, A review of coastal risk management in the UK, *Maritime Engineering* 163, pp. 31–42.

ope, J. and Dean, J.L., 1986. Development of Design Criteria for segmented breakwaters, *Proc. 20th International Conference on Coastal Engineering*, ASCE, USA

Powell, K.A., 1987. *Toe Scour at Seawalls Subject to Wave Action – Literature Review*, Report SR119, Hydraulics Research, Wallingford, UK.

Powell, K.A., 1990. *Predicting Short Term Profile Response for Shingle Beaches*, Report 219, HR Wallingford.

Powell, K.A., 1993. *Dissimilar Sediments: Model Tests of Replenished Beaches Using Widely Graded Sediments*, Report SR 350, HR Wallingford, UK

Powell, K.A. and Whitehouse, R., 1998. The occurence and prediction of scour at coastal structures. *33rd MAFF Conference of River and Coastal Engineers*, DEFRA, UK

Prandle, D., 1978. Residual flows and elevations in the southern North Sea, *Proc. R. Soc. Lond.*, A, 359, pp. 189–228.

Press, W.H., Flannery, B.P., Teukolsky, S.A. and Vetterling, W.T., 1986. *Numerical Recipes: the Art of Scientific Computing*, Cambridge University Press, Cambridge, UK, p. 818.

Prestedge, G.K. and Bosman, D.E., 1994. Sand bypassing at navigation inlets: solution of small scale and large scale problems, *7th Int. Conf. On Beach Preservation Tech.*, American Shore and Beach Presvn. Assn., Tampa, Florida

Pruszak Z., 1993. The analysis of beach profile changes using Dean's method and empirical orthogonal functions, *Coastal Engineering*, 19(3–4), pp. 245–261.

Pruszak, Z. and Rozynski, G., 1998. Variability of multi-bar profiles in terms of random sine functions, *Journal of Waterways, Port, Coastal and Ocean Engineering*, 124(2), pp. 48–56.

Pye, K., 1983. Coastal Dunes, *Processes in Physical Geography*, 7, pp. 531–557.

Pye, K. and Tsoar, H., 1990. *Aeolian Sand and Sand Dunes*, Hyman Ltd., London, UK.

Radder, A.C., 1979. On the parabolic equation method for water wave propagation, *J. Fluid Mech.*, 95, pp. 159–176.

Rakha, K.A., 1998. A Quasi-3D phase-resolving hydrodynamic and sediment transport model. *Coastal Engineering*, 34(3–4), pp. 277–311.

Rakha, K.A., Deigaard, R., Brøker, I., 1997. A phase-resolving cross shore sediment transport model for beach profile evolution. *Coastal Engineering*, 31(1–4), pp. 231–261.

Raudkivi, H., 1990. *Loose Boundary Hydraulics*, 3rd edn, Pergamon, Oxford.

Reeve, D.E., 1992a. Numerical simulation of nearshore residual currents, *Proc. 2nd Int. Conf. On Hydr. and Env. Modelling of Coastal, Estuarine and River Waters*, 1, pp. 55–64.

Reeve, D.E., 1992b. Bathymetric generation of an angular spectrum, *Wave Motion*, 16, pp. 17–228.

Reeve, D.E., 1996. Estimation of extreme Indian monsoon rainfall, *Intl. J. Climatol.*, 16, pp. 105–112.

Reeve, D.E., 1998. Coastal flood risk assessment, *Journal of Waterways, Port, Coastal and Ocean Engineering*, 124(5), pp. 219–228.

Reeve, D.E., 2003. Probabilistic methods for flood risk assessment. In McKee-Smith, J. (ed). *Proc. 28th ICCE*, July 2002, Cardiff, World Scientific, pp. 2324–2334.

Reeve, D.E., 2006. Explicit expression for beach response to non-stationary forcing near a groyne, *Journal of Waterways, Port, Coastal and Ocean Engineering*, 132, pp. 125–132.

Reeve, D.E. and Fleming, C.A., 1997. A statistical-dynamical method for predicting long term coastal evolution, *Coastal Engineering*, 30, pp. 259–280.

Reeve, D.E. and Hiley, R.A., 1992. Numerical prediction of tidal flow in shallow water. In Partridge, P.W. (ed.) *Computer Modelling of Seas and Coastal Regions*, Computational Mechanics Publications. Elsevier Applied Science.

Reeve, D.E., Li, B. and Fleming, C.A., 1996. Validation of storm wave forecasting. *Proc. 31st MAFF Conference of River and Coastal Engineers*, Keele, 3–5 July, pp. 4.3.1–4.3.12.

Reeve, D.E., Li, B. and Fleming, C.A., 1996. Validation of storm wave forecasting, *Proce. of 31st MAFF Conference of River and Coastal Engineers*, Keele, 3–5 July, pp. 4.3.1–4.3.12.

Reeve, D.E., Li, B. and Fleming, C.A., 1999. Classification of beach behaviour using chaos theory. *Proc. 34th MAFF Conference of River and Coastal Engineers*, Keele, 30 June–2 July, pp. 9.3.1–9.3.12.

Reeve, D.E., Li, B., and Thurston, N., 2001. Eigenfunction analysis of decadal fluctuations in sandbank morphology at Great Yarmouth, *Journal Coastal Research*, 18(2), pp. 371–382.

Reeve, D.E. and Spivack, M., 2001. Prediction of long-term coastal evolution using moment equations, *Proc. Coastal Dynamics 2001* ASCE, Lund, Sweden, pp. 723–731.

Reinen-Hamill, R.A. (1997). Numerical modelling of the Canterbury Regional Bight. The use of a shoreline evolution model for management and design purposes. *Pacific Coasts and Ports '97: Proceedings of the 13th Australasian Coastal and Ocean Engineering Conference*, Christchurch, New Zealand, pp. 359–364.

Roelvink, J.A., 1991. Modelling of cross-shore flow and morphology, *Proc. Coastal Sediments, 91*, ASCE, Seattle, pp. 603–617.

Roelvink, J.A. and Broker, I.H., 1993. Cross-shore profile models, *Coastal Engineering*, 21, pp. 163–191.

Rogers J., Hamer B., Brampton A., Challinor S., Glennerster M. and Brenton P. 2010, *Beach Management Manual (Second Edition)*, CIRIA C685, RP 787.

Rosati, J.D., 1990. *Functional Design of Breakwaters for Shore Protection: Empirical Methods*, Tech Rep CERC-90–15, USACE, Waterways Experimental Station, Vicksburg, USA

Rosenblueth, E., 1975. Point estimates for probability moments, *Nat. Acad. Sci.*, 72(10), pp. 3812–3814.

Rossiter, J.R., 1954. The North Sea Storm Surge of 31st January and 1st February 1953. *Phil. Trans. Roy. Soc. London*, A Series, 246, pp. 371–400.

Rouse, H. (ed.), 1950. *Engineering Hydraulics*, John Wiley & Sons, New York.

Royal Society, 1992. *Risk: Analysis, Perception and Management*, Report of a Royal Society Study Group. The Royal Society, London.

Rozynski, G., Larson, M. and Pruszak, Z., 2001. Forced and self organised shoreline response for a beach in the southern Baltic Sea determined through singular spectrum analysis, *Coastal Engineering*, 43, pp. 41–58.

Ruessink, B.G., van Enckevort, I.M.J., Kingston, K.S. and Davidson, M.A., 2000. Analysis of observed two- and three-dimensional nearshore bar behaviour, *Marine Geology*, 169(1–2), pp. 161–183.

Ruggiero, P., List, J., Hanes, D. and Eshleman, J., 2006. Probabilistic shoreline change modeling. *Proceedings of the 30th International Conference on Coastal Engineering*, San Diego, California, USA, pp. 3417–3429.

Sato, S. and Mitsunobu, N., 1991. A numerical model of beach profile change due to random waves, *Proc. Coastal Sediments 1991*, ASCE, Seattle, pp. 674–687.

Saville, T., McClendon, E.W. and Cochran, A.L., 1962. Freeboard allowance for waves in inland reservoirs, *Proc. ASCE*, 18, No. WW2, May.

Schoonees, J.S. and Theron, A.K. 1993. Review of the field-data base for longshore sediment transport. *Coastal Engineering*, 19, 1–25.

Schoonees, J.S. and Theron, A.K., 1994. Accuracy and applicability of the SPM longshore transport formula. *Proc. 24th International Conference on Coastal Engineering*, Kobe, American Society of Civil Engineers, pp. 2595–2609.

Schoonees, J.S. and Theron, A.K., 1995. Evaluation of 10 crosshore sediment transport/morphological models, *Coastal Engineering*, 25, pp. 1–41.

Schoonees, J.S. and Theron, A.K., 1996. Improvement of the most accurate longshore transport formula. *Proc. 25th International Conference on Coastal Engineering*, Orlando, American Society of Civil Engineers, pp. 3652–3665.

Scottish Natural Heritage, 2000. A guide to managing coastal erosion in beach dune systems.

Segur, H., 2007. Waves in shallow water, with emphasis on the tsunami of 2004. In Anjan, K. (ed.) *Tsunami and Nonlinear Waves*, Springer, Berlin Heidelberg

Seymour, R.J., 1977. Estimating wave generation on restricted fetches, *Proc. ASCE*, 103, No. WW2, May.

Shankar, J.N., Cheong, H.-F.and Chan, C.-T., 1997. Boundary fitted grid models for tidal motions in Singapore coastal waters, *J. Hydr. Res.*, 14, pp. 3–19.

Shields, A., 1936. Anwendung der Ahnlichkeitsmechanik und der Turbulenzforschung auf die Geschiebebewegung, Heft 26. *Preuss. Vers. für Wasserbau und Sch~fflau*, Berlin.

Shore Protection Manual (SPM), 1984. USACE, Coastal Engineering Research Centre, Waterways Experimental Station, Vicksburg, USA

Silvester, R. and Hsu, J.R.C., 1997. *Coastal stabilisation*, Advanced Series on Ocean Engineering, 14, World Scientific, Singapore.

Silvester, R., 1974. *Coastal Engineering*, Vols 1 and 2, Elsevier, Oxford.

Silvester, R., 1976. Headland defense of coasts. *Proc. 15th International Conference on Coastal Engineering*, Hawaii, ASCE, Vol. II, pp. 1394–1406.

Simm, J.D. (ed.), Brampton, A.H., Beech, N.W. and Brooke, J.S., 1996. *Beach Management Manual*, CIRIA Report 153, CIRIA, UK

Simons, D.B. and Richardson, E.V., 1961. Forms of Bed Roughness in Alluvial Channels. *Proc. Am. Soc. Civil Engrs.*, vol. 87, no. HY3 p.87.

Smith, R.L., 1986. Extreme value theory based on the largest annual events, *Journal of Hydrology*, 86, pp. 27–43.

Smith, S.D. and Banke, E.G., 1975. Variation of the sea surface drag coefficient with wind speed, *Q. J. Roy. Meteor. Soc.*, 101, pp. 665–673.

Snodgrass, F.E. *et al.*, 1966. Propagation of ocean swell across the Pacific, *Philos. Trans. Roy. Soc. Lond. Ser.* A, 259, pp. 431–497.

Sobey, R.J. and Young, I.R., 1986. Hurricane wind waves – a discrete spectral model, *Journal of Waterways, Port, Coastal and Ocean Engineering*, 112, pp. 370–389.

Soliman, A. and Reeve, D.E., 2003. Numerical study for small freeboard wave overtopping and overflow of sloping sea walls. *Proc. Coastal Structures*, ASCE, Portland, USA, pp. 643–655.

Sorensen, R.M., 1993. *Basic Wave Mechanics for Coastal and Ocean Engineers*, John Wiley & Sons, New York.

Soulsby, R.L., 1994. *Manual of Marine Sands. Report SR 351*, HR Wallingford.

Soulsby, R.L., 1995. Bed shear stresses due to combined waves and currents. In Stive, J.F. et al. (eds) *Advances in Coastal Morphodynamics*. pp. 4–20 to 4–23. Delft Hydraulics, Netherlands, pp. 4.20–4.23.

Soulsby, R.L., 1997. *Dynamics of Marine Sands*. Thomas Telford, London.

Soulsby, R.L. and Whitehouse, R.J.S.W., 1997. Threshold of sediment motion in coastal environments. *Proc. Pacific Coasts and Ports '97 Conf.*, Christchurch, 1, pp. 149–154. University of Canterbury, New Zealand.

Southgate, H.N., 1995. The effects of wave chronology on medium and long term coastal morphology, *Coastal Engineering*, 26, pp. 251–270.

Southgate, H.N. and Brampton, A.H., 2001. *Coastal Morphology Modelling: A Guide to Model Selection and Usage*, Report SR570, H R Wallingford.

Southgate, H.N. and Nairn, R.B., 1993. Deterministic profile modelling of nearshore processes. Part 1: Waves and currents, *Coastal Engineering*, 19, pp. 27–56.

Southgate, H.N., Wijnberg, K.M., Larson, M., Capobianco, M. and Jansen, H., 2003. Analysis of Field Data of Coastal Morphological Evolution over Yearly and Decadal Timescales. Part 2: Non-Linear Techniques, *J. Coastal Research*, 19(4), pp. 776–789.

Spivack, M. and Reeve, D.E., 2000. Source reconstruction in a coastal evolution equation, *J. Comp. Phys.*, Vol. 161, pp. 169–181.

Stauble, D.K. and Kraus, N.C., (eds.), 1993. Beach nourishment engineering and management considerations, Special volume, *Proc. Coastal Zone Management*, ASCE, USA

Steers, J.A., 1960. *The Coastline of England and Wales*, Cambridge University Press, Cambridge, UK.

Stive, M.J.F., and De Vriend, H.J., 1995. Modelling shoreface profile evolution, *Marine Geology*, 126, pp. 235–248.

Stive, M.J.F., Nicholls, R.J. and De Vriend, H.J., 1991. Sea level rise and shore nourishment: a discussion. *Coastal Engineering*, 16(1), pp. 147–163.

Stive M.J.F., Aarninkhof S.G.J., Hamm, L., Hanson, H., Larson, M., Wijnberg, K.M., Nicholls, R.J. and Capobianco, M., 2002. Variability of shore and shoreline evolution. *Coastal Engineering* 47, pp. 211–235.

Sunamura, T, 1989. Sandy beach geomorphology elucidated by laboratory modelling. In Lakhan, V.C., Tenhaile, A.S. (eds), *Applications in Coastal Modelling*, Elsevier, Amsterdam, pp. 159–213.

Sutherland, J. and Gouldby, B., 2002. Vulnerability of coastal defences to climate change. *Water and Maritime Engineering*, 156(WM2), pp. 137–145.

Sverdrup, H.U., 1945. *Oceanography for Meteorologists*, Geo. Allen & Unwin Ltd, London.

Takahasi, R., 1947. On seismic sea waves caused by deformations of the sea bottom, 3rd report, *Bull. Earthq. Res. Inst.*, University of Tokyo, 25, pp. 5–8.

Takens, F., 1981. Detecting strange attractors in turbulenc. In Rand, D.A. and Young, L.-S. (eds) *Dynamical Systems and Turbulence: Lecture Notes in Mathematics*, Vol. 898, Springer, New York, pp. 366–381.

Tanimoto, K., Moto, K., Ishizuka, S., and Goda, Y., 1976. An investigation on design wave force formulae of composite type breakwaters, *Proc. 23rd Japanese Conf. in Coastal Engineering* (in Japanese).

Tanimoto, T., Yagyu, T. and Goda, Y., 1982. Irregular wave tests for composite breakwater foundations. *Proc. 18th International Conference on Coastal Engineering*, ASCE, Vol. 3, pp. 2144–2163.

Tappert, F.D., 1977. The parabolic approximation method, in J.B. Keller and J.S.Papadakis, eds., *Wave propagation and Underwater Acoustics*, Springer, Berlin-Heidelberg, pp. 224–287.

TAW (1997) Technical Report – Erosion Resistance of Grassland as Dike Covering. Technical Advisory Committee for Flood Defence in the Netherlands. Delft 2002.

Thoft-Christensen, P. and Baker, M.J., 1982. *Structural reliability theory and its applications*, Springer-Verlag, Berlin, 267 pp.

Thomas, R.S. and Hall, B., 1992. *Sea Wall Design Guidelines*, CIRIA/Butterworths, London.

Thompson, E.F. and Vincent, C.L., 1983. Prediction of wave height in shallow water, *Proc. Coastal Structures '83*, ASCE, pp. 1000–1008.

Tolman, H. L., 1997. User manual and system documentation of WAVEWATCH-III version 1.15. NOAA / NWS / NCEP / OMB Technical Note 151, 97 pp.

Tolman, H. L., 1999. User manual and system documentation of WAVEWATCH-III version 1.18. NOAA / NWS / NCEP / OMB Technical Note 166, 110 pp.

Tolman, H. L., 2009. User manual and system documentation of WAVEWATCH III version 3.14. NOAA / NWS / NCEP / MMAB Technical Note 276, 194 pp.+ Appendices.

Townend, I.H., 1994. Risk assessment of coastal defences, *Proc. Instn. Civ. Engrs. Water, Maritime and Energy*, 106, pp. 381–384.

Townend, I.H., Fleming, C.A., McLaren, P. and Hunter-Blair, A., 1990. A Regional Study of Coastal Morphology, *Proc. 22nd International Conference on Coastal Engineering*, ASCE, Delft.

Toyoshima, O., 1972. *Coastal Engineering for Practicing Engineers – Beach Erosion*, Morikita Publishing Co., Tokyo, Japan, pp. 227–317. (English translation of Chapter 8, Offshore breakwaters, available through Coastal Eng Res Centre, USACE, Waterways Experimental Station, Vicksburg, USA).

Toyoshima, O., 1974. Design of detached breakwater system, *Proc. 14th International Conference on Coastal Engineering*, ASCE, USA

Tsinker, G., 1986. *Floating Ports*, Gulf Publishing Co., Houston, Texas, USA.

Ursell, F., 1952. Edge waves on a sloping beach, *Proc. Roy. Soc. Lond.*, A, 214, pp. 79–97.

US Army Corps of Engineers, 1984. *Shore Protection Manual*, Coastal Engineering Research Center, Washington.

US Army Corps of Engineers, 2002. *Coastal Engineering Manual* (CEM), http://chl.erdc.usace.army.mil/cem, accessed 2 Aug 2011.

Van der Meer, J.W., 1988. *Rock slopes and gravel beaches under wave attack*, PhD Thesis, Delft Technical University, April 1998. (Published as Delft Hydraulics Commn. No. 396).

Van der Meer, J.W., 1988a. Deterministic and probabilistic design of breakwater armour layers, *Journal of Waterways, Ports, and Coastal and Oceanic Engineering*, 114(1), pp. 510–517.

Van der Meer, J.W., 1988b. *Stability of Cubes, Tetrapods and Accropodes, Design of Breakwaters*, Thomas Telford, London, UK.

Van der Meer, J.W., 1990. Static and dynamic stability of loose materials. In Pilarczyk, K.W. (ed.), *Coastal Protection*, A.A. Balkema, Rotterdam.

Van der Meer, J.W., d' Angremond, K. and Gerding, E., 1995. Toe structure stability of rubble mound breakwaters. *Proc. Adv. Coast. Structures and Breakwaters Conference*, Institution of Civil Engineers, Thomas Telford Publishing, London, UK, pp. 308–321.

Van der Meer, J.W. and Koster, M.J., 1988. Application of computational model on dynamic stability, *Proc. Conf. on Breakwaters – Design of Breakwaters*, Thomas Telford, London.

Van de Meer, J. W., 1993. *Conceptual Design of Rubble Mound Breakwaters*. Publication No 483. WL. Delft Hydraulics, Delft.

Van der Meer, J.W. and Janssen, J.P.F.M., 1995. Wave run-up and wave overtopping at dikes, In Kobayashi, N. and Demirbilek, Z., (eds) *Wave Forces on Inclined and Vertical Wall Structures*, ASCE, New York, pp. 1–27.

Van der Meer, J.W., Tonjes, P. and De Waal, J.P., 1998. A code for dike height design and examination. In Abbott, M. B., Price, W. A. (eds), *Proc. Conf. on Coastlines, Structures and Breakwaters*, Thomas Telford, London.

Van Goor, M.A., Stive, M.J.F., Wang, Z.B. and Zitman, T.J., 2001. Influence of relative sea level rise on coastal inlets and tidal basins, *Proc. Coastal Dynamics 2001*, ASCE, Lund, Sweden, pp. 242–251.

Van Rijn, L.C., 1984. Sediment transport (in 3 parts). *J. Hydraulic Engng.: Part 1*, 110(10), pp. 1431–1456; *Part II*, 110(11), pp. 1613–1641; *Part III*, 110(12), pp. 1733–1754.

Van Rijn, L.C., 1993. *Principles of Sediment Transport in Rivers, Estuaries and Coastal Seas*, Aqua Publications, Amsterdam.

Van Rijn, L.C., Davies, A.G., van de Graaff, J. and Ribberink, J.S., 2001. *Sediment transport Modelling in Marine Coastal Environments*, Aqua Publications, Amsterdam.

Van Wellen, E., Chadwick, A.J. Mason, T., 2000. A review and assessment of longshore sediment transport equations for coarse grained beaches. *Coastal Engineering*, (40) 3, pp. 43–275.

Vautard, R., Yion, P. and Ghil, M., 1992. Singular spectrum analysis: a toolkit for short, noisy, chaotic signals, *Physica D*, 158, pp. 98–126.

Vrijling, J.K., 1982. Probability design method, Eastern Scheldt Storm Surge Barrier, *Proceedings of the Delta Barrier Symposium*, Rotterdam, p 44–49.

Vrijling, J.K. and Meijer, G.J., 1992. Probabilistic coastline position computation, *Coastal Engineering*, 17, pp. 1–23.

Wallingford, H. R., 2000. *The Joint Probability of Waves and Water Levels: JOIN-SEA*, Report SR 537, HR Wallingford, Oxon., UK

Walsh, K.J.E., Betts, H, Church, J, Pittock, A.B., McInnes, K.L., Jackett, D.R. et al., 2004. Using sea level rise projections for urban planning in Australia. *Journal of Coastal Research*, 20(2), pp. 586–598.

Walton, T.L., 1993. Shoreline solution for tapered beach fill, *Journal of Waterways, Port, Coastal and Oceanic Engineering*, ASCE 120(6), pp. 651–655.

Walton, T.L. and Chiu, T.Y., 1979. A review of analytical techniques to solve the sand transport equation and some simplified solutions, *Proc. Coastal Structures*, 1979, ASCE, New York, pp. 809–837.

WAMDI Group, 1988. The WAM model – a third-generation ocean wave prediction model, *J. Phys. Oceanogr.*, 18, pp. 1775–1810.

Wang, P. and Davis, Jr., R.A., 1998. A Beach Profile Model for a Barred Coast – Case Study from Sand Key, West-Central Florida, *J. Coastal Res.*, 14, 3, pp. 981–991.

Wang, B. and Reeve, D.E., 2010. Probabilistic modelling of long-term beach evolution near segmented shore-parallel breakwaters, *Coastal Engineering*, 57, pp. 732–744.

White, W.R., 1972. *Sediment Transport in Channels, a General Function*. Rep. mt. 102, Hydraulics Research, Wallingford.

White, W.R., Paris, E. and Bettess, R. 1980. The frictional characteristics of alluvial streams: a new approach. *Proc. Inst. Civ. Eng.*, Part 2, 69, pp. 737–750.

Whitehouse, R., Balson, P., Brampton, A., Blott, S., *et al.*, 2009. *Characterisation and Prediction of Large-Scale, Long-Term Coastal Geomorphological Behaviours*: Final Science Report: SC060074/SR2. UK Environment Agency, 281 pp.

Wijnberg, K.M. and Terwindt, J.H.J., 1995. Extracting decadal morphological behaviour from high resolution, long term bathymetric surveys along the Holland coast using eigenfunction analysis, *Marine Geology*, 126, pp. 301–330.

Wilson, B.W. and Torum, A., 1968. *The Tsunami of the Alaskan Earthquake, 1964: Engineering Evaluation*, Technical Memorandum 25, US Army Coastal Engineering Center, Ft. Belvoir, VA.

Wilson, K.C., 1989. Friction of wave-induced sheet flow. *Coastal Engineering*, 12: 371–379.

Winant, C.D., Inman, D.L. and Nordstom, C.E. 1975. Description of seasonal beach changes using empirical eigenfunctions, *J. Geophys. Res.*, 80(15), pp. 1979–1986.

Wind, H.G., 1990. Influence functions. *Proceedings of the 21th International Conference on Coastal Engineering*, Costal del Sol-Malaga, Spain, pp. 3281–3294.

Wright, L.D., Short, A.D., Boon, J.D., Kimball, S. and List, J.H., 1987. The morphodynamic effects of incident wave groupiness and tide range on an energetic beach, *Marine Geology*, 74, pp. 1–20.

Yoon, S.B. and Liu, P.L-F., 1989. Interactions of currents and weakly nonlinear water waves in shallow water, *J. of Fluid Mech.*, 205, pp. 397–419.

Young, I.R., 1999. *Wind Generated Ocean Waves*, Elsevier Ocean Engineering Book Series, Vol. 12.

Young, I.R. and Holland, G.J., 1998. *A Multi-Media Atlas of the Oceans Wind and Wave Climate*, Version 2.0, CD-ROM, Pergamon Press.

Zacharioudaki, A. and Reeve, D.E. 2008. Semi-analytical solutions of shoreline response to time varying wave conditions, *ASCE Journal of Waterways, Port, Coastal and Ocean Engineering*, 134(5), pp. 265–274.

Zacharioudaki, A. and Reeve, D.E., 2011. Shoreline evolution under climate change wave scenarios, *Climatic Change*, 108(1–2), pp. 73–105.

Zhang, D.H., Yip, T.L. and Ng, C.-O., 2009. Predicting tsunami arrivals: estimates and policy implications, *Marine Policy*, 33, pp. 643–650.

Zyserman, J.A. and Fredsoe, J., 1994. Data analysis of bed concentration of sediment. *J. Hydraul. Eng. ASCE*, 120(9), pp. 1021–42.

Index